Free Boundary Problems

Theory and Applications

Isabel N. Figueiredo
José F. Rodrigues
Lisa Santos
Editors

Birkhäuser Verlag
Basel · Boston · Berlin

Editors:

Isabel Narra Figueiredo
Departamento de Matemática
Faculdade de Ciências e Tecnologia
Universidade de Coimbra
Apartado 3008
3001-454 Coimbra
Portugal
isabelf@mat.uc.pt

José Francisco Rodrigues
Universidade de Lisboa / CMAF
Av. Prof. Gama Pinto 2
1649-003 Lisboa
Portugal
rodrigue@ptmat.fc.ul.pt

Lisa Santos
Departamento de Matemática
Universidade do Minho
Campus de Gualtar
4710-057 Braga
Portugal
lisa@math.uminho.pt

2000 Mathematics Subject Classification 35-06, 35R35, 80A22, 65Mxx, 65Nxx, 65Z05, 74N20, 74N30, 74N99, 76D27.

Library of Congress Control Number: 2006935948

Bibliographic information published by Die Deutsche Bibliothek
Die Deutsche Bibliothek lists this publication in the Deutsche Nationalbibliografie; detailed bibliographic data is available in the Internet at http://dnb.ddb.de.

ISBN 978-3-7643-7718-2 Birkhäuser Verlag, Basel – Boston – Berlin

Part of Springer Science+Business Media
Printed on acid-free paper produced from chlorine-free pulp TCF ∞
Printed in Germany

ISBN-10: 3-7643-7718-6
ISBN-13: 978-3-7643-7718-2

e-ISBN-10: 3-7643-7719-4
e-ISBN-13: 978-3-7643-7719-9

9 8 7 6 5 4 3 2 1

www.birkhauser.ch

Contents

Preface

This book gathers a collection of refereed articles containing original results reporting the recent original contributions of the lectures and communications presented at the Free Boundary Problems (FBP2005) Conference that took place at the University of Coimbra, Portugal, from 7 to 12 of June 2005. They deal with the Mathematics of a broad class of models and problems involving nonlinear partial differential equations arising in Physics, Engineering, Biology and Finance. Among the main topics, the talks considered free boundary problems in biomedicine, in porous media, in thermodynamic modeling, in fluid mechanics, in image processing, in financial mathematics or in computations for inter-scale problems.

FBP2005 was the 10th Conference of a Series started in 1981 in Montecatini, Italy, that has had a continuous development in the following conferences in Maubuisson, France (1984), Irsee, Germany (1987), Montreal, Canada (1990), Toledo, Spain (1993), Zakopone, Poland (1995), Crete, Greece (1997), Chiba, Japan (1999), Trento, Italy (2002) and will be followed by the next one foreseen to be held in Stockholm, Sweden, in 2008.

In fact, the mathematical analysis and fine properties of solutions and interfaces in free boundary problems have been an active subject in the last three decades and their mathematical understanding continues to be an important interdisciplinary tool for the scientific applications, on one hand, and an intrinsic aspect of the current development of several important mathematical disciplines. This was recognized, in particular, by the *Free Boundary Problems Scientific Programme of the European Science Foundation*, that sponsored three conferences in the nineties in Europe, and is reflected in an electronic newsletter-forum (*FBP-News*, http://fbpnews.org), that started in 2003 and continues to have an important role to promote the exchange of information and ideas between mathematicians interested in this area.

Over 150 participants have gathered during the FBP2005, to present and discuss, in more than 120 talks, the last results on the Mathematics of free boundary problems. The structure of the Conference, advised by a Scientific Committee, combined Main Lectures and Focus Sessions by invitation and was complemented with Focus Discussions and Contribution Talks with selected open proposals by the worldwide scientific community, that constituted almost half of the communications. The conference also integrated in its programme, for the first time, an European Mathematical Society (EMS) Lecture. During the FBP2005 Conference, new people and new problems, with renewed classical subjects, were on stage. This has confirmed that these conferences continue to be an important catalyst for the identification and development of this interdisciplinary mathematical field. They promote, not only in Europe, but all over the world, an interdisciplinary scope in the broadest possible mathematical sense: from experimental observations to modeling, from abstract mathematical analysis to numerical computations.

The credit of the success of the FPB2005 conference is mainly due to the lecturers, the organizers of the focus sessions and all the speakers of the invited and contributed talks, for their valuable contributions. Of course, our acknowledgements also go to the members of the scientific committee, that was constituted by C. Bandle (University of Basel), H. Berestycki (EHESS, Paris), L. Caffarelli (University of Austin, Texas, USA), P. Colli (University of Pavia, Italy), C.J. van Duijn (University of Eindhoven, Netherlands), G. Dziuk (University of Freiburg, Germany), C. Elliott (University of Sussex, UK), A. Fasano (University of Florence, Italy), A. Friedman (University of Ohio, USA), B. Kawohl (University of Koln, Germany), M. Mimura (University of Tokyo, Japan), S. Osher (University of Los Angeles, USA), J.F. Rodrigues (University of Lisbon/CMU Coimbra, Portugal), H. Shahgholian (University of Stockholm, Sweden), J. Sprekels (WIAS Berlin, Germany) and J.L. Vazquez (University Autonoma of Madrid, Spain), as well as to our co-organizer L.N. Vicente (University of Coimbra), the reviewers for performing the evaluation of the articles presented in this book of Proceedings and to K.-H. Hoffmann for accepting it in this Birkhäuser Series. Our thanks also go to the secretariat of the conference, in particular, we wish to acknowledge Rute Andrade for her excellent collaboration, and the Department of Mathematics of the University of Coimbra, for the facilities and active assistance.

Finally, we wish to thank also the important financial support from ESF (European Science Foundation) Scientific Programme (Global) on "Global and Geometrical Aspects of Nonlinear Partial Differential Equations", as well as, the financial support from CMUC (Centro de Matemática da Universidade de Coimbra), CMAF (Centro de Matemática e Aplicações Fundamentais da Universidade de Lisboa), EMS (European Mathematical Society), FLAD (Fundação Luso-Americana) and FCT (Fundação para a Ciência e a Tecnologia).

The Editors

Isabel Narra Figueiredo (Coimbra)
José Francisco Rodrigues (Lisboa)
Lisa Santos (Braga)

International Series of Numerical Mathematics, Vol. 154, 1–9

One-dimensional Shape Memory Alloy Problem with Duhem Type of Hysteresis Operator

Toyohiko Aiki and Takanobu Okazaki

Abstract. In our previous works we have proposed a mathematical model for dynamics of shape memory alloy materials. In the model the relationship between the strain and the stress is given as the generalized stop operator described by the ordinary differential equation including the subdifferential of the indicator function for the closed interval depending on the temperature. Here, we adopt the Duhem type of hysteresis operators as the mathematical description of the relationship in order to deal with the more realistic mathematical model. The aims of this paper are to show our new model and to establish the well-posedness of the model.

Mathematics Subject Classification (2000). Primary 74D10; Secondary 34G25, 35K45, 35Q72.

Keywords. Shape memory alloy, hysteresis, Duhem type.

1. Introduction

In this paper we consider the following system (1.1)–(1.6). The problem denoted by P is to find functions, the displacement u, the temperature field θ and the stress σ on $Q(T) := (0,T) \times (0,1)$, $0 < T < \infty$, satisfying

$$u_{tt} + \gamma u_{xxxx} - \mu u_{xxt} = \sigma_x \quad \text{in } Q(T), \tag{1.1}$$

$$\theta_t - \kappa\theta_{xx} = \sigma u_{xt} + \mu|u_{xt}|^2 \quad \text{in } Q(T), \tag{1.2}$$

$$\sigma_t + \partial I(\theta,\varepsilon;\sigma) \ni g_1(\theta,\varepsilon,\sigma)[\varepsilon_t]^+ - g_2(\theta,\varepsilon,\sigma)[\varepsilon_t]^- \quad \text{in } Q(T), \tag{1.3}$$

$$u(t,0) = u(t,1) = 0, u_{xx}(t,0) = u_{xx}(t,1) = 0 \quad \text{for } 0 < t < T, \tag{1.4}$$

$$\theta_x(t,0) = \theta_x(t,1) = 0 \quad \text{for } 0 < t < T, \tag{1.5}$$

$$u(0) = u_0, u_t(0) = v_0, \theta(0) = \theta_0, \sigma(0) = \sigma_0, \tag{1.6}$$

This work is partially supported by a grant in aid of JSPS ((C)16540146).

where $\varepsilon = u_x$ is the linearized strain, γ, μ and κ are positive constants, and u_0, v_0, θ_0 and σ_0 are initial functions. Also, $I(\theta, \varepsilon; \cdot)$ is the indicator functions of the closed interval $[f_*(\theta, \varepsilon), f^*(\theta, \varepsilon)]$, $\partial I(\theta, \varepsilon; \cdot)$ denotes its subdifferential, where $f_* : R^2 \to R$, $f^* : R^2 \to R$ are given continuous functions with $f_* \leq f^*$ on R^2, and g_1 are g_2 are also given continuous functions on R^3.

In our previous works Aiki-Kenmochi [5], Aiki [1], Aiki-Kadoya-Yoshikawa [4] the system {(1.1), (1.2), (1.4), (1.5), (1.6), (1.7) } was investigated. Here, (1.7) is as follows:

$$\sigma_t + \partial I(\theta, \varepsilon; \sigma) \ni c u_{xt} \quad \text{in } Q(T), \tag{1.7}$$

where c is a positive constant. Also, we quote [5, 1, 4] and Brokate-Sprekels [6] for the physical background for our system. As mentioned in Visintin [12] the differential equation (1.7) is one of characterization for the generalised stop operator. We note that in [4] the well-posedness of the problem without the restriction $\mu^2 > \gamma$, although we assumed this condition in [5, 1]. In that proof by using maximal regularity for complex Ginzburg-Landau equation we could remove the condition.

Next, we give a brief explanation for the Duhem type of hysteresis operators. From the experimental results we know that for shape memory alloy materials the relationship of interior of hysteresis loops is more complicated than one of the stop operator (see Figures 1 and 2).

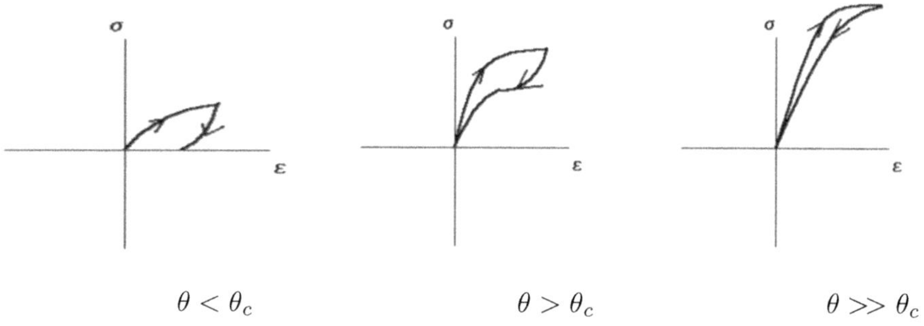

FIGURE 1. Graphs from experiments

Then we adopt the Duhem type of stop operator, which is defined by the ordinary differential equations. For example, the following equation was already introduced in [12]:

$$\sigma_t = g_1(\theta, \varepsilon, \sigma)[\varepsilon_t]^+ - g_2(\theta, \varepsilon, \sigma)[\varepsilon_t]^- .$$

By choosing suitable functions f_*, f^*, g_1 and g_2 we can obtain the graphs which are very close to experimental graphs, numerically (Figure 3). Also, the system including the Duhem type of hysteresis operator was already applied for the magnetization process of ferromagnetic materials and obtained the existence and the uniqueness of a solution to the problem in Aiki-Hoffmann-Okazaki [3]. See [2] for recent works of some mathematical models including hysteresis operators.

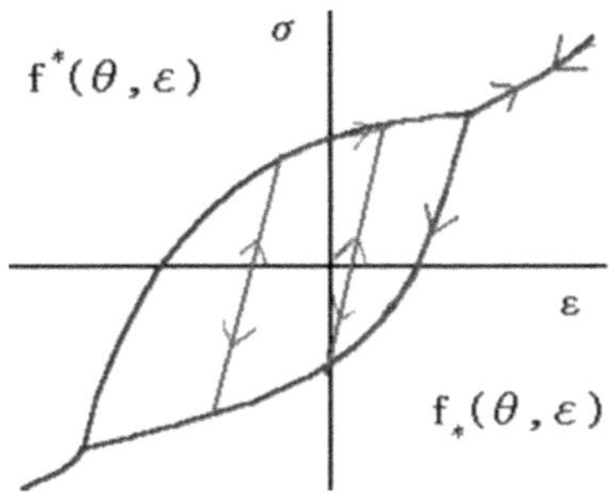

FIGURE 2. Graph of the generalized stop operator

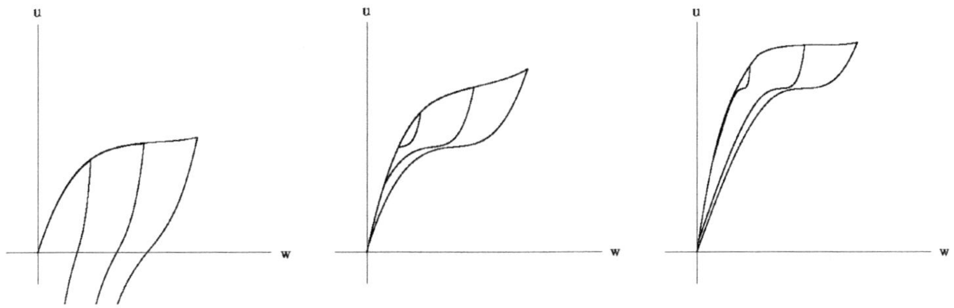

FIGURE 3. Graph from the numerical calculations

At the end of the introduction we show some results concerned with a mathematical model given by the more general hysteretic relations. In a series of papers [8, 9] Krejci and Sprekels studied one-dimensional shape memory models with hysteresis operator of Prandtl-Ishlinskii type, parametrized by the absolute temperature. The problems considered in these papers are more difficult than the one studied in this paper in the sense that in [8] no smoothing viscosity (i.e., $\mu = 0$), and in [9] no smoothing couple stress are assumed (i.e., $\gamma = 0$). Moreover, the above results have been generalized by Krejci, Sprekels and Stefanelli in [10, 11].

Here, we give the advantage and the disadvantage of using the Duhem model for shape memory alloys instead of the Prandtl-Ishlinskii model. The advantage of the Duhem model is to possible to deal with any shape of the load-deformation curves. In case with the Prandtl-Ishlinskii model the initial loading curve must be concave (cf. [7, Section 2]). The disadvantage of the Duhem model is that it is impossible to show thermodynamically consistent at the present time.

2. Main results

The purpose of this section is to give a complete statement for our result. First, we give assumptions for data.

(A1) $f_*, f^* \in C^2(R^2) \cap W^{2,\infty}(R^2)$;

(A2) $g_1 : R^3 \to R$ and $g_2 : R^3 \to R$ are Lipschitz continuous and $g_1, g_2 \in L^\infty(R^3)$;

(A3) $u_0 \in H^4(0,1) \cap H_0^1(0,1)$, $v_0 \in H_0^1(0,1) \cap H^2(0,1)$, $\theta_0 \in H^1(0,1)$, $\sigma_0 \in H^1(0,1)$,

$$f_*(\theta_0, \varepsilon_0) \leq \sigma_0 \leq f^*(\theta_0, \varepsilon_0) \text{ on } (0,1).$$

Next, we define a solution of P as follows:

Definition 2.1. We say that a triplet $\{u, \theta, \sigma\}$ of functions, u, θ and σ is a solution of P on $[0,T]$, $0 < T < \infty$, if and only if the following conditions (S1) and (S2) hold:

(S1) $u \in W^{2,\infty}(0,T; L^2(0,1)) \cap W^{2,2}(0,T; H^1(0,1)) \cap W^{1,\infty}(0,T; H^2(0,1)) \cap W^{1,2}(0,T; H^3(0,1)) \cap L^\infty(0,T; H^4(0,1))$,
$\theta \in W^{1,2}(0,T; L^2(0,1)) \cap L^\infty(0,T; H^1(0,1))$, and
$\sigma \in W^{1,2}(0,T; L^2(0,1)) \cap L^\infty(0,T; H^1(0,1))$.

(S2) (1.1)–(1.6) hold in the usual sense.

This is a main result of this paper.

Theorem 2.2 (Main Theorem). *Assume that* (A1), (A2) *and* (A3) *hold. Then the problem P has a unique solution* $\{u, \theta, \sigma\}$ *on* $[0,T]$ *for any* $T > 0$.

The proof of the uniqueness is given in Section 3 and the proof of the existence is rather long and quite similar to those of [1, 4] so that we omit it.

3. Proof of the uniqueness

In this section we will prove the uniqueness. The proof is similar to those of [1, 4]. Throughout this section we assume (A1)–(A3), and for simplicity we use the notation $H = L^2(0,1)$. Let $T > 0$, $\{u_1, \theta_1, \sigma_1\}$ and $\{u_2, \theta_2, \sigma_2\}$ be solutions of P on $[0,T]$ and put $\varepsilon_i = u_{ix}$, $i = 1, 2$, $u = u_1 - u_2$, $\theta = \theta_1 - \theta_2$, $\sigma = \sigma_1 - \sigma_2$, $\varepsilon = \varepsilon_1 - \varepsilon_2$, and

$$M(s) = \max\{|f_*(\theta_1, \varepsilon_1) - f_*(\theta_2, \varepsilon_2)|_{L^\infty(Q(s))}, |f^*(\theta_1, \varepsilon_1) - f^*(\theta_2, \varepsilon_2)|_{L^\infty(Q(s))}\}$$

for $0 < s \leq T$. Moreover, let $z_1 = \sigma_1 - [\sigma - M(s)]^+$ and $z_2 = \sigma_2 - [\sigma - M(s)]^+$.

For $s \in (0,T]$ we have

$$f_*(\theta_1, \varepsilon_1) \leq z_1 \leq f^*(\theta_1, \varepsilon_1) \text{ and } f_*(\theta_2, \varepsilon_2) \leq z_2 \leq f^*(\theta_2, \varepsilon_2) \text{ a.e. on } Q(s)$$

so that by using the definition of subdifferential it is obvious that

$$\int_0^1 \sigma_{1t}(t)(\sigma_1(t)-z_1(t))dx \le \int_0^1 G_1(t)(\sigma_1(t)-z_1(t))dx \quad \text{for a.e. } t\in[0,s],$$
$$\int_0^1 \sigma_{2t}(t)(\sigma_2(t)-z_2(t))dx \le \int_0^1 G_2(t)(\sigma_2(t)-z_2(t))dx \quad \text{for a.e. } t\in[0,s],$$

where $G_1(t)=(g_1(\theta_1(t),\varepsilon_1(t),\sigma_1(t))[\varepsilon_{1t}(t)]^+ - g_2(\theta_1(t),\varepsilon_1(t),\sigma_1(t))[\varepsilon_{1t}(t)]^-)$, $G_2(t)=(g_1(\theta_2(t),\varepsilon_2(t),\sigma_2(t))[\varepsilon_{2t}(t)]^+ - g_2(\theta_2(t),\varepsilon_2(t),\sigma_2(t))[\varepsilon_{2t}(t)]^-)$. Combining the above two inequalities, we obtain

$$\begin{aligned}&\frac{1}{2}\frac{d}{dt}|[\sigma(t)-M(s)]^+|^2 dx\\ &\quad\le \int_0^1 (G_1(t)-G_2(t))[\sigma(t)-M(s)]^+dx \quad \text{for a.e. } t\in[0,s].\end{aligned} \tag{3.1}$$

Next, (1.1) implies

$$u_{tt}+\gamma u_{xxxx}-\mu u_{txx}=\sigma_x \quad \text{in } Q(T),$$

and we multiply it by u_t. Then, we have

$$\begin{aligned}&\frac{d}{dt}\Big(\frac{1}{2}|u_t(t)|_H^2+\frac{\gamma}{2}|u_{xx}(t)|_H^2\Big)+\mu|u_{tx}(t)|_H^2\\ &\quad=-\int_0^1 \sigma(t)u_{tx}(t)dx \quad \text{for a.e. } t\in[0,T].\end{aligned} \tag{3.2}$$

It follows from (3.1) and (3.2) that

$$\begin{aligned}&\frac{d}{dt}\Big(\frac{1}{2}|[\sigma(t)-M(s)]^+|_H^2+\frac{1}{2}|u_t(t)|_H^2+\frac{\gamma}{2}|u_{xx}(t)|_H^2\Big)+\mu|u_{tx}(t)|_H^2\\
&\le L_g\int_0^1 (|\theta(t)|+|\varepsilon(t)|+|\sigma(t)|([\varepsilon_{1t}(t)]^+ + [\varepsilon_{1t}(t)]^-)[\sigma(t)-M(s)]^+dx\\
&\quad+\int_0^1 |g_1(\theta_2(t),\varepsilon_2(t),\sigma_2(t))||[\varepsilon_{1t}(t)]^+ - [\varepsilon_{2t}(t)]^+|[\sigma(t)-M(s)]^+dx\\
&\quad+\int_0^1 |g_2(\theta_2(t),\varepsilon_2(t),\sigma_2(t))||[\varepsilon_{1t}(t)]^- - [\varepsilon_{2t}(t)]^-|[\sigma(t)-M(s)]^+dx\\
&\quad-\int_0^1 \sigma(t)u_{tx}(t)dx\\
&\le L_g|\varepsilon_{1t}|_{L^\infty(Q(T))}(|\theta(t)|_H+|\varepsilon(t)|_H)|[\sigma(t)-M(s)]^+|_H\\
&\quad+L_g|\varepsilon_{1t}|_{L^\infty(Q(T))}\int_0^1 |\sigma(t)|[\sigma(t)-M(s)]^+dx\\
&\quad+\frac{\mu}{2}|u_{tx}(t)|_H^2+\frac{4\hat{L}_g^2}{\mu}|[\sigma(t)-M(s)]^+|_H^2+\frac{1}{\mu}|\sigma(t)|_H^2 \quad \text{for a.e. } t\in(0,s],\end{aligned}$$

where L_g is a common Lipschitz constant of g_1 and g_2 and $\hat{L}_g = |g_1|_{L^\infty(R^3)} + |g_2|_{L^\infty(R^3)}$, since by Definition 2.1 it holds that $u_{1t} \in L^\infty(0,T;H^2(0,1))$ and $\varepsilon_{1t} = u_{1tx} \in L^\infty(Q(T))$.

By similar calculations we can observe that

$$\begin{aligned}
&\frac{1}{2}\frac{d}{dt}(|[\sigma(t)-M(s)]^+|_H^2 + |[-\sigma(t)-M(s)]^+|_H^2) \\
&\quad + \frac{d}{dt}(|u_t(t)|_H^2 + \gamma|u_{xx}(t)|_H^2) + \mu|u_{tx}(t)|_H^2 \\
&\quad \le C_1(|\theta(t)|_H + |\varepsilon(t)|_H)(|[\sigma(t)-M(s)]^+|_H + |[-\sigma(t)-M(s)]^+|_H) \\
&\quad + C_1\int_0^1 |\sigma(t)|([\sigma(t)-M(s)]^+ + [-\sigma(t)-M(s)]^+)dx + \frac{2}{\mu}|\sigma(t)|_H^2 \\
&\quad + C_1(|[\sigma(t)-M(s)]^+|_H^2 + |[-\sigma(t)-M(s)]^+|_H^2) \quad \text{for a.e. } t \in (0,s],
\end{aligned}$$

where $C_1 = L_g|\varepsilon_{1t}|_{L^\infty(Q(T))} + \frac{4\hat{L}_g^2}{\mu}$. It is easy to see that

$$\begin{aligned}
&\int_0^1 |\sigma(t)|(|[\sigma(t)-M(s)]^+| + |[-\sigma(t)-M(s)]^+|)dx \\
&\qquad \le 2M(s)\sqrt{E(t)} + 2E(t) \qquad \text{for } t \in [0,s],
\end{aligned}$$

where $E(t) = |[\sigma(t)-M(s)]^+|_H^2 + |[-\sigma(t)-M(s)]^+|_H^2$. Thus we see that

$$\begin{aligned}
&\frac{1}{2}\frac{d}{dt}E(t) + \frac{d}{dt}(|u_t(t)|_H^2 + \gamma|u_{xx}(t)|_H^2) + \mu|u_{tx}(t)|_H^2 \\
&\quad \le C_1(|\theta(t)|_H + |\varepsilon(t)|_H)\sqrt{E(t)} + 2C_1(M(s)\sqrt{E(t)} + E(t)) \\
&\qquad + 4M(s)^2 + 4E(t) + C_1E(t) \\
&\quad \le C_2(|\theta(t)|_H + |\varepsilon(t)|_H)\sqrt{E(t)} + C_2M(s)^2 + C_2E(t) \qquad \text{for a.e. } t \in [0,s],
\end{aligned} \tag{3.3}$$

where $C_2 = 2C_1 + 4$.

Now, by (1.2) and routine works we infer that

$$\begin{aligned}
&\frac{1}{2}\frac{d}{dt}|\theta(t)|_H^2 + \frac{\kappa}{2}|\theta_x(t)|_H^2 \\
&\quad \le \int_0^1 (\sigma_1(t)\varepsilon_t(t) + \sigma(t)u_{2xt}(t))\theta(t)dx \\
&\qquad + \mu\int_0^1 |\varepsilon_t(t)|(|u_{1xt}(t)| + |u_{2xt}(t)|)|\theta(t)|dx \\
&\quad \le |\sigma_1|_{L^\infty(Q(T))}|\varepsilon_t(t)|_H|\theta(t)|_H + |\sigma(t)|_H|\varepsilon_{2t}(t)|_{L^\infty(Q(T))}|\theta(t)|_H \\
&\qquad + \mu(|u_{1xt}|_{L^\infty(Q(T))} + |u_{2xt}|_{L^\infty(Q(T))})|\varepsilon_t(t)|_H|\theta(t)|_H \\
&\quad \le C_3(|u_{tx}(t)|_H|\theta(t)|_H + |\sigma(t)|_H|\theta(t)|_H) \\
&\quad \le C_3^2|\theta(t)|_H^2 + \frac{\mu}{4}|u_{tx}(t)|_H^2 + C_3|\theta(t)|_H + 2C_3(M(s)^2 + E(t)) \text{ for a.e. } t \in [0,T],
\end{aligned} \tag{3.4}$$

where $C_3 = |\sigma_1|_{L^\infty(Q(T))} + |\varepsilon_{2t}(t)|_{L^\infty(Q(T))} + \mu(|u_{1xt}|_{L^\infty(Q(T))} + |u_{2xt}|_{L^\infty(Q(T))})$.

Hence, by (3.3) and the above inequality there is a positive constant C_4 such that

$$\frac{d}{dt}E_0(t) + E_1(t) \le C_4(|\theta(t)|_H^2 + M(s)^2 + E(t) + |\varepsilon(t)|_H\sqrt{E(t)})$$

for a.e. $t \in (0, s]$, where

$$E_0(t) = \frac{1}{2}E(t) + |u_t(t)|_H^2 + \gamma|u_{xx}(t)|_H^2 + \frac{1}{2}|\theta(t)|_H^2$$

and

$$E_1(t) = \mu|u_{tx}(t)|_H^2 + \frac{\kappa}{2}|\theta_x(t)|_H^2.$$

In order to give an estimate for $|\theta(t)|_{L^\infty(Q(s))}$ we multiply (1.2) by θ_t. Then, similarly to (3.4), it is clear that

$$\begin{aligned}
&|\theta_t(t)|_H^2 + \frac{\kappa}{2}\frac{d}{dt}|\theta_x(t)|_H^2 \\
&\quad \le C_3(|u_{tx}(t)|_H|\theta_t(t)|_H + |\sigma(t)|_H|\theta_t(t)|_H) \\
&\quad \le C_3^2(|u_{tx}(t)|_H^2 + |\sigma(t)|_H^2) + \frac{1}{2}|\theta_t(t)|_H^3 \qquad \text{for a.e. } t \in [0, T]
\end{aligned}$$

so that

$$\begin{aligned}
&\int_0^t |\theta_\tau(\tau)|_H^2 d\tau + \kappa|\theta_x(t)|_H^2 \\
&\quad \le C_3^2 \int_0^t (|u_{\tau x}(\tau)|_H^2 + |\sigma(\tau)|_H^2) d\tau \\
&\quad \le 2C_3^2 t M(s)^2 + 2C_3^2 \int_0^t |\sigma(\tau)|_H^2 d\tau + C_3^2 \int_0^t |u_{\tau x}(\tau)|_H^2 d\tau \qquad \text{for } 0 \le t \le s.
\end{aligned}$$

Hence, we have

$$\begin{aligned}
&|\theta|_{L^\infty(Q(s))}^2 \\
&\quad \le 2(|\theta|_{L^\infty(0,s;H)}^2 + |\theta_x|_{L^\infty(0,s;H)}^2) \\
&\quad \le 2\sup_{0\le t\le s}|\theta(t)|_H^2 + 2C_3^2 s M(s)^2 + 2C_3^2\Big(\int_0^s |\sigma(\tau)|_H^2 d\tau + \int_0^s |u_{\tau x}(\tau)|_H^2 d\tau\Big).
\end{aligned}$$

On account of this estimate it yields that

$$\begin{aligned}
&M(s)^2 \le 2L_0^2(|\theta(t)|_{L^\infty(Q(s))}^2 + |\varepsilon(t)|_{L^\infty(Q(s))}^2) \\
&\le 2L_0^2\Big(2\sup_{0\le t\le s}|\theta(t)|_H^2 + 2C_3^2 s M(s)^2 + 2C_3^2\Big(\int_0^s |\sigma(\tau)|_H^2 d\tau + \int_0^s |u_{\tau x}(\tau)|_H^2 d\tau\Big)\Big) \\
&\quad + 2L_0^2\Big(\sup_{0\le t\le s}|u_{xx}(t)|_H^2 + \sup_{0\le t\le s}|u(t)|_H^2\Big) \\
&\le C_4\Big(\sup_{0\le t\le s}E_0(t) + \int_0^s (E_0(t) + E_1(t))dt \qquad \text{for } 0 \le s \le T,
\end{aligned}$$

where L_0 is a common Lipschitz constant of f^* and f_* and C_4 is a suitable positive constant.

Moreover, it is easy to see that

$$\int_0^s |\varepsilon(\tau)|_H^2 d\tau \le \int_0^s \Big(\int_0^\tau |u_{xt}(t)|_H dt\Big)^2 d\tau$$
$$\le sT \int_0^s |u_{x\tau}(\tau)|_H^2 d\tau \quad \text{for } 0 \le s \le T.$$

Hence, there is a positive constant C_5 such that

$$\frac{d}{dt}E_0(t) + E_1(t) \le C_5(E_0(t) + M(s)^2) + |\varepsilon(t)|_H^2 \text{ for a.e. } t \in (0, s].$$

Now, applying Gronwall's inequality, the above argument shows that

$$E_0(t) + \int_0^t E_1(\tau)d\tau$$
$$\le C_5 e^{C_5 s}(sM(s)^2 + \int_0^s |\varepsilon(\tau)|_H^2 d\tau)$$
$$\le C_5 e^{C_5 T}(2s \sup_{0\le t\le t} E_0(t) + sT \int_0^s E_1(t)dt) \quad \text{for } 0 \le t \le s \le T.$$

Accordingly, it holds that

$$\sup_{0\le t\le s} E_0(t) + \int_0^s E_1(\tau)d\tau$$
$$\le C_5 e^{C_5 T}(1+T)(2s \sup_{0\le t\le t} E_0(t) + s\int_0^s E_1(t)dt) \quad \text{for } 0 \le s \le T.$$

Here, by choosing a positive number s with $C_5 e^{C_5 T} 2(1+T)s \le \frac{1}{2}$ we conclude that $E_0(t) = E_1(t) = 0$ for $t \in [0, s]$. This implies the uniqueness of a solution. □

Acknowledgement

I am grateful to the referee whose remarks helped to improve this manuscript.

References

[1] T. Aiki, One-dimensional shape memory alloy problems including a hysteresis operator, Funkcial. Ekvac., 46 (2003), 441–469.

[2] T. Aiki, Mathematical models including a hysteresis operator, to appear in *Dissipative phase transitions*, the Series on Advances in Mathematics for Applied Sciences, World Sci. Publishing.

[3] T. Aiki, K.-H. Hoffmann, and T. Okazaki, Well-posedness for a new mathematical model for magnetostrictive thin film multilayers, Advances in Mathematical Sciences and Applications, Vol. 14, (2004) 417–442.

[4] T. Aiki, A. Kadoya and S. Yoshikawa, Hysteresis model for one-dimensional shape memory alloy with small viscosity, in preparation.

[5] T. Aiki and N. Kenmochi, Some models for shape memory alloys, *Mathematical Aspects of Modeling Structure Formation Phenomena*, Gakuto, International Series Mathematical Sciences and Applications, Vol. 17 (2001), 144–162.

[6] M. Brokate and J. Sprekels, *Hysteresis and Phase Transitions*, Springer, Appl. Math. Sci., 121, 1996.

[7] P. Krejci, Hysteresis and periodic solutions of semilinear and quasilinear wave equations, Math. Z., 193 (1986), 247–264.

[8] P. Krejci and J. Sprekels, On a system of nonlinear PDEs with temperature-dependent hysteresis in one-dimensional thermoplasticity, J. Math. Anal. Appl., 209 (1997), 25–46.

[9] P. Krejci and J. Sprekels, Temperature-dependent hysteresis in one-dimensional thermovisco-elastoplasticity, Appl. Math., 43 (1998), 173–205.

[10] P. Krejci, J. Sprekels and U. Stefanelli, Phase-field models with hysteresis in one-dimensional thermoviscoplasticity, SIAM J. Math. Anal., 34 (2002), 409–434.

[11] P. Krejci, J. Sprekels and U. Stefanelli, One-dimensional thermo-visco-plastic processes with hysteresis and phase transitions, Adv. Math. Sci. Appl., 13 (2004), 695–712.

[12] A. Visintin, *Differential Models of Hysteresis*, Appl. Math. Sci., Vol. 111, Springer-Verlag, Berlin, 1993.

Toyohiko Aiki
Department of Mathematics
Faculty of Education
Gifu University
Yanagido 1-1
Gifu 501-1193, Japan
e-mail: `aiki@cc.gifu-u.ac.jp`

Takanobu Okazaki
Department of Mathematics
Faculty of Education
Chiba University
Yayoi-cho, 1-33, Chiba 263-8522, Japan
e-mail: `okazaki@faculty.Chiba-u.jp`

International Series of Numerical Mathematics, Vol. 154, 11–21

Existence and Uniqueness Results for Quasi-linear Elliptic and Parabolic Equations with Nonlinear Boundary Conditions

F. Andreu, N. Igbida, J.M. Mazón and J. Toledo

Abstract. We study the questions of existence and uniqueness of weak and entropy solutions for equations of type $-\text{div } \mathbf{a}(x, Du) + \gamma(u) \ni \phi$, posed in an open bounded subset Ω of $\mathbb{R}^N$, with nonlinear boundary conditions of the form $\mathbf{a}(x, Du)\cdot\eta + \beta(u) \ni \psi$. The nonlinear elliptic operator div $\mathbf{a}(x, Du)$ is modeled on the p-Laplacian operator $\Delta_p(u) = \text{div } (|Du|^{p-2}Du)$, with $p > 1$, γ and β are maximal monotone graphs in $\mathbb{R}^2$ such that $0 \in \gamma(0)$ and $0 \in \beta(0)$, and the data $\phi \in L^1(\Omega)$ and $\psi \in L^1(\partial\Omega)$. We also study existence and uniqueness of weak solutions for a general degenerate elliptic-parabolic problem with nonlinear dynamical boundary conditions. Particular instances of this problem appear in various phenomena with changes of phase like multiphase Stefan problem and in the weak formulation of the mathematical model of the so called Hele Shaw problem.

Mathematics Subject Classification (2000). Primary 35J60; Secondary 35D02.

Keywords. Quasi-linear elliptic equations, Quasi-linear parabolic equations, Stefan problem, Hele Shaw problem, Nonlinear boundary conditions, Nonlinear semigroup theory.

1. Introduction

Let Ω be a bounded domain in $\mathbb{R}^N$ with smooth boundary $\partial\Omega$ and $p > 1$, and let $\mathbf{a} : \Omega \times \mathbb{R}^N \to \mathbb{R}^N$ be a Carathéodory function satisfying

(H_1) there exists $\lambda > 0$ such that $\mathbf{a}(x, \xi) \cdot \xi \geq \lambda|\xi|^p$ for *a.e.* $x \in \Omega$ and for all $\xi \in \mathbb{R}^N$,

(H_2) there exists $\sigma > 0$ and $\theta \in L^{p'}(\Omega)$ such that $|\mathbf{a}(x, \xi)| \leq \sigma(\theta(x) + |\xi|^{p-1})$ for *a.e.* $x \in \Omega$ and for all $\xi \in \mathbb{R}^N$, where $p' = \frac{p}{p-1}$,

(H_3) $(\mathbf{a}(x, \xi_1) - \mathbf{a}(x, \xi_2)) \cdot (\xi_1 - \xi_2) > 0$ for *a.e.* $x \in \Omega$ and for all $\xi_1, \xi_2 \in \mathbb{R}^N$, $\xi_1 \neq \xi_2$.

The hypotheses $(H_1 - H_3)$ are classical in the study of nonlinear operators in divergent form (cf. [31] or [8]). The model example of function $\mathbf{a}$ satisfying these hypotheses is $\mathbf{a}(x,\xi) = |\xi|^{p-2}\xi$. The corresponding operator is the p-Laplacian operator $\Delta_p(u) = \mathrm{div}(|Du|^{p-2}Du)$.

We are interested in the study of existence and uniqueness of weak and entropy solutions for the elliptic problem

$$(S^{\gamma,\beta}_{\phi,\psi}) \quad \begin{cases} -\mathrm{div}\, \mathbf{a}(x,Du) + \gamma(u) \ni \phi & \text{in } \Omega \\ \mathbf{a}(x,Du)\cdot\eta + \beta(u) \ni \psi & \text{on } \partial\Omega, \end{cases}$$

where η is the unit outward normal on $\partial\Omega$, $\psi \in L^1(\partial\Omega)$ and $\phi \in L^1(\Omega)$. The nonlinearities γ and β are maximal monotone graphs in $\mathbb{R}^2$ (see, e.g., [14]) such that $0 \in \gamma(0)$ and $0 \in \beta(0)$. In particular, they may be multivalued and this allows to include the Dirichlet condition (taking β to be the monotone graph D defined by $D(0) = \mathbb{R}$) and the Neumann condition (taking β to be the monotone graph N defined by $N(r) = 0$ for all $r \in \mathbb{R}$) as well as many other nonlinear fluxes on the boundary that occur in some problems in Mechanics and Physics (see, e.g., [20] or [13]). Note also that, since γ may be multivalued, problems of type $(S^{\gamma,\beta}_{\phi,\psi})$ appear in various phenomena with changes of phase like the multiphase Stefan problem (cf [17]) and in the weak formulation of the mathematical model of the so-called Hele Shaw problem (cf. [19] and [22]).

Particular instances of problem $(S^{\gamma,\beta}_{\phi,\psi})$ have been studied in [10], [8], [6] and [2]. The work of Bénilan, Crandall and Sacks [10] was pioneer in this kind of problems. They study problem $(S^{\gamma,\beta}_{\phi,0})$ for any γ and β maximal monotone graphs in $\mathbb{R}^2$ such that $0 \in \gamma(0)$ and $0 \in \beta(0)$, for the Laplacian operator, i.e., for $\mathbf{a}(x,\xi) = \xi$. For nonhomogeneous boundary condition, i.e. $\psi \not\equiv 0$, one can see [27] for ψ in the range of β, and [25, 26] for some particular instances of β and γ. Another important work in the L^1-Theory for p-Laplacian type equations is [8], where problem

$$(D^{\gamma}_{\phi}) \quad \begin{cases} -\mathrm{div}\, \mathbf{a}(x,Du) + \gamma(u) \ni \phi & \text{in } \Omega \\ u = 0 & \text{on } \partial\Omega \end{cases}$$

is studied for any γ maximal monotone graph in $\mathbb{R}^2$ such that $0 \in \gamma(0)$. Following [8], problems $(S^{id,\beta}_{\phi,0})$ and $(S^{id,\beta}_{\phi,\psi})$, where $id(r) = r$ for all $r \in \mathbb{R}$, are studied in [6] and [2], for any β maximal monotone graph in $\mathbb{R}^2$ with closed domain such that $0 \in \beta(0)$.

Our aim is to establish existence and uniqueness of weak and entropy solutions for the general elliptic problem $(S^{\gamma,\beta}_{\phi,\psi})$. The main interest in our work is that we are dealing with general nonlinear operators $-\mathrm{div}\, \mathbf{a}(x,Du)$ with nonhomogeneous boundary conditions and general nonlinearities β and γ. As in [10], a range condition relating the average of ϕ and ψ to the range of β and γ is necessary for existence of weak and entropy solution (see Remark 3.3). However, in contrast to the smooth homogeneous case, $\mathbf{a}$ smooth and $\psi = 0$, for the nonhomogeneous case this range condition is not sufficient for the existence of weak solution. Indeed, in general, the intersection of the domains of β and γ seems to create some obstruction phenomena for the existence of these solutions. In general, even if $D(\beta) = \mathbb{R}$, a weak solution does not exist, as the following example shows. Let γ be such that $D(\gamma) = [0, 1]$, $\beta = \mathbb{R} \times \{0\}$, and let $\phi \in L^1(\Omega)$, $\phi \leq 0$ *a.e.* in Ω, and $\psi \in L^1(\partial\Omega)$, $\psi \leq 0$ *a.e.* in $\partial\Omega$. If there exists $[u, z, w]$ a weak solution of the problem $(S^{\gamma,\beta}_{\phi,\psi})$ (see Definition 3.1), then $z \in \gamma(u)$, therefore $0 \leq u \leq 1$ *a.e.* in Ω, $w = 0$, and it holds that for any $v \in W^{1,p}(\Omega) \cap L^\infty(\Omega)$,

$$\int_\Omega \mathbf{a}(x, Du)Dv + \int_\Omega zv = \int_{\partial\Omega} \psi v + \int_\Omega \phi v.$$

Taking $v = u$, as $u \geq 0$, we get u is constant and

$$\int_\Omega zv = \int_{\partial\Omega} \psi v + \int_\Omega \phi v,$$

for any $v \in W^{1,p}(\Omega) \cap L^\infty(\Omega)$. Consequently, $\phi = z$ *a.e.* in Ω, and ψ must be 0 *a.e.* in $\partial\Omega$.

The main applications we have in mind are the study of doubly nonlinear evolution problems of elliptic-parabolic type and degenerate parabolic problems of Stefan or Hele-Shaw type, with nonhomogeneous boundary conditions and/or dynamical boundary conditions (see [5]). More precisely, we have in mind to study the following degenerate elliptic-parabolic problem with nonlinear dynamical boundary condition

$$P_{\gamma,\beta}(f, g, z_0, w_0) \quad \begin{cases} z_t - \operatorname{div} \mathbf{a}(x, Du) = f, \ z \in \gamma(u), & \text{in } Q_T :=]0, T[\times\Omega \\ w_t + \mathbf{a}(x, Du) \cdot \eta = g, \ w \in \beta(u), & \text{on } S_T :=]0, T[\times\partial\Omega \\ v(0) = v_0 \ \text{ in } \Omega, \ w(0) = w_0 \ \text{ in } \partial\Omega, \end{cases}$$

where $v_0 \in L^1(\Omega)$, $w_0 \in L^1(\partial\Omega)$, $f \in L^1(0, T; L^1(\Omega))$ and $g \in L^1(0, T; L^1(\partial\Omega))$. The dynamical boundary conditions, although not too widely considered in the mathematical literature, are very natural in many mathematical models as heat transfer in a solid in contact with a moving fluid, thermoelasticity, diffusion phenomena, the heat transfer in two phase medium (Stefan problem), problems in fluid dynamics, etc. (cf. [18] or [21] and the reference therein).

Problems of type $P_{\gamma,\beta}(f, g, z_0, w_0)$, that is, the elliptic-parabolic problem with Dirichlet boundary conditions have been studied extensively in the literature (cf. [1], [3], [11], [15], [28] and the references therein). However, with respect to the pure Neumann case, for the multidimensional case, with time-dependent flux g, we only know the paper of Hulshof [23] for the Laplacian operator and γ a uniformly Lipschitz continuous function, $\gamma(r) = 1$ for $r \in \mathbb{R}^+$, $\gamma \in C^1(\mathbb{R}^-)$, $\gamma' > 0$ on $\mathbb{R}^-$ and $\lim_{r\downarrow-\infty} \gamma(r) = 0$; and the paper of Kenmochi [29] also for the Laplace operator and for γ which range is a closed bounded interval. In one space dimension, much more is known (cf. [12] and the references therein).

2. Preliminaries

For a maximal monotone graph ϑ in $\mathbb{R}\times\mathbb{R}$ we shall denote $\vartheta_- := \inf \mathrm{R}(\vartheta)$ and $\vartheta_+ := \sup \mathrm{R}(\vartheta)$, where $\mathrm{R}(\vartheta)$ denotes the range of ϑ. If $0 \in \mathrm{Dom}(\vartheta)$ and ϑ^0 is the main section of ϑ, $j_\vartheta(r) = \int_0^r \vartheta^0(s)ds$ defines a convex l.s.c. function such that $\vartheta = \partial j_\vartheta$. If j_ϑ^* is the Legendre transformation of j_ϑ then $\vartheta^{-1} = \partial j_\vartheta^*$.

In [8], the authors introduce the set

$$\mathcal{T}^{1,p}(\Omega) = \{u : \Omega \longrightarrow \mathbb{R} \text{ measurable such that } T_k(u) \in W^{1,p}(\Omega) \quad \forall k > 0\},$$

where $T_k(s) = \sup(-k, \inf(s,k))$. They also prove that given $u \in \mathcal{T}^{1,p}(\Omega)$, there exists a unique measurable function $\hat{u} : \Omega \to \mathbb{R}^N$ such that

$$DT_k(u) = \hat{u}\chi_{\{|\hat{u}|<k\}} \quad \forall k > 0.$$

This function $\hat{u}$ will be denoted by Du. It is clear that if $u \in W^{1,p}(\Omega)$, then $\hat{u} \in L^p(\Omega)$ and $\hat{u} = Du$ in the usual sense.

As in [6], $\mathcal{T}_{tr}^{1,p}(\Omega)$ denotes the set of functions u in $\mathcal{T}^{1,p}(\Omega)$ satisfying the following conditions, there exists a sequence u_n in $W^{1,p}(\Omega)$ such that

(a) u_n converges to u *a.e.* in Ω,
(b) $DT_k(u_n)$ converges to $DT_k(u)$ in $L^1(\Omega)$ for all $k > 0$,
(c) there exists a measurable function $\tilde{u}$ on $\partial\Omega$, such that u_n converges to $\tilde{u}$ *a.e.* in $\partial\Omega$.

The function $\tilde{u}$ is the trace of u in the generalized sense introduced in [6]. In the sequel, the trace of $u \in \mathcal{T}_{tr}^{1,p}(\Omega)$ on $\partial\Omega$ will be denoted by u.

We say that $\mathbf{a}$ is *smooth* (see [6]) when, for any $\phi \in L^\infty(\Omega)$ such that there exists a bounded weak solution u of the homogeneous Dirichlet problem

$$(D) \qquad \begin{cases} -\operatorname{div} \mathbf{a}(x, Du) = \phi & \text{in } \Omega \\ u = 0 & \text{on } \partial\Omega, \end{cases}$$

there exists $g \in L^1(\partial\Omega)$ such that u is also a weak solution of the Neumann problem

$$(N) \qquad \begin{cases} -\operatorname{div} \mathbf{a}(x, Du) = \phi & \text{in } \Omega \\ \mathbf{a}(x, Du)\cdot\eta = g & \text{on } \partial\Omega. \end{cases}$$

Functions $\mathbf{a}$ corresponding to linear operators with smooth coefficients and p-Laplacian type operators are smooth (see [13] and [30]).

3. The elliptic problem

In this section we give the different concepts of solutions of problem $(S_{\phi,\psi}^{\gamma,\beta})$ and we state the main results obtained in [4].

Definition 3.1. Let $\phi \in L^1(\Omega)$ and $\psi \in L^1(\partial\Omega)$. A triple of functions $[u, z, w] \in W^{1,p}(\Omega) \times L^1(\Omega) \times L^1(\partial\Omega)$ is a *weak solution* of problem $(S^{\gamma,\beta}_{\phi,\psi})$ if $z(x) \in \gamma(u(x))$ *a.e.* in Ω, $w(x) \in \beta(u(x))$ *a.e.* in $\partial\Omega$, and

$$\int_\Omega \mathbf{a}(x, Du) \cdot Dv + \int_\Omega zv + \int_{\partial\Omega} wv = \int_{\partial\Omega} \psi v + \int_\Omega \phi v, \tag{3.1}$$

for all $v \in L^\infty(\Omega) \cap W^{1,p}(\Omega)$.

In general, as it is remarked in [8], for $1 < p \le 2 - \frac{1}{N}$, there exists $f \in L^1(\Omega)$ such that the problem

$$u \in W^{1,1}_{loc}(\Omega), \; u - \Delta_p(u) = f \quad \text{in} \quad \mathcal{D}'(\Omega),$$

has no solution. In [8], to overcome this difficulty and to get uniqueness, a new concept of solution was introduced, named entropy solution. Following these ideas, as in [6] or [2], we introduce the following concept of solution.

Definition 3.2. Let $\phi \in L^1(\Omega)$ and $\psi \in L^1(\partial\Omega)$. A triple of functions $[u, z, w] \in \mathcal{T}^{1,p}_{tr}(\Omega) \times L^1(\Omega) \times L^1(\partial\Omega)$ is an *entropy solution* of problem $(S^{\gamma,\beta}_{\phi,\psi})$ if $z(x) \in \gamma(u(x))$ *a.e.* in Ω, $w(x) \in \beta(u(x))$ *a.e.* in $\partial\Omega$ and

$$\begin{aligned} &\int_\Omega \mathbf{a}(x, Du) \cdot DT_k(u - v) + \int_\Omega zT_k(u - v) + \int_{\partial\Omega} wT_k(u - v) \\ &\quad \le \int_{\partial\Omega} \psi T_k(u - v) + \int_\Omega \phi T_k(u - v) \quad \forall k > 0, \end{aligned} \tag{3.2}$$

for all $v \in L^\infty(\Omega) \cap W^{1,p}(\Omega)$.

Obviously, every weak solution is an entropy solution and an entropy solution with $u \in W^{1,p}(\Omega)$ is a weak solution.

Remark 3.3. If we take $v = T_h(u) \pm 1$ as test functions in (3.2) and let h go to $+\infty$, we get that

$$\int_\Omega z + \int_{\partial\Omega} w = \int_{\partial\Omega} \psi + \int_\Omega \phi.$$

Then necessarily ϕ and ψ must satisfy the following range condition

$$\mathcal{R}^-_{\gamma,\beta} \le \int_{\partial\Omega} \psi + \int_\Omega \phi \le \mathcal{R}^+_{\gamma,\beta},$$

where

$$\mathcal{R}^+_{\gamma,\beta} := \gamma_+|\Omega| + \beta_+|\partial\Omega|, \qquad \mathcal{R}^-_{\gamma,\beta} := \gamma_-|\Omega| + \beta_-|\partial\Omega|.$$

We will write $\mathcal{R}_{\gamma,\beta} :=]\mathcal{R}^-_{\gamma,\beta}, \mathcal{R}^+_{\gamma,\beta}[$ when $\mathcal{R}^-_{\gamma,\beta} < \mathcal{R}^+_{\gamma,\beta}$.

We shall state now the uniqueness result for entropy solutions. Since every weak solution is an entropy solution of problem $(S^{\gamma,\beta}_{\phi,\psi})$, the same result is true for weak solutions.

Theorem 3.4 ([4]). *Let $\phi \in L^1(\Omega)$ and $\psi \in L^1(\partial\Omega)$, and let $[u_1, z_1, w_1]$ and $[u_2, z_2, w_2]$ be entropy solutions of problem $(S^{\gamma,\beta}_{\phi,\psi})$. Then, there exists a constant $c \in \mathbb{R}$ such that*

$$\begin{aligned} u_1 - u_2 = c \quad & a.e. \ in \ \Omega, \\ z_1 - z_2 = 0 \quad & a.e. \ in \ \Omega. \\ w_1 - w_2 = 0 \quad & a.e. \ in \ \partial\Omega. \end{aligned}$$

Moreover, if $c \neq 0$, there exists a constant k such that $z_1 = z_2 = k$.

With respect to the existence of weak solutions we have the following results.

Theorem 3.5 ([4]). *Assume $D(\gamma) = \mathbb{R}$ and $\mathcal{R}^-_{\gamma,\beta} < \mathcal{R}^+_{\gamma,\beta}$. Let $D(\beta) = \mathbb{R}$ or* **a** *smooth.*

(i) *For any $\phi \in L^{p'}(\Omega)$ and $\psi \in L^{p'}(\partial\Omega)$ with*

$$\int_\Omega \phi + \int_{\partial\Omega} \psi \in \mathcal{R}_{\gamma,\beta}, \tag{3.3}$$

there exists a weak solution $[u, z, w]$ of problem $(S^{\gamma,\beta}_{\phi,\psi})$.

(ii) *For any $[u_1, z_1, w_1]$ weak solution of problem $(S^{\gamma,\beta}_{\phi_1,\psi_1})$, $\phi_1 \in L^{p'}(\Omega)$ and $\psi_1 \in L^{p'}(\partial\Omega)$ satisfying (3.3), and any $[u_2, z_2, w_2]$ weak solution of problem $(S^{\gamma,\beta}_{\phi_2,\psi_2})$, $\phi_2 \in L^{p'}(\Omega)$ and $\psi_2 \in L^{p'}(\partial\Omega)$ satisfying (3.3), we have that*

$$\int_\Omega (z_1 - z_2)^+ + \int_{\partial\Omega} (w_1 - w_2)^+ \leq \int_{\partial\Omega} (\psi_1 - \psi_2)^+ + \int_\Omega (\phi_1 - \phi_2)^+.$$

In the case $\mathcal{R}^-_{\gamma,\beta} = \mathcal{R}^+_{\gamma,\beta}$, that is, when $\gamma(r) = \beta(r) = 0$ for any $r \in \mathbb{R}$, existence and uniqueness of weak solutions are also obtained.

Theorem 3.6 ([4]). *For any $\phi \in L^{p'}(\Omega)$ and $\psi \in L^{p'}(\partial\Omega)$ with*

$$\int_\Omega \phi + \int_{\partial\Omega} \psi = 0, \tag{3.4}$$

there exists a unique (up to a constant) weak solution $u \in W^{1,p}(\Omega)$ of the problem

$$\begin{cases} -\text{div } \mathbf{a}(x, Du) = \phi & \text{in } \Omega \\ \mathbf{a}(x, Du) \cdot \eta = \psi & \text{on } \partial\Omega \end{cases}$$

in the sense that

$$\int_\Omega \mathbf{a}(x, Du) \cdot Dv = \int_{\partial\Omega} \psi v + \int_\Omega \phi v,$$

for all $v \in W^{1,p}(\Omega)$.

In order to get the above results, the main idea is to consider the approximated problem

$$(S^{\gamma_{m,n},\beta_{m,n}}_{\phi_{m,n},\psi_{m,n}}) \quad \begin{cases} -\text{div } \mathbf{a}(x, Du) + \gamma_{m,n}(u) \ni \phi_{m,n} & \text{in } \Omega \\ \mathbf{a}(x, Du) \cdot \eta + \beta_{m,n}(u) \ni \psi_{m,n} & \text{on } \partial\Omega, \end{cases}$$

where $\gamma_{m,n}$ and $\beta_{m,n}$ are approximations of γ and β given by

$$\gamma_{m,n}(r) = \gamma(r) + \frac{1}{m}r^+ - \frac{1}{n}r^-$$

and

$$\beta_{m,n}(r) = \beta(r) + \frac{1}{m}r^+ - \frac{1}{n}r^-$$

respectively, $m, n \in \mathbb{N}$, and

$$\phi_{m,n} = \sup\{\inf\{m, \phi\}, -n\}$$

and

$$\psi_{m,n} = \sup\{\inf\{m, \psi\}, -n\},$$

$m, n \in \mathbb{N}$ are approximations of ϕ and ψ, respectively. For these approximated problems we obtain existence of weak solutions with appropriate estimates and monotone properties, which allow us to pass to the limit.

Approximating L^1-data by L^∞-data and using Theorem 3.6, we can get the following result about existence of entropy solutions.

Theorem 3.7 ([4]). *Assume $D(\gamma) = \mathbb{R}$, and $D(\beta) = \mathbb{R}$ or* a *smooth. Let also assume that, if $[0, +\infty[\subset D(\beta)$,*

$$\lim_{k\to+\infty} \gamma^0(k) = +\infty \text{ and } \lim_{k\to+\infty} \beta^0(k) = +\infty, \tag{3.5}$$

and if $]-\infty, 0] \subset D(\beta)$,

$$\lim_{k\to-\infty} \gamma^0(k) = -\infty \text{ and } \lim_{k\to-\infty} \beta^0(k) = -\infty. \tag{3.6}$$

Then,

(i) *for any $\phi \in L^1(\Omega)$ and $\psi \in L^1(\partial\Omega)$, there exists an entropy solution $[u, z, w]$ of problem $(S^{\gamma,\beta}_{\phi,\psi})$.*

(ii) *For any $[u_1, z_1, w_1]$ entropy solution of problem $(S^{\gamma,\beta}_{\phi_1,\psi_1})$, $\phi_1 \in L^1(\Omega)$, $\psi_1 \in L^1(\partial\Omega)$, and any $[u_2, z_2, w_2]$ entropy solution of problem $(S^{\gamma,\beta}_{\phi_2,\psi_2})$, $\phi_2 \in L^1(\Omega)$, $\psi_2 \in L^1(\partial\Omega)$, we have that*

$$\int_\Omega (z_1 - z_2)^+ + \int_{\partial\Omega} (w_1 - w_2)^+ \leq \int_{\partial\Omega} (\psi_1 - \psi_2)^+ + \int_\Omega (\phi_1 - \phi_2)^+.$$

4. The parabolic problem

In this section we give the concept of weak solution for the problem $P_{\gamma,\beta}(f,g,z_0,w_0)$ and we state the existence and uniqueness result for this type of solutions given in [5].

Definition 4.1. Given $f \in L^1(0,T; L^1(\Omega))$, $g \in L^1(0,T; L^1(\partial\Omega))$, $z_0 \in L^1(\Omega)$ and $w_0 \in L^1(\partial\Omega)$, a *weak solution* of $P_{\gamma,\beta}(f, g, z_0, w_0)$ in $[0,T]$ is a couple (z, w) such that $z \in C([0,T] : L^1(\Omega))$, $w \in C([0,T] : L^1(\partial\Omega))$, $z(0) = z_0$, $w(0) = w_0$ and there exists $u \in L^1(0,T;W^{1,p}(\Omega))$ such that $z \in \gamma(u)$ a.e. in Q_T, $w \in \beta(u)$ a.e. on S_T and

$$\begin{aligned} &\frac{d}{dt}\int_\Omega z(t)\xi + \frac{d}{dt}\int_{\partial\Omega} w(t)\xi + \int_\Omega \mathbf{a}(x,Du(t))\cdot D\xi \\ &\quad = \int_\Omega f(t)\xi + \int_{\partial\Omega} g(t)\xi \quad \text{ in } \mathcal{D}'(]0,T[) \end{aligned} \tag{4.1}$$

for any $\xi \in C^1(\overline{\Omega})$.

Recall that even in the case $\beta = 0$, for the Laplacian operator and γ the multivalued Heaviside function (i.e., for the Hele-Shaw problem), existence and uniqueness of weak solutions for this problem is known to be true only if

$$\int_\Omega z_0 + \int_0^t \left(\int_\Omega f + \int_{\partial\Omega} g\right) \in (0,|\Omega|) \quad \text{for any } \ t \in [0,T)$$

(cf., [24] or [29])).

We have the following existence and uniqueness theorem.

Theorem 4.2 ([5]). *Assume* $\mathrm{Dom}(\gamma) = \mathbb{R}$, $\mathcal{R}^-_{\gamma,\beta} < \mathcal{R}^+_{\gamma,\beta}$ *and* $\mathrm{Dom}(\beta) = \mathbb{R}$ *or* **a** *smooth. Let* $T > 0$. *Let* $f \in L^{p'}(0,T;L^{p'}(\Omega))$, $g \in L^{p'}(0,T;L^{p'}(\partial\Omega))$, $z_0 \in L^{p'}(\Omega)$ *and* $w_0 \in L^{p'}(\partial\Omega)$ *such that*

$$\gamma_- \leq z_0 \leq \gamma_+, \quad \beta_- \leq w_0 \leq \beta_+, \tag{4.2}$$

$$\int_\Omega j^*_\gamma(z_0) + \int_\Gamma j^*_\beta(w_0) < +\infty, \tag{4.3}$$

and

$$\mu(t) := \int_\Omega w_0 + \int_{\partial\Omega} z_0 + \int_0^t \left(\int_\Omega f + \int_{\partial\Omega} g\right) \in \mathcal{R}_{\gamma,\beta} \qquad \forall\, t \in [0,T]. \tag{4.4}$$

Then, there exists a weak solution (z,w) *of problem* $P_{\gamma,\beta}(f,g,z_0,w_0)$ *in* $[0,T]$ *such that*

$$\mu(t) = \int_\Omega z(t) + \int_{\partial\Omega} w(t) \qquad \forall\, t \in [0,T]. \tag{4.5}$$

Moreover, the following L^1*-contraction principle holds. For* $i = 1,2$, *let* $f_i \in L^{p'}(0,T;L^{p'}(\Omega))$, $g_i \in L^{p'}(0,T;L^{p'}(\partial\Omega))$, $z_{i0} \in L^{p'}(\Omega)$ *and* $w_{i0} \in L^{p'}(\partial\Omega)$ *satisfying* (4.2), (4.3) *and* (4.4) *for every* i*; and let* (z_i,w_i) *be a weak solution in* $[0,T]$ *of* $P_{\gamma,\beta}(f_i,g_i,z_{i0},w_{i0})$, $i = 1,2$. *Then*

$$\begin{aligned} \int_\Omega (z_1(t) - z_2(t))^+ + \int_{\partial\Omega} (w_1(t) - w_2(t))^+ \leq \int_\Omega (z_{10} - z_{20})^+ + \int_{\partial\Omega} (w_{10} - w_{20})^+ \\ + \int_0^t \int_\Omega (f_1(\tau) - f_2(\tau))^+ \, d\tau + \int_0^t \int_{\partial\Omega} (g_1(\tau) - g_2(\tau))^+ \, d\tau \end{aligned} \tag{4.6}$$

for almost every $t \in]0,T[$. *In particular, problem* $P_{\gamma,\beta}(f,g,z_0,w_0)$ *has a unique weak solution.*

To prove the above theorem we shall use the Nonlinear Semigroups Theory (cf. [7], [9] or [16]). The natural space to study problem $P_{\gamma,\beta}(f,g,z_0,w_0)$ from this point of view is $X = L^1(\Omega) \times L^1(\partial\Omega)$ provided with the natural norm

$$\|(f,g)\| := \|f\|_{L^1(\Omega)} + \|g\|_{L^1(\partial\Omega)}.$$

To rewrite problem $P_{\gamma,\beta}(f,g,z_0,w_0)$ as an abstract Cauchy problem in X, we define the following operator $\mathcal{B}^{\gamma,\beta}$ in X.

$$\mathcal{B}^{\gamma,\beta} := \Big\{((z,w),(\hat{z},\hat{w})) \in X \times X \ : \ \exists u \in W^{1,p}(\Omega) \text{ such that} \\ [u,z,w] \text{ is a weak solution of } (S^{\gamma,\beta}_{z+\hat{z},w+\hat{w}})\Big\}.$$

As a consequence of the results of Section 3, we have $\mathcal{B}^{\gamma,\beta}$ is a T-accretive operator in X, and its closure is $m-T$-accretive in X. Moreover, we have

Theorem 4.3 ([5]). *Under the hypothesis* $\mathrm{Dom}(\gamma) = \mathbb{R}$, *and* $\mathrm{Dom}(\beta) = \mathbb{R}$ *or* a *smooth, we have*

$$\overline{D(\mathcal{B}^{\gamma,\beta})}^{L^1(\Omega)\times L^1(\partial\Omega)} = \left\{(z,w) \in L^1(\Omega) \times L^1(\partial\Omega) : \gamma_- \le z \le \gamma_+,\ \beta_- \le w \le \beta_+\right\}.$$

Then, by the Nonlinear Semigroups Theory, we know that for every $z_0 \in L^{p'}(\Omega)$, $w_0 \in L^{p'}(\partial\Omega)$ and every $f \in L^1(0,T;L^{p'}(\Omega))$, $g \in L^1(0,T;L^{p'}(\partial\Omega))$, satisfying (4.2) and (4.4) there exists a unique mild solution of the abstract Cauchy problem

$$\begin{cases} V'(t) + \mathcal{B}^{\gamma,\beta}(V(t)) \ni (f,g) & t \in (0,T) \\ V(0) = (z_0,w_0). \end{cases} \tag{4.7}$$

In principle, it is not clear how these mild solutions have to be interpreted with respect to the problem $P_{\gamma,\beta}(f,g,z_0,w_0)$. Now, we show that mild-solution are weak solutions of problem $P_{\gamma,\beta}(f,g,z_0,w_0)$ under the hypothesis of Theorem 4.2, which gives the existence part of Theorem 4.2. To get uniqueness, the main difficulties are due to the jumps of γ and β and the non-homogenous boundary conditions. We see that weak solutions are integral solutions ([7]) and consequently mild-solutions. Since $\mathcal{B}^{\gamma,\beta}$ is T-accretive in X, the contraction principle (4.6) is obtained from the general theory.

Acknowledgment

This work has been performed during the visits of the first, third and fourth authors to the Université de Picardie Jules Verne and the visits of the second author to the Universitat de València. They thank these institutions for their support and hospitality. The first, third and fourth authors have been partially supported by PNPGC project, reference BFM2002-01145.

References

[1] H.W. Alt and S. Luckhaus, *Quasilinear elliptic-parabolic differential equations.* Math. Z., **183** (1983), 311–341.

[2] K. Ammar, F. Andreu and J. Toledo, *Quasi-linear elliptic problems in L^1 with non homogeneous boundary conditions.* To appear in Rend. Mat. Univ. Roma.

[3] K. Ammar and P. Wittbold, *Existence of renormalized solutions of degenerate elliptic-parabolic problems.* Proc. Roy. Soc. Edinburg **133A** (2003), 477–496.

[4] F. Andreu, N. Igbida, J.M. Mazón and J. Toledo, *L^1 Existence and Uniqueness Results for Quasi-linear Elliptic Equations with Nonlinear Boundary Conditions.* To appear in Ann. I. H. Poincaré-AN.

[5] F. Andreu, N. Igbida, J.M. Mazón and J. Toledo, *A degenerate elliptic-parabolic problem with nonlinear dynamical boundary conditions.* To appear in Interface and Free Boundaries.

[6] F. Andreu, J.M. Mazón, S. Segura de León and J. Toledo, *Quasi-linear elliptic and parabolic equations in L^1 with nonlinear boundary conditions.* Adv. Math. Sci. Appl., **7** (1997), 183–213.

[7] Ph. Bénilan, *Equations d'évolution dans un espace de Banach quelconque et applications.* Thesis, Univ. Orsay, 1972.

[8] Ph. Bénilan, L. Boccardo, Th. Gallouët, R. Gariepy, M. Pierre and J.L. Vázquez, *An L^1-theory of existence and uniqueness of solutions of nonlinear elliptic equations.* Ann. Scuola Norm. Sup. Pisa Cl. Sci., **22**(2) (1995), 241–273.

[9] Ph. Bénilan, M.G. Crandall and A. Pazy, *Evolution Equations Governed by Accretive Operators.* Book to appear.

[10] Ph. Bénilan, M.G. Crandall and P. Sacks, *Some L^1 existence and dependence results for semilinear elliptic equations under nonlinear boundary conditions.* Appl. Math. Optim., **17** (1986), 203–224.

[11] Ph. Bénilan and P. Wittbold, *On mild and weak solutions of elliptic-parabolic problems.* Adv. Differential Equations **1** (1996), 1053–1073.

[12] M. Bertsch and J. Hulshof. *Regularity results for an elliptic-parabolic free boundary problem.* Trans. Amer. Math. Soc., **297** (1986), 337–350.

[13] H. Brezis, *Problèmes unilatéraux.* J. Math. Pures Appl. **51** (1972), 1–168.

[14] H. Brezis, *Opérateurs Maximaux Monotones et Semi-groupes de Contractions dans les Espaces de Hilbert.* Oxford Univ. Press, Oxford, 1984.

[15] J. Carrillo and P. Wittbold. *Uniqueness of Renormalized Solutions of Degenerate Elliptic-Parabolic Problems.* Journal of Differential Equations, **156** (1999), 93–121.

[16] M.G. Crandall *An introduction to evolution governed by accretive operators.* In *Dynamical System, An International Symposium,* vol. 1 (L. Cesari et al. eds.), Academic Press, New York, 1976, pages 131–165. Dekker, New York, 1991.

[17] J. Crank, *Free and Moving Boundary Problems,* North-Holland, Amsterdam, 1977.

[18] I. Bejenaru, J.I. Diaz, and I. Vrabie, *An abstract approximate controllability result and applications to elliptic and parabolic system with dynamic boundary conditions.* Electronic J. Diff. Eq., 2001(50):1–19.

[19] E. DiBenedetto and A. Friedman, *The ill-posed Hele-Shaw model and the Stefan problem for supercooler water.* Trans. Amer. Math. Soc., **282** (1984), 183–204.

[20] G. Duvaux and J. L. Lions, *Inequalities in Mechanics and Physics*, Springer-Verlag, 1976.

[21] J. Escher, *Quasilinear parabolic systems with dynamical boundary conditions.* Comm. Part. Diff. Equat., **18** (1993), 1309–1364.

[22] C.M. Elliot and V. Janosky, *A variational inequality approach to the Hele-Shaw flow with a moving boundary.* Proc. Roy. Soc. Edinburgh Sect A., **88** (1981), 93–107.

[23] J. Hulshof, *Bounded weak solutions of an elliptic-parabolic Neumann problem.* Trans. Amer. Math. Soc., **303** (1987), 211–227.

[24] N. Igbida, *The mesa-limit of the porous medium equation and the Hele-Shaw problem.* Diff. Int. Equat., **15** (2002), 129–146.

[25] N. Igbida, *The Hele Shaw Problem with Dynamical Boundary Conditions.* Preprint.

[26] N. Igbida, *Nonlinear Heat Equation with Fast/Logarithmic Diffusion.* Preprint.

[27] N. Igbida and M. Kirane, *A degenerate diffusion problem with dynamical boundary conditions.* Math. Ann., **323**(2002), 377–396.

[28] N. Igbida and J.M. Urbano, *Uniqueness for Nonlinear Degenerate Problems.* NoDEA, **10** (2003), 287–307.

[29] N. Kenmochi, *Neumann problem for a class of nonlinear degenerate parabolic equations.* Diff. Int. Equat., **2**(1990), 253–273.

[30] G.M. Lieberman, *Boundary regularity for solutions of degenerate elliptic equations.* Nonlinear Analysis TMA, **12** (1988), 1203–1219.

[31] J. L. Lions, *Quelques méthodes de résolution de problèmes aux limites non linéaires*, Dunod-Gauthier-Vilars, Paris, 1968.

F. Andreu, N. Igbida, J.M. Mazón
Universitat de Valencia
Dr. Moliner 50
E-46100 Burjassot, Spain
e-mail: Fuensanta.Andreu@uv.es, Mazon@uv.es, jjtoledo@uv.es

J. Toledo
LAMFA, CNRS-UMR 6140
Université de Picardie Jules Verne
33 rue Saint Leu
F-80038 Amiens, France
e-mail: noureddine.igbida@u-picardie.fr

International Series of Numerical Mathematics, Vol. 154, 23–32

Finite Time Localized Solutions of Fluid Problems with Anisotropic Dissipation

S. Antontsev and H.B. de Oliveira

Abstract. In this work we consider an incompressible, non-homogeneous, dilatant and viscous fluid for which the stress tensor satisfies a general non-Newtonian law. The new contribution of this work is the consideration of an anisotropic dissipative forces field which depends nonlinearly on the own velocity. We prove that, if the flow of such a fluid is generated by the initial data, then in a finite time the fluid becomes immobile. We, also, prove that, if the flow is stirred by a forces term which vanishes at some instant of time, then the fluid is still for all time grater than that and provided the intensity of the force is suitably small.

Mathematics Subject Classification (2000). 76A05, 76D05, 76E30, 76D03, 35B99.

Keywords. Non-Newtonian Fluids, anisotropic dissipative field, finite time localization effect.

1. Introduction

1.1. Statement of the problem

In this article we consider incompressible and non-homogeneous non-Newtonian fluids. We assume that there are no inner mass sources and the motions are isothermal. These fluids are driven by the following complete system of equations posed in the cylinder $Q = \Omega \times (0,T) \subset \mathbb{R}^N \times \mathbb{R}^+$:

$$\frac{\partial \rho}{\partial t} + \mathbf{u} \cdot \nabla \rho = 0, \quad \operatorname{div} \mathbf{u} = 0; \tag{1.1}$$

$$\rho \left(\frac{\partial \mathbf{u}}{\partial t} + (\mathbf{u} \cdot \nabla)\mathbf{u} \right) = \operatorname{div} \mathbf{S} + \rho\, \mathbf{f}; \tag{1.2}$$

$$\mathbf{S} = -p\mathbf{I} + \mathbf{F}(\mathbf{D}), \quad \mathbf{D} = \frac{1}{2}\left(\nabla \mathbf{u} + \nabla \mathbf{u}^T\right). \tag{1.3}$$

The work of the first author was partially supported by the project FCT-DECONT-UBI.
The work of the second author was partially supported by the project FCT-POCTI-ISFL-209.

In these equations, $\mathbf{u}$, ρ and p are, respectively, velocity, density and pressure in the fluid, and $\mathbf{f}$ is the prescribed mass force. $\mathbf{S}$, $\mathbf{D}$ and $\mathbf{I}$ are, respectively, the stress tensor, the tensor of rate of deformation and the unit tensor. The domain Ω considered here is bounded and its boundary $\partial\Omega$ is assumed to be Lipschitz.

System (1.1)–(1.3) is endowed with the initial and boundary conditions:

$$\mathbf{u}(\mathbf{x},0) = \mathbf{u}_0(\mathbf{x}), \qquad \rho(\mathbf{x},0) = \rho_0(\mathbf{x}) \qquad \text{in} \quad \Omega; \tag{1.4}$$

$$\mathbf{u}(\mathbf{x},t) = \mathbf{0} \quad \text{on} \quad \Gamma_T = \partial\Omega \times (0,T). \tag{1.5}$$

The tensor $\mathbf{F}$ is symmetric and is assumed to satisfy, for all $\mathbf{u} \in \mathbb{R}^N$,

$$\delta\, |\mathbf{D}(\mathbf{u})|^q \leq \mathbf{D}(\mathbf{u}) : \mathbf{F}(\mathbf{u}) \equiv \sum_{i,j=1}^{N} F_{ij} D_{ij}, \quad 0 < \delta = \delta(\rho) < \infty, \ 1 \leq q < \infty. \tag{1.6}$$

Fluids satisfying condition (1.6) are called viscous-plastic if $1 < q < 2$ and dilatant if $q > 2$. Classical Navier-Stokes equations correspond to $q = 2$ and, in this case, for incompressible homogeneous viscous fluids the stress tensor $\mathbf{S}$ has the form $\mathbf{S} = -p\mathbf{I} + 2\mu\mathbf{D}$, where μ is the shear viscosity.

The new contribution of this work, is the consideration of a forces field $\mathbf{f}$ in (1.2) such that

$$\mathbf{f}(\mathbf{x},t,\mathbf{u}) = \mathbf{h}(\mathbf{x},t,\mathbf{u}) + \mathbf{g}(\mathbf{x},t), \tag{1.7}$$

where $\mathbf{g}$ is a given function and $\mathbf{h}$ depends nonlinearly on the velocity field $\mathbf{u}$:

$$-\mathbf{h}(\mathbf{x},t,\mathbf{u}) \cdot \mathbf{u} \geq \sum_{i=1}^{N} \delta_i\, |u_i|^{\sigma_i} \quad \forall\, \mathbf{u} \in \mathbb{R}^N, \qquad \sigma_i \in (1,2), \tag{1.8}$$

for some non-negative constants δ_i, with $i = 1,\ldots,N$. We have the following examples of forces fields $\mathbf{f}$ satisfying (1.7) and (1.8):

$$\text{with } \mathbf{g} \equiv \mathbf{0}, \qquad \mathbf{f}(\mathbf{x},t,\mathbf{u}) = -(\delta_1\, |u_1|^{\sigma_1 - 2} u_1, \ldots, \delta_N\, |u_N|^{\sigma_N - 2} u_N); \tag{1.9}$$

$$\text{with } \mathbf{g} \neq \mathbf{0}, \quad \mathbf{f}(\mathbf{x},t,\mathbf{u}) = -(\delta_1\, |u_1|^{\sigma_1 - 2} u_1, \ldots, \delta_N\, |u_N|^{\sigma_N - 2} u_N) - \mathbf{g}(\mathbf{x},t), \tag{1.10}$$
$$\|\mathbf{g}(\cdot,t)\|_{2,\Omega} \leq C\,(1 - t/t_{\mathbf{g}})_+^{\nu},$$

for some positive constants C, ν, $t_{\mathbf{g}}$ and where $u_+ = \max(0,u)$. Notice the constants $\delta_1,\ldots,\delta_N$ in (1.8) are non-negative and, thus, only one component of the vector field $(\delta_1\, |u_1|^{\sigma_1-2} u_1, \ldots,\ \delta_N\, |u_N|^{\sigma_N-2} u_N)$ appearing in examples (1.9) and (1.10) can be zero.

From the Fluid Mechanics point of view, condition (1.8) means the forces field $\mathbf{f}$ is a feedback term, as one can see from the examples (1.9) and (1.10). This feedback is presented as an anisotropic condition, because the dependence of $\mathbf{f}$ on $\mathbf{u}$ may be different for distinct directions. Moreover, from condition (1.8), we can say the feedback forces field $\mathbf{h}$, and thus $\mathbf{f}$, is dissipative, in order to each component u_k, in all directions x_k where $\delta_k > 0$, for $k = 1,\ldots,N$.

Definition 1.1. We say the weak solutions $(\mathbf{u},\rho)$ of the problem (1.1)–(1.5) possesses the finite time localization property if there exists (a finite time) $t^* \in (0,\infty)$ such that $\mathbf{u}(\mathbf{x},t) = \mathbf{0}$ a.e. in Ω and for all $t \geq t^*$.

1.2. Motivation

In [5, §4.7] was considered, for the first time, the assumption that the forcing term $\mathbf{f}$ in (1.2) to depend on $\mathbf{u}$ and to obey

$$\forall\, \mathbf{u} \in \mathbb{R}^N \quad \delta_0 |\mathbf{u}|^\sigma \leq -\mathbf{f}(x,t,\mathbf{u}) \cdot \mathbf{u}, \quad \sigma \in (1,2), \tag{1.11}$$

with some positive constant δ_0. It was assumed that

$$E(0) = \frac{1}{2} \int_\Omega \rho(x,0) |\mathbf{u}(\mathbf{x},0)|^2 \, d\mathbf{x} < \infty, \qquad 1/M \leq \rho_0 \leq M \equiv Const. \tag{1.12}$$

and were considered weak solutions $(\mathbf{u}, \rho)$ of problem (1.1)–(1.5) such that

$$\mathbf{u} \in \mathrm{L}^\infty(0,T;\mathbf{L}^2(\Omega)) \cap \mathrm{L}^\sigma(0,T;\mathbf{L}^\sigma(\Omega)) \cap \mathrm{L}^2(0,T;\mathbf{W}_0^{1,q}(\Omega)) \quad \text{and} \quad \frac{1}{M} \leq \rho \leq M.$$

First, were considered pseudo-plastic fluids for which (1.6) holds with $\delta = Const.$ and $q \in (2N/(N+2), 2)$. The finite time localization property was proved in the case of $\mathbf{f} = \mathbf{0}$ with t^* expressed by an explicit formulae (see [5, Th. 7.1]). The same property was proved also for a given $t_{\mathbf{f}} > t^*$ if $\|\mathbf{f}(\cdot,t)\|_{p,\Omega} \leq \epsilon \, (1 - t/t_{\mathbf{f}})_+^\nu$ for all $t > 0$. Here $p = Nq/[N(q-1)+q]$, ν is some positive constant and ϵ is a sufficiently small positive constant. Notice that, in the limit case $\epsilon = 0$, we have $\mathbf{f} \equiv \mathbf{0}$. In this case and assuming (1.6), it can be proved the following results (see [5, §4.7] – see also [9, 11]): if $q = 2$, the norm $\|\mathbf{u}\|_{2,\Omega}$ has a time exponential decay, for $t \geq t_{\mathbf{g}}$; if $q > 2$, the norm $\|\mathbf{u}\|_{2,\Omega}$ has a time power decay, for $t \geq t_{\mathbf{g}}$.

Then, were considered pseudo-plastic fluids with vanishing or unbounded density, *i.e.*, assuming that $\|1/\rho_0\|_{m,\Omega} \leq C_1$, $\|\rho_0\|_{M,\Omega} \leq C_2$, and $\min(m, M) > 1$, for some positive constants C_1 and C_2. Under the assumptions that $\sqrt{\rho_0}\, \mathbf{u}_0 \in \mathbf{L}^2(\Omega)$ and (1.6 holds with $\delta = Const.$ and $q \in (2MN/[N(M-1)+2M], 2)$, $M > N/2$, were proved analogous results about finite time localized weak solutions (see [5, Th. 7.2]).

Notice that to obtain these results there is no need to assume condition (1.11) on the forces field $\mathbf{f}$. In these cases, *i.e.*, pseudo-plastic fluids ($q < 2$), the localization effects are determined only by the structure of the tensor $\mathbf{S}$.

Next, was considered the case $q \geq 2$, dilatant ($q > 2$) and Newtonian ($q = 2$) fluids. Under the assumptions that (1.11) and (1.6) with $\delta = \mathrm{Const.}$ and $q \in [2, N)$ hold, was proved the finite time localization property (see [5, §4.7]). In this case, the finite time localization property is determined by the feedback forcing term $\mathbf{f}$.

A few time later, this kind of forces field was improved in planar stationary homogeneous incompressible fluid problems, governed by the classical Navier-Stokes equations. In [1, 2, 3, 4] has been established the property of finite space localization for the velocity $\mathbf{u}$, *i.e.*, has been proved the solutions $\mathbf{u}$ of those problems have compact support in Ω. The isothermal 2-dimensional cases of Stokes and Navier-Stokes problems were considered in a semi-infinite strip $\Omega = (0,\infty) \times (0, L)$, $L = \mathrm{Const.} > 0$, in [1, 2, 3]. There, has been assumed the body forces $\mathbf{f}$ satisfies, for every $\mathbf{u} \in \mathbb{R}^2$, $\mathbf{u} = (u, v)$, and almost all $\mathbf{x} \in \Omega$

$$-\mathbf{f}(\mathbf{x}, \mathbf{u}) \cdot \mathbf{u} \geq \delta \, |u|^\sigma - g(\mathbf{x}),, \qquad \delta = \mathrm{Const.} > 0, \qquad 1 < \sigma < 2,$$

for some function $g \in \mathrm{L}^1(\Omega^{x_g})$, where $\Omega^{x_g} = (0, x_g) \times (0, L)$ and $0 \le x_g < \infty$, such that $g \ge 0$ and $g(\mathbf{x}) = 0$ a.e. in $\Omega_{x_g} = (x_g, \infty) \times (0, L)$. The exothermic version of the Navier-Stokes problem was consider in [4], where it was assumed, for every $\mathbf{u} \in \mathbb{R}^2$, $\mathbf{u} = (u, v)$, $\theta \in [m, M]$ (θ is the temperature), with $m < M$ constants, and almost all $\mathbf{x} \in \Omega$

$$-\mathbf{f}(\mathbf{x}, \theta, \mathbf{u}) \cdot \mathbf{u} \ge \delta\, |u|^{1+\sigma(\theta)} - g(\mathbf{x}, \theta), \qquad \delta = \text{Const.} > 0,$$

for some function $g \in \mathrm{L}^1(\Omega^{x_g} \times \mathbb{R})$ such that $g \ge 0$, $g(\mathbf{x}, \theta) = 0$ a.e. in Ω_{x_g} for every $\theta \in [m, M]$. Here, σ is a Lipschitz continuous function such that $0 < \sigma^- \le \sigma(\theta) \le \sigma^+ < 1$ for every $\theta \in [m, M]$.

In this paper, we are interested in generalize some results of [5, §4] by considering the improvements of the forces fields considered in [1, 2, 3, 4]. Particularly, we will analyse the following very interesting question: does the property of finite time localization, for the weak solutions $\mathbf{u}$ of (1.1)–(1.8), may be assured by dissipation of the forces field $\mathbf{f}$ merely in one direction? Recall that dissipation in order to u_k, in the direction x_k, with $k = 1, \dots, N$, corresponds to assume $\delta_k > 0$ in (1.7)–(1.8). We will prove the weak solutions $(\mathbf{u}, \rho)$ of problem (1.1)–(1.5) are finite time localized, if the stress tensor $\mathbf{S}$ satisfies (1.6) for appropriated values of q and the forces field $\mathbf{f}$ satisfies (1.7)–(1.8).

2. Main results

2.1. Weak formulation

We are interested in weak solutions $(\rho, \mathbf{u})$ to the problem (1.1)–(1.5) such that

$$E(t) + \int_\Omega |\nabla \mathbf{u}|^q \, d\mathbf{x} < \infty, \quad \text{where} \quad E(t) = \frac{1}{2} \int_\Omega \rho(\mathbf{x}, t)\, |\mathbf{u}(\mathbf{x}, t)|^2 \, d\mathbf{x}, \tag{2.1}$$

$$1/M \le \rho \le M, \qquad M \equiv Const. > 0 \tag{2.2}$$

Let us introduce the vector function space

$$\mathbf{J}^r(\Omega) = \left\{ \mathbf{u} \in \mathbf{L}^r(\Omega) : \int_\Omega \mathbf{u} \cdot \nabla \phi \, d\mathbf{x} = 0, \quad \nabla \phi \in \mathbf{L}^{r}(\Omega) \right\}$$

and consider the class of vector functions $\mathbf{u} \in \mathbf{W}_{q,\overline{\sigma}}$, $\mathbf{u} = (u_1, \dots, u_N)$, where

$$\begin{aligned} \mathbf{W}_{q,\overline{\sigma}} =& \{\mathbf{u} \in \mathrm{L}^\infty\left(0, T; \mathbf{L}^2(\Omega)\right) \cap \mathrm{L}^2(0, T; \mathbf{W}_0^{1,q}(\Omega)) : \overline{\sigma} = (\sigma_1, \dots, \sigma_N), \\ & \sigma_i \in (1, 2) \quad \text{and} \quad u_i \in \mathrm{L}^{\sigma_i}(0, T; \mathrm{L}^{\sigma_i}(\Omega)) \quad \forall\, i = 1, \dots, N\}. \end{aligned}$$

Definition 2.1. A pair of functions $(\mathbf{u}, \rho)$ is called a weak solution of problem (1.1)–(1.5), if $\mathbf{u} \in \mathbf{W}_{q,\overline{\sigma}}$, ρ satisfies (2.2) and if the following integral identities:

$$-\int_0^T \int_\Omega \rho \left[\mathbf{u} \cdot \mathbf{\Phi}_t + \mathbf{u} \otimes \mathbf{u} : \nabla \mathbf{\Phi}\right] d\mathbf{x}\, dt \, +$$

$$\int_0^T \int_\Omega \mathbf{F}(\mathbf{u}) : \nabla \mathbf{\Phi} \, d\mathbf{x}\, dt = \int_0^T \int_\Omega \rho\, \mathbf{f} \cdot \mathbf{\Phi} \, d\mathbf{x}\, dt - \int_\Omega \rho_0\, \mathbf{u}_0 \cdot \mathbf{\Phi}(0) \, d\mathbf{x}\,;$$

$$\int_0^T \int_\Omega \rho\left[\varphi_t + (\mathbf{u}\cdot\nabla)\varphi\right] d\mathbf{x}\,dt + \int_\Omega \rho_0\varphi(0)\,d\mathbf{x} = 0\,;$$

are fulfilled for any $\mathbf{\Phi} \in \mathrm{C}^1\left(0,T;\mathbf{J}^1(\Omega)\right)$ and $\varphi \in \mathrm{C}^1\left(0,T;\mathrm{H}^1(\Omega)\right)$ such that $\mathbf{\Phi}(\mathbf{x},T) = 0$ and $\varphi(\mathbf{x},T) = 0$.

According to [5, §4.7] and references therein problem (1.1)–(1.5) has, at least, one weak solution, if the mass force term does not depend on $\mathbf{u}$, *i.e* if we consider $\mathbf{f} = \mathbf{f}(\mathbf{x},t)$ in a suitable functional space. Moreover, at least formally, every weak solution of problem (1.1)–(1.5) satisfies the energy relation

$$\frac{d}{dt}E(t) + \int_\Omega \mathbf{F}(\mathbf{u}) : \mathbf{D}(\mathbf{u})\,d\mathbf{x} = \int_\Omega \rho\mathbf{f}\cdot\mathbf{u}\,d\mathbf{x}, \tag{2.3}$$

where $E(t)$ is given in (2.1). The derivation of (2.3) relies on (1.1)–(1.3), the symmetry of the tensor $\mathbf{F}$, integration-by-parts formulae and boundary condition (1.5).

Now, we assume the forces field $\mathbf{f}$ satisfies (1.7)–(1.8) for a given vector field $\mathbf{g}$. Moreover, we assume the tensor $\mathbf{F}$, given in (1.3), is such that

$$\text{(1.6) holds with } q \geq 2. \tag{2.4}$$

Thus energy relation (2.3), assumptions (2.2) and (2.4), Korn's inequality (see Lemma 2.5) and Sobolev Embedding Theorem, lead us to the estimate

$$\begin{aligned} &\sup_{0\leq t\leq T} E(t) + \int_0^T \int_\Omega \left(\delta\,|\nabla\mathbf{u}|^q + \sum_{i=1}^N \delta_i\,|u_i|^{\sigma_i}\right) d\mathbf{x}\,dt \\ \leq & C(M)\left(E(0) + \int_0^T \left(\int_\Omega |\mathbf{g}(\mathbf{x},t)|^p\,d\mathbf{x}\right)^{\frac{q}{p(q-1)}} dt\right), \end{aligned} \tag{2.5}$$

where $p = Nq/[N(q-1)+q]$. In our further study we assume the existence of, at least, one weak solution of problem (1.1)–(1.5) in the sense of Definition 2.1.

We give here only the ideas of the proof. We consider the three main different cases: $1 \leq q < 2$, $q = 2$ and $q > 2$. For each one of such cases, and for different constitutive laws, there are known existence results for suitable forces given in appropriated function spaces. For $1 \leq q < 2$ and $\mathbf{S} = -p\mathbf{I} + 2\mu\mathbf{D} + \alpha D_{II}^{(q-2)/2}\mathbf{D}$, the existence of a weak solution is proved in [7]. In [6] was proved the existence of a weak solution for $\mathbf{S} = -p\mathbf{I} + 2\mu\mathbf{D}$ (and $q = 2$). Finally, for $\mathbf{S} = -p\mathbf{I} + 2\mu\mathbf{D} + \alpha D_{II}^{(q-2)/2}\mathbf{D}$ and $q \geq 2$, the existence of a weak solution was also proved in [7]. In our problem (1.1)–(1.5), the idea is to use energy estimate (2.5), assumption (2.2), repeating the corresponding arguments of [6, 7] (see also [8, 10]) and to use a fixed point argument. We hope to publish these results elsewhere as soon as possible.

Notice that, according to (2.5), every weak solution satisfies

$$\int_\Omega \left(\delta\,|\nabla\mathbf{u}|^q + \sum_{i=1}^N \delta_i\,|u_i|^{\sigma_i}\right) d\mathbf{x} \in \mathbf{L}^1[0,T].$$

2.2. Finite time localization

We will prove that the weak solutions of (1.1)–(1.5) are finite time localized, if condition (2.4) is satisfied and the forces field $\mathbf{f}$ satisfies (1.7)–(1.8) and exhibits dissipation effect in one direction. For the sake of simplicity, we assume that is $\delta_N = 0$, *i.e.*

$$\text{(1.8) holds with } \delta_N = 0 \text{ and } \delta_j \neq 0 \text{ for all } j \neq N. \tag{2.6}$$

Let us consider the following hypothesis on the domain Ω.

Hypothesis A. The domain Ω is convex, at least, in the x_N direction.

From this assumption, we can say that each line parallel to the x_N axis intersects the boundary $\partial\Omega$ only on two points, say, $\mathbf{x}_N^0 \equiv (x_1^0, \dots, x_{N-1}^0)$ and $\mathbf{x}_N^1 \equiv (x_1^1, \dots, x_{N-1}^1)$, with $x_N^0 \leq x_N^1$.

Theorem 2.2. *Let* $(\mathbf{u}, \rho)$ *be a weak solution of problem* (1.1)–(1.5) *in the sense of Definition* 2.1. *Assume that conditions* (1.12), (2.2), (2.4), (2.6) *and* **Hypothesis A** *are satisfied.*

1. *If* $\mathbf{g}(\mathbf{x}, t) \equiv 0$, *then there exists* $t^* > 0$ *such that* $E(t) = 0$ *for almost all* $t \geq t^*$. *In particular,* $\mathbf{u} \equiv 0$ *in* $Q \cap \{t \geq t^*\}$.
2. *Let* $\mathbf{g} \not\equiv 0$ *satisfy*

$$\|\mathbf{g}(\cdot, t)\|_{p,\Omega} \leq \epsilon \left(1 - \frac{t}{t_{\mathbf{g}}}\right)_+^{\frac{q-1}{q(\mu-1)}}, \qquad p = \frac{Nq}{N(q-1)+q}, \qquad \mu > 1, \tag{2.7}$$

where μ *depends on* p, q, N *and* $\sigma_1, \dots, \sigma_{N-1}$. *Then there exists a constant* $\epsilon_0 > 0$ *such that* $E(t) = 0$ *for almost all* $t \geq t_g$, *if* $\epsilon \geq \epsilon_0 > 0$. *In particular,* $\mathbf{u} \equiv 0$ *in* $Q \cap \{t \geq t_{\mathbf{g}}\}$, *if* $\epsilon \geq \epsilon_0 > 0$.

Remark 2.3. The mechanical sense of this theorem is that if the flow of a non-Newtonian dilatant fluid with the dissipative term satisfying (1.8) with $\mathbf{g} \equiv \mathbf{0}$ is generated by the initial data, then in a finite time the fluid becomes immobile. If the flow is stirred by the source term $\mathbf{g} \neq \mathbf{0}$ which vanishes at the instant $t_{\mathbf{g}}$, then the fluid is still for all $t \geq t_{\mathbf{g}}$ provided the intensity of the source is suitably small.

The main tool to prove Theorem 2.2 are two well-known results, usually denoted in the literature, as the interpolation embedding inequality and the Korn's inequality. For these results see [5, §4.7.2 and Appendix 3] and references therein.

Lemma 2.4 (Interpolation Embedding). *Let* $u \in \mathrm{W}_0^{1,p}(\Omega)$, $1 < p < \infty$. *Then the following interpolation inequality holds:*

$$\|u\|_{q,\Omega} \leq C(\|\nabla u\|_{p,\Omega})^{\theta} (\|u\|_{r,\Omega})^{1-\theta}, \quad \theta = \frac{1/r - 1/q}{1/r - 1/p^*}, \quad p^* = Np/(N-p). \tag{2.8}$$

Here $\;C = ((N-1)/Np^*)^{\theta}, \quad \theta \in [0,1], \quad q \in [r, p^*]$ (*or* $q \in [p^*, r]$), *if* $p < N$;

$C = \max\left((N-1)/Nq, 1 + (p-1)/pr\right)^{\theta}, \quad \theta \in [0,1], \quad q \in [r, \infty)$, *if* $N \leq p$.

If $p > N$, then (2.8) is true for $q = \infty$ with $\theta = Np/[Np + r(p - N)]$ and some constant $C < \infty$ not depending on Ω.

Lemma 2.5 (Korn). *Let $1 \leq p, r < \infty$ and $\mathbf{D}$ be the tensor of rate of deformation. If $\mathbf{u} \in \mathbf{W}_0^{1,p}(\Omega)$, then the following inequality holds*

$$\frac{1}{C}\|\nabla \mathbf{u}\|_{p,\Omega} \leq \|\mathbf{D}(\mathbf{u})\|_{p,\Omega} \leq C\|\nabla \mathbf{u}\|_{p,\Omega}, \quad C = C(p, N, \Omega),\ K = K(p, N, \Omega). \tag{2.9}$$

Proof of Theorem 2.2. We will split this proof into several steps.
Step 1. Applying assumptions (2.6), (2.4) and (2.2) to the energy relation (2.3), and using Korn's inequality (2.9), we obtain

$$\frac{d}{dt}E(t) + CE_{q,\sigma}(t) \leq \int_\Omega \mathbf{u}\cdot\mathbf{g}\,d\mathbf{x},\ E_{q,\sigma}(t) = \int_\Omega \left(|\nabla \mathbf{u}|^q + \sum_{i=1}^{N-1} |u_i|^{\sigma_i}\right) d\mathbf{x}, \tag{2.10}$$

where $C = C_1(K, M, N, \delta_0)$ with $\delta_0 = \min_{i=1,\ldots,N} \delta_i > 0$, K is the Korn's inequality constant and, from (1.8), $1 < \sigma_i < 2$ for all $i = 1, \ldots, N-1$. Notice that $\mathbf{u} \in \mathrm{L}^2(0,T; \mathbf{W}^{1,q}(\Omega))$, $u_i \in \mathrm{L}^{\sigma_i}(0,T; \mathrm{L}^{\sigma_i}(\Omega))$ for all $i = 1, \ldots, N-1$.
Step 2. In this step we establish the following result.

Lemma 2.6. *Let $\mathbf{u} \in \mathrm{L}^2(0,T; \mathbf{W}^{1,q}(\Omega) \cap \mathbf{J}^q(\Omega))$, $u_i \in \mathrm{L}^{\sigma_i}(0,T; \mathrm{L}^{\sigma_i}(\Omega))$ for all $i = 1, \ldots, N-1$. and the* **Hypothesis A** *be fulfilled. Then, for almost all $t \in [0,T]$, the inequality*

$$E(t)^{\frac{1}{\mu}} \leq C\, E_{q,\sigma}(t), \qquad E(t) \equiv \frac{1}{2}\int_\Omega \rho\, |\mathbf{u}|^2\, d\mathbf{x}, \tag{2.11}$$

holds for some positive constants $C = C(\|\mathbf{u}\|_{2,\Omega}, q, N, \sigma_i, \Omega)$, $\mu = \mu(q, N, \sigma_i) > 1$, $i = 1, \ldots, N-1$.

Proof. Without loss of generality, we assume that $\mathbf{u} \in \mathrm{C}^2(\Omega)$ for almost all $t \in [0,T]$. Using the interpolation embedding inequality (2.8), with $q = 2$, $p = q$, $r = \sigma_i$, and Young's inequality, we can write, for any scalar component u_i of vector $\mathbf{u}$, with $i = 1, \ldots, N-1$,

$$\|u_i\|_{2,\Omega}^2 \leq C\left[\int_\Omega \left(|\nabla u_i|^q + |u_i|^{\sigma_i}\right) d\mathbf{x}\right]^{\mu_i}, \qquad \mu_i = 1 + \frac{q(2-\sigma_i)}{q(N+\sigma_i) - N\sigma_i}. \tag{2.12}$$

Notice that the assumptions $1 < \sigma_i < 2$ and $q \geq 2 > \frac{2N}{N+1}$ assure that $\mu_i > 1$ for any $i = 1, \ldots, N-1$. Assuming, without loss of generality, that $\|\mathbf{u}\|_{2,\Omega}^2 \leq 1$, we can rewrite (2.12), and for all $i = 1, \ldots, N-1$, in the form

$$\|u_i\|_{2,\Omega}^2 \leq C\left(E_{q,\sigma}(t)\right)^{\mu_{N-1}}, \qquad \mu_{N-1} = \min_{1\leq i\leq N-1} \mu_i. \tag{2.13}$$

Next we need to derive an analogous estimate for the last component u_N of vector $\mathbf{u}$. Recall that the force term is non-dissipative, in order to u_N, with respect to the x_N direction ($\delta_N = 0$). To establish that, we introduce the hyperplane

$$\Omega(z) = \Omega \cap \{\mathbf{x} = (\mathbf{x}', x_N) \in \mathbb{R}^N : \mathbf{x}' = (x_1, \ldots, x_{N-1}) \quad \text{and} \quad x_N = z\} \subseteq \mathbb{R}^{N-1}.$$

We formally multiply the continuity equation

$$\operatorname{div}\mathbf{u} = 0, \qquad \mathbf{u} = (u_1, \ldots, u_N) \in \mathbf{J}^q(\Omega),$$

by u_N and integrating by parts over $\Omega_(z)$, we came to

$$-\frac{1}{2}\int_{\Omega(z)} \frac{\partial u_N^2}{\partial x_N}\, d\mathbf{x}' = \int_{\Omega(z)} \sum_{i=1}^{N-1} \frac{\partial u_N}{\partial x_i} u_i\, d\mathbf{x}'.$$

But, according to the boundary condition (1.5), we have

$$\frac{1}{2}\frac{\partial}{\partial z}\int_{\Omega(z)} u_N^2\, d\mathbf{x}' = \int_{\Omega(z)} \sum_{i=1}^{N-1} \frac{\partial u_N}{\partial x_i} u_i\, d\mathbf{x}'. \tag{2.14}$$

Let us consider x_N^0 such that $(x_1, \ldots, x_{N-1}, x_N^0) \in \partial\Omega$. Integrating (2.14) with respect to $z \in [x_N^0, x_N]$ and applying, once more, the boundary condition (1.5),

$$\int_{\Omega(z)} u_N^2\, d\mathbf{x}' = 2\int_{x_N^0}^{x_N}\int_{\Omega(z)} \sum_{i=1}^{N-1} \frac{\partial u_N}{\partial x_i} u_i\, d\mathbf{x}'\, dz.$$

We apply Hölder's inequality to obtain

$$\|u_N\|_{2,\Omega(z)}^2 \le C\,\|\nabla u_N\|_{q,\Omega} \sum_{i=1}^{N-1} \|u_i\|_{q',\Omega}\,,$$

where $C = C(N)$. Integrating the last inequality with respect to z and using, again, Hölder's inequality, we achieve to the estimate

$$\|u_N\|_{2,\Omega}^2 \le C\,\|\nabla u_N\|_{q,\Omega} \sum_{i=1}^{N-1} \|u_i\|_{q',\Omega} \le C\,\|\nabla u_N\|_{q,\Omega} \sum_{i=1}^{N-1} \|u_i\|_{2,\Omega}\,,$$

where $C = C(q, N, \Omega)$. Now, applying (2.13) and the definition of $E_{q,\sigma}(t)$ (see 2.10), we came to the inequality

$$\|u_N\|_{2,\Omega}^2 \le C\,(E_{q,\sigma}(t))^{\mu_N}\,, \qquad \mu_N = \frac{1}{q} + \frac{\mu_{N-1}}{2} > 1, \tag{2.15}$$

where $C = C(\|u_i\|_{2,\Omega}\,, q, N, \sigma_i, \Omega)$, $i = 1, \ldots, N-1$. Finally, combining (2.13) and (2.15), we obtain (2.11), where $\mu = \min_{1\le i\le N} \mu_i$. □

Step 3. If $g \equiv 0$, using (2.11) and the energy relation (2.10), we come to the homogeneous ordinary differential inequality

$$\frac{d}{dt}E(t) + CE(t)^{1/\mu} \le 0 \qquad \text{for all } t \ge 0. \tag{2.16}$$

An explicit integration of (2.16) between $t = 0$ and $t \ge t_{\mathbf{g}}$ proves the first assertion.

If $g \not\equiv 0$, we use the estimate

$$\left|\int_\Omega \mathbf{u}\cdot\mathbf{g}\, d\mathbf{x}\right| \le \|\mathbf{g}\|_{p,\Omega}\,\|\mathbf{u}\|_{p',\Omega} \le C\|\mathbf{g}\|_{p,\Omega}\|\nabla\mathbf{u}\|_{q,\Omega} \le \varepsilon\|\nabla\mathbf{u}\|_{q,\Omega} + C(\varepsilon)\|\mathbf{g}\|_{p,\Omega}^{\frac{q}{q-1}}\,,$$

valid for some $\varepsilon \in (0,1)$, and, also, estimate (2.11), assumption (2.7) and the energy relation (2.10), to achieve the nonhomogeneous ordinary differential inequality

$$\frac{d}{dt}E(t) + C_1 E(t)^{1/\mu} \leq C_2 \left(1 - \frac{t}{t_{\mathbf{g}}}\right)_+^{\frac{q-1}{q(\mu-1)}} \qquad \text{for all } t \geq 0.$$

The analysis of this inequality, which have been considered in [5, §1.2], proves the second assertion. □

Example. Let us consider the very interesting particular case of problem (1.1)–(1.5) when $N = 2$ and assume the fluid is homogeneous. The incompressibility condition given in (1.1) and the homogeneity of the fluid allow us to consider problem (1.1)–(1.5) in the form

$$\operatorname{div}\mathbf{u} = 0, \qquad \frac{\partial \mathbf{u}}{\partial t} + (\mathbf{u}\cdot\nabla)\mathbf{u} = \frac{1}{\rho_0}\operatorname{div}\mathbf{S} + \mathbf{f}; \tag{2.17}$$

$$\mathbf{u}(\mathbf{x},0) = \mathbf{u}_0(\mathbf{x}) \quad \text{in} \quad \Omega \qquad \text{and} \qquad \mathbf{u}(\mathbf{x},t) = \mathbf{0} \quad \text{on} \quad \Gamma_T = \partial\Omega \times (0,T). \tag{2.18}$$

where ρ_0 is the constant density of the fluid, the stress tensor $\mathbf{S}$ satisfies (1.3) and the tensor $\mathbf{F}$ satisfies (2.4) with $N = 2$. By classical techniques, we reduce problem (2.17)–(2.18) to the consideration of the problem posed by the following equations

$$-\triangle\psi_t + \psi_x\triangle\psi_y - \psi_y\triangle\psi_x = (\operatorname{div}\mathbf{S})_y - (\operatorname{div}\mathbf{S})_x + \frac{\partial f_2}{\partial x} - \frac{\partial f_2}{\partial y}, \tag{2.19}$$

$$\psi(\mathbf{x},0) = \int_0^y u_0(x,s)\,ds \text{ in } \Omega, \quad \psi = 0 \text{ and } \frac{\partial\psi}{\partial\mathbf{n}} = 0 \text{ on } \Gamma_T = \partial\Omega\times(0,T), \tag{2.20}$$

for the stream function $\psi \equiv \psi(x,y) : \mathbf{u} = (u,v) = (\psi_y, -\psi_x)$. We formally multiply equation (2.19) by a weak solution ψ and integrate by parts over Ω, using (2.20) and (2.4) to obtain

$$\frac{d}{dt}E(t) + CE_{q,\sigma}(t) \leq \int_\Omega (\psi_{\mathbf{y}}, -\psi_{\mathbf{x}})\cdot\mathbf{g}\,d\mathbf{x}, \quad E_{q,\sigma}(t) = \int_\Omega \left(|\mathrm{D}^2\psi|^q + |\psi_y|^{\sigma_1}\right)d\mathbf{x},$$

where $E(t) = 1/2\int_\Omega |\nabla\psi|^2\,d\mathbf{x} \equiv 1/2\int_\Omega |\mathbf{u}|^2\,d\mathbf{x}$. Proceeding as in the proof of Theorem 2.2, we obtain

$$\|\psi_y\|_{2,\Omega}^2 \leq C\left(E_{q,\sigma}(t)\right)^{\mu_1}, \qquad \mu_1 = 1 + \frac{q(2-\sigma_1)}{q(2+\sigma_1)-2\sigma_1} > 1.$$

Then, we multiply the equation $\psi_{yx} - \psi_{xy} = 0$ by $-\psi_x$, where ψ is the stream function associated to a function $\mathbf{u}\in\mathbf{J}^q(\Omega)$, and integrating by parts over $\Omega(z) = \Omega\cap\{(x,y)\in\mathbb{R}^2 : y = z\} \subseteq \mathbb{R}$, we come to

$$-\frac{1}{2}\int_{\Omega(z)} \frac{\partial\psi_x^2}{\partial y}\,dx = \int_{\Omega(z)} \psi_{xx}\psi_x\,dx.$$

Again, proceeding as in the proof of Theorem 2.2, we achieve to the estimate

$$\|\psi_x\|_{2,\Omega}^2 \leq C\left\|\mathrm{D}^2\psi\right\|_{q,\Omega}\|\psi_y\|_{q',\Omega} \leq C\left\|\mathrm{D}^2\psi\right\|_{q,\Omega}\|\psi_y\|_{2,\Omega},$$

where $C = C(N,q,\Omega)$. The rest of the proof follows just in the same manner.

Remark 2.7. The results established in Theorem 2.2 can be extended to unbounded domains satisfying Hypothesis A. The proof is almost the same, we only need to use the known Korn's and interpolation embedding inequalities (Gagliardo-Nirenberg) for these domains. See the papers, cited in [5], by Kondratiev and Oleinik for Korn's inequality in unbounded domains, and by Gagliardo and Nirenberg for the interpolation embedding inequality, also, in unbounded domains.

References

[1] S. Antontsev, J.I. Díaz and H.B. de Oliveira, *On the confinement of a viscous fluid by means of a feedback external field.* C.R. Mecanique, **330** (2002), 797–802.

[2] S. Antontsev, J.I. Díaz and H.B. de Oliveira, *Stopping a viscous fluid by a feedback dissipative field. I. The stationary Stokes problem.*, J. Math. Fluid Mech., no. 4, **6** (2004), 439–461.

[3] S. Antontsev, J.I. Díaz and H.B. de Oliveira, *Stopping a viscous fluid by a feedback dissipative field. II. The stationary Navier-Stokes problem.* Atti Accad. Naz. Lincei Cl. Sci. Fis. Mat. Natur. Rend. Lincei (9) Mat. Appl., no. 3-4, **15** (2004), 257–270.

[4] S. Antontsev, J.I. Díaz and H.B. de Oliveira, *Stopping a viscous fluid by a feedback dissipative field: thermal effects without phase changing,* Progress in Nonlinear Differential Equations and their Applications **61**, Birkhäuser (2005), 1–14.

[5] S. Antontsev, J.I. Díaz and S.I. Shmarev, *Energy methods for free boundary problems,* Progress in Nonlinear Differential Equations and their Applications **48**, Birkhäuser, 2002.

[6] S. Antontsev, A.V. Kazhikhov and V.N. Monakhov, *Boundary value problems in mechanics of nonhomogeneous fluids.* Studies in Mathematics and its Applications **22**, North-Holland, 1990.

[7] E. Fernández-Cara, F. Guillén and R.R. Ortega, *Some theoretical results for viscoplastic and dilatant fluids with variable density.*, Nonlinear Anal., no. 6, **28** (1997), 1079–1100.

[8] E. Fernández-Cara, F. Guillén and R.R. Ortega, *Some theoretical results concerning non-Newtonian fluids of the Oldroyd kind.*, Ann. Scuola Norm. Sup. Pisa Cl. Sci. (4), no. 1, **26** (1998), 1–29.

[9] O.A. Ladyzhenskaya, *The mathematical theory of viscous incompressible flow.* Mathematics and its Applications **2**, Gordon and Breach, 1969.

[10] J. Simon, *Nonhomogeneous viscous incompressible fluids: existence of velocity, density, and pressure.* SIAM J. Math. Anal., no. 5, **21** (1990), 1093–1117.

[11] H. Sohr, *The Navier-Stokes equations.*, Birkhäuser, Basel, 2001.

S. Antontsev
Departamento de Matemática, Universidade da Beira Interior
Convento de Santo António, 6201-001 Covilhã, Portugal
e-mail: `anton@ubi.pt`

H.B. de Oliveira
Faculdade de Ciências e Tecnologia, Universidade do Algarve
Campus de Gambelas, 8005-139 Faro, Portugal
e-mail: `holivei@ualg.pt`

International Series of Numerical Mathematics, Vol. 154, 33–44

Parabolic Equations with Anisotropic Nonstandard Growth Conditions

S. Antontsev and S. Shmarev

Abstract. We study the Dirichlet problem for a class of nonlinear parabolic equations with nonstandard anisotropic growth conditions. Equations of this class generalize the evolutional $p(x,t)$-Laplacian equation. We prove the existence of a bounded weak solution and study its localization (vanishing) properties.

Mathematics Subject Classification (2000). 35K55; 35K65.

Keywords. Nonlinear parabolic equation, nonstandard growth conditions, anisotropic nonlinearity, localization of solutions.

1. Statement of the problem

Let $\Omega \subset \mathbb{R}^n$ be a bounded simple-connected domain and $0 < T < \infty$. We consider the Dirichlet problem for the parabolic equation

$$\begin{aligned} & u_t - \sum_i D_i \left(a_i(z,u)|D_i u|^{p_i(z)-2} D_i u \right) + c(z,u)|u|^{\sigma(z)-2}u = f \quad \text{in } Q, \\ & u = 0 \text{ on } \Gamma, \qquad u(x,0) = u_0(x) \text{ in } \Omega, \end{aligned} \tag{1.1}$$

where $z = (x,t) \in Q \equiv \Omega \times (0,T]$, Γ denotes the lateral boundary of the cylinder Q. Throughout the paper we assume that $p_i(z)$ and $\sigma(z)$ are given measurable in Q functions such that

$$\begin{aligned} & p_i(z) \subseteq \left[\inf_Q p_i, \sup_Q p_i \right] \subset (p_i^-, p_i^+) \subseteq (p^-, p^+), \\ & \sigma(z) \subseteq \left[\inf_Q \sigma, \sup_Q \sigma \right] \subset (\sigma^-, \sigma^+) \end{aligned} \tag{1.2}$$

The first author was partially supported by the research project DECONT, FCT/MCES (Portugal) at the "Centro de Matemática", Universidade da Beira Interior.

The second author was supported by the research grants MTM–2004-05417 (Spain) and HPRN–CT–2002–00274 (EC).

with given finite constants $p_i^{\pm}$, $\sigma^{\pm}$, $p_i^- > 1$, $p^- > 1$, $\sigma^- > 1$. The coefficients a_i and c may depend on x, t, $u(x,t)$ and obey the following conditions: $a_i(z,u)$, $c(z,u)$ are Carathéodory functions, defined for $(z,r) \in \overline{Q} \times \mathbb{R}$ (measurable in z for every $r \in \mathbb{R}$ and continuous in r for a.a. $z \in Q$),

$$\forall (z,r) \in \overline{Q} \times \mathbb{R} \qquad \begin{aligned} &0 < a_0 \le a_i(z,r) \le a_1 < \infty, \\ &0 \le c_0 \le c(z,u) \le c_1 < \infty, \qquad a_0, a_1, c_0, c_1 = \text{const}. \end{aligned} \tag{1.3}$$

Moreover, it is assumed that the exponents $p_i(z)$ and $\sigma(z)$ are continuous in Q with logarithmic module of continuity:

$$\forall\ (z_1,\ z_2) \in Q \quad \sum_i |p_i(z_1) - p_i(z_2)| + |\sigma(z_1) - \sigma(z_2)| \le \omega(|z_1 - z_2|) \tag{1.4}$$

where $\omega(\tau)$ is continuous for $\tau > 0$ and

$$\overline{\lim_{\tau \to 0^+}}\ \omega(\tau) \ln \frac{1}{\tau} = C < +\infty. \tag{1.5}$$

The note is devoted to the study of localization (vanishing) properties of bounded weak solutions to equation (1.1) which degenerates if $|D_i u|^{p_i(z)} = 0$ or becomes singular if $|D_i u|^{p_i(z)} = \infty$. In Section 2 we collect some known facts from the theory of the generalized Lebesgue and Sobolev spaces and introduce the function spaces the solutions of problem (1.1) belong to. In Section 3 we announce the results on the existence of bounded weak solutions of problem (1.1) and give a sketch of proofs. In the rest of the note we study the localization properties of weak solutions: we show that the solutions of problem (1.1) either identically vanish in a finite time (if $p^+ < 2$), or possess the property of finite speed of propagation of disturbances from the initial data (if $p^- > 2$). In the study of the localization properties we use a modification of the method of local energy estimates [8]. Similar properties of solutions of parabolic equations nonlinear with respect to the solution with variable exponents of nonlinearity are obtained in [5, 6]. The localization properties of solutions of elliptic equations with nonstandard growth conditions are studied in [4, 7]. We also refer to the papers [1, 2, 3, 9] for a discussion of the regularity properties of weak solutions of the systems of equations with nonstandard growth conditions (see also the references therein to the previous work on this issue). The continuity properties of solutions of a parabolic equation with variable exponent of nonlinearity are studied in [13].

2. Spaces $L^{p(x)}(\Omega)$ and $W_0^{1,p(x)}(\Omega)$

The definitions of the function spaces and the brief description of their properties presented in this subsection follow [10, 11, 14, 16] (see the review work [12] for the detailed list of references). Let

$$\begin{cases} \Omega \subset \mathbb{R}^n \text{ be a bounded domain, } \partial\Omega \text{ be Lipschitz-continuous,} \\ p(x) \text{ satisfy (1.4) on } \Omega. \end{cases} \tag{2.1}$$

By $L^{p(x)}(\Omega)$ we denote the space of measurable functions $f(x)$ on Ω such that

$$A_{p(\cdot)}(f) = \int_\Omega |f(x)|^{p(x)}\,dx < \infty.$$

The space $L^{p(x)}(\Omega)$ equipped with the norm

$$\|f\|_{p(\cdot)} \equiv \|f\|_{L^{p(x)}(\Omega)} = \inf\left\{\lambda > 0 : \; A_{p(\cdot)}\left(f/\lambda\right) \le 1\right\}$$

becomes a Banach space. The Banach space $W^{1,\,p(x)}(\Omega)$ with $p(x) \in [p^-, p^+] \subset (1,\infty)$ is defined by

$$\begin{aligned} &W^{1,\,p(x)}(\Omega) = \left\{f \in L^{p(x)}(\Omega) : \, |\nabla f| \in L^{p(x)}(\Omega)\right\}, \\ &\|u\|_{W^{1,p(x)}(\Omega)} = \sum_i \|D_i u\|_{p(\cdot)} + \|u\|_{p(\cdot)}. \end{aligned} \tag{2.2}$$

If condition (2.1) is fulfilled, then $C_0^\infty(\Omega)$ is dense in $W_0^{1,\,p(x)}(\Omega)$ and $W_0^{1,\,p(x)}(\Omega)$ can be defined as the closure of $C_0^\infty(\Omega)$ with respect to the norm (2.2). The equivalent norm of $W_0^{1,\,p(x)}$ is defined by $\sum_i \|D_i u\|_{p(\cdot)}$. If $p(x) \in C^0(\overline{\Omega})$, then $W^{1,p(x)}(\Omega)$ is separable and reflexive. If $p(x)$, $q(x) \in C^0(\overline{\Omega})$ and

$$1 < q(x) \le \sup_\Omega q(x) < \inf_\Omega p_*(x) \quad \text{with} \quad p_*(x) = \begin{cases} \dfrac{p(x)n}{n - p(x)} & \text{if } p(x) < n, \\ \infty & \text{if } p(x) > n, \end{cases}$$

then the embedding $W_0^{1,p(x)}(\Omega) \hookrightarrow L^{q(x)}(\Omega)$ is continuous and compact.

The following inequalities hold:

1. $$\min\left(\|f\|_{p(\cdot)}^{p^-}, \|f\|_{p(\cdot)}^{p^+}\right) \le A_{p(\cdot)}(f) \le \max\left(\|f\|_{p(\cdot)}^{p^-}, \|f\|_{p(\cdot)}^{p^+}\right); \tag{2.3}$$

2. Hölder's inequality: for $f \in L^{p(x)}(\Omega)$, $g \in L^{q(x)}(\Omega)$ with

$$\frac{1}{p(x)} + \frac{1}{q(x)} = 1, \quad 1 < p^- \le p(x) \le p^+ < \infty, \quad 1 < q^- \le q(x) \le q^+ < \infty$$

$$\int_\Omega |f\,g|\,dx \le 2\,\|f\|_{p(\cdot)}\,\|g\|_{q(\cdot)}\,. \tag{2.4}$$

3. for every $1 \le q = const < p^-$ we have $\|f\|_q \le C\,\|f\|_{p(\cdot)}$ with the constant $C = 2\,\|1\|_{\frac{p(\cdot)}{p(\cdot)-q}}$.
4. Sobolev's inequality: if conditions (2.1) are fulfilled, there is a constant $C > 0$ such that

$$\forall f \in W_0^{1,p(x)}(\Omega) \quad \|f\|_{p(\cdot)} \le C\,\|\nabla f\|_{p(\cdot)}\,. \tag{2.5}$$

2.1. Anisotropic generalized Sobolev spaces

Let $p_i(x,t)$ and $\sigma(x,t)$ satisfy conditions (1.2) and (1.4). We introduce the Banach space

$$\mathbf{V}_t(\Omega) = \Big\{u|\, u \in L^{\sigma(x,t)}(\Omega) \cap W_0^{1,1}(\Omega), \quad D_i u \in L^{p_i(x,t)}(\Omega)\Big\},$$
$$\|u\|_{\mathbf{V}_t} = \|u\|_{\sigma(\cdot,t),\Omega} + \sum_{i=1}^{n} \|D_i u\|_{p_i(\cdot,t),\Omega}$$

and its dual $\mathbf{V}_t'(\Omega)$. For every $t \in [0,T]$ $\mathbf{V}_t(\Omega) \subset \mathbf{X} = W_0^{1,p^-}(\Omega) \cap L^{\sigma^-}(\Omega)$, so that $\mathbf{V}_t(\Omega)$ is reflexive and separable as a closed subspace of $\mathbf{X}$. Denote $\mathbf{p}(z) = (p_1(z), \dots, p_n(z))$ and set

$$A_{\mathbf{p}(\cdot),Q}(\nabla u) = \sum_{i=1}^{n} \int_Q |D_i u|^{p_i(x,t)}\, dxdt.$$

The following counterpart of (2.3) holds:

$$\begin{aligned}\min\Big\{\sum_i \|D_i u\|^{p^+}_{p_i(\cdot),Q}, \sum_i \|D_i u\|^{p^-}_{p_i(\cdot),Q}\Big\} &\le A_{\mathbf{p}(\cdot),Q}(\nabla u) \\ &\le \max\Big\{\sum_i \|D_i u\|^{p^-}_{p_i(\cdot),Q}, \sum_i \|D_i u\|^{p^+}_{p_i(\cdot),Q}\Big\}.\end{aligned} \tag{2.6}$$

By $\mathbf{W}(Q)$ we define the Banach space

$$\mathbf{W}(Q) = \Big\{u(x,t)|u \in L^{\sigma(x,t)}(Q),\ D_i u \in L^{p_i(x,t)}(Q),\ u = 0 \text{ on } \Gamma\Big\},$$
$$\|u\|_{\mathbf{W}(Q)} = \sum_i \|D_i u\|_{L^{p_i(x,t)}(Q)} + \|u\|_{L^{\sigma(x,t)}(Q)}$$

with the dual $\mathbf{W}'(Q) = \Big\{w|\, w \in L^{\sigma'(x,t)}(Q) \cap L^{p_i'(x,t)}(0,T; W^{-1,p_i'(x,t)}(\Omega))\Big\}$. Here and throughout the text we use the notation $1/s + 1/s' = 1$.

3. Existence of bounded weak solution

The solution of problem (1.1) is understood in the following sense.

Definition 3.1. *A locally integrable function $u(x,t) \in \mathbf{W}(Q)$ is called* weak solution *of problem* (1.1) *if for every test-function $\zeta \in L^{\sigma(x,t)}(Q) \cap L^\infty\big(0,T; L^2(\Omega)\big)$ such that $\zeta = 0$ on $\partial\Omega \times (0,T)$, $D_i\zeta \in L^{p_i(x,t)}(Q)$, $\zeta_t \in L^2(Q)$, and every $t_1, t_2 \in [0,T]$ the following identity holds:*

$$\int_{t_1}^{t_2}\int_\Omega \Big(u\zeta_t - \sum_i a_i |D_i u|^{p_i - 2} D_i u\, D_i\zeta - c|u|^{\sigma-2} u\,\zeta + f\zeta\Big)\, dz = \int_\Omega u\zeta dx\Big|_{t_1}^{t_2}. \tag{3.1}$$

Theorem 3.1. *Let conditions* (1.2), (1.3) *and* (1.4) *be fulfilled,* $p^- > \dfrac{2n}{n+2}$, $c_0 \geq 0$, $f \in L^2(Q)$ *and* $u_0 \in L^2(\Omega)$. *Then problem* (1.1) *has at least one weak solution* $u \in \mathbf{W}$ *satisfying the estimate*

$$\|u\|^2_{L^\infty(0,T;L^2(\Omega))} + \int\limits_Q \left(a_0 \sum_i |D_i u|^{p_i} + c_0 |u|^\sigma \right) dz \leq M \left[\|u_0\|^2_{L^2(\Omega)} + \|f\|^2_{2,Q} \right]$$

with a universal constant M *not depending on* T.

Theorem 3.2. *Let* $u_0 \in L^\infty(\Omega)$ *and* $\|f(\cdot,t)\|_{\infty,\Omega} \in L^1(0,T)$. *Then the weak solution of problem* (1.1) *satisfies the estimate*

$$\operatorname*{ess\ sup}_{t \in (0,T)} \|u(\cdot,t)\|_{\infty,\Omega} \leq \|u_0\|_{\infty,\Omega} + \int_0^T \|f(\cdot,\tau)\|_{\infty,\Omega}\, d\tau.$$

The strict limitation on the volume of this note prevents us from presenting here the detailed proofs. We limit ourselves by a very short sketch of the arguments leaving the detailed proofs for future publications. Theorem 3.1 is proved by constructing a sequence of Galerkin's approximations: since $C_0^\infty(\Omega)$ is dense in $W_0^{1,p(x,t)}(\Omega)$ for every $t \in (0,T)$, we may construct the sequence of approximate solutions $u^N(x,t) = \sum_{k=1}^N u_k^N(t)\psi_k(x)$ following the classical scheme given in [15, Ch.3]. To justify the limit passage as $N \to \infty$ we rely on the monotonicity of the elliptic part of equation (1.1): let $\mathcal{M}(s) = |s|^{p-2}s$, then $\forall\, \xi, \eta \in \mathbb{R}^n$

$$(\mathcal{M}(\xi) - \mathcal{M}(\eta))\,(\xi - \eta) \geq \begin{cases} 2^{-p}\,|\xi - \eta|^p & \text{if } 2 \leq p < \infty, \\ (p-1)\,|\xi - \eta|^2\,(|\xi|^p + |\eta|^p)^{\frac{p-2}{p}} & \text{if } 1 < p < 2. \end{cases}$$

To prove Theorem 3.2 we take for the test-function in (3.1) the function u_k^{2m-1} with $m \in \mathbb{N}$ and

$$u_k = \min\{|u|,\, k\} \operatorname{sign} u \equiv \begin{cases} k & \text{if } u > k, \\ u & \text{if } |u| \leq k, \\ -k & \text{if } u < -k. \end{cases}$$

Estimate (3.2) follows then after a suitable choice of the parameter k, depending on $\|u_0\|_{\infty,\Omega}$, and the limit passage as $m \to \infty$ (see [6]).

4. Vanishing in a finite time

For the sake of presentation, in this section we deal with the equation (1.1) with isotropic nonlinearity. The general anisotropic case is studied likewise and does not present any principal difficulty.

Theorem 4.1. *Let* $u(z)$ *be a weak solution of problem* (1.1) *with* $p_i(z) = p(z)$, *and let conditions* (1.2) , (1.3) *and* (1.4) *be fulfilled.*

1. *Let* $f \equiv 0$ *and either*

$$c_0 \geq 0, \quad \max\left\{1, \frac{2n}{n+2}\right\} < p^- \leq p_i(z) = p(z) \leq p^+ < 2, \tag{4.1}$$

or

$$\left\{\begin{array}{l} c_0 > 0, \quad \sigma^+ < 2 \leq p^+, \\ \dfrac{p^+\,\sigma^+\,[p^-(n+\sigma^-) - n\sigma^-]}{np^-(2-\sigma^-)\sigma^+ + [(2+n)p^- - 2n)]\,\sigma^- p^+} \equiv \nu \in (1/2, 1). \end{array}\right. \tag{4.2}$$

Then $u(z) = 0$ *in* $Q \cap \{t \geq t^*\}$ *with some finite* t^* *depending on* $\|u_0\|_{2,\Omega}$, n, $p^\pm$, $\sigma^\pm$, a_0, c_0.

2. *Let either* (4.1), *or* (4.2) *holds. Assume that* $f \not\equiv 0$ *and*

$$\|f(\cdot,t)\|_{2,\Omega}^{\frac{2\nu}{2\nu-1}} \leq \varepsilon\left[1 - \frac{t}{t_f}\right]_+^{\frac{\nu}{1-\nu}}, \quad t_f > t^*, \tag{4.3}$$

with $\epsilon = const > 0$, *and* $\nu = p^+/2 < 1$ *if* (4.1) *is fulfilled,* ν *defined by* (4.2) *otherwise. Then there exists a constant* $\varepsilon_0 > 0$ *such that* $u(z) = 0$ *in* $Q \cap \{t \geq t_f > t^*\}$, *provided that* $\varepsilon_0 \geq \varepsilon > 0$.

Proof. Let us introduce the functions

$$E(t) = \|u\|_{2,\Omega}^2, \quad I(t) = \int_\Omega f u\,dx, \quad Q(t) = \sum_i \int_\Omega \left(a_i\,|D_i u|^{p(z)} + c\,|u|^{\sigma(z)}\right) dx.$$

Choosing the solution $u(z)$ for the test-function in (3.1), we obtain the energy relation

$$\frac{1}{2}E'(t) + Q(t) = I(t). \tag{4.4}$$

It is easy to see that Q, $I \in L^1(0,T)$, and

$$Q_0(t) \equiv \int_\Omega \left(a_0\,|\nabla u|^{p(z)} + c_0\,|u|^{\sigma(z)}\right) dx \leq Q(t) \leq \int_\Omega \left(a_1\,|\nabla u|^{p(z)} + c_1\,|u|^{\sigma(z)}\right) dx.$$

Without loss of generality we may assume that $E(t) \leq 1$ for a.e. $t \in (0,T)$. Using the interpolation inequalities in the Sobolev spaces of functions defined on Ω and depending on t as a parameter, we have that for a.e. $t \in (0,T)$ the inequality

$$E^{\nu_*(t)}(t) \leq \min\left\{E^{p_*^-(t)/2}(t),\ E^{p_*^+(t)/2}(t)\right\} \leq C\int_\Omega |\nabla u(z)|^{p(z)}\,dx \tag{4.5}$$

holds in which C is a constant depending only on $p^\pm$, n, Ω, and $\nu_*(t) = p_*^+(t)/2 \leq 1$ if condition (4.1) is fulfilled. If condition (4.2) holds, then

$$E^{\nu_*(t)}(t) \leq C\int_\Omega \left(|\nabla u(z)|^{p(z)} + |u(z)|^{\sigma(z)}\right) dx \tag{4.6}$$

with the exponent

$$\nu_*(t) = \frac{p_*^+ \, \sigma_*^+ \, [p_*^- (n + \sigma_*^-) - n\sigma_*^-]}{np_*^- (2 - \sigma_*^-)\sigma_*^+ + \left[(2+n)p_*^- - 2n)\right] \sigma_*^- p_*^+} \in (1/2, 1). \tag{4.7}$$

Gathering (4.4) with (4.5) (or (4.6)) and making use of Young's inequality, we arrive at the inequality

$$\frac{1}{2} E'(t) + C\, E^{\nu_*(t)} \le \frac{C}{2} E^{\nu_*(t)} + C\, \|f(\cdot, t)\|_{2,\Omega}^{\nu_*(t)/(2\nu_*(t)-1)}.$$

Applying (4.3), we obtain the nonhomogeneous ordinary differential inequality

$$\frac{1}{2} E'(t) + C\, E^{\nu_*(t)} \le C' \left(\varepsilon \left[1 - \frac{t}{t_f} \right]_+^{\frac{\nu}{1-\nu}} \right)^{\frac{2\nu-1}{2\nu} \frac{\nu_*(t)}{2\nu_*(t)-1}}. \tag{4.8}$$

Because of the assumption $E(t) \le 1$, (4.8) can be strengthen as follows:

$$\frac{1}{2} E'(t) + C\, E^{\nu} \le C' \left(\varepsilon \left[1 - \frac{t}{t_f} \right]_+^{\frac{\nu}{1-\nu}} \right) \quad \text{with } \nu = \sup_{(0,T)} \nu_*(t). \tag{4.9}$$

If $\epsilon = 0$, the straightforward integration of this inequality gives the estimate

$$E^{1-\nu}(t) \le E^{1-\nu}(0) - (1-\nu)Ct, \tag{4.10}$$

whence the former assertion of the theorem with $t^* = E^{1-\nu}(0)/(1-\nu)C$. The latter assertion ($\varepsilon > 0$) follows like in [8, Ch.1,Lemma 2.1] via comparison with a suitable barrier function. □

Remark 4.1. *The second assertion of the theorem means that under the imposed conditions every weak solution of problem* (1.1) *vanishes simultaneously with the source term at the same instant* t_f.

5. Space localization of solutions

Let us denote

$$B_\rho(x_0) = \{x | \, |x - x_0| < \rho\}, \quad S_\rho(x_0) = \partial B_\rho(x_0), \quad Q_\rho(x_0) = B_\rho(x_0) \times (0, T).$$

In this section we study the property of finite speed of propagation of disturbances for the local weak solutions of equation (1.1) in the cylinder $Q_{\rho_0}(x_0) = B_{\rho_0}(x_0) \times (0, T)$, $\overline{B}_{\rho_0} \subset \Omega$, regardless of the boundary conditions on $\partial\Omega$. Throughout the section we assume that the total energy of the solution under study is finite: $\forall\, t \in (0, T)$

$$D(\rho_0, t) = \sup_{(0,\, t)} \|u(\cdot, \tau)\|_{L^2(B_{\rho_0})}^2 + \int_0^t \int_{B_{\rho_0}} (|\nabla u|^p + c_0 |u|^\sigma)\, dxdt \le C, \tag{5.1}$$

with some constant C not depending on t. By local weak solution we mean the following:

Definition 5.1. *A measurable function $u(x,t)$ is called local weak solution of equation* (1.1) *if for every $Q_\rho(x_0) \subset Q$*

1. $u \in \mathbf{W}(Q_\rho(x_0)) \cap L^\infty(0,T;L^2(B_\rho(x_0)))$, $u_t \in \mathbf{W}'(Q_\rho(x_0))$,
2. *for every test-function ζ such that $\zeta \in \mathbf{W}(Q_\rho(x_0))$, $\zeta_t \in \mathbf{W}'(Q_\rho(x_0))$, $\zeta = 0$ outside $Q_\rho(x_0)$, the integral identity holds:*

$$\int_\Omega u\,\zeta\,dx\Big|_0^T + \int_Q \Big[-u\,\zeta_t + \sum_i a_i|D_iu|^{p_i-2}D_iu\,D_i\zeta + c|u|^{\sigma-2}u\,\zeta\Big]\,dz = \int_Q f\,\zeta\,dz.$$

Obviously, the weak solution of problem (1.1) $u \in \mathbf{W}(Q)$ is a local weak solution in the sense of this definition.

5.1. The basic energy relation

Let $x_0 \in \Omega$ be an arbitrary fixed point.

Theorem 5.1. *Let $\dfrac{1}{p^-} - \dfrac{1}{p^+} < \dfrac{1}{n}$. Then for every $B_\rho(x_0) \subset \Omega$, $t \in (0,T]$ the local weak solution of equation* (1.1) *satisfies the energy relation*

$$\begin{aligned}\frac{1}{2}\int_{B_\rho(x_0)} u^2(x,\tau)\,dx\Big|_{\tau=0}^{\tau=t} &+ \int_0^t\int_{B_{\rho_0}(x_0)}\Big(\sum_i a_i|D_iu|^{p_i} + c\,|u|^\sigma - f\,u\Big)\,dxdt \\ &= -\sum_i\int_0^t\int_{S_\rho(x_0)} u\,a_i|D_iu|^{p_i-2}D_iu\cdot\nu_i\,dS\,dt \equiv I(\rho,t),\end{aligned} \tag{5.2}$$

where ν is the unit outer normal to $S_\rho(x_0)$.

Proof. We will need the following assertion.

Lemma 5.1. *Let $A_{\mathbf{p}(\cdot)}(\nabla u) \le K$. If $\dfrac{1}{p^-} - \dfrac{1}{p^+} < \dfrac{1}{n}$, then $I(r) \in L^1(0,\rho)$.*

Proof. For the sake of brevity, let us denote $\omega = B_\rho(x_0) \times (0,T)$. By (2.4)

$$\int_0^\rho |I(r)|\,dr \le 2a_0 \sum_i \|u\|_{p_i(\cdot),\omega}\,\|D_iu\|_{p_i(\cdot),\omega}\,.$$

According to the embedding theorem $\|u\|_{p_i(\cdot),\omega} \le C\,\|\nabla u\|_{p^-,\omega}$, provided that $p_i(x) < \dfrac{np^-}{n-p^-}$. Finally, for every $i = 1,\dots,n$

$$\begin{aligned}\|D_iu\|^{p^-}_{p^-(\cdot),\omega} &= \int_\omega |D_iu|^{p^-}\,dxdt \le 2\,\||D_iu|^{p^-}\|_{\frac{p_i(\cdot)}{p^-},\omega}\|1\|_{\frac{p_i}{p_i-p^-},\omega} \\ &\le 2\,\|1\|_{\frac{p_i}{p_i-p^-},\omega}\max\Big\{A^{p^-/p_i^-}_{p_i(\cdot),\omega}(D_iu),\ A^{p^-/p_i^+}_{p_i(\cdot),\omega}(D_iu)\Big\} \le C(|\Omega|,T,p_i^\pm,p^\pm,K). \qquad \square\end{aligned}$$

The rest of the proof follows the proof of Lemma 2.1 in [8, Ch.3]: $\square$

5.2. Finite speed of propagation of disturbances from the data

Theorem 5.2 (Finite speed of propagation). *Let conditions* (1.2), (1.3), (1.4) *and the conditions of Theorem* 5.1 *be fulfilled. Assume that either*

$$c_0 = 0, \quad \textit{and} \quad 2 < p^- \le p(x,t), \tag{5.3}$$

or

$$c_0 > 0, \quad \sigma^+ < p^-, \quad \max\left\{1, \frac{2n}{n+2}\right\} < p^- \le p^+ \le p(x,t) \le 2, \tag{5.4}$$

and that $f = 0$ in the cylinder $Q_{\rho_0}(x_0) = B_{\rho_0}(x_0) \times (0,T)$. Then every local weak solution $u(z)$ of equation (1.1) *in $Q_{\rho_0}(x_0)$, satisfying* (5.1), *possesses the property of finite speed of propagation: $u(x,t) = 0$ in $x \in B_{\rho(t)}(x_0)$ with $0 \le t \le t_* < T$ and $\rho(t)$ given by the formula*

$$\rho^{1+\beta}(t) = \rho_0^{1+\beta} - Ct^\lambda D^{1-\nu}(\rho_0, t) \tag{5.5}$$

with some positive constants C, ν, and λ, β which depend on the constants in conditions (1.2), (1.3) *and* (1.4).

Remark 5.1. *Since the function $\rho(t)$ defined by* (5.5) *is monotone decreasing, the set $B_{\rho(t)}(x_0)$ is nonempty for small t.*

Let us now assume that there exists $\rho_1 > 0$ such that $B_{\rho_1}(x_0) \subset \Omega$ and that for some $\rho_0 \in (0, \rho_1)$

$$u_0(x) \equiv 0 \quad x \in B_{\rho_0}(x_0),\ f(x,t) \equiv 0 \quad \text{in } Q_{\rho_0(x_0)}, \tag{5.6}$$

$$\|u_0\|^2_{2,B_\rho(x_0)} + \|f\|^2_{2,Q_\rho(x_0)} \le \varepsilon(\rho - \rho_0)_+^{1/(1-\nu)}, \tag{5.7}$$

for all $\rho \in [\rho_0, \rho_1]$, $D(\rho_1, T) < \infty$, with the positive constant ν defined below, and some $\varepsilon > 0$. This assumption means that the functions $u_0(x)$ and $f(z)$ are sufficiently "flat" near the boundaries of their supports.

Theorem 5.3 (The waiting time effect). *Let conditions* (5.3) *or* (5.4)) *and* (5.6), (5.7) *hold, and the conditions of Theorem* 5.1 *be fulfilled. Then every weak local solution $u(z)$ of equation* (1.1) *possesses the waiting time property: there exists a positive constant $t^* \le T$ such that $u(x,t) = 0$ in $B_{\rho_0}(x_0) \times [0, t^*]$.*

Sketch of proof. For the sake of simplicity we consider the case (5.3) of Theorem 5.2 and assume that $f \equiv 0$. Let us introduce the energy functions

$$E(\rho,t) = \int_0^t \int_{B_\rho(x_0)} |\nabla u|^{p(z)} dz, \quad b(\rho,t) = \|u(\cdot,\iota)\|^2_{2,B_\rho(x_0)}, \tag{5.8}$$

$$\overline{E}(\rho,t) = \max_{\tau \le t} E(\rho,\tau), \quad \overline{b}(\rho,t) = \max_{\tau \le t} \|u(\cdot,\tau)\|^2_{2,B_\rho(x_0)} \tag{5.9}$$

for which

$$E_\rho = \int_0^t \int_{S_\rho(x_0)} |\nabla u|^p \, dx dt, \quad E_t = \int_{B_\rho(x_0)} |\nabla u|^p \, dx. \tag{5.10}$$

We recall that due to the regularity of weak solutions stated in Theorem 5.1 the functions E_t, E_ρ and $E_{t\rho}$ are well defined in the corresponding functional spaces. Let us consider first the cylinder $B_{\rho_0}(x_0) \times (0, T^*)$, $\rho_0 > 0, T^* > 0$ assuming that

$$p^+ - p^- \leq \varepsilon(\rho_0, T^*) \tag{5.11}$$

with a sufficiently small $\varepsilon > 0$. Using the interpolation inequalities (see [4], formulas (5.15), (5.22) with $\sigma^\pm = 2$), and then following the proof of [8, Theorem 2.1, p. 133] we estimate the right-hand side $I(\rho, t)$ of (5.2) as follows:

$$|I(\rho,t)| \leq C \int_0^t \max\left\{E_{t\rho}^{(p^+-1)/p^+},\ E_{t\rho}^{(p^--1)/p^-}\right\} \|u\|_{p^+,S_\rho} dt, \tag{5.12}$$

$$\begin{aligned} \|u\|_{p^+,S_\rho(x_0)} &\leq C(\|\nabla u\|_{p^-,B_\rho(x_0)} + \rho^{-\delta} b^{\frac{1}{2}})^\theta b^{\frac{1-\theta}{2}} \\ &\leq C\left(\max\left\{E_t^{1/p^+}, E_t^{1/p^-}\right\} + \rho^{-\delta} b^{\frac{1}{2}}\right)^\theta b^{\frac{1-\theta}{2}}, \end{aligned} \tag{5.13}$$

where

$$\theta = \frac{p^-}{p^+} \frac{n(p^+-2)+2}{n(p^--2)+2p^-} < 1, \quad \delta = \frac{n(p^--2)+2p^-}{2p^-} > 1. \tag{5.14}$$

Substituting (5.12), (5.13) into (5.2) we arrive at the inequality

$$b + E \leq C\bar{b}^{\frac{1-\theta}{2}} \int_0^t \max\left\{E_{t\rho}^{1-\frac{1}{p^+}},\ E_{t\rho}^{1-\frac{1}{p^-}}\right\} \left(\max\left\{E_t^{\frac{\theta}{p^+}},\ E_t^{\frac{\theta}{p^-}}\right\}] + \rho^{-\delta\theta} b^{\frac{\theta}{2}}\right) d\tau.$$

Not loosing generality we may assume that $\overline{E}(\rho_0, T) \leq D(\rho_0, T) \leq 1$ and $T \leq 1$. Applying the integral representations

$$\int_0^t E_{t\rho} dt = E_\rho(\rho, t), \qquad \int_0^t E_t dt = E(\rho, t),$$

and using Hölder's inequality, we derive the inequality

$$(\bar{b} + E)^\mu \leq C t^\kappa \rho^{-\delta\theta} \max\left\{E_\rho^{1-1/p^+},\ E_\rho^{1-1/p^-}\right\},$$

with the exponents

$$\mu = 1 - \frac{\theta}{p^+} - \frac{1-\theta}{2}, \qquad \kappa = \min\left\{\frac{1}{p^+}\left(1 - \frac{\theta p^+}{p^-}\right), \frac{1-\theta}{p^-}\right\}.$$

Since $\overline{E} \leq 1$, this inequality leads to the ordinary nonlinear differential inequality for the energy function $\overline{E}$:

$$\overline{E}^\nu \leq (\bar{b} + \overline{E})^\nu \leq C t_*^\lambda \rho^{-\beta} \overline{E}_\rho \tag{5.15}$$

with the exponents of nonlinearity

$$\nu = \frac{p^-}{p^- - 1}\left(\frac{1}{2} + \theta\left(\frac{p^+-2}{2p^+}\right)\right) < 1, \qquad \lambda = \frac{\kappa p^+}{p^+ - 1}, \qquad \beta = \delta\theta \frac{p^+}{p^+ - 1}.$$

The function $\overline{E}$ is considered as a function of the variable ρ depending on t as a parameter. Notice that due to condition (5.11) on the oscillation of $p(z)$ the inequality $\mu < 1$ with variable $p(z)$ immediately follows if $\mu < 1$ in the special case when $p^+ = p^-$, which much easier to check. The required estimate (5.5) follows now after integration of the differential inequality in the limits (ρ, ρ_0). To complete the proof, we take a ball of an arbitrary radius ρ_0, take a finite covering of this ball with balls of small radius ρ' such that the oscillation condition (5.11) is fulfilled, and then repeat the previous arguments in every of these balls.

In the case of Theorem 5.3 the same proceeding leads to the nonhomogeneous ordinary differential inequality

$$\overline{E}^{\nu} \leq Ct_*^{\lambda}\rho^{-\beta}\overline{E}_{\rho} + \varepsilon\,[\rho - \rho_0]_+^{\frac{\nu}{1-\nu}}\,,\ \rho \in (\rho_0, \rho_1). \tag{5.16}$$

The analysis of this inequality is based on [8, Ch.1, Lemma 2.4] (see also [4, Lemma 5.3]). □

Remark 5.2. *The conclusions about the space-and-time localization properties of solutions to problem* (1.1) *are based on the analysis of the nonlinear ordinary differential inequalities for the energy functions. When dealing with these inequalities we always reduced them, by means of some suitable assumptions, to the nonlinear inequalities with constant exponents of nonlinearity, which are already studied* (*see, for instance, inequality* (4.8) *and its counterpart* (4.9)). *The study of the properties of functions satisfying the nonlinear ordinary differential inequalities with variable exponents of nonlinearity is still an open question.*

References

[1] E. Acerbi and G. Mingione, *Regularity results for a class of functionals with non-standard growth*, Arch. Rational Mech. Anal., 156 (2001), pp. 121–140.

[2] E. Acerbi and G. Mingione, *Regularity results for stationary electro-rheological fluids*, Arch. Rational Mech. Anal., 164 (2002), pp. 213–259.

[3] E. Acerbi, G. Mingione, and G.A. Seregin, *Regularity results for parabolic systems related to a class of non-Newtonian fluids*, Ann. Inst. H. Poincaré Anal. Non Linéaire, 21 (2004), pp. 25–60.

[4] S. Antontsev and S. Shmarev, *Elliptic equations and systems with nonstandard growth conditions: existence, uniqueness and localization properties of solutions*, Nonlinear Analysis Serie A: Theory and Methods (to appear).

[5] S. Antontsev and S. Shmarev, *Existence and uniqueness of solutions of degenerate parabolic equations with variable exponents of nonlinearity*, Preprint 3, Departamento de Matemática, Universidade da Beira Interior, 2004, 16 pp., www.ubi.pt/externos/noe.html. (To appear in "Fundamental and Applied Mathematics", 2005.)

[6] ——, *A model porous medium equation with variables exponent of nonlinearity: existence, uniqueness and localization properties of solutions*, Nonlinear Analysis Serie A: Theory and Methods, 60 (2005), pp. 515–545.

[7] ——, *On localization of solutions of elliptic equations with nonhomogeneous anisotropic degeneracy*, Siberian Math. Journal, 46(5) (2005), pp. 963–984.

[8] S. Antontsev, J.I. Díaz, and S. Shmarev, *Energy Methods for Free Boundary Problems:Applications to Non-linear PDEs and Fluid Mechanics*, Birkhäuser, Boston, 2002. Progress in Nonlinear Differential Equations and Their Applications, Vol. 48.

[9] S. Antontsev and V.V. Zhikov, *Higher integrability for parabolic equations of $p(x,t)$-Laplacian type*, Advances in Differential Equations, 10(9), pp. 1053–1080, (2005).

[10] L. Diening, *Riesz potential and Sobolev embeddings on generalized Lebesgue and Sobolev spaces $L^{p(\cdot)}$ and $W^{k,p(\cdot)}$*, Preprint 22/2002-15.07.2002, Albert-Ludwigs-University Freiburg, 2002. pp.1–13.

[11] D. Edmunds and J. Rákosník, *Sobolev embeddings with variable exponent*, Studia Math., 143(3) (2000), pp. 267–293.

[12] P. Harjulehto and P. Hästö, *An overview of variable exponent Lebesgue and Sobolev spaces*, in Future trends in geometric function theory, vol. 92 of Rep. Univ. Jyväskylä Dep. Math. Stat., Univ. Jyväskylä, Jyväskylä, 2003, pp. 85–93.

[13] E. Henriques and J.M. Urbano, *Intrinsic scaling for PDE's with an exponential nonlinearity*, Preprint 04-33, Universidade de Coimbra, Departamento de Matemática, 2004.

[14] O. Kováčik and J. Rákosník, *On spaces $L^{p(x)}$ and $W^{k,p(x)}$*, Czechoslovak Math. J., 41 (1991), pp. 592–618.

[15] O. A. Ladyženskaja, V.A. Solonnikov, and N.N. Ural'tseva, *Linear and quasilinear equations of parabolic type*, American Mathematical Society, Providence, R.I., 1967. Translated from the Russian by S. Smith. Translations of Mathematical Monographs, Vol. 23.

[16] J. Musielak, *Orlicz spaces and modular spaces*, vol. 1034 of Lecture Notes in Mathematics, Springer-Verlag, Berlin, 1983.

S. Antontsev and S. Shmarev
Departamento de Matemática
Universidade da Beira Interior, Portugal
e-mail: anton@ubi.pt

S. Shmarev
Departaménto de Matematicas
Universidad de Oviedo, España
e-mail: shmarev@orion.ciencias.uniovi.es

International Series of Numerical Mathematics, Vol. 154, 45–54

Parabolic Systems with the Unknown Dependent Constraints Arising in Phase Transitions

Masayasu Aso, Michel Frémond and Nobuyuki Kenmochi

Abstract. We consider a system of nonlinear parabolic PDEs which includes a constraint on the time-derivative depending on the unknowns. This system is a mathematical model for irreversible phase transitions. In our phase transition model, the constraint $p := p(\theta, w)$ is a function of the temperature θ and the order parameter (state variable) w and it is imposed on the velocity of the order parameter, for instance, in such a way that $p(\theta, w) \leq w_t \leq p(\theta, w) +$ (a positive constant). We give an existence result of the problem.

Mathematics Subject Classification (2000). Primary 35K45; Secondary 35K50.

Keywords. System of nonlinear parabolic PDEs, irreversible phase transitions.

1. Introduction

The irreversible phase change is very often observed in solid-liquid systems, for instance, in the solidification process of eggs; in fact, once eggs are solidified in high temperature, their states never return to raw ones even if they are put in cold water. It is called that the phase transition is irreversible.

In this paper we consider the following system:

$$\theta_t + w_t - \nu\Delta\theta = h(x,t) \text{ in } Q := \Omega \times (0,T), \tag{1.1}$$

$$w_t + \alpha(w_t - p(\theta, w)) - \kappa\Delta w + \beta(w) \ni f(\theta, w) \text{ in } Q, \tag{1.2}$$

$$\frac{\partial\theta}{\partial n} = \frac{\partial w}{\partial n} = 0 \text{ on } \Sigma := \Gamma \times (0,T), \tag{1.3}$$

$$\theta(\cdot, 0) = \theta_0, \ w(\cdot, 0) = w_0 \text{ on } \Omega, \tag{1.4}$$

where Ω is a bounded domain in $\mathbf{R}^3$ with smooth boundary Γ, $0 < T < \infty$; ν and κ are positive constants; α and β are maximal monotone graphs in $\mathbf{R} \times \mathbf{R}$; $p(\cdot,\cdot)$ is a function of C^2-class on $\mathbf{R} \times \mathbf{R}$ and $f(\cdot,\cdot)$ is a Lipschitz continuous function on $\mathbf{R} \times \mathbf{R}$. Moreover h is a function on Q, and θ_0 and w_0 are functions on Ω, which are prescribed as the data. We denote by (P) the system of (1.1)–(1.4).

In this paper, we suppose for α, β, $p(\cdot,\cdot)$ and $f(\cdot,\cdot)$ that

(A) $0 \in \alpha(0)$, $D(\alpha) = [0, N_0]$ or $[0, N_0)$ for some finite positive number N_0, where $D(\alpha)$ is the domain of α.

(B) β is the subdifferential of the indicator function of the interval $(-\infty, 1]$, namely

$$\beta(r) = \begin{cases} \emptyset & \text{for } r > 1, \\ [0,\infty) & \text{for } r = 1, \\ \{0\} & \text{for } r < 1. \end{cases}$$

(C) All of the first- and second-order partial derivatives of p are continuous and bounded on $\mathbf{R} \times \mathbf{R}$, $p \geq 0$ on $\mathbf{R} \times \mathbf{R}$ and

$$p(\theta, w) = 0, \ \forall \theta \in \mathbf{R}, \ \forall w \in \mathbf{R} \text{ with } w \geq 1; \tag{1.5}$$

note here that $1 = \sup D(\beta)$.

(D) f is Lipschitz continuous and bounded on $\mathbf{R} \times \mathbf{R}$.

We note here that the term $\alpha(w_t - p(\theta, w))$ requires automatically $0 \leq p(\theta, w) \leq w_t \leq p(\theta, w) + N_0$, which is a velocity constraint depending on the unknowns θ and w; in the context of irreversible phase transition in solid-liquid systems, the unknown θ is temperature and w, $0 \leq w \leq 1$, is the volume fraction of solid in the system under consideration. In this paper we give an existence result for problem (P). We refer for related works on irreversible phase transition, for instance, to [7, 9] in the case of prescribed constraints.

2. Main result

System (P) is treated under the following assumptions on the data h, θ_0, w_0:

$$h \in W^{1,2}(0,T;L^2(\Omega)) \cap L^\infty(Q), \ \theta_0, w_0 \in H^2(\Omega), \ \frac{\partial \theta_0}{\partial n} = \frac{\partial w_0}{\partial n} = 0 \text{ on } \Gamma. \tag{2.1}$$

For simplicity, we assume that the diffusion coefficients ν and κ are equal to 1, and use the notation $-\Delta_0$ to indicate the Laplace operator $-\Delta$ with homogeneous Neumann boundary condition. Now we give the definition of solution of (P) (with $\nu = 1$ and $\kappa = 1$).

Definition 2.1. A pair of functions $\{\theta, w\}$ is called a solution of (P), if it satisfies the following conditions (1)–(4):

(1) θ, $w \in W^{1,2}(0,T;L^2(\Omega)) \cap L^\infty(0,T;H^1(\Omega)) \cap L^2(0,T;H^2(\Omega))$.

(2) $\theta'(t) + w'(t) - \Delta_0 \theta(t) = h(t)$ in $L^2(\Omega)$ for a.e. $t \in (0,T)$, where θ' and w' denote the time-derivatives of θ and w, respectively.

(3) There exist functions $\xi, \eta \in L^2(0,T;L^2(\Omega))$ with $\xi \in \alpha(w' - p(\theta, w))$ and $\eta \in \beta(w)$ a.e. in Q such that

$$w'(t) + \xi(t) - \Delta_0 w(t) + \eta(t) = f(\theta(t), w(t)) \text{ in } L^2(\Omega) \text{ for a.e. } t \in (0,T).$$

(4) $\theta(0) = \theta_0$ and $w(0) = w_0$ in $L^2(\Omega)$.

Our existence result is stated as follows:

Theorem 2.1. *In addition to assumptions* (A), (B), (C) *and* (D), *suppose that* (2.1) *holds. Then problem* (P) *has at least one solution* $\{\theta, w\}$ *in the sense of Definition* 2.1 *such that*

$$w' \in L^\infty(Q),\ \theta,\ w \in W^{1,2}(0,T;H^1(\Omega)) \cap L^\infty(0,T;H^2(\Omega)).$$

Remark 2.1. As far as the existence (in the sense of Definition 2.1) of a solution of (P) is concerned, we can replace the conditions for h and θ_0 by weaker ones

$$h \in L^\infty(Q),\ \theta_0 \in H^1(\Omega) \cap L^\infty(\Omega). \tag{2.2}$$

To prove the above theorem we first consider approximate problems $(P_{\mu,\lambda})$ of the following form:

$$(\theta + J_\mu w)' - \Delta_0 \theta = h(t,x) \text{ in } Q, \tag{2.3}$$

$$w' + \alpha(w' - p(\theta, J_\mu w)) + A_\mu w + \beta_\lambda(w) \ni f(\theta, w) \text{ in } Q, \tag{2.4}$$

$$\theta(\cdot,0) = \theta_0,\ w(\cdot,0) = w_0 \text{ in } \Omega, \tag{2.5}$$

where A_μ and J_μ, $0 < \mu \le 1$, and β_λ, $0 < \lambda \le 1$, are defined as follows:

(1) For each $0 < \mu \le 1$, φ_μ is the Moreau-Yosida regularization of the function

$$\varphi(v) := \begin{cases} \dfrac{1}{2}\displaystyle\int_\Omega |\nabla v|^2 dx, & \text{if } v \in H^1(\Omega), \\ \infty, & \text{otherwise,} \end{cases}$$

namely,

$$\varphi_\mu(v) = \inf_{z \in L^2(\Omega)} \left\{ \frac{1}{2\mu}|z - v|^2_{L^2(\Omega)} + \varphi(z) \right\},\ \forall v \in L^2(\Omega),$$

and A_μ is the subdifferential of φ_μ in $L^2(\Omega)$, that is $A_\mu = \partial\varphi_\mu$.

(2) For each $0 < \mu \le 1$, J_μ is the resolvent of the subdifferential $\partial\varphi$ $(= -\Delta_0)$ in $L^2(\Omega)$, namely, $J_\mu := (I + \mu\partial\varphi)^{-1}$.

(3) For each $0 < \lambda \le 1$, β_λ is the Moreau-Yosida regularization of β in $\mathbf{R}$, namely

$$\beta_\lambda(r) := \frac{r - (I + \lambda\beta)^{-1} r}{\lambda} \text{ for all } r \in \mathbf{R}.$$

It is well known that $A_\mu := -\Delta_0 J_\mu$ and it converges to $-\Delta_0$ in $L^2(\Omega)$ as $\mu \to 0$, and β_λ converges to β in $\mathbf{R} \times \mathbf{R}$ as $\lambda \to 0$ in the sense of graph (cf. [5, 8, 10]).

So far as the approximate problems $(P_{\mu,\lambda})$ are concerned, the existence of their solutions is shown in a way similar to that in [1–4]; in fact, $(P_{\mu,\lambda})$ has at least one solution $\{\theta_{\mu,\lambda}, w_{\mu,\lambda}\}$ satisfying that

$$\theta_{\mu,\lambda} \in W^{1,2}(0,T;H^1(\Omega)) \cap L^\infty(0,T;H^2(\Omega)) \text{ and } w_{\mu,\lambda} \in W^{1,2}(0,T;L^2(\Omega)).$$

A solution of our problem (P) is constructed as a limit of approximate solutions $\{\theta_{\mu,\lambda}, w_{\mu,\lambda}\}$ as μ and λ tend to zero.

3. Uniform estimates

In this section, we give some uniform estimates for $\{\theta_{\mu,\lambda}, w_{\mu,\lambda}\}$.

We use hereafter notations C_i, M_i, $i \in \mathbf{N}$, to indicate positive constants in inequalities which we derive in this section; C_i denotes a positive constant which does not depend on any of the data h, θ_0, w_0 and parameters μ, $\lambda \in (0,1]$ (but may depend in general on p, f), and also M_i denotes a positive constant which is independent of μ, $\lambda \in (0,1]$. Moreover, we denote by $M_i(\lambda)$, $i \in \mathbf{N}$, a positive constant which depends only on $\lambda \in (0,1]$, but not on $\mu \in (0,1]$.

Lemma 3.1.

(i) *The following inequality holds:*

$$|\theta_{\mu,\lambda}|_{L^\infty(Q)}, |w_{\mu,\lambda}|_{L^\infty(Q)}, |w'_{\mu,\lambda}|_{L^\infty(Q)} \leq M_1 + M_1 T, \ \forall \mu, \ \lambda \in (0,1],$$

where $M_1 = |\theta_0|_{L^\infty(\Omega)} + |w_0|_{L^\infty(\Omega)} + |h|_{L^\infty(Q)} + |p|_{L^\infty(\mathbf{R}^2)} + N_0$.

(ii) *There exists a positive constant M_2 independent of $\mu, \lambda \in (0,1]$ such that*

$$\begin{aligned}&|A_\mu p(\theta_{\mu,\lambda}, J_\mu w_{\mu,\lambda})|^2_{L^2(\Omega)} \leq |\Delta_0 p(\theta_{\mu,\lambda}, J_\mu w_{\mu,\lambda})|^2_{L^2(\Omega)}\\ &\leq M_2\left(|\Delta_0\theta_{\mu,\lambda}|^2_{L^2(\Omega)} + |A_\mu w_{\mu,\lambda}|^2_{L^2(\Omega)} + |\nabla\theta_{\mu,\lambda}|^2_{L^2(\Omega)} + \varphi_\mu(w_{\mu,\lambda}) + 1\right)\end{aligned}$$

a.e. on $(0,T)$.

Proof. The inequalities for $w_{\mu,\lambda}$ and $w'_{\mu,\lambda}$ in (i) are trivial by assumption (A), and the one for $\theta_{\mu,\lambda}$ is immediately obtained by comparing it with the function $q(t) := M_1 + M_1 t$. In fact, it is enough to multiply $(\theta_{\mu,\lambda} - q)' - \Delta_0(\theta_{\mu,\lambda} - q) = h - J_\mu w'_{\mu,\lambda} - M_1 (\leq 0)$ by $[\theta_{\mu,\lambda} - q]^+$ and use the usual comparison technic.

Now we prove the inequality in (ii). Since all of the second-order partial derivatives of p are bounded by assumption (C), it follows that

$$\begin{aligned}|A_\mu p(\theta_{\mu,\lambda}, J_\mu w_{\mu,\lambda})|^2_{L^2(\Omega)} &\leq |\Delta_0 p(\theta_{\mu,\lambda}, J_\mu w_{\mu,\lambda})|^2_{L^2(\Omega)} \qquad (3.1)\\ \leq M_3\Big(|\nabla\theta_{\mu,\lambda}|^4_{L^4(\Omega)} &+ |\nabla J_\mu w_{\mu,\lambda}|^4_{L^4(\Omega)} + |\Delta_0\theta_{\mu,\lambda}|^2_{L^2(\Omega)} + |A_\mu w_{\mu,\lambda}|^2_{L^2(\Omega)}\Big)\end{aligned}$$

for a.e. on $(0,T)$. By the Gagliardo-Nirenberg interpolation inequality (cf. [11]) and the result of (i), the following inequalities hold:

$$\begin{aligned}|\nabla\theta_{\mu,\lambda}|^4_{L^4(\Omega)} &\leq C_1 |\theta_{\mu,\lambda}|^2_{H^2(\Omega)} |\theta_{\mu,\lambda}|^2_{L^\infty(\Omega)}\\ &\leq C_2\left(|\theta_{\mu,\lambda}|^2_{L^2(\Omega)} + |\nabla\theta_{\mu,\lambda}|^2_{L^2(\Omega)} + |\Delta_0\theta_{\mu,\lambda}|^2_{L^2(\Omega)}\right)|\theta_{\mu,\lambda}|^2_{L^\infty(\Omega)}\\ &\leq M_4\left(|\Delta_0\theta_{\mu,\lambda}|^2_{L^2(\Omega)} + |\nabla\theta_{\mu,\lambda}|^2_{L^2(\Omega)} + 1\right).\end{aligned}$$

Similarly, we get

$$\begin{aligned}|\nabla J_\mu w_{\mu,\lambda}|^4_{L^4(\Omega)} &\leq C_3 |J_\mu w_{\mu,\lambda}|^2_{H^2(\Omega)} |J_\mu w_{\mu,\lambda}|^2_{L^\infty(\Omega)}\\ &\leq M_5\left(|A_\mu w_{\mu,\lambda}|^2_{L^2(\Omega)} + \varphi_\mu(w_{\mu,\lambda}) + 1\right)\end{aligned}$$

for a.e. on $(0,T)$. In virtue of (3.1) together with the above inequalities we find a required constant M_2. □

Lemma 3.2. *There exists a positive constant $M_1(\lambda)$ depending only on $\lambda \in (0,1]$ (and on the data p, f, h, θ_0 and w_0 as well, but not on $\mu \in (0,1]$) such that*

$$\sup_{t\in[0,T]} \left\{ |\theta_{\mu,\lambda}|^2_{H^2(\Omega)} + |A_\mu w_{\mu,\lambda}|^2_{L^2(\Omega)} + \varphi_\mu(w_{\mu,\lambda}) + \int_\Omega \hat{\beta}_\lambda(w_{\mu,\lambda})dx \right\}$$

$$+|\theta'_{\mu,\lambda}|^2_{L^2(0,T;L^2(\Omega))} + |\nabla\theta'_{\mu,\lambda}|^2_{L^2(0,T;L^2(\Omega))} + \int_0^T \varphi_\mu(w'_{\mu,\lambda})dt \le M_1(\lambda),$$

where $\hat{\beta}$ is a primitive of β.

Proof. Compute: (2.3) $\times\, \theta'_{\mu,\lambda}$, (2.3) $\times\, (-\Delta_0\theta'_{\mu,\lambda})$, (2.4) $\times \left(w'_{\mu,\lambda} - p(\theta_{\mu,\lambda}, J_\mu w_{\mu,\lambda})\right)$ and (2.4) $\times\, A_\mu(w'_{\mu,\lambda} - p(\theta_{\mu,\lambda}, J_\mu w_{\mu,\lambda}))$. Then, with the help of Lemma 3.1, the Schwarz's inequality and some other relations such as $|w'_{\mu,\lambda}| \le |p|_{L^\infty(\mathbf{R}^2)} + N_0$ a.e. on Q and

$$(\xi, A_\mu(w'_{\mu,\lambda} - p(\theta_{\mu,\lambda}, J_\mu w_{\mu,\lambda}))_{L^2(\Omega)} \ge 0, \ \forall \xi \in \alpha(w'_{\mu,\lambda} - p(\theta_{\mu,\lambda}, J_\mu w_{\mu,\lambda})),$$

we have respectively (cf. [4]):

(1) $\frac{1}{2}|\theta'_{\mu,\lambda}|^2_{L^2(\Omega)} + \frac{1}{2}\frac{d}{dt}|\nabla\theta_{\mu,\lambda}|^2_{L^2(\Omega)} \le |h|^2_{L^2(\Omega)} + C_4,$

(2) $\frac{1}{2}|\nabla\theta'_{\mu,\lambda}|^2_{L^2(\Omega)} + \frac{d}{dt}E^0_{\mu,\lambda} \le \varphi_\mu(w'_{\mu,\lambda}) + \frac{1}{2}|\Delta_0\theta_{\mu,\lambda}|^2_{L^2(\Omega)} + \frac{1}{2}|h'|^2_{L^2(\Omega)},$
where $E^0_{\mu,\lambda}(\cdot) := \frac{1}{2}|\Delta_0\theta_{\mu,\lambda}|^2_{L^2(\Omega)} - (h, -\Delta_0\theta_{\mu,\lambda})_{L^2(\Omega)},$

(3) $\frac{d}{dt}E^1_{\mu,\lambda} \le M_2(\lambda)|A_\mu w_{\mu,\lambda}|^2_{L^2(\Omega)} + C_5,$
where $E^1_{\mu,\lambda}(\cdot) := \varphi_\mu(w_{\mu,\lambda}) + \int_\Omega \hat{\beta}_\lambda(w_{\mu,\lambda})dx,$

(4) $$\varphi_\mu(w'_{\mu,\lambda}) + \frac{d}{dt}E^2_{\mu,\lambda} \le \frac{1}{8}|\theta'_{\mu,\lambda}|^2_{L^2(\Omega)} + M_3(\lambda)\left(|\Delta_0\theta_{\mu,\lambda}|^2_{L^2(\Omega)} + |A_\mu w_{\mu,\lambda}|^2_{L^2(\Omega)}\right.$$
$$\left. +|\nabla\theta_{\mu,\lambda}|^2_{L^2(\Omega)} + \varphi_\mu(w_{\mu,\lambda}) + 1\right),$$

where $E^2_{\mu,\lambda}(\cdot) := \frac{1}{2}|A_\mu w_{\mu,\lambda}|^2_{L^2(\Omega)} + (\beta_\lambda(w_{\mu,\lambda}) - f(\theta_{\mu,\lambda}, w_{\mu,\lambda}), A_\mu w_{\mu,\lambda})_{L^2(\Omega)}.$

Now, by (1) + (2) + (3) + (4) $\times$ 2 we observe that

$$\frac{1}{4}|\theta'_{\mu,\lambda}|^2_{L^2(\Omega)} + \frac{1}{2}|\nabla\theta'_{\mu,\lambda}|^2_{L^2(\Omega)} + \varphi_\mu(w'_{\mu,\lambda}) + \frac{d}{dt}E_{\mu,\lambda} \tag{3.2}$$
$$\le M_3(\lambda)E_{\mu,\lambda} + M_6(||h||^2_{L^2(\Omega)} + 1),$$

where $||h||_{L^2(\Omega)} := |h|_{L^2(\Omega)} + |h'|_{L^2(\Omega)}$ and $E_{\mu,\lambda}(\cdot) := \frac{1}{2}|\nabla\theta_{\mu,\lambda}|^2_{L^2(\Omega)} + E^0_{\mu,\lambda}(\cdot) + E^1_{\mu,\lambda}(\cdot) + 2E^2_{\mu,\lambda}(\cdot)$. Applying the Gronwall's inequality to (3.2), we obtain the required inequality with a certain constant $M_1(\lambda)$. □

On account of estimates in Lemmas 3.1 and 3.2, for each fixed $\lambda \in (0,1]$ we find a sequence $\{\mu_n\} \subset (0,1]$ with $\mu_n \to 0$ as $n \to \infty$ and functions θ_λ, w_λ such that

$$\begin{cases} \theta_{\mu_n,\lambda} \to \theta_\lambda & \text{in } C([0,T];H^1(\Omega)), \\ J_{\mu_n} w_{\mu_n,\lambda} \to w_\lambda & \text{in } C([0,T];H^1(\Omega)), \end{cases} \tag{3.3}$$

as $n \to \infty$, and

$$\begin{cases} w_{\mu_n,\lambda} \to w_\lambda & \text{in } C([0,T];L^2(\Omega)), \\ \theta'_{\mu_n,\lambda} \to \theta'_\lambda & \text{weakly in } L^2(0,T;H^1(\Omega)), \\ J_{\mu_n} w'_{\mu_n,\lambda} \to w'_\lambda & \text{weakly in } L^2(0,T;H^1(\Omega)), \\ -\Delta_0 \theta_{\mu_n,\lambda} \to -\Delta_0 \theta_\lambda & \text{weakly in } L^2(0,T;L^2(\Omega)), \\ w'_{\mu_n,\lambda} \to w'_\lambda & \text{weakly in } L^2(0,T;L^2(\Omega)), \\ A_{\mu_n} w_{\mu_n,\lambda} \to -\Delta_0 w_\lambda & \text{weakly in } L^2(0,T;L^2(\Omega)), \\ \beta_\lambda(w_{\mu_n,\lambda}) \to \beta_\lambda(w_\lambda) & \text{in } C([0,T];L^2(\Omega)), \\ f(\theta_{\mu_n,\lambda}, w_{\mu_n,\lambda}) \to f(\theta_\lambda, w_\lambda) & \text{in } C([0,T];L^2(\Omega)), \\ p(\theta_{\mu_n,\lambda}, J_{\mu_n} w_{\mu_n,\lambda}) \to p(\theta_\lambda, w_\lambda) & \text{in } C([0,T];L^2(\Omega)), \end{cases} \tag{3.4}$$

as $n \to \infty$. Also, put

$$\xi_{\mu,\lambda} := -w'_{\mu,\lambda} - A_\mu w_{\mu,\lambda} - \beta_\lambda(w_{\mu,\lambda}) + f(\theta_{\mu,\lambda}, w_{\mu,\lambda}).$$

Then, $\xi_{\mu,\lambda} \in \alpha(w'_{\mu,\lambda} - p(\theta_{\mu,\lambda}, J_\mu w_{\mu,\lambda}))$ a.e. on Q, and it follows from the above convergences that

$$\xi_{\mu_n,\lambda} \to -w'_\lambda + \Delta_0 w_\lambda - \beta_\lambda(w_\lambda) + f(\theta_\lambda, w_\lambda) =: \xi_\lambda \text{ weakly in } L^2(0,T;L^2(\Omega)), \tag{3.5}$$

as $n \to \infty$. Moreover it is easy to see that

$$\begin{aligned} &\limsup_{n\to\infty} \int_0^T (\xi_{\mu_n,\lambda}, w'_{\mu_n,\lambda} - p(\theta_{\mu_n,\lambda}, J_{\mu_n} w_{\mu_n,\lambda}))_{L^2(\Omega)} dt \\ &\qquad \le \int_0^T (\xi_\lambda, w'_\lambda - p(\theta_\lambda, w_\lambda))_{L^2(\Omega)} dt. \end{aligned} \tag{3.6}$$

By the maximal monotonicity of α, (3.5) and (3.6) show that $\xi_\lambda \in \alpha(w'_\lambda - p(\theta_\lambda, w_\lambda))$ a.e. on Q. Consequently, passing to the limit in (2.3)–(2.5) as $n \to \infty$, we see that $\theta_\lambda,\ w_\lambda \in W^{1,2}(0,T;H^1(\Omega)) \cap L^\infty(0,T;H^2(\Omega))$ and the pair $\{\theta_\lambda, w_\lambda\}$ is a solution of problem $(P_\lambda) := \{(3.7)\text{-}(3.10)\}$:

$$\theta'_\lambda + w'_\lambda - \Delta_0 \theta_\lambda = h \text{ in } L^2(\Omega), \text{a.e. on } (0,T), \tag{3.7}$$

$$w'_\lambda + \xi_\lambda - \Delta_0 w_\lambda + \beta_\lambda(w_\lambda) = f(\theta_\lambda, w_\lambda) \text{ in } L^2(\Omega), \text{a.e. on } (0,T), \tag{3.8}$$

$$\xi_\lambda \in \alpha(w'_\lambda - p(\theta_\lambda, w_\lambda)) \text{ a.e. in } Q, \tag{3.9}$$

$$\theta_\lambda(0) = \theta_0,\ w_\lambda(0) = w_0 \text{ in } L^2(\Omega). \tag{3.10}$$

Moreover we have the following uniform estimates of $\{\theta_\lambda, w_\lambda\}$ with respect to $\lambda \in (0,1]$.

Lemma 3.3. *There is a positive constant M_7, independent of $\lambda \in (0,1]$, such that*

$$\begin{aligned} &|\theta_\lambda|_{W^{1,2}(0,T;H^1(\Omega))} + |w_\lambda|_{W^{1,2}(0,T;H^1(\Omega))} + |\beta_\lambda(w_\lambda)|_{L^\infty(0,T;L^2(\Omega))} \\ &\qquad + |\theta_\lambda|_{L^\infty(0,T;H^2(\Omega))} + |w_\lambda|_{L^\infty(0,T;H^2(\Omega))} \le M_7 \end{aligned} \tag{3.11}$$

Proof. Let $\{\theta_{\mu,\lambda}, w_{\mu,\lambda}\}$ be again the solution of $(P_{\mu,\lambda})$. Then, in addition to estimates (1) - (4) in the proof of Lemma 3.2, we have by multiplying (2.4) by $A_\mu(w'_{\mu,\lambda} - p(\theta_{\mu,\lambda}, J_\mu w_{\mu,\lambda})) + \frac{d}{dt}\beta_\lambda(w_{\mu,\lambda})$ that

$$(5)\ \varphi_\mu(w'_{\mu,\lambda}) + \frac{d}{dt}E^3_{\mu,\lambda} + (\xi_{\mu,\lambda}, \beta'_\lambda(w_{\mu,\lambda})p(\theta_{\mu,\lambda}, J_\mu w_{\mu,\lambda}))_{L^2(\Omega)}$$
$$\leq \frac{1}{8}|\theta'_{\mu,\lambda}|^2_{L^2(\Omega)} + M_8\left(|A_\mu w_{\mu,\lambda} + \beta_\lambda(w_{\mu,\lambda})|^2_{L^2(\Omega)} + |\Delta_0\theta_{\mu,\lambda}|^2_{L^2(\Omega)}\right.$$
$$\left. + |\nabla\theta_{\mu,\lambda}|^2_{L^2(\Omega)} + \varphi_\mu(w_{\mu,\lambda}) + 1\right),$$

where $E^3_{\mu,\lambda}(\cdot) := \frac{1}{2}|A_\mu w_{\mu,\lambda} + \beta_\lambda(w_{\mu,\lambda})|^2_{L^2(\Omega)}$
$$- (A_\mu w_{\mu,\lambda} + \beta_\lambda(w_{\mu,\lambda}), f(\theta_{\mu,\lambda}, w_{\mu,\lambda}))_{L^2(\Omega)}.$$

In the above computation we used some known inequalities such as

$$(\xi_{\mu,\lambda}, \beta'_\lambda(w_{\mu,\lambda})w'_{\mu,\lambda})_{L^2(\Omega)} \geq (\xi_{\mu,\lambda}, \beta'_\lambda(w_{\mu,\lambda})p(\theta_{\mu,\lambda}, J_\mu w_{\mu,\lambda}))_{L^2(\Omega)},$$
$$(\beta_\lambda(w_{\mu,\lambda}), A_\mu w_{\mu,\lambda})_{L^2(\Omega)} \geq 0,$$
$$|A_\mu w_{\mu,\lambda} + \beta_\lambda(w_{\mu,\lambda})|_{L^2(\Omega)} \geq |A_\mu w_{\mu,\lambda}|_{L^2(\Omega)}, |\beta_\lambda(w_{\mu,\lambda})|_{L^2(\Omega)}.$$

Now, we compute (1) + (2) + (5) $\times$ 2 to get

$$G_{\mu,\lambda} + \frac{d}{dt}\tilde{E}_{\mu,\lambda} + 2\left(\xi_{\mu,\lambda}, \beta'_\lambda(w_{\mu,\lambda})p(\theta_{\mu,\lambda}, J_\mu w_{\mu,\lambda})\right)_{L^2(\Omega)} \tag{3.12}$$
$$\leq M_9\tilde{E}_{\mu,\lambda} + M_{10}(||h||^2_{L^2(\Omega)} + 1),$$

where $\tilde{E}_{\mu,\lambda}(\cdot) := \frac{1}{2}|\nabla\theta_{\mu,\lambda}|^2_{L^2(\Omega)} + E^0_{\mu,\lambda}(\cdot) + 2E^3_{\mu,\lambda}(\cdot)$ and $G_{\mu,\lambda}(\cdot) := \frac{1}{4}|\theta'_{\mu,\lambda}|^2_{L^2(\Omega)} + \frac{1}{2}|\nabla\theta'_{\mu,\lambda}|^2_{L^2(\Omega)} + \varphi_\mu(w'_{\mu,\lambda})$. Also, we derive from (3.12) that

$$\mathrm{e}^{-M_9t}G_{\mu,\lambda} + \frac{d}{dt}\left\{\mathrm{e}^{-M_9t}\tilde{E}_{\mu,\lambda}\right\} + 2\mathrm{e}^{-M_9t}(\xi_{\mu,\lambda}, \beta'_\lambda(w_{\mu,\lambda})p(\theta_{\mu,\lambda}, J_\mu w_{\mu,\lambda}))_{L^2(\Omega)} \tag{3.13}$$
$$\leq M_{10}\mathrm{e}^{-M_9t}(||h||^2_{L^2(\Omega)} + 1).$$

For each $s \in [0,T]$, integrating (3.13) on $[0,s]$ yields that

$$\int_0^s \mathrm{e}^{-M_9t}G_{\mu,\lambda}dt + \mathrm{e}^{-M_9s}\tilde{E}_{\mu,\lambda}(s) + 2\int_0^s \mathrm{e}^{-M_9t}(\xi_{\mu,\lambda}, \beta'_\lambda(w_{\mu,\lambda})p(\theta_{\mu,\lambda}, J_\mu w_{\mu,\lambda}))_{L^2(\Omega)}dt$$
$$\leq \tilde{E}_{\mu,\lambda}(0) + M_{11}\int_0^s \mathrm{e}^{-M_9t}(||h||^2_{L^2(\Omega)} + 1)dt. \tag{3.14}$$

Here we observe that $\tilde{E}_{\mu,\lambda}(0)$ is bounded in $\mu, \lambda \in (0,1]$ and

$$\lim_{n\to\infty}\int_0^s \mathrm{e}^{-M_9t}(\xi_{\mu_n,\lambda}, \beta'_\lambda(w_{\mu_n,\lambda})p(\theta_{\mu_n,\lambda}, J_{\mu_n}w_{\mu_n,\lambda}))_{L^2(\Omega)}dt$$
$$= \int_0^s \mathrm{e}^{-M_9t}(\xi_\lambda, \beta'_\lambda(w_\lambda)p(\theta_\lambda, w_\lambda))_{L^2(\Omega)}dt = 0,$$

since $\beta'_\lambda(w_\lambda)p(\theta_\lambda, w_\lambda) = 0$ a.e. on Q. Hence, taking the limit in (3.14) with $\mu = \mu_n$, we have

$$\int_0^s e^{-M_9 t} G_\lambda dt + e^{-M_9 s}\tilde{E}_\lambda(s) \le \tilde{E}_\lambda(0) + M_{11}\int_0^s e^{-M_9 t}(||h||^2_{L^2(\Omega)} + 1)dt \quad (3.15)$$

for all $s \in [0, T]$, where

$$\begin{aligned} \tilde{E}_\lambda(\cdot) &:= \frac{1}{2}|\theta_\lambda|^2_{H^1(\Omega)} - (h, -\Delta_0\theta_\lambda)_{L^2(\Omega)} + |-\Delta_0 w_\lambda + \beta_\lambda(w_\lambda)|^2_{L^2(\Omega)} \\ &\quad -2(-\Delta_0 w_\lambda + \beta_\lambda(w_\lambda), f(\theta_\lambda, w_\lambda))_{L^2(\Omega)}, \\ G_\lambda(\cdot) &:= \frac{1}{4}|\theta'_\lambda|^2_{L^2(\Omega)} + \frac{1}{2}|\nabla\theta'_\lambda|^2_{L^2(\Omega)} + \varphi(w'_\lambda). \end{aligned}$$

Since $\{\tilde{E}_\lambda(0)\}_{\lambda\in(0,1]}$ is bounded, this inequality implies that $\tilde{E}_\lambda$ is uniformly bounded on $[0, T]$ with respect to $\lambda \in (0, 1]$. From this fact together with (3.15) we have an inequality which is required in the lemma. □

4. Proof of Theorem 2.1

We are now in a position to give a proof of Theorem 2.1.

Proof of Theorem 2.1 Let $\{\theta_\lambda, w_\lambda\}$ be a solution of (P_λ), $\lambda \in (0, 1]$, for which (3.11) in Lemma 3.3 holds, and put $\xi_\lambda := -w'_\lambda + \Delta_0 w_\lambda - \beta_\lambda(w_\lambda) + f(\theta_\lambda, w_\lambda)$ again.

First, by the uniform estimates (3.11) we can choose a sequence $\{\lambda_n\} \subset (0, 1]$ tending to 0 as $n \to \infty$ with functions $\theta, w \in W^{1,2}(0, T; H^1(\Omega)) \cap L^\infty(0, T; H^2(\Omega))$ and $\xi,\ \eta \in L^\infty(0, T; L^2(\Omega))$ such that

$$\theta_n := \theta_{\lambda_n} \to \theta,\ w_n := w_{\lambda_n} \to w \text{ in } C([0, T]; H^1(\Omega)) \text{ as } n \to \infty.$$

and

$$\xi_n := \xi_{\lambda_n} \to \xi,\ \beta_{\lambda_n}(w_n) \to \eta \text{ weakly-star in } L^\infty(0, T; L^2(\Omega)) \text{ as } n \to \infty;$$

note that the following convergences hold, too:

$$\begin{cases} \theta'_n \to \theta' & \text{weakly in } L^2(0, T; H^1(\Omega)), \\ w'_n \to w' & \text{weakly in } L^2(0, T; H^1(\Omega)), \\ -\Delta_0\theta_n \to -\Delta_0\theta & \text{weakly in } L^2(0, T; L^2(\Omega)), \\ -\Delta_0 w_n \to -\Delta_0 w & \text{weakly in } L^2(0, T; L^2(\Omega)), \\ f(\theta_n, w_n) \to f(\theta, w) & \text{in } C([0, T]; L^2(\Omega)), \\ p(\theta_n, w_n) \to p(\theta, w) & \text{in } C([0, T]; L^2(\Omega)), \end{cases} \quad (4.1)$$

as $n \to \infty$. From the convergences in (4.1) it follows that

$$\theta' + w' - \Delta_0\theta = h \text{ in } L^2(\Omega),\ \text{ a.e. on } (0, T),$$

$$w' + \xi - \Delta_0 w + \eta = f(\theta, w) \text{ in } L^2(\Omega),\ \text{ a.e. on } (0, T),$$

$$\theta(0) = \theta_0,\ w(0) = w_0 \text{ in } L^2(\Omega).$$

Moreover, by the demiclosedness of β (cf. [6]), we have $\eta \in \beta(w)$ a.e. on Q. Therefore, to complete the proof of Theorem 2.1 it is enough to show that $\xi \in \alpha(w' - p(\theta, w))$ a.e. on Q, which is derived as follows. We obverse that

$$\begin{aligned}
&\limsup_{n\to\infty} \int_0^T (\xi_n, w_n' - p(\theta_n, w_n))_{L^2(\Omega)} dt \\
= &\limsup_{n\to\infty} \int_0^T (-w_n' + \Delta_0 w_n - \beta_{\lambda_n}(w_n) + f(\theta_n, w_n), w_n' - p(\theta_n, w_n))_{L^2(\Omega)} dt \\
= &\limsup_{n\to\infty} \Big\{ -|w_n'|^2_{L^2(0,T;L^2(\Omega))} - \frac{1}{2}|\nabla w_n(T)|^2_{L^2(\Omega)} - \int_\Omega \hat{\beta}_{\lambda_n}(w_n(T)) dx \\
&+ \frac{1}{2}|\nabla w_0|^2_{L^2(\Omega)} + \int_\Omega \hat{\beta}_{\lambda_n}(w_0) dx \\
&- \int_0^T (-w_n' + \Delta_0 w_n - \beta_{\lambda_n}(w_n) + f(\theta_n, w_n), p(\theta_n, w_n))_{L^2(\Omega)} dt \Big\} \\
\leq &-|w'|^2_{L^2(0,T;L^2(\Omega))} - \frac{1}{2}|\nabla w(T)|^2_{L^2(\Omega)} - \int_\Omega \hat{\beta}(w(T)) dx + \frac{1}{2}|\nabla w_0|^2_{L^2(\Omega)} \\
&+ \int_\Omega \hat{\beta}(w_0) dx - \int_0^T (-w' + \Delta_0 w - \eta + f(\theta, w), p(\theta, w))_{L^2(\Omega)} dt \\
= &\int_0^T (-w' + \Delta_0 w - \eta + f(\theta, w), w' - p(\theta, w))_{L^2(\Omega)} dt \\
= &\int_0^T (\xi, w' - p(\theta, w))_{L^2(\Omega)} dt
\end{aligned}$$

By the maximal monotonicity of α the above inequalities imply that $\xi \in \alpha(w' - p(\theta, w))$ a.e. on Q. □

Remark 4.1. Under the weaker condition (2.2), a solution $\{\theta, w\}$ of (P) can be constructed by using energy inequalities (1), (3), (4) (in Lemma 3.2) and (5) (in Lemma 3.3). In this case, θ has less regularity

$$\theta \in W^{1,2}(0,T; L^2(\Omega)) \cap L^\infty(0,T; H^1(\Omega)) \cap L^2(0,T; H^2(\Omega)).$$

References

[1] M. Aso, A nonlinear system for irreversible phase change with constraint, Adv. Math. Sci. Appl., **15** (2005), 451–466.

[2] M. Aso, M. Frémond and N. Kenmochi, Phase change problems with temperature dependent constraints for the volume fraction velocities, Nonlinear Anal., **60** (2005), 1003–1023.

[3] M. Aso, M. Frémond and N. Kenmochi, Quasi-variational evolution problems for irreversible phase change, pp. 517–525 *in Proceedings of International Conference on Nonlinear Partial Differential Equations and Their Applications,* ed. N. Kenmochi,

M. Ôtani and S. Zheng, GAKUTO Intern. Ser. Math. Sci. Appl., **Vol. 20**, Gakkotosho, Tokyo, 2004.

[4] M. Aso and N. Kenmochi, *A class of doubly nonlinear quasi-variational evolution problems*, to appear in GAKUTO Intern. Ser. Math. Sci. Appl., **Vol. 23** (2005).

[5] H. Attouch, *Variational Convergence for Functions and Operators*, Applicable Mathematics Series, Pitman Advanced Publishing Program, Boston, 1984.

[6] V. Barbu, *Nonlinear semigroups and differential equations in Banach spaces*, Noordhoff, Leyden, 1976.

[7] G. Bonfanti, M. Frémond and F. Luterotti, Global solution to a nonlinear system for irreversible phase changes, Adv. Math. Sci. Appl., **10** (2000), 1–24.

[8] H. Brézis, *Opérateurs Maximaux Monotones et Semi-groupes de contractions dans les espaces de Hilbert*, Math. Studies **5**, North-Holland, Amsterdam, 1973.

[9] P. Laurençot, G. Schimperna and U. Stefanelli, Global existence of a strong solution to the one-dimensional full model for irreversible phase transitions, J. Math. Anal. Appl., **271** (2002), 426–442.

[10] U. Mosco, Convergence of convex sets and of solutions of variational inequalities, Adv. Math., **3** (1969), 510–585.

[11] E. Zeidler, *Nonlinear Functional Analysis and Its Applications. II/A. Linear Monotone Operators,* Springer-Verlag, New York/Berlin, 1990.

Masayasu Aso
Department of Mathematics
Faculty of Education
Chiba University
1-33 Yayoi-chō, Inage-ku, Chiba 263-8522 Japan
e-mail: masayasu.aso@faculty.chiba-u.jp

Michel Frémond
Laboratoire Lagrange, LCPC
58 Boulevard Lefebvre
F-75732 Paris Cedex 15, France
e-mail: fremond@lcpc.fr

Nobuyuki Kenmochi
Department of Mathematics
Faculty of Education
Chiba University
1-33 Yayoi-chō, Inage-ku, Chiba 263-8522 Japan
e-mail: kenmochi@faculty.chiba-u.jp

International Series of Numerical Mathematics, Vol. 154, 55–64

The N-membranes Problem with Neumann Type Boundary Condition

A. Azevedo, J.F. Rodrigues and L. Santos

Abstract. We consider the problem of finding the equilibrium position of N membranes constrained not to pass through each other, under prescribed volumic forces and boundary tensions. This model corresponds to solve variationally a N-system for linear second order elliptic equations with sequential constraints. We obtain interior and boundary Lewy-Stampacchia type inequalities for the respective solution and we establish the conditions for stability in measure of the interior contact zones of the membranes.

Mathematics Subject Classification (2000). Primary 35R35; Secondary 35J50.

Keywords. Variational inequalities, Lewy-Stampacchia inequalities, coincidence sets.

1. Introduction

Let Ω be a bounded open subset of $\mathbb{R}^d$ with Lipschitz boundary Γ. Denote by $\boldsymbol{u} = (u_1, \ldots, u_N)$ the equilibrium displacements of N ($N \geq 2$) elastic membranes, each one constrained not to pass through the others, subject to external volumic forces $\mathbf{f} = (f_1, \ldots, f_N)$ and boundary tensions $\mathbf{g} = (g_1, \ldots, g_N)$. The problem consists of minimizing the energy functional

$$E(\boldsymbol{u}) = \int_\Omega \left(\frac{1}{2} \left(a(\boldsymbol{u}, \boldsymbol{u}) + c\, \boldsymbol{u} \cdot \boldsymbol{u} \right) - \boldsymbol{f} \cdot \boldsymbol{u} \right) + \int_\Gamma \left(\frac{1}{2} b\, \boldsymbol{u} \cdot \boldsymbol{u} - \boldsymbol{g} \cdot \boldsymbol{u} \right), \tag{1.1}$$

in the convex set

$$\mathbb{K}_N = \left\{ \boldsymbol{v} = (v_1, \ldots, v_N) \in \left[H^1(\Omega) \right]^N : v_1 \geq \cdots \geq v_N \text{ a.e. in } \Omega \right\}, \tag{1.2}$$

This work was partially supported by FCT (Fundação para a Ciência e Tecnologia).

where $a(\boldsymbol{u},\boldsymbol{v}) = \sum_{k=1}^{N} a(u_k, v_k)$, with $a(u,v) = a_{ij} u_{x_i} v_{x_j}$ (using the summation convention for $i,j = 1,\dots,d$) and $\boldsymbol{u}\cdot\boldsymbol{v}$ denotes the usual internal product between $\boldsymbol{u}$ and $\boldsymbol{v}$.

The N-membranes problem attached to rigid supports was considered in [3] for N linear coercive elliptic operators of second order and extended in [1] to quasilinear operators, with smooth coefficients of p-Laplacian type. For general linear second order elliptic operators with measurable coefficients, see also [2].

Although Neumann boundary type problems can also be considered for more general operators, for simplicity, here we assume

$$\begin{cases} a_{ij} \in L^\infty(\Omega),\ a_{ij} = a_{ji}, \quad \exists \nu > 0\, \forall \xi \in \mathbb{R}^d \quad a_{ij}\xi_i\xi_j \geq \nu|\xi|^2, \\ c \in L^\infty(\Omega),\ b \in L^\infty(\Gamma), \quad c \geq c_0 \geq 0,\ b \geq b_0 \geq 0, \quad c_0 + b_0 > 0. \end{cases} \tag{1.3}$$

$$\begin{cases} f_1,\dots,f_N \in L^p(\Omega), \quad g_1,\dots,g_N \in L^q(\Gamma), \\ p \geq \frac{2d}{d+2} \quad \text{if } d \geq 3, \quad p > 1 \text{ if } d = 2, \\ q \geq \frac{2(d-1)}{d} \quad \text{if } d \geq 3, \quad q > 1 \text{ if } d = 2. \end{cases} \tag{1.4}$$

Here we use $\bigvee$ and $\bigwedge$ for the supremum and infimum, respectively, of two or more functions

$$\bigvee_{k=1}^{N} \xi_k = \sup\{\xi_1,\dots,\xi_N\}, \qquad \bigwedge_{k=1}^{N} \xi_k = \inf\{\xi_1,\dots,\xi_N\},$$

and, accordingly, we set $\xi^+ = \xi \vee 0$ and $\xi^- = -(\xi \wedge 0)$.

The minimization problem (1.1)–(1.2) is equivalent to the variational inequality

$$\begin{cases} \boldsymbol{u} \in \mathbb{K}_N : \\ \displaystyle\int_\Omega \big(a(\boldsymbol{u}, \boldsymbol{v} - \boldsymbol{u}) + c\,\boldsymbol{u}\cdot(\boldsymbol{v} - \boldsymbol{u})\big) + \int_\Gamma b\,\boldsymbol{u}\cdot(\boldsymbol{v} - \boldsymbol{u}) \\ \qquad\qquad \geq \displaystyle\int_\Omega \boldsymbol{f}\cdot(\boldsymbol{v} - \boldsymbol{u}) + \int_\Gamma \boldsymbol{g}\cdot(\boldsymbol{v} - \boldsymbol{u}), \qquad \forall \boldsymbol{v} \in \mathbb{K}_N. \end{cases} \tag{1.5}$$

For $N = 2$ this problem can be considered, when the solution is known, as two one obstacle problems. For $N \geq 3$, the upper and the lower membranes are of this type, but each membrane in between may be considered a solution of a two obstacles problem. This last problem corresponds to a variational inequality with the convex set given in the form

$$\mathbb{K}_\psi^\varphi = \{\xi \in H^1(\Omega) : \psi \leq \xi \leq \varphi \text{ a.e. in } \Omega\},$$

where the given obstacles are such that $\psi \leq \varphi$. For two obstacles, the Lewy--Stampacchia inequalities for the solution v are

$$f \wedge A\varphi \leq Av \leq f \vee A\psi \ \text{ a.e. in } \Omega, \qquad g \wedge B\varphi \leq Bv \leq g \vee B\psi \ \text{ a.e. on } \Gamma, \tag{1.6}$$

where A and B denote the associated differential and boundary operators, respectively,

$$Av = -\left(a_{ij}v_{x_i}\right)_{x_j} + cv, \qquad \text{in } \Omega, \tag{1.7}$$

$$Bv = a_{ij}v_{x_i}n_j + bv, \qquad \text{on } \Gamma, \tag{1.8}$$

$(n_1, \ldots, n_d)$ denoting the unit outward normal vector to Γ.

The iteration of these inequalities yields the new set of N inequalities for the solution $\boldsymbol{u}$ of the N-membranes problem, both in Ω and on Γ

$$\bigwedge_{k=1}^{l} f_k \le Au_l \le \bigvee_{k=l}^{N} f_k, \qquad \text{a.e. in } \Omega, \qquad l = 1, \ldots, N, \tag{1.9}$$

$$\bigwedge_{k=1}^{l} g_k \le Bu_l \le \bigvee_{k=l}^{N} g_k, \qquad \text{a.e. on } \Gamma, \qquad l = 1, \ldots, N, \tag{1.10}$$

which allows to reduce the regularity of the solutions to the corresponding regularity of a system of equations, as shown in the next section. In particular, in the following special cases:

- $f_1 = \cdots = f_N = f$, the solution $\boldsymbol{u}$ of the variational inequality (1.5) satisfies the system of N equations $Au_k = f$ a.e. in Ω, $k = 1, \ldots, N$;
- $g_1 = \cdots = g_N = g$, the solution $\boldsymbol{u}$ of the variational inequality (1.5) satisfies the Neumann boundary conditions $Bu_k = g$ a.e. on Γ, $i = 1, \ldots, N$, although in the general case we only can say that $\boldsymbol{u}$ satisfies Signorini type boundary conditions.

Another interesting result is the stability of the $\frac{N(N-1)}{2}$ coincidence sets

$$I_{k,l} = \{x \in \Omega : u_k(x) = \cdots = u_l(x) \text{ for a.e. } x \in \Omega\}, \qquad 1 \le k < l \le N, \tag{1.11}$$

the sets of contact of $l - k + 1$ consecutive membranes. Given a subset A of Ω, we denote by χ_A (the characteristic function of A), i.e., $\chi_A(x) = 1$ if $x \in A$ and $\chi_A(x) = 0$ if $x \in \Omega \setminus A$. As we have shown in [1] this is a consequence of writing the solution of (1.5) as the solution of a semilinear system involving the characteristic functions $\chi_{I_{k,l}}$. We exemplify the argument in the simple cases $N = 2, 3$.

For $N = 2$ there is only one possible coincidence set, the contact of u_1 with u_2. If the two forces associated with the two membranes are almost everywhere different in Ω ($f_1 \ne f_2$ a.e. in Ω), then the characteristic function $\chi_{I_{1,2}}$ of $I_{1,2}$ is easily shown to converge strongly in any $L^s(\Omega)$, $1 < s < \infty$, for variations of the forces in $L^p(\Omega)$.

For $N = 3$ there are three possible coincidence sets, the sets $I_{1,2}$, $I_{2,3}$ and $I_{1,3} = I_{1,2} \cap I_{2,3}$. Setting $\chi_{k,l} = \chi_{I_{k,l}}$, $1 \le k < l \le 3$, the characteristic functions $\chi_{k,l}$ of the sets $I_{k,l}$ are shown to converge strongly in any $L^s(\Omega)$, $1 < s < \infty$, for variations of the forces f_1, f_2 and f_3 in $L^p(\Omega)$, as long as

$$f_1 \ne f_2, \qquad f_2 \ne f_3, \qquad f_1 \ne \frac{1}{2}(f_2 + f_3), \qquad \frac{1}{2}(f_1 + f_2) \ne f_3. \tag{1.12}$$

This is a consequence of the fact that the solution $\boldsymbol{u}$ of (1.5) satisfies the system a.e. in Ω,

$$\begin{cases} Au_1 = f_1 + \frac{1}{2}(f_2 - f_1)\chi_{1,2} + \frac{1}{6}(2f_3 - f_2 - f_1)\chi_{1,3} \\ Au_2 = f_2 - \frac{1}{2}(f_2 - f_1)\chi_{1,2} + \frac{1}{2}(f_3 - f_2)\chi_{2,3} + \frac{1}{6}(2f_2 - f_1 - f_3)\chi_{1,3} \\ Au_3 = f_3 - \frac{1}{2}(f_3 - f_2)\chi_{2,3} + \frac{1}{6}(2f_1 - f_2 - f_3)\chi_{1,3}. \end{cases} \tag{1.13}$$

Notice that the system (1.13) contains the case $N = 2$, that reduces only to the two first equations of this system, with $I_{2,3} = \emptyset$ (so $\chi_{2,3} = \chi_{1,3} = 0$). Even in the more complicated situation of $N > 3$, the stability result can still be extended in the interior of Ω as we show in Section 3. However, the corresponding stability result on the boundary Γ is an open question. In this paper we have chosen to present only the Neumann case when $\Gamma = \partial\Omega$, but all the results are still valid, with simple adaptations, for the mixed problem where $\partial\Omega = \Gamma_0 \cup \Gamma_1$, with Dirichlet data on Γ_0 and Neumann data on Γ_1 (see [7], for instance).

2. The Lewy-Stampacchia inequalities

We begin this section recalling a theorem for the double obstacle problem:

Theorem 2.1. *Suppose that $\psi_1, \psi_2 \in H^1(\Omega)$, $f \in L^p(\Omega)$, $g \in L^q(\Gamma)$, p, q defined as in (1.4). Let u be the solution of the variational inequality*

$$\int_\Omega \Big(a(u, v-u) + cu(v-u)\Big) + \int_\Gamma b(v-u) \geq \int_\Omega f(v-u) + \int_\Gamma g(v-u), \tag{2.1}$$

with the assumptions (1.3), *in the convex set*

$$\mathbb{K}_{\psi_1}^{\psi_2} = \{v \in H^1(\Omega) : \psi_1 \leq v \leq \psi_2 \text{ a.e. in } \Omega\}. \tag{2.2}$$

If $(A\psi_1 - f)^+$, $(A\psi_2 - f)^- \in L^p(\Omega)$ and $(B\psi_1 - g)^+$, $(B\psi_2 - g)^- \in L^q(\Gamma)$, then

$$f \wedge A\psi_1 \leq Au \leq f \vee A\psi_2, \qquad \text{a.e. in } \Omega, \tag{2.3}$$

$$g \wedge B\psi_1 \leq Bu \leq g \vee B\psi_2, \qquad \text{a.e. on } \Gamma. \tag{2.4}$$

Proof. The proof of this theorem is a simple adaptation of the arguments used for the one obstacle problem with Neumann boundary condition (see, for instance, [9] or [7]). □

Remark 2.2. We observe that both the lower and the upper one obstacle variational inequalities (2.1) in the convex sets

$$\mathbb{K}_{\psi_1} = \{v \in H^1(\Omega) : v \geq \psi_1 \text{ a.e. in } \Omega\}$$

and

$$\mathbb{K}^{\psi_2} = \{v \in H^1(\Omega) : v \leq \psi_2 \text{ a.e. in } \Omega\},$$

can be regarded as particular cases of the double obstacle problem, corresponding formally to $\psi_2 = +\infty$ and $\psi_1 = -\infty$, respectively.

Given N functions $\varphi_1, \ldots, \varphi_N$, we define, for $1 \leq k < l \leq N$, the average of $\varphi_k, \ldots, \varphi_l$ as

$$\langle \varphi \rangle_{k,l} = \frac{\varphi_k + \cdots + \varphi_l}{l - k + 1}. \tag{2.5}$$

Denote

$$\xi_0 = \max\left\{ \langle f \rangle_{1,k} : k = 1, \ldots, N \right\}, \qquad \eta_0 = \max\left\{ \langle g \rangle_{1,k} : k = 1, \ldots, N \right\} \tag{2.6}$$

and, for $k = 1, \ldots, N$,

$$\xi_k = k\big(\xi_0 - \langle f \rangle_{1,k}\big) \qquad \eta_k = k\big(\eta_0 - \langle g \rangle_{1,k}\big). \tag{2.7}$$

We may approximate the solution of (1.5) by the solution of the penalized problem given by the semilinear system with Neumann boundary conditions, for $k = 1, \ldots, N$,

$$\begin{cases} Au_k^\varepsilon + \xi_k \theta_\varepsilon(u_k^\varepsilon - u_{k+1}^\varepsilon) - \xi_{k-1}\theta_\varepsilon(u_{k-1}^\varepsilon - u_k^\varepsilon) = f_k & \text{in } \Omega, \\ Bu_k^\varepsilon + \eta_k \theta_\varepsilon(u_k^\varepsilon - u_{k+1}^\varepsilon) - \eta_{k-1}\theta_\varepsilon(u_{k-1}^\varepsilon - u_k^\varepsilon) = g_k & \text{on } \Gamma, \end{cases} \tag{2.8}$$

with the conventions $u_0^\varepsilon = +\infty$, $u_{N+1}^\varepsilon = -\infty$, where for $\varepsilon > 0$, θ_ε is defined by $\theta_\varepsilon(s) = -1$ if $s \leq -\varepsilon$, $\theta_\varepsilon(s) = \frac{s}{\varepsilon}$, if $-\varepsilon < s < 0$ and $\theta_\varepsilon(s) = 0$ for $s \geq 0$.

Proposition 2.3. *With the assumptions* (1.3) *and* (1.4), *problem* (2.8) *has a unique solution* $(u_1^\varepsilon, \ldots, u_N^\varepsilon)$, *bounded independently of* ε *in* $\left[H^1(\Omega)\right]^N$. *Besides that,* $A\boldsymbol{u}^\varepsilon$ *and* $B\boldsymbol{u}^\varepsilon$ *are bounded independently of* ε *in* $[L^p(\Omega)]^N$ *and in* $[L^q(\Gamma)]^N$, *respectively.*

Proof. Consider the monotone operator

$$\begin{aligned} \langle \Psi_\varepsilon(\boldsymbol{v}), \boldsymbol{w} \rangle &= \sum_{k=1}^N \int_\Omega \Big(\xi_k \theta_\varepsilon(v_k - v_{k+1}) - \xi_{k-1}\theta_\varepsilon(v_{k-1} - v_k) \Big) w_k \\ &\quad + \sum_{k=1}^N \int_\Gamma \Big(\eta_k \theta_\varepsilon(v_k - v_{k+1}) - \eta_{k-1}\theta_\varepsilon(v_{k-1} - v_k) \Big) w_k. \end{aligned} \tag{2.9}$$

The problem (2.8) is equivalent to the semilinear variational problem

$$\begin{cases} \boldsymbol{u}^\varepsilon \in \left[H^1(\Omega)\right]^N : \\ \displaystyle\int_\Omega (a(\boldsymbol{u}^\varepsilon, \boldsymbol{v}) + c\,\boldsymbol{u}^\varepsilon \cdot \boldsymbol{v}) + \int_\Gamma b\,\boldsymbol{u}^\varepsilon \cdot \boldsymbol{v} + \langle \Psi_\varepsilon(\boldsymbol{u}^\varepsilon), \boldsymbol{v} \rangle \\ \qquad = \displaystyle\int_\Omega \boldsymbol{f} \cdot \boldsymbol{v} + \int_\Gamma \boldsymbol{g} \cdot \boldsymbol{v}, \qquad \forall \boldsymbol{v} \in \left[H^1(\Omega)\right]^N \end{cases} \tag{2.10}$$

and this problem has a unique solution, by standard monotone methods.

Since

$$A\boldsymbol{u}^\varepsilon = \boldsymbol{f} - \big(\xi_k \theta_\varepsilon(u_k^\varepsilon - u_{k+1}^\varepsilon) - \xi_{k-1}\theta_\varepsilon(u_{k-1}^\varepsilon - u_k^\varepsilon)\big)_{k=1,\ldots,N},$$

$-1 \leq \theta_\varepsilon \leq 0$ and $\boldsymbol{f}, \boldsymbol{\xi} \in \left[L^p(\Omega)\right]^N$, it follows that $\{A\boldsymbol{u}^\varepsilon : 0 < \varepsilon < 1\}$ belongs to a bounded subset of $\left[L^p(\Omega)\right]^N$. Analogously, after integration by parts, the set $\{B\boldsymbol{u}^\varepsilon : 0 < \varepsilon < 1\}$ is bounded in $\left[L^q(\Gamma)\right]^N$. □

Proposition 2.4. *Under the assumptions* (1.3) *and* (1.4), *let* $\boldsymbol{u}^\varepsilon$ *be the solution of problem* (2.8) *and* $\boldsymbol{u}$ *the solution of the variational inequality* (1.5). *Then*

$$u_k^\varepsilon \leq u_{k-1}^\varepsilon + \varepsilon, \qquad k = 2, \dots, N, \tag{2.11}$$

and, when $\varepsilon \to 0$,

$$\boldsymbol{u}^\varepsilon \longrightarrow \boldsymbol{u} \qquad in \ \left[H^1(\Omega)\right]^N,$$

$$A\boldsymbol{u}^\varepsilon \longrightarrow A\boldsymbol{u} \ in \ \left[L^p(\Omega)\right]^N\text{-weak}, \qquad B\boldsymbol{u}^\varepsilon \longrightarrow B\boldsymbol{u} \ in \ \left[L^q(\Gamma)\right]^N\text{-weak}.$$

Proof. We begin noticing that,

$$\xi_k \geq 0 \ \ (k \geq 1), \qquad \left(\xi_{k-1} - \xi_{k-2}\right) - \left(\xi_k - \xi_{k-1}\right) = f_k - f_{k-1} \ \ (k \geq 2),$$

$$\eta_k \geq 0 \ \ (k \geq 1), \qquad \left(\eta_{k-1} - \eta_{k-2}\right) - \left(\eta_k - \eta_{k-1}\right) = g_k - g_{k-1} \ \ (k \geq 2).$$

To prove (2.11), we multiply the k−th equation of (2.8) by $(u_k^\varepsilon - u_{k-1}^\varepsilon - \varepsilon)^+$ and integrate on Ω. Using that $\theta_\varepsilon(u_{k-1}^\varepsilon - u_k^\varepsilon)(u_k^\varepsilon - u_{k-1}^\varepsilon - \varepsilon)^+ = -(u_k^\varepsilon - u_{k-1}^\varepsilon - \varepsilon)^+$ and $\theta_\varepsilon(u_k^\varepsilon - u_{k+1}^\varepsilon) \geq -1$, we obtain

$$\begin{aligned} &\int_\Omega A\, u_k^\varepsilon (u_k^\varepsilon - u_{k-1}^\varepsilon - \varepsilon)^+ \\ &\qquad \leq \int_\Omega \left[f_k + \xi_k - \xi_{k-1}\right](u_k^\varepsilon - u_{k-1}^\varepsilon - \varepsilon)^+ \\ &\qquad\quad + \int_\Gamma \left[g_k + \eta_k - \eta_{k-1}\right](u_k^\varepsilon - u_{k-1}^\varepsilon - \varepsilon)^+. \end{aligned} \tag{2.12}$$

With similar arguments, if we multiply, for $k \geq 2$, the $(k-1)$−th equation of (2.8) by $(u_k^\varepsilon - u_{k-1}^\varepsilon - \varepsilon)^+$ and integrate on Ω we obtain,

$$\begin{aligned} &\int_\Omega A\, u_{k-1}^\varepsilon (u_k^\varepsilon - u_{k-1}^\varepsilon - \varepsilon)^+ \\ &\qquad \geq \int_\Omega \left[f_{k-1} + \xi_{k-1} - \xi_{k-2}\right](u_k^\varepsilon - u_{k-1}^\varepsilon - \varepsilon)^+ \\ &\qquad\quad + \int_\Gamma \left[g_{k-1} + \eta_{k-1} - \eta_{k-2}\right](u_k^\varepsilon - u_{k-1}^\varepsilon - \varepsilon)^+. \end{aligned} \tag{2.13}$$

Subtracting equation (2.13) from (2.12), using the assumptions (1.3), the conclusion (2.11) follows.

The strong convergence in $\left[H^1(\Omega)\right]^N$ of $\boldsymbol{u}^\varepsilon$ to the solution $\boldsymbol{u}$ of the variational inequality (1.5), when $\varepsilon \to 0$, follows by a standard argument.

The uniform boundedeness of $\{A\boldsymbol{u}^\varepsilon : 0 < \varepsilon < 1\}$ in $\left[L^p(\Omega)\right]^N$ implies the weak convergence of $A\boldsymbol{u}^\varepsilon$ to $A\boldsymbol{u}$ in $\left[L^p(\Omega)\right]^N$, and, analogously, the boundedeness of $\{B\boldsymbol{u}^\varepsilon : 0 < \varepsilon < 1\}$ in $\left[L^q(\Gamma)\right]^N$ implies the weak convergence of $B\boldsymbol{u}^\varepsilon$ to $B\boldsymbol{u}$ in $\left[L^q(\Gamma)\right]^N$. □

We are now able to prove the following result:

Theorem 2.5. *Under the assumptions* (1.3) *and* (1.4), *the solution* $\boldsymbol{u}$ *of the problem* (1.5) *satisfies the following Lewy-Stampacchia type inequalities*

$$\left.\begin{array}{rcccl} f_1 & \leq & A\,u_1 & \leq & f_1\vee\cdots\vee f_N \\ f_1\wedge f_2 & \leq & A\,u_2 & \leq & f_2\vee\cdots\vee f_N \\ & & \vdots & & \\ f_1\wedge\cdots\wedge f_{N-1} & \leq & A\,u_{N-1} & \leq & f_{N-1}\vee f_N \\ f_1\wedge\cdots\wedge f_N & \leq & A\,u_N & \leq & f_N \end{array}\right\} \quad a.e.\ in\ \Omega \tag{2.14}$$

and

$$\left.\begin{array}{rcccl} g_1 & \leq & B\,u_1 & \leq & g_1\vee\cdots\vee g_N \\ g_1\wedge g_2 & \leq & B\,u_2 & \leq & g_2\vee\cdots\vee g_N \\ & & \vdots & & \\ g_1\wedge\cdots\wedge g_{N-1} & \leq & B\,u_{N-1} & \leq & g_{N-1}\vee g_N \\ g_1\wedge\cdots\wedge g_N & \leq & B\,u_N & \leq & g_N \end{array}\right\} \quad a.e.\ on\ \Gamma. \tag{2.15}$$

Proof. If $(v, u_2, \dots, u_N) \in \mathbb{K}_N$, with $v \in \mathbb{K}_{u_2}$, we see that $u_1 \in \mathbb{K}_{u_2}$ solves the variational inequality (1.5) with $f = f_1$. Observing that $Au_2 \in L^p(\Omega)$ and that $Bu_2 \in L^q(\Gamma)$, by (2.3) and (2.4) we have

$$f_1 \leq A\,u_1 \leq f_1 \vee A\,u_2 \quad \text{a.e. in } \Omega$$

$$g_1 \leq B\,u_1 \leq g_1 \vee B\,u_2 \quad \text{a.e. in } \Gamma.$$

Since $u_k \in \mathbb{K}_{u_{k+1}}^{u_{k-1}}$ solves the two obstacles problem (2.1) with $f = f_k$, $k = 2, \dots, N-1$, and satisfies, by (2.3) and (2.4),

$$f_k \wedge A\,u_{k-1} \leq A\,u_k \leq f_k \vee A\,u_{k+1} \quad \text{a.e. in } \Omega,$$

$$g_k \wedge B\,u_{k-1} \leq B\,u_k \leq g_k \vee B\,u_{k+1} \quad \text{a.e. in } \Gamma.$$

As $u_N \in \mathbb{K}^{u_{N-1}}$ satisfies

$$f_N \wedge A\,u_{N-1} \leq A\,u_N \leq f_N \quad \text{a.e. on } \Omega,$$

$$g_N \wedge B\,u_{N-1} \leq B\,u_N \leq g_N \quad \text{a.e. on } \Gamma,$$

(2.14) and (2.15) are easily obtained by simple iterations. □

Remark 2.6. The Lewy-Stamppachia inequalities appeared first in [6] for the obstacle problem with Dirichlet boundary conditions and were extended to the Neumann case in [5] (see also [9] and [8]).

From (2.14) and (2.15) the following corollary is immediate:

Corollary 2.7. *Let* $\boldsymbol{u}$ *be the solution of the variational inequality* (1.5). *We have*

if $\boldsymbol{f} = (f, \dots, f)$, *then* $A\boldsymbol{u} = \boldsymbol{f}$ *in* Ω, *if* $\boldsymbol{g} = (g, \dots, g)$, *then* $B\boldsymbol{u} = \boldsymbol{g}$ *on* Γ.

From the linear elliptic regularity theory (see [4] or [8], for instance) we have

Corollary 2.8. *Under the assumptions* (1.3) *and* (1.4), *the solution* $\boldsymbol{u}$ *of* (1.5) *is in* $\left[C^{0,\alpha}(\overline{\Omega})\right]^N$, *for some* $0 < \alpha < 1$. *Besides that, if* $a_{ij} \in C^{0,1}(\overline{\Omega})$ *then* $\boldsymbol{u} \in \left[W^{2,p}_{loc}(\Omega)\right]^N$ *and* $\boldsymbol{u} \in \left[C^{1,\beta}(\Omega)\right]^N$ *if* $0 < \beta = 1 - \frac{d}{p} < 1$; *if in addition* $\Gamma \in C^{1,1}$, $b \in C^{0,1}(\Gamma)$ *and* $\boldsymbol{f} \in \left[L^2(\Omega)\right]^N$, $\boldsymbol{g} \in \left[L^2(\Gamma)\right]^N$ *then* $\boldsymbol{u} \in \left[W^{3/2,2}(\Omega)\right]^N$; *finally, if also* $g_1 = \cdots = g_N \in W^{1-\frac{1}{p},p}(\Gamma)$, *then* $\boldsymbol{u} \in \left[W^{2,p}(\Omega)\right]^N$.

3. The stability of the coincidence sets

Let $\boldsymbol{u_n}$ be the solution of the N-membranes problem (1.5), under the assumptions (1.3), with given data $\boldsymbol{f_n}$ and $\boldsymbol{g_n}$ satisfying (1.4). Assuming that $\boldsymbol{f_n}$ converges to $\boldsymbol{f}$ in $[L^p(\Omega)]^N$ and that $\boldsymbol{g_n}$ converges to $\boldsymbol{g}$ in $[L^q(\Gamma)]^N$, we shall extend now the following stability result in $L^s(\Omega)$ $(1 \leq s < \infty)$ of [1] for the corresponding coincidence sets (defined in (1.11)),

$$\chi_{\{u_k^n=\cdots=u_l^n\}} \xrightarrow[n]{} \chi_{\{u_k=\cdots=u_l\}}, \qquad \text{for } 1 \leq k < l \leq N.$$

Recalling the inequalities (2.14), $\boldsymbol{Au} = \boldsymbol{F}$ a.e. in Ω, for some function $\boldsymbol{F} \in [L^p(\Omega)]^N$, as in Lemma 2 of [8], we have

$$Au_k = Au_{k+1} \qquad \text{a.e. in } \{x \in \Omega : u_k(x) = u_{k+1}(x)\}$$

and so we can characterize a.e. in Ω each F_k in terms of f_l and the characteristic functions $\chi_{\{u_r=\cdots=u_s\}}$, $1 \leq l \leq N$, $1 \leq r < s \leq N$.

In what follows, we use, as before, the convention, $u_0 = +\infty$ and $u_{N+1} = -\infty$. We define the following sets

$$\Theta_{k,l} = \{x \in \Omega : u_{k-1}(x) > u_k(x) = \cdots = u_l(x) > u_{l+1}(x)\}, \tag{3.1}$$

the sets of contact of exactly the membranes $u_k, \ldots, u_l$.

Proposition 3.1. *If* $k, l \in \mathbb{N}$ *are such that* $1 \leq k \leq l \leq N$, *we have*

1. $Au_r = \begin{cases} \langle f \rangle_{k,l} & \text{a.e. in } \Theta_{k,l} \quad \text{if } r \in \{k, \ldots, l\}, \\ f_r & \text{a.e. in } \Theta_{k,l} \quad \text{if } r \notin \{k, \ldots, l\}. \end{cases}$
2. *If* $k < l$ *then for all* $r \in \{k, \ldots, l\}$ $\langle f \rangle_{r+1,l} \geq \langle f \rangle_{k,r}$ *a.e. in* $\Theta_{k,l}$.

Proof. Because of the regularity result $\boldsymbol{Au} \in [L^p(\Omega)]^N$, the proof of this proposition is the same as for the case with boundary Dirichlet condition, done in [1], since it was done locally at a.e. point $x \in \Omega$. □

Remark 3.2. It is well known that a necessary condition for existing contact in the case of two membranes u_1 and u_2, subject to external forces f_1 and f_2 respectively, is that $f_2 \geq f_1$. Depending on the boundary conditions, this condition may be (or not) sufficient for contact.

We would like to emphasize that condition 2. of the preceding proposition is a necessary condition for the first $r-k$ membranes ($k<r\leq l$) to be in contact with the other $l-r+1$ membranes. We can interpret physically the condition 2. by regarding the first $r-k$ membranes as one membrane where a force with the intensity of the average of the forces $f_k,\ldots,f_r$ is applied and all the other $l-r+1$ as another one where it was applied a force with the intensity equal to the average of the remaining forces $f_{r+1},\ldots,f_l$.

As for the boundary Dirichlet condition case, we may characterize the variational inequality (1.5) as a system of N equations, coupled through the characteristic functions of the coincidence sets $I_{k,l}$. In (1.13) we presented the system for $N=3$, containing as a special case $N=2$. The next theorem presents the general case.

Theorem 3.3. *Under the assumptions* (1.3), *let* $\boldsymbol{u}$ *be the solution of the problem* (1.5) *with data* $\boldsymbol{f}$ *and* $\boldsymbol{g}$ *satisfying* (1.4). *Then*

$$Au_r = f_r + \sum_{1\leq k<l\leq N,\ k\leq r\leq l} b_r^{k,l}\,\chi_{k,l} \quad a.e.\ in\ \Omega, \tag{3.2}$$

where

$$b_r^{k,l} = b_r^{k,l}[f] = \begin{cases} \langle f\rangle_{k,l} - \langle f\rangle_{k,l-1} & if\ r=l \\ \langle f\rangle_{k,l} - \langle f\rangle_{k+1,l} & if\ r=k \\ \frac{2}{(l-k)(l-k+1)}\left(\langle f\rangle_{k+1,l-1} - \frac{1}{2}(f_k+f_l)\right) & if\ k<r<l. \end{cases}$$

Also exactly as in [1], using the variational convergence $\boldsymbol{u_n}\to\boldsymbol{u}$ in $\left[H^1(\Omega)\right]^N$, we may prove the continuous dependence of the coincidence sets with respect to the external data.

Theorem 3.4. *Assuming* (1.3) *and given* $n\in\mathbb{N}$, *let* $\boldsymbol{u_n}$ *denote the solution of problem* (1.5) *with given data* $\boldsymbol{f_n}\in\left[L^p(\Omega)\right]^N$, $\boldsymbol{g_n}\in\left[L^q(\Gamma)\right]^N$, *with* p,q *as in* (1.4).

Suppose that

$$\boldsymbol{f_n}\xrightarrow[n]{}\boldsymbol{f}\ \ in\ \ \left[L^p(\Omega)\right]^N, \qquad \boldsymbol{g_n}\xrightarrow[n]{}\boldsymbol{g}\ \ in\ \ \left[L^q(\Gamma)\right]^N.$$

Then

$$\boldsymbol{u_n}\xrightarrow[n]{}\boldsymbol{u}\ \ in\ \ \left[H^1(\Omega)\right]^N. \tag{3.3}$$

If, in addition, the limit forces satisfy

$$\langle f\rangle_{k,r}\neq\langle f\rangle_{r+1,l} \qquad for\ all\ k,r,l\in\{1,\ldots,N\}\ with\ k\leq r<l, \tag{3.4}$$

then, for any $1\leq s<\infty$, $\forall\,k,l\in\{1,\ldots,N\}$, $k<l$,

$$\chi_{\{u_k^n=\cdots=u_l^n\}}\xrightarrow[n]{}\chi_{\{u_k=\cdots=u_l\}} \qquad in\ L^s(\Omega). \tag{3.5}$$

Remark 3.5. The condition (3.4) for the stability of the coincidence sets for $N=2$ is simply $f_2\neq f_1$ and for $N=3$, the condition (1.12) (see [2] for a direct proof).

Remark 3.6. It would be interesting to prove a condition analogous to the system (3.2) for the boundary operator B (under additional regularity of the solution $\boldsymbol{u}$), i.e., to find sufficient conditions for some coefficients $\gamma_r^{j,k}$ involving the averages $\langle g\rangle_{k,l}$ such that, if $\hat{I}_{k,l} = \{x \in \Gamma : u_k(x) = \cdots = u_l(x)\}$, then

$$Bu_r = g_r + \sum_{1\le k<l\le N,\ k<r<l} \gamma_r^{k,l}\, \chi_{\hat{I}_{k,l}} \quad \text{a.e. on } \Gamma.$$

References

[1] Azevedo, A. & Rodrigues, J.F. & Santos, L. *The N-membranes problem for quasilinear degenerate systems* Interfaces and Free Boundaries Volume **7** Issue 3 (2005) 319–337.

[2] Azevedo, A. & Rodrigues, J.F. & Santos, L. *Remarks on the two and three membranes problem* (Taiwan 2004) Elliptic and parabolic problems: recent advances, Chen, C. C. & Chipot, M. & Lin, C. S. (Eds.), World Scientific Singapore (2005) 19–33.

[3] Chipot, M. & Vergara-Cafarelli, G. *The N-membranes problem* Appl. Math. Optim. **13** n$^{\rm o}$3 (1985) 231–249.

[4] Gilbard, D. & Trudinger, N.S. *Elliptic partial differential equations of second order* 2nd edition, Springer-Verlag, Berlin, 1983.

[5] Hanouzet, B. & Joly, J.L. *Méthodes d'ordre dans l'interprétation de certains inéquations variationnelles et applications* J. Functional Analysis **34** (1979) 217–249 (see also C. R. Ac. Sci. Paris **281** (1975) 373–376).

[6] Lewy, H. & Stampacchia, G. *On the smoothness of superharmonics which solve a minimum problem* J. Analyse Math. **23** (1970) 227–236.

[7] Rodrigues, J.F. *Obstacle problems in mathematical physics* North Holland, Amsterdam, 1987.

[8] Rodrigues, J.F. *Stability remarks to the obstacle problem for the p-Laplacian type equations* Calc. Var. Partial Differential Equations **23** no. 1 (2005) 51–65

[9] Troianiello, G.M. *Elliptic differential equations and obstacle problems* Plenum Press, New York, 1987.

[10] Vergara-Caffarelli, G. *Regolarità di un problema di disequazioni variazionali relativo a due membrane* Atti Accad. Naz. Lincei Rend. Cl. Sci. Fis. Mat. Natur. (8) **50** (1971) 659–662.

A. Azevedo and L. Santos
University of Minho
Campus de Gualtar
4700–030 Braga, Portugal
e-mail: `assis@math.uminho.pt`
e-mail: `lisa@math.uminho.pt`

J.F. Rodrigues
CMUC/University of Coimbra and University of Lisbon
Av. Prof. Gama Pinto, 2
1649–003 Lisboa, Portugal
e-mail: `rodrigue@fc.ul.pt`

International Series of Numerical Mathematics, Vol. 154, 65–74

Modelling, Analysis and Simulation of Bioreactive Multicomponent Transport

Markus Bause and Willi Merz

Abstract. In this work we present a bioreactive multicomponent model that incorporates relevant hydraulic, chemical and biological processes of contaminant transport and degradation in the subsurface. Our latest results for the existence, uniqueness and regularity of solutions to the model equations are summarized; cf. [4, 9]. The basic idea of the proof of regularity is sketched briefly. Moreover, our numerical discretization scheme that has proved its capability of approximating reliably and efficiently solutions of the mathematical model is described shortly, and an error estimate is given; cf. [2, 3]. Finally, to illustrate our approach of modelling and simulating bioreactive transport in the subsurface, the movement and expansion of a m-xylene plume is studied numerically under realistic field-scale assumptions.

1. Introduction

In particular in industrialized countries, groundwater and soil pollution has become a major environmental threat. In many cases groundwater and soil contain a mixture of organic and anorganic substances. Usually, the contamination itself is hardly accessible in the subsurface. But, fortunately, biodegradation tends to attenuate at least some contaminants during groundwater transport. However, its potential is difficult to predict. Mathematical models and numerical simulations can be used to predict the long-term evaluation of contaminant plumes and help to design remediation techniques for field scale problems.

In the sequel, a mathematical model incorporating relevant processes of contaminant transport and (bio-)degradation in the subsurface is presented, the mathematical properties of its solutions are given and a reliable and efficient approximation scheme for numerical simulations is proposed. Finally, a realistic aquifer contamination scenario is investigated numerically.

2. Mathematical model

Microbial degradation mechanisms in soil-groundwater systems are mostly based on redox reactions (i.e., electron transfers), and the biomass acts as a catalyst for these reactions. The activity and dynamic of the biomass is dependent on the availability of an organic substrate, acting as the electron donor, and an electron acceptor. In this paper, the aerobic degradation of a single substrate is considered only, since we focus here on analyzing the mathematical properties of solutions to the governing equations and proposing an accurate numerical approximation scheme. The main principles hold equally for multiple electron donors and acceptors. For modelling of multicomponent bioreactive transport processes we refer to [11] and the references therein.

In this work, biomass growth is described by Monod-type kinetics. A detailed discussion of this modelling approach can be found, for instance, in [5, 11, 13]. The governing equations for the electron donor c_D, electron acceptor c_A and immobile biomass c_X are respectively given by

$$\begin{aligned} \partial_t(\Theta c_D) - \nabla\cdot(D_D\nabla c_D - qc_D) &= -\mu\,, \\ \partial_t(\Theta c_A) - \nabla\cdot(D_A\nabla c_A - qc_A) &= -\alpha_{A/D}\mu\,, \\ \partial_t c_X + k_d c_X &= \tfrac{Y}{\Theta}\left(1 - \tfrac{c_X}{c_{X_{\max}}}\right)\mu\,, \end{aligned} \tag{1}$$

$$\mu = \Theta\mu_{\max}\, c_X \,\frac{c_D}{K_D + c_D}\,\frac{K_{ID}}{K_{ID} + c_D}\,\frac{c_A}{K_A + c_A}\,\frac{K_{IA}}{K_{IA} + c_A}\,.$$

We consider solving the Monod model (1) of partial and ordinary differential equations over $Q_T = \Omega \times (0,T)$, $T > 0$, where $\Omega \subset \mathbb{R}^d$, $d = 2, 3$, is a bounded domain. The equations (1) are supplied with the initial conditions

$$c_D(\cdot,0) = c_{D,0}\,, \quad c_A(\cdot,0) = c_{A,0}\,, \quad c_X(\cdot,0) = c_{X,0} \tag{2}$$

in Ω, and the non-homogeneous Dirichlet and Neumann boundary conditions

$$c_i = g_i \quad \text{on } \Sigma_{T,D}\,, \qquad D_i\nabla c_i\cdot\nu = h_i \quad \text{on } \Sigma_{T,N}\,, \tag{3}$$

for $i = D, A$, where $\Sigma_{T,D} = \Gamma_D \times (0,T)$ and $\Sigma_{T,N} = \Gamma_N \times (0,T)$. Here, Γ_D and Γ_N denote the portions of the boundary $\partial\Omega = \Gamma_D \cup \Gamma_N$ where Dirichlet and Neumann boundary conditions are prescribed, respectively.

In equations (1), the parameter $\alpha_{A/D}$, k_d, Y, $c_{X_{\max}}$, $\mu_{\max}$, K_D, K_{ID}, K_A and K_{IA} are given constants. In our numerical example (cf. Sec. 5), saturated groundwater flow is considered and Θ is assumed to be a constant, too. We refer to [2, 5, 13] for further details of the model equations and parameter. For our theoretical results of Sec. 3 and 4 the diffusion-dispersion matrix $D = D(x)$ and the velocity vector $q = q(x)$ are assumed to be given sufficiently smooth functions. For the sake of physical realism, in our computational studies the Scheidegger parametrization (cf. [12])

$$D_i = (\Theta d_i + \beta_t|q|)I + (\beta_l - \beta_t)\frac{q\otimes q}{|q|}\,, \quad i = D, A\,, \tag{4}$$

of the diffusion-dispersion tensor is used. In (4), the parameter d_i, β_t and β_l are given constants and I is the identity matrix. Further, the velocity field q (volumetric flux) in (1) and (4) is obtained by solving the standard single phase Darcy flow problem, along with appropriate boundary conditions,

$$\nabla \cdot q = 0\,, \qquad q + K_s \nabla(\psi + z) = 0\,. \tag{5}$$

Here, ψ is the pressure head, q is the (Darcy) flux, $K_s = K_s(x)$ is the hydraulic permeability in the saturated zone and z denotes the height against the gravitational direction. For numerical simulations of contaminant transport and biodegradation scenarios also in the vadose zone we refer to [2].

3. Existence and regularity of solutions

Due to the relevance of the Monod model (1) for hydrological and environmental studies, as well as for civil and environmental engineering, and due to a lack in its mathematical analysis, we found it interesting enough to study whether the set of equations (1)–(3) admits unique (global) solutions and, moreover, whether solutions of higher order regularity exist under appropriate assumptions on the data and boundary of Ω. From the point of view of numerical computations, regularity results are of particular importance for the development of approximation schemes and the application of higher order discretization techniques; cf. [2, 4] and Sec. 4.

As regards the existence and uniqueness of solutions to the model equations (1)–(3) the following result was proved; cf. [9].

Theorem 3.1. *Let $\Omega \subset \mathbb{R}^2$ be a bounded domain. Then there exists a unique nonnegative solution*

$$c_D, c_A \in W_p^{2,1}(Q_T)\,,\ p > 2\,, \qquad c_X \in C^1([0,T]; C(\overline{\Omega})) \tag{6}$$

to the Monod model (1)–(3) *for any given $T \in (0, \infty)$.*

In (6), the notation of the space-time function spaces $W_p^{l,l/2}(Q_T)$ is standard; cf., e.g., [8]. We note that even the nonnegativeness of the species' concentrations c_D, c_A and c_X is ensured. The proof is based on the fixed point theorem of Leray–Schauder and can be carried over to three-dimensional domains $\Omega \subset \mathbb{R}^3$.

Now we address ourselves to solutions of (1)–(3) of higher regularity than it has been established in (6). Our regularity result is summarized in Theorem 3.2. In the sequel we sketch briefly the basic steps of its proof. For further details we refer to [4].

We consider equations (1)–(3) and, for brevity, assume that the coefficient functions D_i and q are constant. The proofs can directly be carried over to the case of space and time dependent functions D_i and q. Since the existence of a unique, nonnegative solution is given by Theorem 3.1 and the coupling of the equations (1) is through the lower order terms only, it suffices to study a single transport equation in (1) in the sequel. Precisely, we consider the coupled system of equations

$$\partial_t u - \nabla \cdot (D\nabla u - qu) = -\mu\,, \quad \partial_t c_X + ac_X + bc_X^2 = 0\,, \tag{7}$$

with a positive scalar $D \in \mathbb{R}$ and functions $a, b, \mu : Q_T \mapsto \mathbb{R}$. We prescribe the initial and, for simplicity, the Dirichlet boundary conditions

$$u(\cdot, 0) = u_0\,, \quad c_X(\cdot, 0) = c_{X,0} \text{ in } \Omega\,, \quad u = \bar{u} \text{ on } \Sigma_T\,. \tag{8}$$

We make the following assumptions on the data of the problem:

$$\begin{aligned} \partial\Omega \in C^4\,, \quad u_0\,,\, c_{X,0} \in W_2^{\frac{10}{3}}(\Omega)\,, \quad u_0\,,\, c_{X,0} \geq 0 \text{ in } \Omega\,, \\ \bar{u} \in W_2^{\frac{7}{2},\frac{7\cdot 3}{4\cdot 2}}(\Sigma_T)\,, \quad \bar{u} \geq 0 \text{ on } \Sigma_T\,. \end{aligned} \tag{9}$$

The assumption on $\partial\Omega$, that $\partial\Omega \in C^4$, is stronger than needed. Below, an elliptic regularity result is required that is satisfied if $\partial\Omega \in C^4$, but is also satisfied by certain other classes of domains. Moreover, we suppose that the second order compatibility conditions are satisfied.

Next, we specify the regularity assumptions on the functions a, b, μ. We compare (7) with the original system (1) to see that a, b, μ are composed in terms of the solution to (1)–(3). Due to (6), we directly get that

$$a, b \in W_p^{2,1}(Q_T)\,, \quad \text{for } p > 2\,. \tag{10}$$

The ordinary differential equation in (7) is of Bernoulli-type such that c_X may be represented explicitly in terms of a and b. It turns out that

$$c_X \in W_p^{2,2}(Q_T) \cap W_p^1(0, T; W_p^2(\Omega))\,, \quad \text{for } p > 2\,, \tag{11}$$

and this in fact yields that, also,

$$\mu \in W_p^{2,1}(Q_T)\,, \quad \text{for } p > 2\,. \tag{12}$$

Finally, results of elliptic regularity theory then yield the following theorem.

Theorem 3.2. *Let* (9) *and the second order compatibility conditions be satisfied. Then, the solutions c_D, c_A of* (1)–(3) *satisfy*

$$c_D\,, c_A \in W_2^{4,3}(Q_T) \cap W_2^1(0, T; W_2^4(\Omega))\,. \tag{13}$$

A detailed proof of Theorem 3.2 can be found in [4]. Along with (11), the result (13) shows that the Monod model (1)–(3) admits solutions of higher order regularity under suitable assumptions on the data and boundary of Ω. Together with our computational experiences, this observation brings us to use higher-order discretizations for the numerical solution of (1)–(3). In (11) and (13), we have established a slightly higher-order regularity than we will in fact need to derive optimal order error estimates (cf. Sec. 4) for our numerical approach.

4. Numerical approximation of solutions

We shall now briefly describe our discretization techniques that we proposed in [2], analyzed in [4] and apply here to solve numerically the equations (1)–(3); cf. Sec. 5. For the spatial discretization we use conforming finite element methods. Let $\mathcal{T}_h = \{K\}$ be a finite decomposition of mesh size h of $\overline{\Omega}$ into triangles or tetrahedrons, respectively. The approximation spaces V_h for the electron donor and acceptor C_i, $i = D, A$, and X_h for the biomass C_X are defined as $V_h = \{C_i \in C(\overline{\Omega}) \mid C_{i|K} \in P_2(K) \text{ for } K \in \mathcal{T}_h\} \cap \mathring{W}^1_{2,\Gamma_D}$ and $X_h = \{C_X \in C(\overline{\Omega}) \mid C_{X|K} \in P_2(K) \text{ for } K \in \mathcal{T}_h\}$, where $P_j(K)$, $j \in \mathbb{N}$, denotes the space of continuous polynomials of maximum degree j. Further, $\mathring{W}^1_{2,\Gamma_D} = \{\phi \in W^1_2(\Omega) \mid \phi = 0 \text{ on } \Gamma_D\}$. For the temporal discretization of equations (1)–(3) we use the two step backward differentiation formula BDF_2 which is of second-order accuracy and has excellent stability properties; cf. [6].

Let us suppose that coefficients $\{Q, D_i\} \in W^1_\infty(\Omega) \times L^\infty(\Omega)$, $i = A, D$, are prescribed where Q may be the velocity field itself or a suitable approximation of it. Let P_{Z_h} denote the L^2-projection onto the finite element space Z_h. The discretization of the Monod model (1)–(3) by the Galerkin method and BDF_2 then reads as:

Set $C_i^0 = P_{V_h} c_{i,0}$ and $C_X^0 = P_{X_h} c_X^0$. For all time steps $n = 0, \ldots, \mathcal{N} - 2$ compute approximations $C_i^{n+2} \in V_h$, $i = D, A$, and $C_X^{n+2} \in X_h$ by solving

$$\begin{aligned}
&\gamma_{n+2}\,\langle \Theta\, C_i^{n+2}, V_i\rangle - \gamma_{n+1}\,\langle \Theta\, C_i^{n+1}, V_i\rangle + \gamma_n\,\langle \Theta\, C_i^{n}, V_i\rangle \\
&+\tau_{n+1}\langle Q\cdot\nabla C_i^{n+2}, V_i\rangle + \tau_{n+1}\langle D_i(t_{n+2})\nabla C_i^{n+2}, \nabla V_i\rangle \\
&+\tau_{n+1}\langle(\nabla\cdot Q)C_i^{n+2}, V_i\rangle = -\tau_{n+1}\langle \alpha_i\,\mathcal{U}^{n+2}, V_i\rangle + \tau_{n+1}\langle h_i, V_i\rangle_{\Gamma_N}\,,
\end{aligned} \tag{14}$$

for all $V_i \in V_h$, $i = D, A$, and

$$\begin{aligned}
&\gamma_{n+2}\, C_X^{n+2} - \gamma_{n+1}\, C_X^{n+1} + \gamma_n\, C_X^{n} \\
&\qquad + \tau_{n+1} k_d\, C_X^{n+2} = \tau_{n+1}\frac{Y}{\Theta}\left(1 - \frac{C_X^{n+2}}{c_{X_{\max}}}\right)\mathcal{U}^{n+2}
\end{aligned} \tag{15}$$

for all nodes $(x_j)_{j=1,\ldots,M}$ associated with degrees of freedom of C_X^{n+2}, where

$$\mathcal{U}^{n+2} = \Theta\,\mu_{\max}\, C_X^{n+2}\,\frac{C_D^{n+2}}{K_D + C_D^{n+2}}\,\frac{K_{ID}}{K_{ID} + C_D^{n+2}}\,\frac{C_A^{n+2}}{K_A + C_A^{n+2}}\,\frac{K_{IA}}{K_{IA} + C_A^{n+2}}\,.$$

By $\langle\cdot,\cdot\rangle$ and $\langle\cdot,\cdot\rangle_{\Gamma_N}$ we denote the standard L^2 inner product in $L^2(\Omega)$ and $L^2(\Gamma_N)$, respectively. Further, in (14) and (15) we use the abbreviations $\gamma_{n+2} = 1 + \tau_{n+1}/(\tau_{n+1} + \tau_n)$, $\gamma_{n+1} = 1 + \tau_{n+1}/\tau_n$ and $\gamma_n = \tau_{n+1}^2/((\tau_{n+1} + \tau_n)\tau_n)$ as well as $\alpha_D = 1$ and $\alpha_A = \alpha_{A/D}$. In our computations we will consider non-vanishing Dirichlet boundary values g_i, $i = D, A$, in (3). For short, the variational formulation (14) is given for homogeneous Dirichlet boundary conditions only. Nonhomogeneous boundary values are incorporated by standard techniques.

Finally, let us address the treatment of the groundwater flow problem (5). Generally, to calculate the (Darcy) velocity vector q of the model equations (1), we prefer to use mixed finite element methods due to their conservation properties and flux approximation as part of the formulation itself; cf. [1] and the references therein. However, to ensure optimal order convergence properties of the species C_A, C_D and C_X in (14) and (15), a higher order approximation Q of the flux q is needed. Precisely, in [2] it was shown for a model transport problem that a third order accurate flux approximation Q in fact leads to optimal, third order L^2-convergence of the piecewise quadratic approximation of the species. Since the algorithmic complexity of higher order mixed finite element methods is relatively large, for the considered approach (14), (15) to the Monod model we currently still use a conforming approximation of the flow problem (5) by cubic finite elements which yields in fact a third order accurate approximation Q of the velocity vector q for (14). For the future we plan to use higher order mixed methods.

Now we shall show briefly that the regularity result that we have established in Theorem 3.2 in fact enables us to derive optimal order error estimates for the numerical approach (14), (15). To reduce technical overhead and focus on key principles, we consider for simplicity the nonlinear model transport problem

$$\partial_t u + q \cdot \nabla u - \nabla \cdot (D \nabla u) + r \frac{u^2}{1+u^2} = f\,, \quad \text{in } \Omega\,, \text{ for } t > 0\,, \tag{16}$$

along with homogeneous Dirichlet boundary conditions $u = 0$ on $\partial\Omega$ for $t > 0$, and the initial condition $u(\cdot, 0) = u_0$ in Ω. We note that equation (16) contains a reactive term of Monod type and, thus, is of the same structure as the first and second of the equations (1). In (16), $D = D(x) \in \mathbb{R}^{d,d}$, $d = 2, 3$, is a symmetric and uniformly positive definite and bounded matrix.

In particular, we shall concentrate on the spatial discretization error. This is based on the observation made in [2, 7, 10] that in the context of bioreactive transport the spatial discretization requires particular care. Schemes with much numerical diffusion produce an artificial mixing of the species which may lead to an overestimation of the biodegradation potential. Let now $u_h : [0, T] \mapsto V_h$ denote the semidiscrete finite element approximation of the solution u of (16). Then we have the following result; cf. [4].

Theorem 4.1. *For $t \in [0, T]$ there holds*

$$\|u(t) - u_h(t)\| + h\|\nabla(u(t) - u_h(t))\| \leq C h^3\,,$$

where $C = C(\|u\|_{C([0,T];W_2^3(\Omega))}, \|\partial_t u\|_{L^1(0,T;W_2^3(\Omega))})$ is independent of h and bounded above (montonically) in terms of $\|u\|_{C([0,T];W_2^3(\Omega))}$, $\|\partial_t u\|_{L^1(0,T;W_2^3(\Omega))}$.

Together with Theorem 3.2 and along with Sobolev embedding results (cf. [8]), Theorem 4.1 thus yields an optimal order error estimate for the proposed finite element approximation of the Monod model (1).

Θ	0.33	$\alpha_{A/D}$	2.16	$\mu_{\max}$	1.13	K_{IA}	∞
d_A, d_D	7.4e−5	k_d	0.025	K_D	0.79	K_s	0.045
β_t	0.03	Y	0.52	K_A	0.1		
β_l	0.36	$c_{X_{\max}}$	1.0	K_{ID}	91.7		

TABLE 1. Transport- and biodegradation parameters for the test problem.

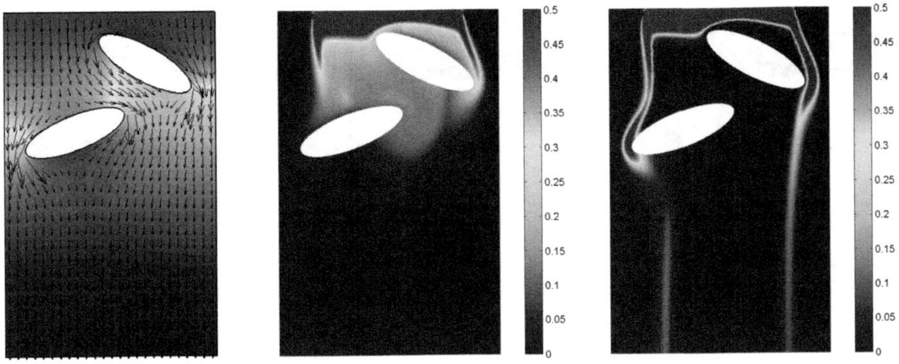

FIGURE 1. Profile of pressure head ψ and flow field q (*left*) and concentration of biomass at $T = 44$ (*center*) and $T = 300$ (*right*).

5. Numerical example

To illustrate our approach of modelling and simulating bioreactive transport in the subsurface, we shall now present a computational study of a realistic groundwater contamination scenario by a m-xylene plume and, thereby, provide valuable insights into the complex interactions of biological, chemical and physical processes that are involved in natural attenuation phenomena.

Our computational domain $\Omega = (0,6) \times (0,10)$ is a section of the aquifer with two impermeable heterogeneities of elliptic form inside that are obtained by rotating two ellipses with center in $(2, 6.5)^\top$ and $(4, 8.5)^\top$, respectively, both with semi-axes equal to 1.5 and 0.5, by $\pi/8$ and $-\pi/6$, respectively; cf. Fig. 1. The groundwater flow field is modelled by the Darcy flow problem (5). For the Monod model (1)–(3) we use the field-measured and laboratory-derived input parameters that are given in [13] and summarized in Table 1. The boundary conditions for the hydraulic head $p = \psi + z$ with $z \equiv x_2$ are

$$p(x) = 0 \quad \text{for } x \in [0,6] \times \{0\}, \quad p(x) = 20 \quad \text{for } x \in [0,6] \times \{10\}$$

and $K_s \nabla p \cdot \nu = 0$ else. The initial conditions for the species are

$$c_D(x,0) = 0.0, \quad c_A(x,0) = 2.0, \quad c_X(x,0) = 0.03 \quad \text{for } x \in \Omega.$$

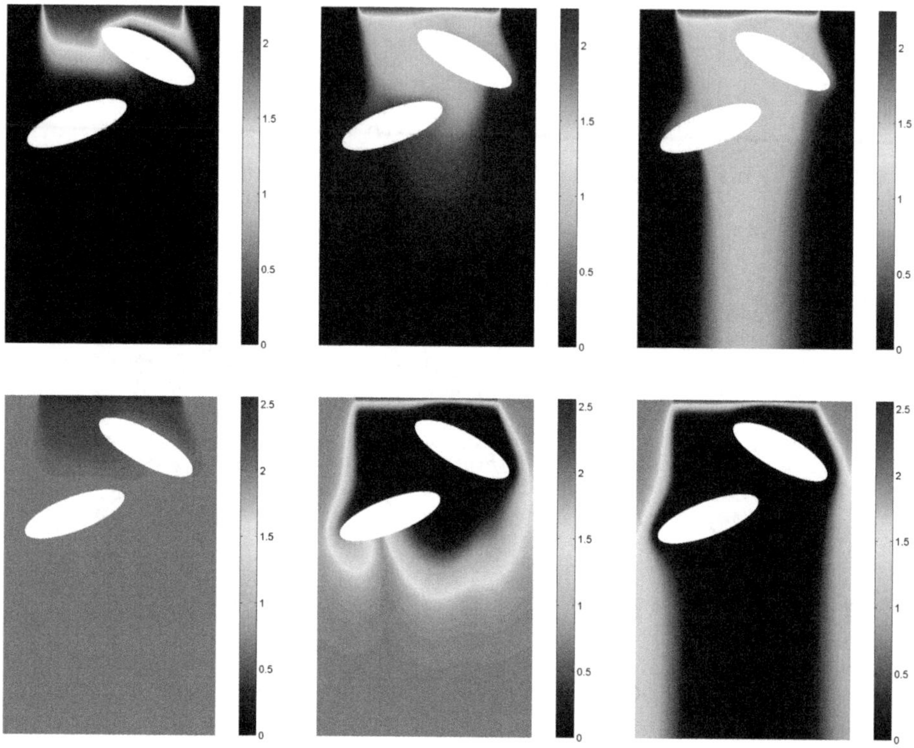

FIGURE 2. Concentration of electron donor m-xylene (*top*) and electron acceptor oxygen (*bottom*) at $T = 10$ (*left*), $T = 44$ (*center*) and $T = 300$ (*right*).

The contaminant and electron donor m-xylene is injected at the upper inflow boundary. Precisely, for $t > 0$ we prescribe the boundary conditions

$$c_D(x,t) = 2.24\,, \quad c_A(x,t) = 2.55 \quad \text{for } x \in [1,5] \times \{10\}\,,$$
$$c_D(x,t) = 0.0\,, \quad c_A(x,t) = 2.0 \quad \text{for } x \in ([0,1] \times \{10\}) \cup ([5,6] \times \{10\})$$

and $D_i \nabla c_i \cdot \nu = 0$, $i = D, A$, for $x \in \partial\Omega \backslash ([0,6] \times \{10\})$.

The calculated pressure head ψ and flow field $q = -K_s \nabla(\psi + z)$ are visualized in Fig. 1. The concentration profiles of the contaminant m-xylene and the electron acceptor (oxygen) at time $T = 10, 44$ and 300 are shown in Fig. 2. The corresponding biomass concentrations can be found in Fig. 1. For visualization purposes we have restricted the color range of the biomass to the interval [0,0.5].

Figs. 1 and 2 show that the contaminant is transported by the flow field to the lower boundary of Ω. Simultaneously, it is degraded by a reaction between electron donor, acceptor and biomass. However, the reaction is restricted to those regions of Ω where the concentrations of the species are sufficiently large. Besides a thin layer close to the inflow boundary where the involved substances (contaminant m-xylene and oxygen) are injected, it is basically the interface between the electron donor (contaminant) and the surrounding region where still enough electron acceptor (oxygen) is available. If a numerical method with much inherent artificial (i.e., numerical) diffusion would be used to solve the model equations (1), then the interface between the electron donor and acceptor would smear out and the reaction would take place in the larger region. Consequently, the contaminant would be degraded too fast. This might lead to a completely wrong prediction of the natural attenuation potential for the considered site; cf. [2, 4, 5, 7, 10] for further details. As shown in [2], the higher order discretization techniques proposed in Sec. 4 help to reduce significantly the amount of inherent numerical diffusion and prevent wrong predictions.

References

[1] M. Bause, P. Knabner, *Computation of variably saturated subsurface flow by adaptive mixed hybrid finite element methods*, Adv. Water Resour., **27**:565–581, 2004.

[2] M. Bause, P. Knabner, *Numerical simulation of contaminant biodegradation by higher order methods and adaptive time stepping*, Comput. Visual. Sci., **7**:61–78, 2004.

[3] M. Bause, *Computational study of field scale BTEX transport and biodegration in the subsurface.* In M. Feistauer et al. (eds.), Numerical Mathematics and Advanced Applications (ENUMATH 2003), 112–122, Springer, Berlin, 2004.

[4] M. Bause, W. Merz, *Higher order regularity and approximation of solutions to the Monod biodegradation model*, Appl. Numer. Math., **55**:154–172, 2005.

[5] O.A. Cirpka, E.O. Frind, R. Helmig, *Numerical simulation of biodegradation controlled by transverse mixing*, J. Contam. Hydrol., **40**:159–182, 1999.

[6] E. Hairer, S.P. Norsett, G. Wanner, *Solving Ordinary Differential Equations I: Nonstiff Problems*, Springer, Berlin, 1987.

[7] R. Klöfkorn, D. Kröner, M. Ohlberger, *Local adaptive methods for convection dominated problems*, Internat. J. Numer. Methods Fluids, **40**:79–91, 2002.

[8] O.A. Ladyženskaya, V.A. Solonnikov, N.N. Ural'ceva, *Linear and Quasilinear Equations of Parabolic Type*, Am. Math. Soc., Providence, RI, 1968.

[9] W. Merz, *Global existence result of the monod model*, Adv. Math. Sci. Appl., to appear in: **15** no. 2, 2005.

[10] M. Ohlberger, C. Rohde, *Adaptive finite volume approximations of weakly coupled convection dominated parabolic systems*, IMA J. Numer. Anal., **22**:253–280, 2002.

[11] A. Prechtel, *Modelling and Efficient Simulation of Hydrogeochemical Multicomponent Transport Problems by Process-Preserving Decoupling Techniques*, Shaker, Aachen, 2005.

[12] A.E. Scheidegger, *General theory of dispersion in porous media*, J. Geophys. Res., **66**:3273–3278, 1961.

[13] M. Schirmer, J.W. Molson, E.O. Frind, J.F. Barker, *Biodegradation modelling of a dissolved gasoline plume applying independent laboratory and field parameters*, J. Contam. Hydrol., **46**:339–374, 2000.

Markus Bause and Willi Merz
Institut für Angewandte Mathematik
Universität Erlangen-Nürnberg
Martensstraße 3
D-91058 Erlangen, Germany
e-mail: bause@am.uni-erlangen.de
e-mail: merz@am.uni-erlangen.de

International Series of Numerical Mathematics, Vol. 154, 75–84

Asymptotic Properties of the Nitzberg-Mumford Variational Model for Segmentation with Depth

Giovanni Bellettini and Riccardo March

Abstract. We consider the Nitzberg-Mumford variational formulation of the segmentation with depth problem. This is an image segmentation model that allows regions to overlap in order to take into account occlusions between different objects. The model gives rise to a variational problem with free boundaries. We discuss some qualitative properties of the Nitzberg-Mumford functional within the framework of the relaxation methods of the Calculus of Variations. We try to characterize minimizing segmentations of images made up of smooth overlapping regions, when the weight of the fidelity term in the functional becomes large. This should give some theoretical information about the capability of the model to reconstruct both occluded boundaries and depth order.

Mathematics Subject Classification (2000). 46E30,49J45,49Q20.

Keywords. Variational methods, relaxation of functionals, computer vision, image segmentation.

1. Introduction

In computer vision theory the segmentation problem consists in finding the homogeneous regions of an image, which are supposed to correspond to meaningful parts of objects in the three-dimensional world. Several variational models have been proposed in recent years which deal with the image segmentation problem [1, 7, 8]. In particular Nitzberg and Mumford [10, 11] proposed a variational model for segmentation with depth that allows regions to overlap to take into account the partial occlusion of farther objects by those that are nearer. The minimization of the functional should give the shape of the objects in an image, the reconstruction of the occluded boundaries, and the ordering of the objects in space. Some analytical and numerical studies can be found in [3, 4, 6, 11].

The Nitzberg-Mumford model incorporates (partially) the way that an image derives from a two-dimensional projection of a three-dimensional visual scene. The luminous intensity (grey level) of the image is represented by a function $g \in L^\infty(\mathbf{R}^2)$ with compact support. The regions of the segmentation constitute a finite collection of subsets of the plane

$$\{E_1, \dots, E_n\}, \qquad E_i \subset \mathbf{R}^2 \ \forall i,$$

and the sets E_i may overlap. A partial ordering $<$ between the regions represents the *relative depth* information:

$$E_i < E_j \qquad \text{means } E_i \text{ occludes } E_j. \tag{1.1}$$

A segmentation is then an ordered family of regions $\{E_i\}_{i=1}^n$. The *visible part* of the region E_i is the set E_i' given by

$$E_1' = E_1, \qquad E_i' = E_i \setminus \bigcup_{j=1}^{i-1} E_j \quad \text{for } i = 2, \dots, n.$$

The set E_{n+1}' is the background region:

$$E_{n+1}' = \mathbf{R}^2 \setminus \bigcup_{i=1}^{n} E_i.$$

Nitzberg and Mumford [10] gave a variational formulation of the segmentation problem by looking for a minimizer of the following functional:

$$\mathcal{G}(n, E_1, \dots, E_n) = \mu \sum_{i=1}^{n+1} \int_{E_i'} (c_i - g)^2 dx + \sum_{i=1}^{n} \int_{\partial E_i} [1 + \psi(\kappa_i)] d\mathcal{H}^1, \tag{1.2}$$

where $\mu > 0$ is a weight, c_i is the mean of g on E_i':

$$c_i = \frac{1}{\mathrm{meas}(E_i')} \int_{E_i'} g dx \quad \forall i = 1, \dots, n, \qquad c_{n+1} = 0,$$

$\mathcal{H}^1$ is the one-dimensional Hausdorff measure, and $\kappa_i(x)$ is the curvature of ∂E_i at x (the sets E_i are supposed to have sufficiently smooth boundaries). We choose the function $\psi : \mathbf{R} \to [0, +\infty[$ even, convex, and of class $\mathcal{C}^1$ (see [10, 11] for more details). The functional $\mathcal{G}$ is minimized over the number of regions n and the ordered families of regions $\{E_1, \dots, E_n\}$.

In the present paper we will concentrate on the case

$$\psi(\kappa) = |\kappa|^p, \qquad p > 1.$$

Since the number n of sets is not known in advance, an interesting mathematical problem is to understand, under suitable assumptions on the image g, if a minimizing n is equal to the number of shapes which are actually present in the image g. Moreover it is interesting to know if a minimizing segmentation matches the depth order embedded in the data. The answers to such questions should give some theoretical information about the reconstruction capabilities of the model. Here, we begin to look for an answer when the image g is piecewise constant with a finite number of values, and $\mu \to +\infty$.

We conclude the introduction by recalling that the proofs of the results here presented will appear in a forthcoming paper, where a variational model that enables the reconstruction of interwoven objects will also be considered.

2. The asymptotic problem

The building block of the energy (1.2) is the functional

$$\mathcal{F}(E) = \int_{\partial E} [1 + |\kappa|^p] \, d\mathcal{H}^1,$$

which is well defined if $E \in \mathcal{C}^2(\mathbf{R}^2)$, where $\mathcal{C}^2(\mathbf{R}^2)$ denotes the class of all bounded open subsets of $\mathbf{R}^2$ of class $\mathcal{C}^2$.

If we apply the direct method of the Calculus of Variations to the problem of minimizing the functional $\mathcal{G}$, we are led to consider sequences of sets $\{E_h\}_h$ on which the functional $\mathcal{F}$ is uniformly bounded. If the sets are contained in a ball independent of h such sequences are compact in $L^1(\mathbf{R}^2)$, and it can be proved [2] that the functional $\mathcal{F}$ is lower semicontinuous on the class $\mathcal{C}^2(\mathbf{R}^2)$ with respect to the convergence in $L^1(\mathbf{R}^2)$ (i.e., the L^1 convergence of the characteristic functions of sets). However, since in general the limit sets are not of class $\mathcal{C}^2$, it is necessary to extend the functional $\mathcal{F}$ to the set of all Lebesgue measurable subsets of $\mathbf{R}^2$, in such a way that the extended functional $\overline{\mathcal{F}}$ is still lower semicontinuous [2].

If we set $\mathcal{F}(E) = +\infty$ if $E \notin \mathcal{C}^2(\mathbf{R}^2)$, by using the relaxation method of the Calculus of Variations, the natural candidate for the functional $\overline{\mathcal{F}}$ is the lower semicontinuous envelope of $\mathcal{F}$ with respect to the $L^1(\mathbf{R}^2)$–topology, i.e.,

$$\overline{\mathcal{F}}(E) = \inf \left\{ \liminf_{h \to +\infty} \mathcal{F}(E_h) : E_h \to E \text{ in } L^1(\mathbf{R}^2) \right\}.$$

If $E \in \mathcal{C}^2(\mathbf{R}^2)$ then $\overline{\mathcal{F}}(E) = \mathcal{F}(E)$. In [2] it has been proved that there exist sets E such that $\overline{\mathcal{F}}(E) < +\infty$, the boundary of which is not smooth. In particular, if ∂E is smooth except at an even number of cusp points, then $\overline{\mathcal{F}}(E)$ is finite. The characterization of the sets E with $\overline{\mathcal{F}}(E) < +\infty$ has been given in [5].

We now denote by $\overline{\mathcal{G}}$ the lower semicontinuous envelope of $\mathcal{G}$ with respect to the $[L^1(\mathbf{R}^2)]^n$ product topology (for a fixed n). Because of the continuity of the terms $\int_{E_i'} (c_i - g)^2 dx$ we have

$$\overline{\mathcal{G}}(n, E_1, \ldots, E_n) = \mu \sum_{i=1}^{n+1} \int_{E_i'} (c_i - g)^2 dx + \sum_{i=1}^{n} \overline{\mathcal{F}}(E_i).$$

The following is an existence result for the functional $\overline{\mathcal{G}}$.

Theorem 2.1. [Existence] *There exists a finite family of sets $\{E_1, \ldots, E_n\}$, such that $\overline{\mathcal{F}}(E_i) < +\infty$ for any $i \in \{1, \ldots, n\}$, which minimizes the functional $\overline{\mathcal{G}}$.*

Moreover, it can be proved that the number n of sets minimizing $\overline{\mathcal{G}}$ is uniformly bounded with respect to μ. We are now in a position to define the asymptotic variational problem. Let g be a piecewise constant function assuming a finite set of values. We find that, as $\mu \to +\infty$, the sets which minimize $\overline{\mathcal{G}}$ converge in the $[L^1(\mathbf{R}^2)]^n$ product topology, up to a subsequence, to a family of n sets $\{\tilde{E}_1, \ldots, \tilde{E}_n\}$ such that

$$c_i = g(x) \quad \text{if } x \in \tilde{E}_i' \quad \forall i = 1, \ldots, n.$$

Moreover, if the sets $\tilde{E}_1, \ldots, \tilde{E}_n$ are of class $\mathcal{C}^2(\mathbf{R}^2)$, we have

$$J_g \subseteq \bigcup_{i=1}^{n} \partial \tilde{E}_i \ ,$$

where J_g denotes the jump set of g, i.e., the visible portions of the boundaries. Then we study the following variational problem: minimize the functional

$$\mathcal{G}_\infty(n, E_1, \ldots, E_n) = \sum_{i=1}^{n} \overline{\mathcal{F}}(E_i),$$

over the number of sets n and the families of sets $\{E_1, \ldots, E_n\}$, under the constraint

$$J_g \subseteq \bigcup_{i=1}^{n} \partial E_i. \tag{2.1}$$

The set J_g is the datum of the problem and the number n of sets is among the unknowns. The existence of minimizers of $\mathcal{G}_\infty$ can be proved from Theorem 2.1 and the uniform boundedness of n with respect to μ.

3. Energy functionals on systems of curves

In this section we need functionals defined on systems of curves of class $W^{2,p}$. We call regular curve any function $\gamma : [0,1] \to \mathbf{R}^2$ such that $|\dot{\gamma}| \neq 0$ in $[0,1]$. By $(\gamma) = \{\gamma(t) : t \in [0,1]\}$ we denote the trace of γ and by $l(\gamma)$ its length.

We define the functional

$$\mathcal{F}(\gamma) = \int_0^{l(\gamma)} [1 + |\kappa|^p] ds = l(\gamma) + l(\gamma)^{1-2p} \int_0^1 \left| \frac{d^2\gamma}{dt^2} \right|^p dt,$$

where s denotes the arclength parameter.

We call system of curves a finite family $\Gamma = \{\gamma^1, \ldots, \gamma^m\}$ of closed regular curves of class $W^{2,p}$ such that $|\dot{\gamma}^i|$ is constant on $[0,1]$ for any $i = 1, \ldots, m$. The trace (Γ) of a system of curves Γ is defined as $\cup_{i=1}^m (\gamma^i)$.

If Γ is a system of curves we define the functional

$$\mathcal{F}(\Gamma) = \sum_{i=1}^{m} \mathcal{F}(\gamma^i).$$

If E is a set of class $\mathcal{C}^2(\mathbf{R}^2)$ and Γ is a parametrization of ∂E, then we have

$$\partial E = (\Gamma), \qquad \mathcal{F}(E) = \mathcal{F}(\Gamma).$$

If E is not of class $\mathcal{C}^2(\mathbf{R}^2)$, but $\overline{\mathcal{F}}(E) < +\infty$, then it can be proved [2, 5] that E is open and there exists a system Γ of curves such that

$$\partial E \subseteq (\Gamma), \qquad \overline{\mathcal{F}}(E) = \mathcal{F}(\Gamma). \tag{3.1}$$

We say that a point $q \in (\Gamma)$ is a self-intersection point of Γ if, for any neighborhood U_q of q, $(\Gamma) \cap U_q$ cannot be written as the graph of a $W^{2,p}$ function.

Let $q = \gamma^i(t_1) = \gamma^j(t_2)$ for some $\gamma^i, \gamma^j \in \Gamma$ and $t_1, t_2 \in [0,1]$. We say that at q there is a tangential self-intersection if

$$\frac{d\gamma^i(t_1)}{dt} \quad \text{and} \quad \frac{d\gamma^j(t_2)}{dt} \qquad \text{are parallel.}$$

We say that at q there is a transversal self-intersection if

$$\frac{d\gamma^i(t_1)}{dt} \quad \text{and} \quad \frac{d\gamma^j(t_2)}{dt} \qquad \text{are not parallel.}$$

At a point q there may be both tangential and transversal self-intersections.

We say that two systems of curves Γ and $\widehat{\Gamma}$ are equivalent if

$$(\widehat{\Gamma}) = (\Gamma), \qquad \mathcal{F}(\widehat{\Gamma}) = \mathcal{F}(\Gamma).$$

The following is an approximation result for systems of curves, which is useful in order to estimate the optimal number n of sets E_i.

Theorem 3.1. [Density] *Let Γ be a system of curves. Then there exist a system of curves $\widehat{\Gamma}$ equivalent to Γ and a sequence $\{\Gamma_h\}_h$ of systems of curves such that*

(i) *Γ_h has a finite number of tangential self-intersections for any h;*
(ii) *$\Gamma_h \to \widehat{\Gamma}$ strongly in $W^{2,p}$;*
(iii) *$(\Gamma_h) \subseteq (\Gamma)$ for any h.*

The density theorem permits us to resort to systems of curves with a finite number of tangential self-intersections.

4. Estimate of the number of sets

In this section we give some estimates of the number n of sets minimizing the energy $\mathcal{G}_\infty$. We will need some assumptions on the image g in order to achieve the desired results.

Let $A_1, \ldots, A_N$ denote N sets of class $\mathcal{C}^2(\mathbf{R}^2)$ such that for any $i, j \in \{1, \ldots, N\}$, with $i \neq j$, the boundaries ∂A_i and ∂A_j intersect transversally at a finite number of points (see Figure 1). Moreover, let the sets be ordered according to the ordering relation (1.1):

$$A_1 < A_2 < \cdots < A_N.$$

The function g is assumed piecewise constant and it is defined by

$$g = \sum_{i=1}^{N} c_i \chi_{A_i'}, \tag{4.1}$$

where the c_i are positive constants, so that g is constant on the visible parts A_i' of the regions.

The visible portion of the boundary of the region A_i is defined by

$$(\partial A_1)' = \partial A_1, \qquad (\partial A_i)' = \partial A_i \setminus \bigcup_{j=1}^{i-1} A_j \quad \text{for } i = 2, \ldots, N.$$

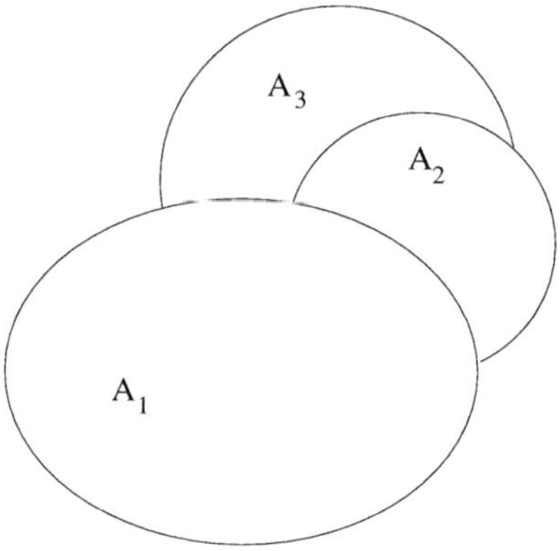

FIGURE 1. The sets A_i in the image g.

By the assumptions on the sets A_i, the visible portion of ∂A_i consists of a finite number of $\mathcal{C}^2$ arcs terminating in a finite set $\mathcal{M}_i$ of endpoints. We set

$$J_g = \partial A_1 \bigcup_{i=2}^{N} (\partial A_i)', \qquad \mathcal{M} = \bigcup_{i=2}^{N} \mathcal{M}_i.$$

We now study the relation between the number n minimizing the functional $\mathcal{G}_\infty$ and the number N of the regions which are actually present in the image g.

Let $\{E_1, \ldots, E_n\}$ be a minimizer of $\mathcal{G}_\infty$; then, using (2.1) and (3.1), there exist n systems of curves Γ^i, $i = 1, \ldots, n$, such that

$$J_g \subseteq \bigcup_{i=1}^{n} (\Gamma^i), \qquad \inf \mathcal{G}_\infty = \sum_{i=1}^{n} \mathcal{F}(\Gamma^i) \,. \tag{4.2}$$

We then define

$$\Gamma = \{\Gamma^1, \ldots, \Gamma^n\},$$

so that $J_g \subseteq (\Gamma)$.

Let Σ denote the set of all finite families $\sigma = \{\gamma^1, \ldots, \gamma^m\}$ of $W^{2,p}$ curves which connect pairwise all the endpoints in $\mathcal{M}$ in such a way that for any point $p \in \mathcal{M}_i$, the tangent vectors of σ and $(\partial A_i)'$ in p are parallel, for any $i \in \{2, \ldots, N\}$.

We have the following lemma.

Lemma 4.1. *If Γ has a finite number of tangential self-intersections then there exists $\sigma \in \Sigma$ such that*

$$\mathcal{F}(\Gamma) \geq \int_{J_g} [1 + |\kappa|^p] d\mathcal{H}^1 + \mathcal{F}(\sigma). \tag{4.3}$$

The proof of the lemma is based on the following argument. It can be proved [2] that each system of curves Γ^i has not transversal self-intersections. Using this fact, the property $J_g \subseteq (\Gamma)$, the hypothesis of a finite number of tangential self-intersections of Γ, and the assumptions on the image g, a proof by induction (based on the finiteness hypothesis) shows that Γ not only covers J_g, but also connects pairwise the endpoints in $\mathcal{M}$ by means of additional $W^{2,p}$ curves. Hence, there exists a family $\sigma \in \Sigma$ such that

$$J_g \bigcup (\sigma) \subseteq (\Gamma),$$

and the estimate (4.3) is then obtained by a covering argument.

We are now in a position to state the following energy estimate.

Theorem 4.2. [Energy estimate] *If* $J_g \subseteq \bigcup_{i=1}^{n} \partial E_i$, *with* $\overline{\mathcal{F}}(E_i) < +\infty$ *for any* $i \in \{1, \dots, n\}$, *then*

$$\inf \mathcal{G}_\infty \geq \int_{J_g} [1 + |\kappa|^p] d\mathcal{H}^1 + \inf_{\sigma \in \Sigma} \mathcal{F}(\sigma). \tag{4.4}$$

The proof of the estimate is based on the density result of Theorem 3.1, which permits us to remove from Lemma 4.1 the hypothesis of a finite number of tangential self-intersections of Γ. Then, using (4.2), Lemma 4.1 and taking the infimum over Σ, we have

$$\mathcal{G}_\infty(n, E_1, \dots, E_n) = \mathcal{F}(\Gamma) \geq \int_{J_g} [1 + |\kappa|^p] d\mathcal{H}^1 + \inf_{\sigma \in \Sigma} \mathcal{F}(\sigma),$$

from which the estimate (4.4) follows.

Now we may construct $\widehat{N}$ sets $\widehat{E}_1, \dots, \widehat{E}_{\widehat{N}}$, with $W^{2,p}$ boundary, by connecting pairwise the endpoints in $\mathcal{M}$ by means of curves which minimize the energy

$$\int_0^{l(\gamma)} [1 + |\kappa|^p] ds$$

with given tangent vectors in $\mathcal{M}$ (see Figure 2). In the case $p = 2$ the minimizing curves are called *elastica* because of their application to the theory of flexible inextensible rods [9].

Then we have

$$J_g \subseteq \bigcup_{i=1}^{\widehat{N}} \partial \widehat{E}_i, \qquad \inf \mathcal{G}_\infty \leq \sum_{i=1}^{\widehat{N}} \overline{\mathcal{F}}(\widehat{E}_i).$$

In particular, there exists a finite family of curves $\widehat{\sigma} \in \Sigma$ such that

$$\bigcup_{i=1}^{\widehat{N}} \partial \widehat{E}_i = J_g \bigcup (\widehat{\sigma}), \qquad \text{and} \qquad \inf \mathcal{G}_\infty \leq \int_{J_g} [1 + |\kappa|^p] d\mathcal{H}^1 + \mathcal{F}(\widehat{\sigma}).$$

If $\widehat{\sigma}$ minimizes $\mathcal{F}(\sigma)$ in Σ then, using Theorem 4.2, it follows

$$\inf \mathcal{G}_\infty = \int_{J_g} [1 + |\kappa|^p] d\mathcal{H}^1 + \mathcal{F}(\widehat{\sigma}) \ ,$$

so that the family of sets $\{\widehat{E}_1, \dots, \widehat{E}_{\widehat{N}}\}$ minimizes $\mathcal{G}_\infty$.

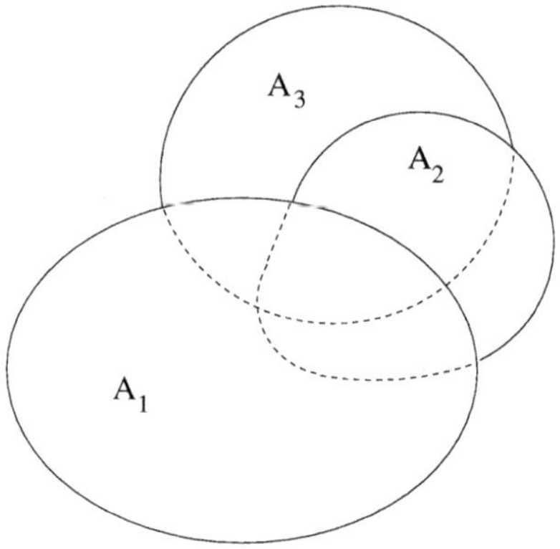

FIGURE 2. Endpoints in $\mathcal{M}$ connected by elastica curves.

Hence, in this case the asymptotic functional $\mathcal{G}_\infty$ is minimized by a family of sets with cardinality $n = \widehat{N}$ and the number of regions that are actually present in the image g is reconstructed. Note that the minimizer $\widehat{\sigma}$ of $\mathcal{F}(\sigma)$ in Σ may be such that $\widehat{N} \neq N$. This may happen in the following case. Suppose that connecting the endpoints of the visible boundary $(\partial A_i)'$ of the region A_i, a set $\widehat{E}_i$ is obtained. Analogously, a set $\widehat{E}_j$ is obtained connecting the endpoints of $(\partial A_j)'$ for some $j \neq i$. However, it may happen that the endpoints of $(\partial A_i)'$ and $(\partial A_j)'$ can be connected by forming a single set $\widehat{E}_{i,j}$ such that

$$\overline{\mathcal{F}}(\widehat{E}_{i,j}) < \overline{\mathcal{F}}(\widehat{E}_i) + \overline{\mathcal{F}}(\widehat{E}_j).$$

This case corresponds to a family of curves $\widehat{\sigma} \in \Sigma$ such that $\widehat{N} < N$.

However, the assumption that there exists a family $\widehat{\sigma}$ with the above properties and minimizing $\mathcal{F}(\sigma)$ in Σ, is not satisfied in general. We are able to improve such a result in the following way.

Let $\Sigma_0 \subseteq \Sigma$ be the subset of the families σ such that there exists a finite family of sets $\{E_1, \ldots, E_M\}$, with $\overline{\mathcal{F}}(E_i) < +\infty$ for any $i \in \{1, \ldots, M\}$ and $J_g \subseteq \cup_{i=1}^M \partial E_i$, such that

$$\sum_{i=1}^{M} \overline{\mathcal{F}}(E_i) = \int_{J_g} [1 + |\kappa|^p] d\mathcal{H}^1 + \mathcal{F}(\sigma). \tag{4.5}$$

The case considered so far corresponds to $\widehat{\sigma} \in \Sigma_0$ and $M = \widehat{N}$.

We have the following lemma.

Lemma 4.3. *If $\sigma \in \Sigma \setminus \Sigma_0$ and $E_1, \ldots, E_M$ are such that*

$$J_g \bigcup (\sigma) \subseteq \bigcup_{i=1}^{M} \partial E_i,$$

then

$$\sum_{i=1}^{M} \overline{\mathcal{F}}(E_i) \geq \int_{J_g} [1 + |\kappa|^p] d\mathcal{H}^1 + \mathcal{F}(\sigma) + c_0\pi, \tag{4.6}$$

where $c_0 = [(p/p')^{1/p} + (p'/p)^{1/p'}]$, *with* $1/p + 1/p' = 1$.

The proof of the lemma is based on the following argument. If $\overline{\mathcal{F}}(E_i) < +\infty$ for any i, by using (3.1) there exists a system of curves Γ such that

$$\bigcup_{i=1}^{M} \partial E_i \subseteq (\Gamma), \qquad \sum_{i=1}^{M} \overline{\mathcal{F}}(E_i) = \mathcal{F}(\Gamma).$$

Assume first that Γ has a finite number of tangential self-intersections. Then, if $\sigma \notin \Sigma_0$, we prove by means of an inductive method (based on the finiteness assumption) that the set $(\Gamma) \setminus (J_g \cup (\sigma))$ contains at least a closed curve. Hence, by the Hölder inequality the energy of a closed curve is greater than or equal to the constant $c_0\pi$, and the inequality (4.6) then follows by a covering argument. Then the hypothesis of a finite number of tangential self-intersections of Γ is removed by means of the density result of Theorem 3.1.

Then, using Theorem 4.2, Lemma 4.3 and (4.5), we obtain the following result.

Proposition 4.4. *If the inequality*

$$\inf_{\sigma \in \Sigma_0} \mathcal{F}(\sigma) \leq \inf_{\sigma \in \Sigma} \mathcal{F}(\sigma) + c_0\pi \tag{4.7}$$

holds, then there exist a family of curves $\sigma \in \Sigma_0$ *and sets* $E_1, \ldots, E_M$, *with* $\overline{\mathcal{F}}(E_i) < +\infty$ *and* $J_g \subseteq \cup_{i=1}^{M} \partial E_i$, *such that*

$$\sum_{i=1}^{M} \overline{\mathcal{F}}(E_i) = \inf \mathcal{G}_\infty,$$

and

$$\sum_{i=1}^{M} \overline{\mathcal{F}}(E_i) = \int_{J_g} [1 + |\kappa|^p] d\mathcal{H}^1 + \mathcal{F}(\sigma).$$

If the assumption (4.7) of Proposition 4.4 is satisfied, it follows that the number M of regions is reconstructed from the image g. Then, the family of sets $\{E_1, \ldots, E_M\}$ can be endowed with the ordering relation

$$E_1 < E_2 < \cdots < E_M\ ,$$

as follows

$$\partial E_1 \subseteq J_g, \qquad \partial E_i \setminus \bigcup_{j=1}^{i-1} E_j \subseteq J_g \quad \text{for } i = 2, \ldots, M\ .$$

Remark 4.5. If $N = 2$ in the definition (4.1) of g and $\mathcal{M}$ consists of two points, then the assumption (4.7) is unnecessary, $\mathcal{G}_\infty$ is minimized by $n = 2$ and the two endpoints of $\mathcal{M}$ are connected by an elastica.

References

[1] G. Aubert and P. Kornprobst, *Mathematical Problems in Image Processing*, Springer, Applied Mathematical Sciences 147, 2002.

[2] G. Bellettini, G. Dal Maso and M. Paolini, *Semicontinuity and relaxation properties of a curvature depending functional in 2D.* Ann. Scuola Norm. Sup. Pisa (4) **20** (1993), 247–299.

[3] G. Bellettini and M. Paolini, *Variational properties of an image segmentation functional depending on contours curvature.* Adv. Math. Sci. Appl. **5** (1995), 681–715.

[4] G. Bellettini and R. March, *An image segmentation variational model with free discontinuities and contour curvature.* Math. Mod. Meth. Appl. Sci. **14** (2004), 1–45.

[5] G. Bellettini and L. Mugnai, *Characterization and representation of the lower semicontinuous envelope of the elastica functional.* Ann. Inst. Henri Poincaré, Anal. Non Linéaire **21** (2004), 839–880.

[6] S. Esedoglu and R. March, *Segmentation with depth but without detecting junctions.* J. Math. Imaging Vision **18** (2003), 7–15.

[7] J.M. Morel and S. Solimini, *Variational Methods in Image Segmentation*, Vol. 14 of Progress in Nonlinear Differential Equations and Their Applications, Birkhäuser, Basel, 1995.

[8] D. Mumford and J. Shah, *Optimal approximations by piecewise smooth functions and associated variational problems.* Comm. Pure Appl. Math. **42** (1989), 577–685.

[9] D. Mumford, *Elastica and computer vision.* Algebraic Geometry and Applications, C. Bajaj ed., Springer-Verlag, Heidelberg, 1992.

[10] M. Nitzberg and D. Mumford, *The* 2.1-*D sketch.* Proceedings of the Third International Conference on Computer Vision, Osaka, 1990.

[11] M. Nitzberg, D. Mumford and T. Shiota, *Filtering, Segmentation and Depth*, Springer, Lecture Notes in Computer Science 662, 1993.

Giovanni Bellettini
Dipartimento di Matematica
Università di Roma "Tor Vergata"
Via della Ricerca Scientifica
I-00133 Roma, Italy
e-mail: `belletti@mat.uniroma2.it`

Riccardo March
Istituto per le Applicazioni del Calcolo, CNR
Viale del Policlinico 137
I-00161 Roma, Italy
e-mail: `r.march@iac.cnr.it`

International Series of Numerical Mathematics, Vol. 154, 85–94

The ∞-Laplacian First Eigenvalue Problem

Marino Belloni

Abstract. We review some results about the first eigenvalue of the infinity Laplacian operator and its first eigenfunctions in a general norm context. Those results are obtained in collaboration with several authors: V. Ferone, P. Juutinen and B. Kawohl (see [BFK], [BK1], [BJK] and [BK2]). In Section 5 we make some remarks on the simplicity of the first eigenvalue of Δ_∞: this will be the object of a joint work with A. Wagner (see [BW]).

Mathematics Subject Classification (2000). Primary 35P30; Secondary 35J70, 49L25, 49R50.

Keywords. Nonlinear eigenvalue problems, Degenerate elliptic equation, Variational methods, Viscosity solutions.

1. Introduction

Imagine a nonlinear elastic membrane, fixed on a boundary $\partial\Omega$ of a plane domain Ω. If $u(x)$ denotes its vertical displacement, and if its deformation energy is given by $\int_\Omega |\nabla u|^p \, dx$, then a minimizer of the Rayleigh quotient

$$\int_\Omega |\nabla u|^p \, dx \; / \; \int_\Omega |u|^p \, dx$$

on $W_0^{1,p}(\Omega)$ satisfies the Euler-Lagrange equation

$$-\Delta_p u = \lambda_p \, |u|^{p-2} u \quad \text{in } \Omega, \tag{1.1}$$

where $\Delta_p u = \operatorname{div}(|\nabla u|^{p-2} \nabla u)$ is the well-known p-Laplace operator. This eigenvalue problem has been extensively studied in the literature, see [L3]. A somewhat strange recent result is that (as $p \to \infty$) the limit equation reads

$$\min \{ \, |\nabla u| - \Lambda_\infty u, \; -\Delta_\infty u \, \} = 0. \tag{1.2}$$

Here $\Delta_\infty u = \sum_{i,j} u_{x_i} u_{x_j} u_{x_i x_j}$ is the ∞-Laplacian, $\Lambda_\infty = \lim_{p\to\infty} \Lambda_p$ where $\Lambda_p = \lambda_p^{1/p}$ (see [JLM1, FIN]).

Now suppose that the membrane is not isotropic. It is for instance woven out of elastic strings like a piece of material. Then the deformation energy can

This work was completed with the support of the research project *Calcolo delle Variazioni e Teoria Geometrica della Misura*.

be anisotropic, see [BK2, BJK]. We are mainly interested in generalizing the result on eigenfunctions for the p-Laplacian to the situation, where $\Omega \subset \mathbb{R}^n$ is no longer equipped with the Euclidean norm, but instead with a general norm $|\cdot|$, for instance with $|x| = (\sum_{i=1}^n |x_i|^q)^{1/q}$ and $q \in (1, \infty)$. In that case a Lipschitz continuous function $u : \Omega \mapsto \mathbb{R}$ (in a convex domain Ω) has Lipschitz constant $L = \sup_{z \in \Omega} |\nabla u(z)|^*$, where $|\cdot|^*$ denotes the dual norm to $|\cdot|$, because $|u(x) - u(y)| \leq L\,|x - y|$ with this L. Then we study the asymptotic behavior of the first eigenvalue (eigenfunctions) when $p \to \infty$. The case when $p \to 1$ and the corresponding limiting problem for the first eigenvalue is not considered, see [KF, KLR].

It is well known, that the infinite-Laplacian operator Δ_∞ is closely related to finding a minimal Lipschitz extension of a given function $\phi \in C^{0,1}(\partial\Omega)$ into Ω: see Section 2. In [BFK] the eigenvalue problem was carried over to a general norm and studied for finite p, while in [BK2] the eigenvalue problem was investigated first for finite p and the special non-Euclidean norm $|x| = (\sum_{i=1}^n |x_i|^{p'})^{1/p'}$ with p' conjugate to p, and then for the limit $p \to \infty$. In [BJK] the eigenvalue problem was investigated for general strictly convex norm $|x|$, and then for the limit $p \to \infty$.

This paper is organized as follows.

In Section 2 we introduce the ∞-Laplacian operator and we survey some old and recent results.

In Section 3 we introduce the first eigenvalue of the operator Δ_∞, and we survey some results obtained in [BJK, BFK, BK1] and [BK2].

In Section 4 we expose some examples related to the results quoted in Section 3 (see [BJK, BK2]).

In Section 5 we expose some unpublished material on the simplicity of the first eigenvalue obtained in [BW], a joint work in preparation with A. Wagner.

2. The ∞-Laplacian operator Δ_∞

Suppose $f \in C^{0,1}(\partial\Omega)$, where $\Omega \subseteq \mathbb{R}^n$ is an open set, having $L(f, \partial\Omega)$ as the least constant for which

$$|f(x) - f(y)| \leq L|x - y|, \qquad \forall x, y \in \partial\Omega$$

holds. Is it possible to find $u \in C^{0,1}(\Omega)$ (a *Minimal Lipschitz Extension*, M.L.E. shortly) such that

(i) $u(x) = f(x)$ for all $x \in \partial\Omega$;
(ii) $L(u, \Omega) = L(f, \partial\Omega)$?

Such an extension u exists (see [McS, W]) and satisfies $\Lambda(f)(x) \leq u(x) \leq \Psi(f)(x)$ for every $x \in \Omega$, where

$$\begin{cases} \Lambda(f)(x) &= \sup_{y \in \partial\Omega} (f(y) - L(f, \partial\Omega)|x - y|)\,, \\ \Psi(f)(x) &= \inf_{y \in \partial\Omega} (f(y) + L(f, \partial\Omega)|x - y|)\,. \end{cases}$$

The functions $\Lambda(f)$, $\Psi(f)$ are itself M.L.E., and then there exist infinitely many functions satisfying (i), (ii). Also, it can happen that u satisfies (i) and (ii) but an inequality like $L(u,V) > L(u,\partial V)$ holds for some $V \subset\subset \Omega$!

Example. $(n = 1)$ Consider $\Omega =]-4,0[\cup]0,4[$, and $f \in C^{0,1}(\partial\Omega)$ defined as: $f(-4) = f(0) = 0$ and $f(4) = 4$. We have immediately that $L(f,\partial\Omega) = 1$, and if we construct $v = \Lambda(f)$ (exercise!), the lower M.L.E., it is evident that for $V =]-3,-1[$ we have $L(v,\partial V) = 0 < 1 = L(v,V)$ (see also [ACJ] where a similar example is considered).

The following definition is a way to recover such an instability (see Aronsson [A1], and also [ACJ]).

Definition 2.1. A function u is an *Absolutely Minimal Lipschitz Extension* if u satisfy (i), (ii) and
(iii) $L(u,V) = L(f,\partial V)$ for every $V \subset\subset \Omega$.

More precisely, Aronsson defined the class of Absolutely Minimal Lipschitz (A.M.L. shortly) functions as the class of continuous functions in Ω satisfying (iii), i.e., it is not necessary to be an extension to be Absolutely minimal (see [ACJ], where this approach is stressed). Aronsson proved that A.M.L. functions must satisfy the Dirichlet boundary value problem

$$\begin{cases} -\Delta_\infty u(x) = 0 & x \in \Omega \\ u(x) = f(x) & x \in \partial\Omega \end{cases} \tag{2.1}$$

where $\Delta_\infty u = \nabla u \nabla(|\nabla u|^2)$ is the ∞-Laplacian (the formal limit as $p \to \infty$ of p-Laplacian operator). In 1993 Jensen proved that (2.1) has a unique viscosity solution. This result was known for C^2 solutions of (2.1) (see [A1]); unfortunately, as shown by Aronsson itself [A2], there exist ∞-harmonic functions not of class C^2: for example $u(x_1,x_2) = |x_1|^{4/3} - |x_2|^{4/3}$. For the definitions and relevant properties of viscosity solutions, see [CIL]. Other uniqueness proof were given by Barles and Busca [BB] still using a viscosity approach and, more recently, by Aronsson, Crandall and Juutinen [ACJ] where the proof is obtained via the Comparison by Cones approach introduced in [CEG] (see also [CDP, GPP] for some generalizations of the Comparison with Cones to more general functionals). Recently O. Savin [S] found the C^1 regularity for the solutions of $\Delta_\infty u = 0$ in 2-dimensional domains. In [ACJ] some results on minimal Lipschitz extensions are generalized from the Euclidean to a general norm, see also [Wu]. The infinity-Laplacian operator plays also an important role in problems of optimal transportation. For technical reasons it is often approximated by p-Laplacians with large p, see for instance [EG], [BDP]. Another very active fields connected with the infinity-Laplacian operator is the supremal calculus in L^∞, see [BJW] and the references therein.

3. The first eigenvalue of Δ_∞

If we minimize the functional

$$I_p(v) = \int_\Omega (|\nabla u|^*)^p \; dx \quad \text{on} \quad K := \{ \, v \in W_0^{1,p}(\Omega) \mid ||v||_{L^p(\Omega)} = 1 \, \}, \tag{3.1}$$

then via standard arguments (see [BFK]) a minimizer u_p exists for every $p > 1$ and it is a weak solution to the equation

$$-Q_p u := -\text{div}\left((|\nabla u_p|^*)^{p-2} J(\nabla u_p)\right) = \lambda_p |u_p|^{p-2} u_p \; , \tag{3.2}$$

that is

$$\int_\Omega (|\nabla u_p|^*)^{p-2} \left\langle J(\nabla u_p), \nabla v \right\rangle \, dx = \lambda_p \int_\Omega |u_p|^{p-2} u \cdot v \; dx \tag{3.3}$$

for any $v \in W_0^{1,p}(\Omega)$. Here $\lambda_p = I_p(u_p)$ and

$$J_i(\xi) := \frac{\partial}{\partial \xi_i} \left(\frac{(|\xi|^*)^2}{2} \right) . \tag{3.4}$$

Clearly (3.4) is well defined as long as the dual norm $H(\eta) = |\eta|^*$ is of class $C^1(\mathbb{R}^n \setminus \{0\})$. Recall that (3.4) is well defined (and single-valued) if and only if the norm $|\cdot|$ is strictly convex, i.e., if its unit sphere $\{x : \; |x| = 1\}$ contains no nontrivial line segments, see [Z] p. 400. Note further that in this case $J(0) = 0$ and that for the Euclidean norm the duality map reduces to the identity $J(\nabla u) = \nabla u$. Note finally that $\Lambda_p := \lambda_p^{1/p}$ is the minimum of the Rayleigh quotient

$$R_p(v) := \frac{\left(\int_\Omega \; (|\nabla v|^*)^p \; dx\right)^{1/p}}{||v||_p} \tag{3.5}$$

on $W_0^{1,p}(\Omega) \setminus \{0\}$.

Theorem 3.1. *Suppose that $H(\eta) = |\eta|^*$ is of class $C^1(\mathbb{R}^n \setminus \{0\})$ or that the norm $|\cdot|$ is strictly convex. Then for every $p \in [2, \infty)$, the nonnegative minimizer u_p of* (3.1) *is unique, positive and of class $C^{1,\alpha}$. It solves* (3.2) *in the weak sense of* (3.3).

For a proof, see [BFK] (see also [L1, L2]).

The function distance to the boundary $\delta(x) := \inf_{y \in \partial\Omega} |x - y|$ is Lipschitz continuous, satisfies $|\nabla \delta(x)|^* = 1$ almost everywhere in Ω and it is equal to zero on the boundary of Ω. We have then for every $\varphi \in W_0^{1,\infty}(\Omega)$ and $y \in \partial\Omega$

$$|\varphi(x)| = |\varphi(x) - \varphi(y)| \leq || \; |\nabla \varphi|^* ||_\infty \delta(x),$$

which implies

$$\frac{1}{||\delta||_\infty} \leq \frac{|| \; |\nabla \varphi|^* ||_\infty}{||\varphi||_\infty}. \tag{3.6}$$

Now let us define

$$\Lambda_\infty := \frac{|| \; |\nabla \delta|^* ||_\infty}{||\delta||_\infty} \left(= \frac{1}{||\delta||_\infty} \right). \tag{3.7}$$

Therefore Λ_∞ is a geometric quantity related to Ω. It is the inverse of the radius of the largest (in general non-Euclidean) ball inside Ω. We can now prove the following Theorem, which explains the analytic meaning of Λ_∞.

Theorem 3.2. *The following limit holds*

$$\left(\lim_{p\to\infty} \lambda_p^{1/p} =\right) \lim_{p\to\infty} \Lambda_p = \Lambda_\infty.$$

Here $\Lambda_p = R_p(u_p)$ *and the Rayleigh quotient* R_p *is given by* (3.5).

For a proof, see [BJK] (see also [JLM1], where the Euclidean case is considered). For a Γ-convergence's proof, see [BW] (see [DM] for the definitions and relevant properties of Γ-convergence).

Now we state the theorem which characterize the limit eigenvalue equation.

Theorem 3.3. *If* $H(\eta) := |\eta|^*$ *is of class* $C^2(\mathbb{R}^n \setminus \{0\})$ *then every cluster point* u_∞ *of the sequence* $\{u_p\}$ *is a viscosity solution of the equation*

$$F_\infty(u, \nabla u, D^2 u) = \min\ \{\ |\nabla u|^* - \Lambda_\infty u,\ -Q_\infty u\} = 0$$

with $Q_\infty u = \langle D^2 u J(\nabla u), J(\nabla u)\rangle$ *representing the infinite-Laplacian in the Finsler metric.*

For a proof, see [BJK] (see also [JLM1]; in [BK2] we consider the case when H fails to be of class $C^2(\mathbb{R}^n \setminus \{0\})$). In [BW] we are trying to obtain this eigenvalue equation starting from a more variational approach (see Section 5 for more details).

The function F_∞, in the setting of viscosity solutions given in [CIL], results degenerate elliptic but not proper. Therefore a comparison result cannot be obtained via standard theory. Thanks to the change of variable $w_\infty = \log u_\infty$, we arrive at the following equation

$$G_\infty(\nabla w, D^2 w) := \min\ \left\{\ |\nabla w|^* - \Lambda_\infty,\ -Q_\infty w - (|\nabla w|^*)^4\right\} = 0, \qquad (3.8)$$

where Q_∞ is defined as before. Equation (3.8), in contrast to $F_\infty = 0$, is now proper and then we can obtain the following comparison result.

Theorem 3.4. *Let* Ω *be a bounded domain, and suppose that* u *is a uniformly continuous viscosity subsolution and* v *a uniformly continuous viscosity supersolution of* (3.8) *in* Ω. *Then the following equality holds:*

$$\sup_{x\in\overline{\Omega}}(u(x) - v(x)) = \sup_{x\in\partial\Omega}(u(x) - v(x)). \qquad (3.9)$$

For a proof see [BJK] (see also [JLM1]).

It is well known that for any $1 < p < \infty$, the eigenvalue λ_p can be characterized by the property that $\lambda = \lambda_p$ is the only real number for which the equation

$$-\mathrm{div}\left((|\nabla u_p|^*)^{p-2} J(\nabla u_p)\right) = \lambda |u_p|^{p-2} u_p$$

has a continuous positive solution with zero boundary value. Also Λ_∞ has an analogous characterization.

Theorem 3.5. *Let Ω be any bounded domain and suppose that the norm $|\cdot|$ is of class $C^2(\mathbb{R}^n \setminus \{0\})$. If u is a continuous positive viscosity solution in Ω of*

$$\min\{|\nabla u|^* - \Lambda u, -Q_\infty u\} = 0$$

with zero boundary value, then $\Lambda = \Lambda_\infty$.

For a proof, see [BJK] (see also [JLM1]).

4. Examples and observations

Example. If the norm under consideration for $x \in \Omega$ is the usual ℓ_q- norm, i.e., for $|x| = (\sum_{i=1}^n |x_i|^q)^{1/q}$ with $q \in (1, \infty)$, the duality map according to (3.4) is easily calculated as

$$J_i(y) = (|y|_{q'})^{2-q'} |y_i|^{q'-2} y_i,$$

with $q' = q/(q-1)$ as conjugate exponent. Notice that this differs from the J in [ACJ], Example 5.2. Then the p-Laplace operator in this Finsler metric is explicitly given by, see [BFK]

$$Q_p u = \sum_{i=1}^n \frac{\partial}{\partial x_i} \left(|\nabla u|_{q'}^{p-q'} \left| \frac{\partial u}{\partial x_i} \right|^{q'-2} \frac{\partial u}{\partial x_i} \right).$$

The ∞-Laplace operator in the same Finsler metric is explicitly given by

$$Q_\infty u = |\nabla u|_{q'}^{4-2q'} \sum_{i,j=1}^n \left(\frac{\partial^2 u}{\partial x_i x_j} \left| \frac{\partial u}{\partial x_i} \right|^{q'-2} \frac{\partial u}{\partial x_i} \left| \frac{\partial u}{\partial x_j} \right|^{q'-2} \frac{\partial u}{\partial x_j} \right)$$

and for $q = 2$ this expression shrinks down to the customary

$$\Delta_\infty u = \sum_{i,j=1}^n \frac{\partial^2 u}{\partial x_i x_j} \frac{\partial u}{\partial x_i} \frac{\partial u}{\partial x_j}.$$

Remark 4.1. It should be remarked that the distance function minimizes the Rayleigh quotient R_∞, but that $\delta(x)$ is in general not a viscosity solution of the limiting eigenvalue problem, unless Ω is a "ball" in the Finsler metric, see [JLM1], [JLM2], [BK2] and [BJK].

Remark 4.2. In general, if the domain is not a "*ball*", there exist infinitely many minimizers of the Rayleigh quotient R_∞. For example, see the function (5.2) (see [FIN]).

Remark 4.3. If Ω is a "ball" in $\mathbb{R}^n$ and $p = n$, then all the level sets of solutions to (3.2)

$$-Q_n u = \lambda_n |u|^{n-2} u$$

are similar "balls", see [BFK].

Remark 4.4. The smoothness assumption made on the dual spheres in[BJK] is violated if the underlying norm is the ℓ_1 or ℓ_∞ norm. However, the pde $-Q_p = 1$ and its limit as $p \to \infty$ was studied even in this case in [IK], see also [K1], [BDM], [Ja] and [IL] for the case of the Euclidean norm and for variants of this problem. The eigenvalue problem is studied in [BK2]

Remark 4.5. Clearly the eigenvalue λ_p depends on Ω. There is an analogue of the Faber-Krahn inequality which states that among all domains of given volume $\lambda_p(\Omega)$ becomes minimal if Ω is a "ball" in the Finsler metric: see [BFK].

Example. Consider the square $S := \{(x, y) \mid \max\{|x|, |y|\} < 1\}$. Assume that the distance is given by $d(x, y) = |x_1 - y_1| + |x_2 - y_2|$. This function does not satisfy the regularity property stated in the previous section. We know that the function distance to the boundary

$$\delta_S(x, y) = d((x, y), \partial S) = \min\{1 - |x|, 1 - |y|\} \qquad \forall (x, y) \in S \tag{4.1}$$

is a minimizer for the Rayleigh quotient $R_\infty = \max\{u_x, u_y\}/||u||_\infty$. But δ_S does not solve the eigenvalue equation

$$\min\{\ \max\{|u_x|, |u_y|\}\ - u,\ -\tilde{\Delta}_\infty u\} = 0, \tag{4.2}$$

where $\tilde{\Delta}_\infty u(x_0) := \sum_{i \in I(\nabla u(x_0))} |u_{x_i}(x_0)| u_{x_i, x_i}(x_0)$ is the pseudo$-\infty$-Laplacian (see [BK2]). We recall that $I(\xi) = \{k\ :\ |\xi_k| = \max_i |\xi_i|\}$.
On the other hand, the function

$$u_\infty(x, y) = (1 - |x|)(1 - |y|).$$

is a minimizer of the Rayleigh quotient and also a viscosity solution of (4.2). We obtain this function as the limit of the sequence of eigenfunctions u_p of the pseudo$-p$-Laplacian. For the details, see [BK2].

5. Uniqueness

For $p = \infty$ the uniqueness of the first eigenfunction is a completely open problem. In fact Theorem 3.4 is just a *local* comparison result: this is because, as observed, just one function can be set equal to 0 on the boundary of Ω. Clearly, when Ω is a *"ball"* (relative to the metric) then the first eigenfunction is unique because the function $d(x, \partial\Omega$ is the unique minimizer of

$$\min\left\{R_\infty(u, \Omega) = \frac{||\nabla u||_{\infty,\Omega}}{||u||_{\infty,\Omega}}\ :\ u \in W_0^{1,\infty}(\Omega)\right\}. \tag{5.1}$$

If Ω is a *stadium* (or an *annulus*) then (see [BW]) the first eigenfunction is still unique and is the function distance to the boundary: the proof is obtained by a careful use of the comparison result (Theorem 3.4) together with the observation that there exists a point $x_0 \in \Omega$ and a minimizer (see [FIN]) as follows

$$C_{x_0}(x) = \begin{cases} \Lambda_\infty^{-1} - d(x, x_0), & \text{if } d(x, x_0) < ||\delta||_\infty \\ 0, & \text{otherwise,} \end{cases} \qquad x \in \Omega. \tag{5.2}$$

If $\{x \in \Omega : \delta(x) = ||\delta||_\infty\} \neq \{x \in \Omega : \delta(x)$ not differentiable$\}$, then (see [BW]) the function δ fails to be a solution of the ∞-eigenvalue equation.

As stated in Theorem 3.1, for every $p \in (1, \infty)$ the first eigenvalue Λ_p is simple, and the conjecture is that Λ_∞ is simple too.

In [BW] we follow a more variational approach to the minimum problem (5.1). First we define (see [BW]) a local minimizer of (5.1) as follows.

Definition 5.1. $u \in W_0^{1,\infty}(\Omega)$ is a local minimizer of (5.1) if for all open sets $V \subset \Omega$ and all $\hat{v} \in W_0^{1,\infty}(\Omega)$, where

$$\hat{v}(x) := \begin{cases} u & : \quad x \in \Omega \setminus V \\ v & : \quad x \in V, \end{cases}$$

for some $v \in W_0^{1,\infty}(\Omega)$ and $\|\hat{v}\|_\infty = \|u\|_\infty$, there holds

$$\|Du\|_\infty \leq \|D\hat{v}\|_\infty.$$

Usually a global minimizer is a local minimizer, but Definition 5.1 reverses the rules: local implies global. Incidentally, this is a big problems in the L^∞ calculus. We are still working on this approach, and for example we are able to show the following theorem.

Theorem 5.2. *If u_∞ is a local minimizer for* (5.1), *then* $|\nabla u_\infty| - \Lambda_\infty u_\infty \geq 0$ *in the viscosity sense, where* $\Lambda_\infty = R_\infty(u_\infty, \Omega)$.

Acknowledgment

The author would like to thank the referee for the careful reading of the manuscript.

References

[A1] G. Aronsson: *Extension of functions satisfying Lipschitz conditions*, Ark. Mat. **6** (1967), 551–561.

[A2] G. Aronsson: *On the partial differential equation* $u_x{}^2 u_{xx} + 2u_x u_y u_{xy} + u_y{}^2 u_{yy} = 0$, Ark. Mat. **7** (1968), 395–425.

[ACJ] G. Aronsson, M.G. Crandall & P. Juutinen: *A tour of the theory of absolutely minimizing functions*, Bull. Amer. Math. Soc. (N.S.) **41** (2004), no. 4, 439–505.

[BB] G. Barles & J. Busca: *Existence and comparison results for fully nonlinear degenerate elliptic equations without zeroth-order term*, Comm. Partial Differential Equations **26** (2001), no. 11-12, 2323–2337.

[BJW] E.N. Barron, R. Jensen & C.Y. Wang: *The Euler equation and absolute minimizers of L^∞ functionals*, Arch. Ration. Mech. Anal. **157** (2001), no. 4, 255–283.

[BFK] M. Belloni, V. Ferone & B. Kawohl: *Isoperimetric inequalities, Wulff shape and related questions for strongly nonlinear elliptic operators*, J. Appl. Math. Phys. (ZAMP), **54** (2003) pp. 771–783.

[BJK] M. Belloni, P. Juutinen & B. Kawohl: *The p-Laplace eigenvalue problem and viscosity solutions as $p \to \infty$ in a Finsler metric*, J. Europ. Math. Soc., to appear.

[BK1] M. Belloni & B. Kawohl: *A direct uniqueness proof for equations involving the p-Laplace operator*, Manuscripta Math. **109** (2002), no. 2, 229–231.

[BK2] M. Belloni & B. Kawohl: *The pseudo-p-Laplace eigenvalue problem and viscosity solutions as $p \to \infty$*, ESAIM Control Optim. Calc. Var. **10** (2004), no. 1, 28–52.

[BW] M. Belloni, & A. Wagner: *manuscript*.

[BDM] T. Bhattacharya, E. DiBenedetto & J. Manfredi: *Limits as $p \to \infty$ of $\Delta_p u_p = f$ and related extremal problems*, Some topics in nonlinear PDEs (Turin, 1989). Rend. Sem. Mat. Univ. Politec. Torino 1989, Special Issue, 15–68 (1991).

[BDP] G. Buttazzo & L. De Pascale: *Optimal Shapes and Masses, and Optimal Transportation Problems*, Lecture Notes in Math., 1813, Springer, Berlin (2003), 11–51.

[CDP] T. Champion & L. De Pascale: *A principle of comparison with distance functions for absolute minimizers*, preprint (2004) (see http://cvgmt.sns.it/cgi/get.cgi/papers/chadep04/).

[CIL] M.G. Crandall; H. Ishii & P.L. Lions: *User's guide to viscosity solutions of second order partial differential equations*, Bull. Amer. Math. Soc. (N.S.) **27** (1992), no. 1, 1–67.

[CE] M.G. Crandall & L.C. Evans: *A remark on infinity harmonic functions*, Proceedings of the USA-Chile Workshop on Nonlinear Analysis (Via del Mar-Valparaiso, 2000), 123–129 (electronic), Electron. J. Differ. Equ. Conf., 6, Southwest Texas State Univ., San Marcos, TX, 2001.

[CEG] M.G. Crandall, L.C. Evans & R.F. Gariepy: *Optimal Lipschitz extensions and the infinity Laplacian.*, Calc. Var. Partial Differential Equations **13** (2001), no. 2, 123–139.

[DM] G. Dal Maso: *An introduction to Γ-convergence*, Progress in Nonlinear Differential Equations and their Applications, 8. Birkhäuser Boston, Inc., Boston, MA (1993).

[EG] L.C. Evans & W. Gangbo:*Differential equations methods for the Monge-Kantorovich mass transfer problem*, Mem. Am. Math. Soc. **653** (1999) pp. 1–66.

[FIN] N. Fukagai, M. Ito & K. Narukawa: *Limit as $p \to \infty$ of p-Laplace eigenvalue problems and L^∞-inequality of the Poincaré type*, Differential Integral Equations **12** (1999), no. 2, 183–206.

[GPP] A. Garroni, M. Ponsiglione & F. Prinari: *From 1-homogeneous supremal functionals to difference quotients: relaxation and Gamma-convergence*, Calc. Var. (to appear).

[IK] T. Ishibashi & S. Koike: *On fully nonlinear pdes derived from variational problems of L^p-norms*, SIAM J. Math. Anal. **33** (2001) pp. 545–569.

[IL] H. Ishii & P. Loreti: *Limits of solutions of p-Laplace equations as p goes to infinity and related variational problems*, SIAM J. Math. Anal., to appear (see also http://www.edu.waseda.ac.jp/ ishii/).

[J] R. Jensen: *Uniqueness of Lipschitz extensions: minimizing the sup norm of the gradient*, Arch. Rational Mech. Anal. **123** (1993), no. 1, 51–74.

[Ja] U. Janfalk: *Behaviour in the limit, as $p \to \infty$, of minimizers of functionals involving p-Dirichlet integrals*, SIAM J. Math. Anal. **27** (1996) pp. 341–360.

[K1] B. Kawohl: *A family of torsional creep problems*, J. reine angew. Math. **410** (1990) 1–22.

[KF] B. Kawohl & V. Fridman: *Isoperimetric estimates for the first eigenvalue of the p-Laplace operator and the Cheeger constant*, Comment. Math. Univ. Carolin. **44** (2003), no. 4, 659–667.

[KLR] B. Kawohl & T. Lachand-Robert: *Characterization of Cheeger sets for convex subsets of the plane*, Pacific J. Math., to appear.

[JLM1] P. Juutinen, P. Lindqvist, J. Manfredi: *The ∞-eigenvalue problem*, Arch. Ration. Mech. Anal. **148** (1999), no. 2, 89–105.

[JLM2] P. Juutinen, P. Lindqvist & J. Manfredi: *The infinity Laplacian: examples and observations*, Papers on analysis, 207–217, Rep. Univ. Jyväskylä Dep. Math. Stat., 83, Univ. Jyväskylä, Jyväskylä, 2001.

[L1] P. Lindqvist: *On the equation* $\operatorname{div}(|\nabla u|^{p-2}\nabla u)+\lambda|u|^{p-2}u=0$, Proc. Amer. Math. Soc. **109** (1990), no. 1, 157–164.

[L2] P. Lindqvist: *Addendum: "On the equation* $\operatorname{div}(|\nabla u|^{p-2}\nabla u)+\lambda|u|^{p-2}u=0$", Proc. Amer. Math. Soc. **116** (1992), no. 2, 583–584.

[L3] P. Lindqvist: *On a nonlinear eigenvalue problem*, Fall School in Analysis (Jyväskylä, 1994), Report, 68, Univ. Jyväskylä, Jyväskylä, (1995) 33–54.

[McS] E.J. McShane: *On the extension of the range of a function*, Bull. Amer. Math. Soc. **40** (1934), 837–842.

[S] O. Savin: *C^1 regularity for infinity harmonic functions in two dimensions*, preprint (2004).

[W] H. Whitney: *Analytic extensions of differentiable functions defined in closed sets*, Trans. Amer. Math. Soc. **36** (1934), no. 1, 63–89.

[Wu] Y. Wu: *Absolute minimizers in Finsler metric*, PhD Thesis (1995), Berkeley.

[Z] E. Zeidler: *Nonlinear Functional Analysis and Applications III, Variational Methods and Optimization*, Springer Verlag, Heidelberg (1984).

Marino Belloni
Dipartimento di Matematica
Università di Parma
Via Parco delle Scienze, 53/A
I-43100 Parma, Italy
e-mail: marino.belloni@unipr.it

International Series of Numerical Mathematics, Vol. 154, 95–106

Comparison of Two Algorithms to Solve the Fixed-strike Amerasian Options Pricing Problem

Alfredo Bermúdez, María Rodríguez-Nogueiras and Carlos Vázquez

Abstract. Amerasian options pricing problems are formulated, using Black-Scholes and Merton methodology, as unilateral obstacle problems for degenerate parabolic convection-diffusion-reaction operators. We mainly focus on the numerical solution of these problems and we compare two algorithms based on the augmented Lagrange formulation. Moreover, we use higher-order Lagrange-Galerkin methods for the time-space discretization. Finally, numerical results show the performance of the proposed methods.

Keywords. Black-Scholes models, Amerasian options, Lagrange-Galerkin method, Duality algorithms, Active set methods.

In this work we consider the numerical solution of the Amerasian options (with continuous arithmetic averaging and fixed strike) pricing problem. These options are *path-dependent* financial derivatives whose *payoffs* (i.e., their values at the end of the contract) somehow depend on an averaging price of another financial product called the *underlying asset* during a period of time. Following Black-Scholes and Merton's methodology, the value of an Asian option of American type, or Amerasian option, solves a two-dimensional linear complementarity problem for a strongly degenerate parabolic differential operator, with no diffusion in one of the spatial directions (see [12]).

In the literature regarding the European counterpart pricing problem some changes of variable were proposed to reduce the spatial dimension in one (see [21, 25]). Nevertheless, these techniques cannot be applied either to the American case, or to more general problems as those with share-dependent volatility, where it is mandatory to solve the two-dimensional problem. Regarding now its numerical solution, a forward shooting algorithm is proposed in [2] and finite volume methods with high-order nonlinear flux limiter for the convective terms, combined with a penalty method for the inequality constraints, are applied in [27]. In [18] an implicit finite element method combined with a PSOR procedure and operator regularization is proposed. Recently, in [20] TVD and WENO discretizations have been applied. In [13] authors solve numerically the Amerasian option problem, under jump diffusion models, governed by integro-partial differential equations. They apply a splitting technique that solves the transport equation in the average direction with a Lagrangian scheme, and solves one-dimensional Black-Scholes equations in the asset direction, with second-order implicit finite differences.

In the present work we address the numerical solution of unilateral-obstacle problems arising when pricing Amerasian options. The proposed algorithm results from the combination of higher-order characteristics methods for time discretization, higher-order finite elements for space discretization and two different algorithms for the free boundary problem (the *Bermúdez-Moreno algorithm* (*BM*), and the *Augmented Lagrangian Active Set method* (*ALAS*)). This work is a continuation of some previous ones where the numerical solution of financial pricing problems by Lagrange-Galerkin methods and iterative algorithms were addressed. In [8] a numerical algorithm consisting of combining the iterative algorithm introduced in [3] with first-order Lagrange-Galerkin methods to solve general early exercise two factor pricing problems is applied. In [7] higher-order Lagrange-Galerkin methods for Eurasian options are used. They have been extended to the general constrained case in [6] where the *BM* algorithm has been employed. The main novelty here is the use of a more recently proposed algorithm, *ALAS* algorithm [17], to face with the nonlinearity of the pricing problem.

Notice that the here proposed numerical methodology is quite general and can be applied to general two factor products as, for instance, convertible bonds (see [1]). However, we have taken into account the specific features of Asian options pricing problems, when optimizing, for instance, some algebraic computations.

In the following, we first formulate the mathematical problem; secondly we recall the basic features of the Lagrange-Galerkin discretization; thirdly, we describe both the *BM* algorithm and the *ALAS* algorithm and give some "a priori" comparison. Finally, we show some numerical results and final conclusions.

1. Mathematical formulation

The option pricing complementarity problem can be formulated as follows [6]:

Find $V : (x_1, x_2, \tau) \in \Omega \times [0,T] \longrightarrow \mathbb{R}$, *satisfying*

$$\mathcal{L}[V] \leq 0, \quad V \geq \Lambda, \quad \mathcal{L}[V]\,(V-\Lambda) = 0 \quad \textit{in } \Omega \times (0,T), \tag{1.1}$$

subjected to the initial condition

$$V(x_1, x_2, 0) = \Lambda(x_1, x_2) \quad \textit{for } (x_1, x_2) \in \Omega, \tag{1.2}$$

and to the boundary condition

$$\frac{\partial V}{\partial x_1}(x_1, x_2, \tau) = g(x_1, x_2, \tau) \quad \textit{on } \Gamma_{1,+} \times (0,T). \tag{1.3}$$

In the previous formulation, $\Omega := (0, x_1^*) \times (0, x_2^*)$, $\Gamma = \partial\Omega$, and $\Gamma_{1,+} = \Gamma \cap \{x_1 = x_1^*\}$, for x_1^* and x_2^* "large enough" positive numbers. Moreover, the linear operator $\mathcal{L}$ is

$$\mathcal{L}[\phi] = \frac{\partial \phi}{\partial \tau} - \mathsf{Div}(\mathbf{A}\ \nabla\phi) + \mathbf{v} \cdot \nabla\phi + l\phi, \tag{1.4}$$

for ϕ defined in $\Omega \times (0, T)$, and

$$\mathbf{A} = \begin{pmatrix} \frac{1}{2}\sigma^2 x_1^2 & 0 \\ 0 & 0 \end{pmatrix}, \quad \mathbf{v} = \begin{pmatrix} (\sigma^2 - r + d_0)x_1 \\ \dfrac{x_2 - x_1}{T - \tau} \end{pmatrix}, \quad l = r. \tag{1.5}$$

Moreover,

$$\Lambda(x_1, x_2) = (x_2 - K)_+, \tag{1.6}$$

and

$$g(x_1, x_2, \tau) := \begin{cases} \dfrac{\tau}{T} e^{-r\tau}, & \text{if} \quad 0 < x_2 < K\dfrac{T}{T - \tau}, \\ \dfrac{e^{-d_0\tau} - e^{-r\tau}}{T(r - d_0)} & \text{if} \quad K\dfrac{T}{T - \tau} < x_2 < x_2^*. \end{cases} \tag{1.7}$$

The unknown function V denotes the value of the option, as a function of the underlying asset value, x_1, the averaging variable, x_2, and the time to maturity, τ. Moreover, the averaging interval is $[0, T]$, and the strike is denoted by $K > 0$. The positive financial parameters, r, σ, d_0 are the interest rate of the economy, and the volatility and the dividend yield of the underlying asset, respectively.

Remark 1.1. Notice that the complementarity problem (1.1)–(1.3) admits the corresponding variational inequality formulation (see, for instance, [6]) which is usually associated to obstacle-like problems.

Remark 1.2. Using Black-Scholes and Merton techniques the Amerasian options pricing problem is firstly posed in an unbounded spatial domain. Keeping in mind the numerical solution of the problem by using finite elements, a truncation of the spatial domain is needed. For the sake of simplicity we have directly formulated the problem in a bounded domain with its corresponding boundary conditions. For more details concerning this process see, for instance [8, 6, 19].

The following propositions concern the noncoincidence region of the free boundary problem and have been stated in [8, 19] by using some financial arguments.

Proposition 1.3. *At points (x_1, x_2, t) such that $x_2 \leq K$ at time t , the function V does not reach the obstacle.*

Proposition 1.4. *At points (x_1, x_2, t) such that $x_1 > (1 + rt)\,\dfrac{1}{t - T_i} x_2$ the function V does not reach the obstacle.*

The iterative algorithms we propose for the numerical solution of (1.1)–(1.3) are based on the Lagrange formulation. It refers to the fact that the inequality involving the operator $\mathcal{L}$ is replaced by an equality by means of an appropriate Lagrange variable or multiplier to be called P. This mixed formulation appears when dealing with duality methods for solving obstacle problems (see [14], for example). Thus, problem (1.1)–(1.3) is equivalent to the following mixed formulation:

Find V and $P : \Omega \times [0,T] \longrightarrow \mathbb{R}$ satisfying the partial differential equation

$$\frac{\partial V}{\partial \tau} - \mathsf{Div}\,(\mathbf{A}\nabla V) + \mathbf{v}\cdot\nabla V + lV + P = 0 \quad \text{in } \Omega \times (0,T)\,, \tag{1.8}$$

the complementarity conditions

$$V \geq \Lambda, \quad P \leq 0, \quad (V-\Lambda)P = 0 \quad \text{in } \Omega \times (0,T), \tag{1.9}$$

and initial and boundary conditions (1.2)–(1.3).

2. Higher-order Lagrange-Galerkin for the unconstrained problem

Very often, in differential equations for pricing financial products the diffusion is quite small relative to the convection for some regions of the domain and/or for particular values of the parameters. This is reinforced for Asian options, as there is no diffusion in one of the spatial dimensions. In such circumstances numerical schemes present difficulties. A possible upwinding scheme that leads to symmetric and stable approximations of the transport PDE, reducing temporal errors and allowing for large timesteps without loss of accuracy, is the characteristics method for time discretization. An application in finance of the classical characteristics method has been already developed in [24, 1, 8, 13]. Moreover, while most papers and books on financial derivatives employ finite differences for the numerical solution, the use of finite elements has several advantages. Firstly, unstructured meshes can be convenient to refine at some parts of the domain as, for instance, near free boundaries or where the initial condition is less smooth. Secondly, it provides greater flexibility for changing final or boundary conditions as well as incorporating inequality constraints, if necessary.

In the present work, which mainly focus on the free-boundary aspect, we use the second-order Lagrange-Galerkin method analyzed in [22, 4, 5] for time-space discretization. The application of this general method for convection-diffusion-reaction PDEs to Asian options has been explained in [6]. Particularly, we consider $\mathcal{Q}_h^2$ finite element spaces,

$$\mathcal{Q}_h^2 \quad := \quad \{f \in \mathrm{C}^0(\overline{\Omega}), f|_K \in Q^2(K) \forall K \in \mathcal{T}_h\}, \tag{2.1}$$

where $Q^2(K)$ is the space of polynomials of degree less than or equal to two in each variable, and we use Simpson rule for numerical integration. So, if we consider a finite elements space of dimension N_{dof}, we have to solve a $N_{dof} \times N_{dof}$ system

$$\mathsf{M}_h V_h^n = b_h^{n-1}, \tag{2.2}$$

at each time step, where matrix M_h does not depend on time. This fact allows us to compute its Choleski factorization only once. Moreover, if meshes satisfy the condition that their edges are parallel to the axis and an adequate mesh numbering is used, this matrix is block diagonal. (See [6, 19] for details.)

3. Numerical treatment of the free boundary problem

The most common method of handling the early exercise condition (which leads to the free boundary problem) in numerical finance is simply to advance the discrete solution over a timestep ignoring the restriction and then to make its projection on the set of constraints (see, for example, [11]). In the case of a single factor (American vanilla put pricing problem, for instance) the algebraic linear complementary problem is commonly solved using a projected iteration method (PSOR) that captures the unknown exercise boundary at each time step (see Wilmott [26]). In [10] a multigrid method to accelerate convergence of the basic relaxation method is suggested and in [24] a Uzawa's algorithm to better capture the free boundary is used. Moreover, in [15] an implicit penalty method for pricing American options is proposed. Authors show that, when variable timestep is used, quadratic convergence is achieved. The drawbacks of projected relaxation methods are that their rates of convergence depend on the choice of the relaxation parameter, they deteriorate when the meshes are refined and, moreover, they do not take into account the domain decomposition given by the free boundary. In this section we describe two algorithms for which the developed regularization does not introduce any further source of error as penalty methods do: the Bermúdez-Moreno algorithm and the Augmented Lagrangian Active Set method.

3.1. The Bermúdez-Moreno iterative algorithm (BM)

This method has been introduced in [3] for solving elliptic variational inequalities. It consists of approximating the solution of the variational inequality by a sequence of solutions of variational equalities. While this method has been extensively applied to solve free boundary problems in computational mechanics, its application to price financial derivatives has been recently proposed [8].

In order to apply the duality method proposed in [3], we introduce a new Lagrange multiplier, Q, in terms of a parameter $\omega > 0$, by

$$Q := P - \omega V. \tag{3.1}$$

Then, condition (1.9) can be equivalently formulated as

$$Q(\mathbf{x},t) \in G^{\omega}(V(\mathbf{x},t)) \qquad \text{a.e. in } \Omega^* \times (0,T),$$

where $G^{\omega} := G - \omega I$, I is the identity function and G denotes the following multi-valued maximal monotone operator (see [9]):

$$G(Y) = \begin{cases} \emptyset & \text{if} \quad Y < \Lambda, \\ (-\infty, 0] & \text{if} \quad Y = \Lambda, \\ 0 & \text{if} \quad \Lambda < Y, \end{cases} . \tag{3.2}$$

We recall that, if B is a maximal monotone operator in a Hilbert space then its *resolvent* operator is the single-valued contraction $J_\lambda = (I + \lambda B)^{-1}$ and its *Yosida regularization* is the Lipschitz-continuous mapping $G_\lambda = \lambda^{-1}(I - J_\lambda)$, where λ is any positive real number (see for instance [9]).

The following equivalence is straightforward:

$$p \in B(v) \Leftrightarrow p = B_\lambda(v + \lambda p). \tag{3.3}$$

In fact, a similar equivalence holds for operator $B^\omega := B - \omega I$, for any $\lambda < 1/\omega$. In the particular case of $B = G$ given by (3.2), the Yosida regularization of G^ω is

$$G_\lambda^\omega(Y) = \begin{cases} \dfrac{Y - \Lambda}{\lambda} & \text{if} \quad Y < (1 - \omega\lambda)\Lambda, \\ -\dfrac{\omega}{1 - \omega\lambda} Y & \text{if} \quad Y \geq (1 - \omega\lambda)\Lambda, \end{cases} \tag{3.4}$$

and equivalence (3.3) becomes

$$Q = G_\lambda^\omega(V + \lambda Q). \tag{3.5}$$

The above developments lead to consider the following algorithm:

1. *Initialization:* Q_0 is arbitrarily given.
2. *Iteration m:* Q_m is known.
   (a) Compute V_{m+1} by solving

$$\begin{cases} \dfrac{\partial V_{m+1}}{\partial \tau} - \mathsf{Div}\,(\mathbf{A}\nabla V_{m+1}) + \mathbf{v} \cdot \nabla V_{m+1} & & \\ \qquad +(r + \omega)V_{m+1} + Q_m = 0 & & \text{in } \Omega^* \times (0, T), \\ V_{m+1}(x_1, x_2, 0) = \Lambda(x_1, x_2) & & \text{in } \Omega^*, \\ \dfrac{\partial V_{m+1}}{\partial x_1}(x_1, x_2, \tau) = g(x_1, x_2, \tau) & & \text{on } \Gamma_{1,+}^* \times (0, T). \end{cases} \tag{3.6}$$

   (b) Update Lagrange multiplier Q by

$$Q_{m+1} = \mu G_\lambda^\omega [V_{m+1} + \lambda Q_m] + (1 - \mu)Q_m \quad \text{in } \Omega^* \times (0, T), \tag{3.7}$$

   where μ is a relaxation parameter, $\mu \in (0, 1]$.

Remark 3.1. We emphasize that, since (3.5) is completely equivalent to (3.1), this algorithm does not introduce any further source of error as penalty methods do.

Convergence results in [3] can be easily adapted to our degenerate case. In particular, one can show convergence as far as ω and λ are chosen such that $\omega\lambda = 1/2$ but, unfortunately, the speed of convergence depends on these parameters.

3.2. Augmented Lagrangian Active set method (ALAS)

The ALAS algorithm proposed in [17] is here applied to the fully discretized in time and space mixed formulation (1.8)–(1.9). In this method the basic iteration of the active set consists of two steps. In the first one the domain is decomposed into active and inactive parts (depending on whether the constraint "acts" or not), and then, a reduced linear system associated to the inactive part is solved. We use the algorithm for unilateral problems, which is based on the augmented Lagrangian formulation. Some a priori known properties of our particular problem are taken into account in order to improve the performance of this method.

First, for any decomposition $\mathrm{N} = \mathrm{I} \cup \mathrm{J}$, where $\mathrm{N} := \{1, 2, \ldots, N_{dof}\}$, let us denote by $[\mathsf{M}_h]_{\mathrm{II}}$ the principal minor of matrix M_h and by $[\mathsf{M}_h]_{\mathrm{IJ}}$ the codiagonal block indexed by I and J. Thus, for each mesh time t_n, the ALAS algorithm computes not only V_h^n and P_h^n but also a decomposition $N = \mathrm{J}^n \cup \mathrm{I}^n$ such that

$$\begin{aligned} \mathsf{M}_h V_h^n + P_h^n &= b_h^{n-1}, \\ [P_h^n]_j + \beta [V_h^n - \Lambda]_j &\leq 0 \qquad \forall j \in \mathrm{J}^n, \\ [P_h^n]_i &= 0 \qquad \forall i \in \mathrm{I}^n, \end{aligned} \tag{3.8}$$

for any positive constant β. In the above, I^n and J^n are, respectively, the *inactive* and the *active* sets at time t_n. More precisely, the iterative algorithm builds sequences $\left\{V_{h,m}^n\right\}_m$, $\left\{P_{h,m}^n\right\}_m$, $\{\mathrm{I}_m^n\}_m$ and $\{\mathrm{J}_m^n\}_m$, converging to V_h^n, P_h^n, I^n and J^n, by means of the following steps:

1. Initialize $V_{h,0}^n$ and $P_{h,0}^n \leq 0$. Choose $\beta > 0$. Set $m = 0$.
2. Compute

$$\begin{aligned} Q_{h,m}^n &= \min\left\{0, P_{h,m}^n + \beta\left(V_{h,m}^n - \Lambda\right)\right\}, \\ \mathrm{J}_m^n &= \left\{j \in \mathrm{N}, \left[Q_{h,m}^n\right]_j < 0\right\}, \\ \mathrm{I}_m^n &= \{i \in \mathrm{N}, \left[Q_{h,m}^n\right]_i = 0\}. \end{aligned}$$

3. If $m \geq 1$ and $J_m^n = J_{m-1}^n$ then convergence is achieved. Stop.
4. Let V and P be the solution of the linear system

$$\begin{aligned} &\mathsf{M}_h V + P = b^{n-1}, \\ &P = 0 \text{ on } \mathrm{I}_m^n \text{ and } V = \Lambda \text{ on } \mathrm{J}_m^n. \end{aligned} \tag{3.9}$$

Set $V_{h,m+1}^n = V, P_{h,m+1}^n = \min\{0, P\}, m = m + 1$ and go to 2.

It is important to notice that, instead of solving the full linear system in (3.9), for $\mathrm{I} = \mathrm{I}_m^n$ and $\mathrm{J} = \mathrm{J}_m^n$ the following reduced one on the inactive set is solved:

$$\begin{aligned} [\mathsf{M}_h]_{\mathrm{II}} [V]_{\mathrm{I}} &= \left[b^{n-1}\right]_{\mathrm{I}} - [\mathsf{M}_h]_{\mathrm{IJ}} [\Lambda]_{\mathrm{J}}, \\ [V]_{\mathrm{J}} &= [\Lambda]_{\mathrm{J}}, \\ P &= b^{n-1} - \mathsf{M}_h V. \end{aligned} \tag{3.10}$$

Remark 3.2. In a unilateral obstacle problem, the parameter β only influences the first iteration.

In [17], authors proved convergence of the algorithm in a finite number of steps for a Stieltjes matrix (i.e., a real symmetric positive definite matrix with negative off-diagonal entries [23]) and a suitable initialization. They also proved that $\mathrm{I}_m \subset \mathrm{I}_{m+1}$. Nevertheless, a Stieltjes matrix can be only obtained for linear elements but never for "our" quadratic elements because we have some positive off-diagonal entries coming from the stiffness matrix (actually we use a lumped mass matrix). However, we have obtained good results by using ALAS algorithm with quadratic finite elements and the following particular additional features:

- The algorithm is initialized as proposed in [17]:
$$V^n_{h,0} = \Lambda \quad \text{and} \quad P^n_{h,0} = b^n - \mathsf{M}_h V^n_{h,0}.$$
- We compute the set
$$\mathrm{I}^n_* := \{i, \in \mathrm{N},\ \mathbf{x}^i = (x^i_1, x^i_2) \text{ is a mesh node with } x^i_2 < K,\quad x^i_1 > (1 + r(t_n - T_i)x^i_2\},$$
and impose that $\mathrm{I}^n_* \subset \mathrm{I}^n_m$ for every m (using Propositions 1.3 and 1.4).
- We do not assume monotonicity with respect to m for the sets $\{\mathrm{I}^n_m\}$.

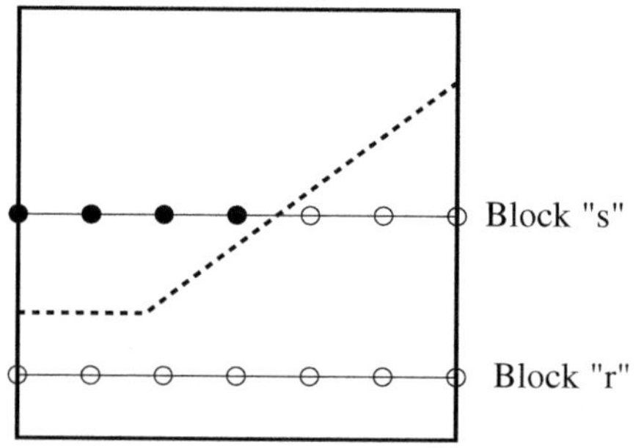

FIGURE 1. Spatial domain of solution for the Amerasian call options pricing problem, separating the active from the inactive set. Two sets of FE nodes with the same x_2 coordinate are represented, and the nodes inside the active set are filled.

Special care has to be taken for an efficient solution of the linear system when using the ALAS algorithm. Meshes with edges parallel to the axis and with suitable mesh numbering have already been used in the BM algorithm. The fact that in the ALAS algorithm only an incomplete linear system is solved requires a deeper study. More precisely, by ordering the nodes from right to left and from bottom to top, we are led to a matrix with N_{x_2} blocks of dimension N_{x_1}. In other words, each set of nodes with the same x_2 coordinate gives rise to a block in the matrix. Thus, for each block either all of the nodes are inside the inactive set (the case of Block "r" in Figure 1) or only the first $n(x_2)$ nodes (with $n(x_2) \leq N_{x_1}$) belong to the inactive set (the case of Block "s" in Figure 1). The main point is that also for the ALAS algorithm we develop the factorization of the (complete) matrix only once outside the time loop and the iterative algorithm loop, and, at each iteration, we solve N_{x_2} systems of variable dimension (less or equal than N_{x_1}).

A "general" comparison between the two iterative algorithms is not practical because the performance of this second algorithm is very problem-dependent.

For example, the larger the active set, the more efficient the second algorithm is. Nevertheless, we can establish the following a priori comments related to the comparison of the two algorithms when applied to our particular problem. They will be completed when showing the numerical results in the next section:

- Linear systems in the ALAS algorithm are smaller than those in the BM algorithm.
- The ALAS algorithm uses some a priori known data about the inactive set.
- The BM algorithm is strongly parameter (ω) dependent, whereas the (β) parameter appearing in ALAS algorithm only influences the first iteration.
- ALAS algorithm can be interpreted as a semi-smooth Newton method [16], and thus it exhibits a super-linear convergence rate.

4. Numerical results

To complete our analysis, we compare the performance of the BM and the ALAS algorithms for three sets of parameters. In Table 1 we show the results corresponding to the computational domain $\Omega^* = (0, x_1^*) \times (0, x_2^*)$, with $x_1^* = x_2^* = 3K$. Uniform meshes in time and non-uniform meshes in space have been used. If N_{ts} denotes the number of time steps and $N_{dof} = N_{x_1} = N_{x_2}$ denotes the number of d.o.f. in each spatial direction, we introduce the following notation for the meshes:

	Mesh-1	Mesh-2	Mesh-3	Mesh-4
N_{ts}	34	67	133	265
N_{dof}	67	133	265	529

Regarding Table 1 we can conclude that the ALAS algorithm is more efficient than the BM one because the computational time is smaller for the same meshes and leads to analogous results. Notice that the main difference between these algorithms is the solution of the linear systems and the updating of the Lagrange multiplier (or active/inactive sets), whereas the computation and factorization of the full matrix, etc. ... are common to both algorithms. Thus, since the algebraic part of the program (in particular, the resolution of the already factorized linear systems) has been optimized in both cases, the difference in time is lower than the difference in the number of iterations.

On the other hand, the mean number of iterations is clearly less in the ALAS algorithm than in the BM one. To this respect, it is interesting to note that, in the ALAS algorithm, parameter β seems not to influence the number of iterations (as it is claimed in [17]). However, the mean number of iterations in the BM algorithm crucially depends on the choice of parameter ω. Furthermore, we have observed that the number of iterations as a function of parameter ω is a convex function, and that its minimum depends not only on the test data (i.e., volatility, time to maturity, etc) but also on the mesh parameters. For this reason, we have first searched the value of ω giving the minimum number of iterations for each test data and mesh. This optimal choice is used in Table 1. In Figure 2 we show the mean number of iterations in the BM algorithm as a function of ω for the four mesh refinements and the data set $\sigma = 0.2, T_f - T_i = 1, r = 0.1, d_0 = 0, K = 100$.

$\sigma = 0.4,\ T_f - T_i = 0.5$								
	Mesh-1		Mesh-2		Mesh-3		Mesh-4	
	BM	ALAS	BM	ALAS	BM	ALAS	BM	ALAS
Parameter	1.5E2	1.E5	4.E2	1.E5	1.15E3	1.E5	1.9E3	1.E5
N_{it}	23	4	26	5	31	6	33	9
Op. Value	8.5528	8.5503	8.5310	8.5299	8.5220	8.5232	8.5222	8.5223
Time	4	3	38	24	411	271	4689	3560

$\sigma = 0.1,\ T_f - T_i = 0.25$								
	Mesh-1		Mesh-2		Mesh-3		Mesh-4	
	BM	ALAS	BM	ALAS	BM	ALAS	BM	ALAS
Parameter	1.5E2	1.E5	3.25E2	1.E5	7.E2	1.E5	1.15E3	1.E5
N_{it}	10	3	11	3	12	3	13	3
Op. Value	1.9759	1.9560	1.9529	1.9308	1.9511	1.9348	1.9590	1.9590
Time	4	3	38	30	413	349	5259	4685

$\sigma = 0.2,\ T_f - T_i = 1.$								
	Mesh-1		Mesh-2		Mesh-3		Mesh-4	
	BM	ALAS	BM	ALAS	BM	ALAS	BM	ALAS
Parameter	7.5E1	1.E5	1.5E2	1.E5	4.5E2	1.E5	1.05E3	1.E5
N_{it}	19	3	20	4	25	5	24	7
Op. Value	7.5475E	7.5445E	7.5441	7.5425E	7.5405	7.5406	7.5386	7.5417
Time	5	3	43	31	474	355	5589	4737

TABLE 1. Results comparing BM and ALAS algorithms applied to the Amerasian problem with data $r = 0.1, d_0 = 0, K = 100$. The parameters (Parameter) are λ for BM and β for ALAS. N_{it} denotes the mean number of iterations and option value (Op. Value) corresponds to $(S, M, t) = (100, 100, T_i)$. The computation time (Time) is in measured in seconds.

Acknowledgment

The first and second authors have been supported by MCYT (Spain) Project VEM2003-20069-C03-02. The second author was also supported by the 'Ministerio de Educación, Cultura y Deporte' (Spain).

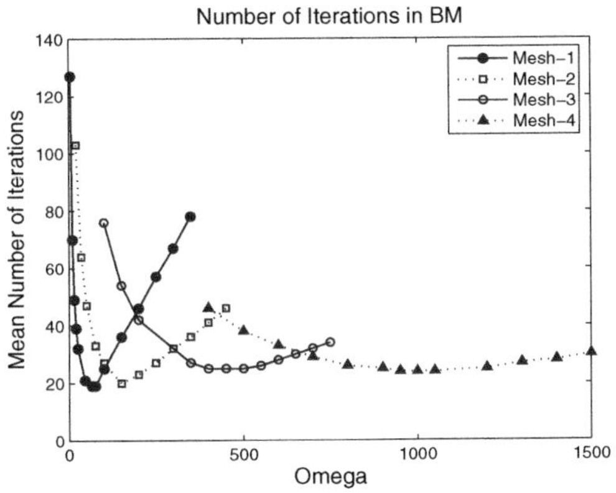

FIGURE 2. Mean number of iterations taken for the BM algorithm convergence as a function of parameter ω for different meshes and for data $\sigma = 0.2, T_f - T_i = 1., r = 0.1, d_0 = 0, K = 100$.

References

[1] G. Barone-Adesi, Ana Bermúdez and J. Hatgioannides, *Two-factor convertible bonds valuation using the method of characteristics/finite elements*, J. Econom. Dynam. Control **27** (2003), no. 10, 1801–1831.

[2] J. Barraquand and T. Pudet, *Pricing of American path-dependent contingent claims*, Math. Finance **6** (1996), no. 1, 17–51.

[3] A. Bermúdez and C. Moreno, *Duality methods for solving variational inequalities*, Comput. Math. Appl. **7** (1981), no. 1, 43–58.

[4] A. Bermúdez, M.R. Nogueiras, and C. Vázquez, *Numerical analysis of convection-diffusion-reaction problems with higher order characteristics/finite elements. Part I: Time discretization*, (to appear in SIAM J. Numer. Anal.).

[5] ______, *Numerical analysis of convection-diffusion-reaction problems with higher order characteristics/finite elements. Part II: Fully discretized scheme and quadrature formulas*, (to appear in SIAM J. Numer. Anal.).

[6] ______, *Numerical solution of variational inequalities for pricing Asian options by higher order Lagrange-Galerkin methods*, Appl. Numer. Math. **56** (2006), no. 10-11, 1256–1270.

[7] ______, *Numerical methods for pricing fixed strike Eurasian options with continuous arithmetic averaging*, CD-ROM Proceedings of Stochastics and Finance, Lisbon, 2004.

[8] Ana Bermúdez and M.R. Nogueiras, *Numerical solution of two-factor models for valuation of financial derivatives*, Math. Mod. Meth. Appl. Sci. **14** (2004), 2, 295–327.

[9] H. Brezis, *Analyse fonctionnelle*, Collection Mathématiques Appliquées pour la Maîtrise, Masson, Paris, 1983.

[10] N. Clarke and K. Parrot, *Multigrid american option pricing with stochastic volatility*, App. Math. Finance **6** (1994), no. 6, 177–195, ICOSAHOM'92 (Montpellier, 1992).

[11] L. Clelow and C. Strickland, *Implementing derivative models*, Jonh Wiley & Sons, Nonsei, 1998, Méthodes Mathématiques de l'Informatique, No. 6.

[12] J.N. Dewynne and P.Wilmott, *Asian options as linear complementarity problems: analysis and finite difference solutions*, Advances in Futures and Options Research **8** (1995), 145–173.

[13] Y. d'Halluin, P.A. Forsyth, and G. Labahn, *A semi-Lagrangian approach for American Asian options under jump diffusion*, to appear in SIAM J. Sci. Comp. (2005).

[14] P. Faurre, *Analyse numérique*, École Polytechnique, Palaiseau, 1988, Notes d'optimisation.

[15] P.A. Forsyth and K.R. Vetzal, *Quadratic convergence for valuing American options using a penalty method*, SIAM J. Sci. Comput. **23** (2002), no. 6, 2095–2122.

[16] K. Ito and K. Kunisch, *Semi-smooth Newton methods for state-constrained optimal control problems*, Systems Control Lett. **50** (2003), no. 3, 221–228.

[17] T. Kärkkäinen, K. Kunisch, and P. Tarvainen, *Augmented Lagrangian active set methods for obstacle problems*, J. Optim. Theory Appl. **119** (2003), no. 3, 499–533.

[18] M.D. Marcozzi, *On the valuation of Asian options by variational methods*, SIAM J. Sci. Comput. **24** (2003), no. 4, 1124–1140.

[19] M.R. Nogueiras, *Numerical Analysis of Second Order Lagrange-Galerkin Schemes. Application to Option Pricing Problems.*, Ph.D. thesis, Universidade de Santiago de Compostela, 2005.

[20] C.W. Oosterlee, J.C. Frisch, and F.J. Gaspar, *TVD, WENO and blended BDF discretizations for Asian options*, Comput. Vis. Sci. **6** (2004), no. 2-3, 131–138.

[21] L.C.G. Rogers and Z. Shi, *The value of an Asian option*, J. Appl. Probab. **32** (1995), no. 4, 1077–1088.

[22] H. Rui and M. Tabata, *A second order characteristic finite element scheme for convection-diffusion problems*, Numer. Math. **92** (2002), no. 1, 161–177.

[23] R.S. Varga, *Matrix iterative analysis*, Prentice-Hall Inc., Englewood Cliffs, N.J., 1962.

[24] C. Vázquez, *An upwind numerical approach for an American and European option pricing model*, Appl. Math. Comput. **97** (1998), no. 2-3, 273–286.

[25] J. Vecer, *A new PDE approach for pricing arithmetic average Asian options*, J. Comp. Finance (2001), 105–113.

[26] P. Wilmott, *Derivatives. the theory and practice of financial engineering*, John Wiley & Sons, Chichester, 1999.

[27] R. Zvan, P.A. Forsyth, and K. Vetzal, *Robust numerical methods for PDE of Asian options*, J. Comput. Finance **1** (1998), no. 1, 39–78.

Alfredo Bermúdez
Dep. de Matemática Aplicada.Univ. Santiago de Compostela
e-mail: `mabermud@usc.es`

María Rodríguez-Nogueiras
Dep. de Matemáticas. Univ. da Coruña
e-mail: `marianog@usc.es`

Carlos Vázquez
Dep. de Matemáticas. Univ. da Coruña
e-mail: `carlosv@udc.es`

International Series of Numerical Mathematics, Vol. 154, 107–116

Nonlinear Diffusion Models for Self-gravitating Particles

Piotr Biler and Robert Stańczy

Abstract. This paper deals with parabolic-elliptic systems of drift-diffusion type modelling gravitational interaction of particles. The main feature is presence of a nonlinear diffusion describing physically relevant density-pressure relations. We study the existence of solutions of the evolution problem, and recall results on the existence of steady states, and the blow up of solutions in cases when drift prevails the diffusion.

Mathematics Subject Classification (2000). Primary 35K60; Secondary 35Q, 35B40, 82C21.

Keywords. Nonlinear diffusion, drift-diffusion systems, mean field equations, local existence of solutions, steady states, blow up of solutions.

1. Introduction

Our aim is to describe some recent results on drift-diffusion equations coupled with the Poisson equation

$$n_t = \nabla \cdot \left(D_* \left(\nabla p + n \nabla \varphi \right) \right), \tag{1.1}$$

$$\Delta \varphi = n. \tag{1.2}$$

These systems are models for the evolution of the density of particles $n = n(x,t) \geq 0$ defined for $(x,t) \in \Omega \times \mathbb{R}^+$, $\Omega \subset \mathbb{R}^d$. The particles attract each other gravitationally through the Newtonian potential $\varphi = \varphi(x,t)$ generated by themselves. The pressure $p \geq 0$ is a sufficiently regular ($\mathcal{C}^2(\mathbb{R}^+ \times \mathbb{R}^+)$) function of the density n and the temperature $\vartheta > 0$

$$p = p(n, \vartheta). \tag{1.3}$$

The coefficient $D_* > 0$ above may depend on n, ϑ, φ, x, … . The natural boundary condition on $\partial\Omega$ for (1.1)–(1.2) is the nonlinear no-flux one

$$(\nabla p + n \nabla \varphi) \cdot \overline{\nu} = 0, \tag{1.4}$$

where $\overline{\nu}$ is the unit normal vector to $\partial\Omega$. The potential φ satisfies either

$$\varphi = E_d * n, \tag{1.5}$$

$E_d(x) = -((d-2)\sigma_d)^{-1}|x|^{2-d}$ being the fundamental solution of the Laplacian in $\mathbb{R}^d$, $d \geq 3$, or the homogeneous Dirichlet boundary condition

$$\varphi_{|\partial\Omega} = 0, \tag{1.6}$$

which is mathematically somewhat simpler. In the case of radially symmetric solutions (1.1)–(1.4), (1.6) is equivalent to (1.1)–(1.4), (1.5): the solutions of (1.5) and (1.6) differ by a constant, cf. the discussion of this issue in [4, 5, 1].

Systems for self-gravitating particles may describe concentration and collapse phenomena, manifesting themselves by, e.g., a finite (or infinite) time blow up of solutions for certain initial data.

Systems like (1.1)–(1.2) or (1.1) with the equation $\Delta\varphi = -n$ for the potential have been considered beginning with Nernst–Planck (1895) and Debye–Hückel (1926) models for the dynamics of ions in a solute (i.e., repulsing particles). Then, they have been used in semiconductor theory and plasma physics. Models for attracting particles are studied at least from early nineties of the twentieth century: [18] and [4]. The papers of P.-H. Chavanis and his collaborators (see [9] and numerous references therein) are another source of those *hydrodynamical* models in astrophysics, with an immediate reference to kinetic equations. Besides the statistical mechanics, systems of the form (1.1)–(1.2) appear also in modelling of chemotaxis phenomena, generalizing the classical approach which involved the linear diffusion with $p(n) = \text{const}\, n$. These biological models, supplemented with the homogeneous Neumann conditions for n and φ are applicable in a description of concentration of either cells or microorganisms due to chemical agents. We refer the reader to [12] for a comprehensive review of these aspects of parabolic-elliptic systems like (1.1)–(1.2), and to [17] for very recent results on related systems with nonlinear diffusion in $\mathbb{R}^N$.

Such *mean field* systems can be studied either in the *canonical ensemble* (i.e., the *isothermal* setting), when $\vartheta = $ const is fixed, or in the *microcanonical ensemble* (the *nonisothermal* case) with a variable temperature: either $\vartheta = \vartheta(t)$ or $\vartheta = \vartheta(x,t)$. Then, the energy balance is described either by a relation like

$$E = \frac{d}{2}\int_\Omega p \,\mathrm{d}x + \frac{1}{2}\int_\Omega n\varphi \,\mathrm{d}x = \text{const} \tag{1.7}$$

(which, for a given n, defines $\vartheta = \vartheta(t)$ in an implicit way) in the former case, or by an evolution partial differential equation for $\vartheta = \vartheta(x,t)$ as was in Streater's models in [2] (much more complicated mathematically).

In order to determine the evolution, we also impose an initial condition

$$n(x,0) = n_0(x) \geq 0, \tag{1.8}$$

and either $\vartheta_0 > 0$ or a value of E which is to be conserved. As a consequence of (1.4), total mass

$$M = \int_\Omega n(x,t)\,\mathrm{d}x \tag{1.9}$$

is conserved during the evolution. Moreover, sufficiently regular solutions of the evolution problem with $n_0 \geq 0$ remain positive.

The main mathematical questions concerning the systems are the following:

- existence, nonexistence and multiplicity of steady states, either for given M, ϑ or for M, E fixed,
- local in time existence of solutions of the evolution problem,
- asymptotics of global in time solutions,
- possibility of finite time blow up of solutions (corresponding to either a gravitational collapse or an explosion),
- spreading of compactly supported initial data when Ω is unbounded (and related free boundary problems).

Two systems with particular density-pressure relations have been extensively studied: for Brownian (or Maxwell–Boltzmann) and Fermi–Dirac particles.

The model of self-gravitating Brownian particles, which consists of (1.1)–(1.2), with (2.1) below, supplemented by (1.7), has been considered in [9, 11] for radially symmetric solutions (n, φ), and in [10, 6] without this symmetry assumption. Studies of the corresponding isothermal problem with $\vartheta \equiv 1$ had been conducted earlier, see, e.g., [18, 4]. We refer the reader also to [1, 5] for stationary solutions for the Brownian particles models. The main issues are:

- gravitational collapse is possible for $d \geq 2$ in the isothermal model, and for $d \geq 3$ in the nonisothermal model,
- the existence of steady states with prescribed mass and energy in $d \geq 3$ dimensions is controlled by the parameter E/M^2 which should be large enough (in $d = 2$ they exist for each $M > 0$ and each $E \in \mathbb{R}$).

Since the Fermi–Dirac model, see (2.3) below, involves nonlinear diffusion, even local in time existence of solutions is much harder to establish than in the Brownian (linear diffusion) case, see [3] where in the isothermal case a specific choice of the coefficient D_* has been considered. In particular:

- structure of the set of steady states with given M and ϑ is different (but less complicated) than in the Brownian case ([9]),
- the existence of steady states with given mass and energy is controlled by the parameter $\min\left\{\frac{E}{M^2}, \frac{E}{M^{1+2/d}}\right\}$ ([7, 16]),
- gravitational collapses cannot occur in $d \leq 3$ dimensions in the isothermal case ([3]),
- the gravitational collapse is possible for $d \geq 4$ for suitable initial data in the nonisothermal case ([7]).

We will study in this paper local in time existence of solutions – a rather difficult question in such a general setting with (1.3), mainly because of nonlinear boundary conditions (1.4) for n and possible degeneracies of diffusion. This subject which has not been considered in [8], where a thorough discussion of the following topics can be found:

- examples of density-pressure relations more general than Maxwell–Boltzmann and Fermi–Dirac,
- entropy functionals, entropy production rates and estimates of the energy,

- existence of steady states with prescribed mass and large energy,
- nonexistence of global in time solutions of (1.1)–(1.2) with general density-pressure relations (1.3), $D_* = 1$, one of the conditions (1.5), (1.6), and either negative initial entropy or low energy, and thus, *a fortiori*, nonexistence of steady states for arbitrary D_*,
- continuation of local in time solutions of (1.1)–(1.2) with polytropic density-pressure relations (cf. (2.4) below).

It is worth noting that while "local" results on the existence of steady states (i.e., for a small range of control parameters M, ϑ, E) are quite similar for general $p = p(n, \vartheta)$, the global structure of the set of steady states (existence of multiple steady states, bifurcations, end-points of branches of solutions, etc.) is rather sensitive to variations of the form of p in (1.3), cf. results for the Maxwell–Boltzmann and Fermi–Dirac models. The problem is relatively well understood in the case of the Boltzmann model, see [1] and references therein. There are some numerical results in the case of radially symmetric solutions of the Fermi–Dirac model in the ball of $\mathbb{R}^3$, cf. [9]. Three different approaches are useful when studying the existence and multiplicity of stationary solutions of the system (1.1)–(1.2), namely: monotonicity methods (as, e.g., in [13, 14]), variational methods (as, e.g., in [3, 7, 8, 15]), and integral equations (as, e.g., in [8, 15, 16]).

Results on the nonexistence of steady states and, more generally, nonexistence of solutions of the evolution problem defined for all $t \geq 0$ are proved using either the moment method (for the boundary condition (1.5)) or Pohozaev–Rellich type identities (for (1.6)). These are results on the isothermal problem with the pressure p asymptotically resembling a polytropic relation (2.4), and with quite general density-pressure relations in the microcanonical setting, see [8, Sec. 5] and [7, Sec. 2]. Similar results for chemotaxis systems with nonlinear diffusion are in [17].

Thus, weakly nonlinear diffusion (e.g., that in (2.4) with small $\gamma > 0$) is not strong enough to prevent from a blow up of solutions, at least for initial data of negative energy. On the other hand, strongly nonlinear diffusion (e.g., (2.4) with relatively large γ) guarantees the continuation of local in time solutions to the global ones. Indeed, for $p(s) \sim \kappa s^{1+\gamma}$ with $\gamma > 1 - 2/d$, using the entropy $W_{\rm iso}$ (3.1), one proves in [8, Sec. 6] that $\sup_{t>0} \int_\Omega p(x,t)\,dx < \infty$ with a bound depending on the initial data. Then, $\sup_{\delta \leq t \leq T} |n(t)|_{L^q(\Omega)} < \infty$ is proved by an iteration process for any $1 < q < \infty$ and $0 < \delta < T < \infty$, with a bound depending also on δ and T. Since for $\gamma \leq 1 - 2/d$ the blow up of solutions, caused by $\int_\Omega n(x,t)|x|^2\,dx \to 0$ as $t \nearrow T_{\max}$, is accompanied by the unboundedness of $L^q(\Omega)$ norms as $t \nearrow T_{\max}$ ([8, Sec. 5]), one can conclude that such phenomena do not occur in the complementary range of polytropic exponents γ, $\gamma > 1 - 2/d$, i.e., for diffusion terms $\Delta p(n)$ in (1.1) strong enough.

Notation. In the sequel $|\,.\,|_q$ will denote the $L^q(\Omega)$ norm. All inessential constants which may vary from line to line will be denoted by C.

2. Examples of diffusions

Without entering into the kinetic theory considerations, cf. [9] and [8], we recall here the most common density-pressure relations encountered in the (classical and quantum) statistical mechanics and generalize these examples.

Example. For Maxwell–Boltzmann distributions

$$p_{\mathrm{MB}}(n, \vartheta) = \vartheta n \tag{2.1}$$

holds, which is a consequence of the Maxwellian form of distributions $f = f(x, v, t)$ maximizing the Boltzmann entropy $-\int_{\mathbb{R}^d} f \log f \, \mathrm{d}v$ under the local density constraint $n(x, t) = \int_{\mathbb{R}^d} f \, \mathrm{d}v$. This classical Boltzmann relation and $D_* = \mathrm{const}$ lead to *linear* Brownian diffusion term Δn in (1.1).

Example. Similarly, Fermi–Dirac distributions maximize the entropy

$$S = -\int_{\mathbb{R}^d} \left(\frac{f}{\eta_0} \log \frac{f}{\eta_0} + \left(1 - \frac{f}{\eta_0}\right) \log\left(1 - \frac{f}{\eta_0}\right)\right) \mathrm{d}v$$

whose form *a priori* prevents from the overcrowding of particles at any point (x, v, t): $0 \le f \le \eta_0$. Then one has, see, e.g., [3, (1.1)–(1.3)],

$$n = \eta_0 2^{d/2-1} \sigma_d \vartheta^{d/2} I_{d/2-1}(\lambda), \quad p = \eta_0 2^{d/2} \frac{\sigma_d}{d} \vartheta^{d/2+1} I_{d/2}(\lambda),$$

where I_α denotes the Fermi integral of order $\alpha > -1$ defined for $\lambda > 0$ by

$$I_\alpha(\lambda) = \int_0^\infty \frac{y^\alpha \, \mathrm{d}y}{\lambda \mathrm{e}^y + 1}. \tag{2.2}$$

Hence

$$p_{\mathrm{FD}}(n, \vartheta) = \frac{\mu}{d} \vartheta^{d/2+1} \left(I_{d/2} \circ I_{d/2-1}^{-1} \right) \left(\frac{2}{\mu} \frac{n}{\vartheta^{d/2}} \right) \tag{2.3}$$

for a constant $\mu > 0$, which leads to a *nonlinear* diffusion in (1.1). Properties of Fermi integrals (2.2) (convexity, asymptotics, etc.) relevant to study the system (1.1)–(1.2), (2.3) are collected in [3, Sec. 2] and [7, Sec. 5].

An analogous construction is used to define Bose–Einstein distributions whose properties, however, much differ from those of Fermi–Dirac ones, see [8, Sec. 2].

Example. Polytropes are classical equations of state of a gas

$$p_{1+\gamma}(n, \vartheta) = \kappa_\gamma \vartheta^{1-\gamma d/2} n^{1+\gamma} \tag{2.4}$$

with $0 < \gamma < 2/d$, and a polytropic constant κ_γ. These are densities corresponding to distributions maximizing the Rényi–Tsallis entropy $\frac{-1}{q-1} \int_{\mathbb{R}^d} (f^q - f) \, \mathrm{d}v$, where $q = 1 + 1/(1/\gamma - d/2) \in (1, \infty)$. The limit value $\gamma = 2/d$ leads to the pressure

$$p_{1+2/d}(n, \vartheta) = \kappa_{2/d} n^{1+2/d} \tag{2.5}$$

independent of ϑ. The limit case $\gamma \searrow 0$ corresponds to the Boltzmann density-pressure relation (2.1). The polytropic relations define evolution equations with nonlinear diffusions as, e.g., in the porous media equation.

In all the examples above the scaling relation $p(\vartheta^{d/2}n,\vartheta) = \vartheta^{d/2+1}p(n,1)$ is satisfied for each $\vartheta > 0$, and thus the self-similar form of the pressure

$$p(n,\vartheta) = \vartheta^{d/2+1}P\left(\frac{n}{\vartheta^{d/2}}\right) \tag{2.6}$$

follows with a sufficiently regular function P defined on $\mathbb{R}^+$.

In all physically relevant examples the function P behaves like power functions of s near 0 and near ∞: $c_0 s^{1+\gamma_0}$, $c_\infty s^{1+\gamma_\infty}$ with some special exponents γ_0, $\gamma_\infty \geq 0$, and constants c_0, $c_\infty > 0$, respectively. For instance, $\gamma_0 = 0$, $\gamma_\infty = 2/d$ for the Fermi–Dirac pressure (2.3).

Remark 2.1. Given E and the instantaneous value of $n = n(x,t)$, the physically justifiable property $\frac{\partial p}{\partial \vartheta} > 0$ permits us to define the temperature $\vartheta = \vartheta(t)$ (and thus the pressure $p = p(n,\vartheta)$) in a unique way. If $p \in \mathcal{C}^1$ has the self-similar form (2.6) and $P(s)/s^{1+2/d}$ is strictly decreasing, then $\frac{\partial p}{\partial \vartheta} > 0$. Indeed, differentiating we get $\left(\frac{d}{2}+1\right)P(s) > \frac{d}{2}sP'(s)$ whence

$$0 < \frac{\partial p}{\partial \vartheta} = \left(\frac{d}{2}+1\right)\vartheta^{d/2}P\left(\frac{n}{\vartheta^{d/2}}\right) - \frac{d}{2}nP'\left(\frac{n}{\vartheta^{d/2}}\right).$$

Moreover, $P(s) \leq C_{\max}s^{1+2/d}$ for any $s \geq 1$, and $P(s) \geq C_{\min}s^{1+2/d}$ for any $s \leq 1$ and some constants $C_{\min}$, $C_{\max} > 0$.

3. Useful estimates for entropies and energies

In the isothermal setting the function

$$W_{\mathrm{iso}} = \frac{1}{\vartheta}\int_\Omega \left(\vartheta nh - p + \frac{1}{2}n\varphi\right)\,\mathrm{d}x \tag{3.1}$$

is a (neg)entropy for the problem (1.1)–(1.2), (1.4), either (1.5) or (1.6), and (1.8): $\frac{\mathrm{d}}{\mathrm{d}t}W_{\mathrm{iso}} = -\int_\Omega \vartheta n D_* \left|\nabla\left(h+\frac{\varphi}{\vartheta}\right)\right|^2\,\mathrm{d}x \leq 0$. Here the function h is defined (up to a constant depending on the temperature) for an arbitrary increasing $\mathcal{C}^1$ function p of $n > 0$ and for fixed $\vartheta > 0$ by the relation $\frac{\partial h}{\partial n} = \frac{1}{\vartheta n}\frac{\partial p}{\partial n}$.

In the nonisothermal setting the existence of a nontrivial entropy needs an assumption on the structure of p in (1.3), i.e., on the dependence on ϑ. A simple sufficient condition is (2.6) which implies that h has the self-similar form $h(n,\vartheta) = H\left(\frac{n}{\vartheta^{d/2}}\right)$ with a function H satisfying $H'(s) = P'(s)/s$. Then the function

$$W = \int_\Omega \left(nh - \left(\frac{d}{2}+1\right)\frac{p}{\vartheta}\right)\,\mathrm{d}x \tag{3.2}$$

is an entropy for the problem (1.1)–(1.2), (1.4), (1.8), one of the boundary condition (1.5) or (1.6), together with the energy relation (1.7). Moreover, the following production of entropy formula holds: $\frac{\mathrm{d}}{\mathrm{d}t}W = -\int_\Omega \vartheta n D_* \left|\nabla\left(h+\frac{\varphi}{\vartheta}\right)\right|^2\,\mathrm{d}x \leq 0$, see [8, Sec. 3]. Due to the minus sign before the pressure term and the boundary conditions (1.4), the entropy W does not provide, in general, substantial *a priori* estimates for n and φ. However, see [7, Lemma 3.6] and the proof of Theorem 4.1 below for *a priori* estimates for ϑ.

In some cases, the energy relation (1.7) leads to interesting *a priori* estimates.

Lemma 3.1. *If $d = 3$, $P(s) \ge \varepsilon s^{1+\gamma}$ for some $\varepsilon > 0$, $\gamma = 2/d$, and all $s \ge 0$, then the total energy (1.7) controls the thermal energy $\frac{d}{2}\int_\Omega p\,\mathrm{d}x$ and the absolute value of the potential energy $\frac{1}{2}\left|\int_\Omega n\varphi\,\mathrm{d}x\right|$ from above. More precisely,*

$$\left|\int_\Omega n\varphi\,\mathrm{d}x\right| \le C|n|_q^2 \le C|n|_{1+\gamma}^{2r}|n|_1^{2-2r} \tag{3.3}$$

with $1/q = r/(1+\gamma) + (1-r) = (d+2)/(2d)$, so that for each $\delta > 0$

$$E + CM^{1+\nu} \ge \left(\varepsilon\frac{d}{2} - \delta\right)\int_\Omega n^{1+2/d}\,\mathrm{d}x, \tag{3.4}$$

and for each $0 < c_0 < d/2$ there exists $C = C(c_0, \Omega)$ such that

$$E \ge c_0\int_\Omega p\,\mathrm{d}x + \left|\int_\Omega n\varphi\,\mathrm{d}x\right| - CM^{1+\nu}, \tag{3.5}$$

where $\nu = 2\gamma/(\gamma d + 2 - d) = 4/(d(4-d))$.

Lemma 3.2. *If $d > 2$, $P(s) \ge \varepsilon s^{1+\gamma}$ for some $\varepsilon > 0$, $\gamma > 1 - 2/d$ and all $s \ge 0$, then the counterpart of the estimate (3.3) reads*

$$\left|\int_\Omega n\varphi\,\mathrm{d}x\right| \le C|n|_{1+\gamma}^{2r}|n|_1^{2-2r} \le c|n|_{1+\gamma}^{1+\gamma}\Theta^{1-\gamma d/2} + CM^{1+\nu}\Theta^{-\kappa} \tag{3.6}$$

with any $c > 0$, any parameter $\Theta > 0$ and some constant $C = C(c, \Omega)$ independent of Θ, $\nu = 2\gamma/(\gamma d + 2 - d)$, $\kappa = (1 - \gamma d/2)(d-2)/(\gamma d + 2 - d)$.
Now, for each $\delta > 0$ the estimate corresponding to (3.4) is

$$E + CM^{1+\nu}\vartheta^{-\kappa} \ge \left(\varepsilon\frac{d}{2} - \delta\right)\int_\Omega n^{1+\gamma}\vartheta^{1-\gamma d/2}\,\mathrm{d}x. \tag{3.7}$$

Moreover, for each $0 < c_0 < d/2$ the following is true

$$E \ge c_0\int_\Omega p\,\mathrm{d}x + \left|\int_\Omega n\varphi\,\mathrm{d}x\right| - CM^{1+\nu}\vartheta^{-\kappa}. \tag{3.8}$$

If $\gamma = 1 - 2/d$ such an estimate is meaningful for small values of mass M only, namely the following estimate holds

$$E \ge \left(\frac{d}{2} - c_0 M^{2/d}\right)\int_\Omega p\,\mathrm{d}x + \left|\int_\Omega n\varphi\,\mathrm{d}x\right| - C. \tag{3.9}$$

Since $\gamma \le 2/d$ in physical situations, (3.8) will be of main interest for $d = 3$ and (3.9) for $d = 4$.

The proofs of the above lemmas, similar to [7, Lemma 3.3], [8, Lemma 3.2], involve standard Sobolev–Gagliardo–Nirenberg and Hölder inequalities.

Remark 3.3. The results in Lemmas 3.1 and 3.2 apply to the case of polytropic density-pressure relations and the Fermi–Dirac model: $\gamma = 2/d$, $d \le 3$, arbitrary

$M > 0$, and $d = 4$, small $M > 0$, see [7, (29)]. They are also valid (with different proofs) for the Maxwell–Boltzmann case: $d = 2$, $\gamma = 0$ and small M, see [1, 6].

Remark 3.4. From Lemma 3.2 follows, by the Jensen inequality, the estimate

$$E + CM^{1+\nu}\vartheta^{-\kappa} \geq \beta M^{1+\gamma}|\Omega|^{-\gamma}\vartheta^{1-\gamma d/2}$$

with a $\beta > 0$, which readily implies an *a priori* upper bound for the temperature ϑ. Indeed, since $\gamma < 2/d$, the right-hand side increases to ∞ when $\vartheta \to \infty$ while the left-hand side is bounded.

4. Results on the existence of solutions

The first step of the construction of local in time solutions of the nonisothermal evolution problem consists in the analysis of the isothermal problem with $\vartheta =$ const. This has been done for the Maxwell–Boltzmann and Fermi–Dirac cases in [4] and [3, Sec. 3], respectively. In the latter publication, solutions have been constructed as limits of solutions of approximated parabolic problems where the Poisson equation has been replaced by the penalized parabolic equation

$$\frac{1}{k}\varphi_t - \Delta\varphi + u = 0, \quad k > 0. \tag{4.1}$$

Then, the approach via energy estimates and the passage to the limit $k \to \infty$ has led to the solution of the original parabolic-elliptic system. The next step is an analysis of the problem with a given (continuous) temperature $\vartheta = \vartheta(t) \in (0, \infty)$, $t \in [0, T]$, see, e.g., [10] and for a slightly another approach [6]. The final step of the construction of solutions for the nonisothermal evolution problem is to look for a fixed point of the operator

$$\mathcal{T} : \vartheta \longmapsto \Theta. \tag{4.2}$$

Here ϑ, $\Theta \in \mathcal{C}[0, T]$, $\Theta(t)$ is an instantaneous temperature determined by an energy relation involving both temperatures: the old one $\vartheta(t)$ (appearing implicitly in n, φ that solve (1.1)–(1.2)) and the new one $\Theta(t)$ (appearing in the pressure term, cf. (4.3) below). Usually, this needs Schauder type arguments, so one should prove invariance (i.e., *a priori* estimates on ϑ) and compactness properties (e.g., a bound on $\dot{\Theta}$) of the operator $\mathcal{T}$. Here, we present the *a priori* estimates on ϑ. The other details of the construction will appear in a more comprehensive forthcoming paper. Evidently, assumptions guaranteeing (global in time) existence of solutions of the evolution problem are, in a sense, complementary to those implying finite time blow up for $D_* = 1$, e.g., assumptions on negativity of E, cf. [8, Sec. 5].

The theorem below covers the case of three-dimensional nonisothermal Fermi–Dirac model.

Theorem 4.1. *Assume that the convex function $P \in \mathcal{C}^1$ satisfies*

(i) $P(s)/s^{1+2/d} \searrow \varepsilon > 0$,
(ii) $\liminf_{s\searrow 0} P(s)/s > 0$.

Moreover, let the data satisfy the condition $W(0) < 0$ *and*

(iii) $W(0)/M < \liminf_{s\to\infty} R(s)/s \equiv \ell$,

where $R(s) = sH(s) - \left(\frac{d}{2}+1\right)P(s)$ *is the entropy density in* (3.2). *Then the temperature* ϑ *for local in time solution satisfies the a priori estimate* $a \le \vartheta \le b$ *for some constants* $0 < a < b < \infty$.

Proof. For $\vartheta \in \mathcal{C}([0,T];(0,\infty))$ we consider the map $\Theta := \mathcal{T}(\vartheta)$ as the new temperature defined for each moment $t \in [0,T]$ by the energy relation

$$E = \frac{d}{2}\int_\Omega \Theta^{1+d/2}P(n\Theta^{-d/2})\,\mathrm{d}x + \frac{1}{2}\int_\Omega n\varphi\,\mathrm{d}x. \tag{4.3}$$

Thanks to the assumption (i) the operator $\mathcal{T}$ is well defined by the argument in Remark 2.1. First, we prove that $\mathcal{T}(\vartheta)$ is bounded from above uniformly in ϑ. Using the lower estimate of the pressure (ii), i.e., $P(s) \ge c_1 s$ for some $c_1 > 0$, and the estimate of the potential energy (3.8), the upper bound for the new temperature Θ follows. Namely,

$$\frac{3}{2}E + \frac{C}{2}M^{1+\nu} \ge E - \frac{1}{2}\int_\Omega n\varphi\,\mathrm{d}x = \frac{d}{2}\int_\Omega \Theta^{1+d/2}P(n\Theta^{-d/2})\,\mathrm{d}x \ge \frac{d}{2}\Theta M c_1$$

holds, and thus we arrive at $\Theta \le b := (dc_1)^{-1}\left(\frac{3E}{M} + CM^\nu\right)$.

Now, we prove the second part of the claim, i.e., we estimate the temperature $\vartheta = \mathcal{T}(\vartheta)$ from below for initial data satisfying (iii), i.e., $(1-\delta)W(0)/M < \ell$ for some $\delta > 0$. Taking $K = K(\delta)$ sufficiently large to have $R(s)/s > (1-\delta)W(0)/M$ for all $s \ge K$ we obtain

$$\begin{aligned} 0 > W(0) \ge W(t) &= \left(\int_{\{\frac{n}{\vartheta^{d/2}}\ge K\}} + \int_{\{\frac{n}{\vartheta^{d/2}}<K\}}\right)\vartheta^{d/2}R\left(\frac{n}{\vartheta^{d/2}}\right)\mathrm{d}x \\ &\ge M(1-\delta)W(0)/M - C(\delta)|\Omega|\vartheta^{d/2} \end{aligned}$$

with $-C(\delta) \le \inf_{0\le s\le K} R(s)$ for some $C(\delta) > 0$. This leads to the required bound for ϑ: $\vartheta \ge a > 0$, since $\vartheta^{d/2} > -\delta W(0)/(C(\delta)|\Omega|) > 0$.

Observe that if $\ell = \lim_{s\to\infty} R(s)/s$, then $\ell = -\frac{d}{2}\lim_{s\to\infty} s^{1+2/d}\left(s^{-2/d}H(s)\right)'$. Moreover, the condition $W(0) < 0$ is satisfied for initial data if $M^{2/d} \gg \vartheta_0$ holds.

Acknowledgment

The preparation of this paper was partially supported by the KBN (MNI) grant 2/P03A/002/24.

References

[1] P. Biler, J. Dolbeault, M.J. Esteban, P.A. Markowich, T. Nadzieja, *Steady states for Streater's energy-transport models of self-gravitating particles.* "Transport in Transition Regimes", N. Ben Abdallah, A. Arnold, P. Degond et al., eds., Springer IMA Volumes in Mathematics and Its Applications **135**, New York: Springer (2004), 37–56.

[2] P. Biler, A. Krzywicki, T. Nadzieja, *Self-interaction of Brownian particles coupled with thermodynamic processes.* Rep. Math. Phys. **42** (1998), 359–372.

[3] P. Biler, Ph. Laurençot, T. Nadzieja, *On an evolution system describing self-gravitating Fermi–Dirac particles.* Adv. Diff. Eq. **9** (2004), 563–586.

[4] P. Biler, T. Nadzieja, *Existence and nonexistence of solutions for a model of gravitational interaction of particles, I.* Colloq. Math. **66** (1994), 319–334.

[5] P. Biler, T. Nadzieja, *Structure of steady states for Streater's energy-transport models of gravitating particles.* Topol. Methods Nonlinear Anal. **19** (2002), 283–301.

[6] P. Biler, T. Nadzieja, *Global and exploding solutions in a model of self-gravitating systems.* Rep. Math. Phys. **52** (2003), 205–225.

[7] P. Biler, T. Nadzieja, R. Stańczy, *Nonisothermal systems of self-interacting Fermi–Dirac particles.* "Nonlocal Elliptic and Parabolic Problems", P. Biler, G. Karch, T. Nadzieja, eds., Banach Center Publications **66**, Warsaw: Polish Acad. Sci. (2004), 61–78.

[8] P. Biler, R. Stańczy, *Parabolic-elliptic systems with general density-pressure relations.* "Variational Problems and Related Topics", RIMS, Kyoto, M. Misawa, T. Suzuki, eds., Sūrikaisekikenkyūsho Kōkyūroku **1405** (2004), 31–53.

[9] P.-H. Chavanis, *Generalized kinetic equations and effective thermodynamics.* "Nonlocal Elliptic and Parabolic Problems", P. Biler, G. Karch, T. Nadzieja, eds., Banach Center Publications **66**, Warsaw: Polish Acad. Sci. (2004), 79–101.

[10] C.J. van Duijn, I.A. Guerra, M.A. Peletier, *Global existence conditions for a non-local problem arising in statistical mechanics.* Adv. Diff. Eq. **9** (2004), 133–158.

[11] I.A. Guerra, T. Nadzieja, *Convergence to stationary solutions in a model of self-gravitating systems.* Colloq. Math. **98** (2003), 42–47.

[12] D. Horstmann, *From 1970 until present: The Keller–Segel model in chemotaxis and its consequences I, II.* Jahresber. Deutsch. Math.-Verein. **105** (2003), 103–165, and **106** (2004), 51–69.

[13] A. Raczyński, *On a nonlocal elliptic problem.* Appl. Math. (Warsaw) **26** (1999), 107–119.

[14] A. Raczyński, *Steady states for polytropic equation of self-gravitating gas.* Math. Meth. Appl. Sci. **28** (2005), 1881–1896.

[15] R. Stańczy, *Steady states for a system describing self-gravitating Fermi–Dirac particles.* Diff. Int. Eq. **18** (2005), 567–582.

[16] R. Stańczy, *Self-attracting Fermi–Dirac particles in canonical and microcanonical setting.* Math. Meth. Appl. Sci. **28** (2005), 975–990.

[17] Y. Sugiyama, *Global existence in the supercritical cases and finite time blow-up in the subcritical cases to degenerate Keller–Segel systems.* 1–21, preprint.

[18] G. Wolansky, *On steady distributions of self-attracting clusters under friction and fluctuations*, Arch. Rational Mech. Anal. **119** (1992), 355–391.

Piotr Biler and Robert Stańczy
Mathematical Institute, University of Wrocław
pl. Grunwaldzki 2/4, PL-50-384 Wrocław, Poland
e-mail: biler@math.uni.wroc.pl, stanczr@math.uni.wroc.pl

International Series of Numerical Mathematics, Vol. 154, 117–124

Existence, Uniqueness and an Explicit Solution for a One-Phase Stefan Problem for a Non-classical Heat Equation

Adriana C. Briozzo and Domingo A. Tarzia

Abstract. Existence and uniqueness, local in time, of the solution of a one-phase Stefan problem for a non-classical heat equation for a semi-infinite material is obtained by using the Friedman-Rubinstein integral representation method through an equivalent system of two Volterra integral equations. Moreover, an explicit solution of a similarity type is presented for a non-classical heat source depending on time and heat flux on the fixed face $x = 0$.

Mathematics Subject Classification (2000). Primary 35R35, 80A22, 35C05; Secondary 35K20, 35K55, 45G15, 35C15.

Keywords. Stefan problem, Non-classical heat equation, Free boundary problems, Similarity solution, Nonlinear heat sources, System of Volterra integral equations.

1. Introduction

The one-phase Stefan problem for a semi-infinite material is a free boundary problem for the classical heat equation which requires the determination of the temperature distribution u of the liquid phase (melting problem) or of the solid phase (solidification problem), and the evolution of the free boundary $x = s(t)$. Phase-change problems appear frequently in industrial processes and other problems of technological interest [2, 4, 6, 9, 12]. A large bibliography on the subject was given in [20].

Non-classical heat conduction problem for a semi-infinite material was studied in [3, 5, 10, 22, 23], e.g., problems of the type

$$\begin{array}{ll} u_t - u_{xx} = -F(u_x(0,t)), & x > 0,\ t > 0, \\ u(0,t) = 0, & t > 0 \\ u(x,0) = h(x), & x > 0 \end{array} \tag{1.1}$$

It was supported by CONICET PIP No. 3579 and ANPCYT PICT No. 03-11165.

where $h(x), x > 0$, and $F(V), V \in \mathbb{R}$, are continuous functions. The function F, henceforth referred as control function, is assumed to fulfill the following condition

$$(H1) \qquad F(0) = 0.$$

As it was observed in [22, 23] the heat flux $w(x,t) = u_x(x,t)$ for problem (1.1) satisfies a classical heat conduction problem with a nonlinear convective condition at $x = 0$, which can be written in the form

$$\begin{cases} w_t - w_{xx} = 0, & x > 0,\ t > 0, \\ w_x(0,t) = F(w(0,t)), & t > 0, \\ w(x,0) = h^{'}(x) \geq 0, & x > 0. \end{cases} \tag{1.2}$$

The literature concerning problem (1.2) has constantly increased from the appearance of the papers [13, 15, 17]. In [21] a one-phase Stefan problem for a non-classical heat equation for a semi-infinite material was presented. The free boundary problem consists in determining the temperature $u = u(x,t)$ and the free boundary $x = s(t)$ with a control function F which depends on the evolution of the heat flux at the boundary $x = 0$, satisfying the following conditions

$$\begin{cases} u_t - u_{xx} = -F(u_x(0,t)), & 0 < x < s(t), 0 < t < T, \\ u(0,t) = f(t) \geq 0, & 0 < t < T, \\ u(s(t),t) = 0, u_x(s(t),t) = -\dot{s}(t), & 0 < t < T, \\ u(x,0) = h(x) \geq 0, & 0 \leq x \leq b = s(0)\ (b > 0). \end{cases} \tag{1.3}$$

In Section 2 we present a result on the local existence and uniqueness in time of the solution of the one-phase Stefan problem (1.3) for a non-classical heat equation with temperature boundary condition at the fixed face $x = 0$. First, we prove that the free boundary problem (1.3) is equivalent to a system of two Volterra integral equations (2.4)–(2.5) [8, 14] following the Friedman-Rubinstein's method given in [7, 18](see also [19]). Then, we prove that the problem (2.4)–(2.5) has a unique local solution in time by using the Banach contraction theorem.

In Section 3 we show an explicit solution of a similarity type for a one-phase Stefan problem for a non classical control function F which depends on time and heat flux on the fixed face $x = 0$.

2. Existence and uniqueness of the non-classical free boundary problem

We have the following equivalence:

Theorem 2.1. *The solution of the free boundary problem* (1.3) *is given by*

$$\begin{aligned} u(x,t) \quad &= \quad \int_0^b G(x,t;\xi,0)h(\xi)d\xi + \int_0^t G_\xi(x,t;0,\tau)f(\tau)d\tau \\ &+ \int_0^t G(x,t;s(\tau),\tau)v(\tau)d\tau - \iint_{D(t)} G(x,t;\xi,\tau)F(V(\tau))d\xi d\tau\ , \end{aligned} \tag{2.1}$$

$$s(t) = b - \int_0^t v(\tau)d\tau \tag{2.2}$$

where $D(t) = \{(x,\tau)/\ 0 < x < s(\tau), 0 < \tau < t\}$, *with* $f \in C^1[0,T)$, $h \in C^1[0,b]$, $h(b) = 0$, $h(0) = f(0)$, F *is a Lipschitz function over* $C^0[0,T]$, *and the functions* $v \in C^0[0,T]$, $V \in C^0[0,T]$ *defined by*

$$v(t) = u_x(s(t),t) \quad , \quad V(t) = u_x(0,t) \tag{2.3}$$

must satisfy the following system of two Volterra integral equations

$$\begin{aligned} v(t) &= 2\int_0^b N(s(t),t;\xi,0)h'(\xi)d\xi \\ &-2\int_0^t N(s(t),t;0,\tau)\,\dot{f}(\tau)d\tau + 2\int_0^t G_x(s(t),t;s(\tau),\tau)v(\tau)d\tau \\ &+2\int_0^t \left[N(s(t),t;s(\tau),\tau) - N(s(t),t;0,\tau)\right]F(V(\tau))d\tau\ , \end{aligned} \tag{2.4}$$

$$\begin{aligned} V(t) &= \int_0^b N(0,t;\xi,0)h'(\xi)d\xi \\ &-\int_0^t N(0,t;0,\tau)\,\dot{f}(\tau)d\tau + \int_0^t G_x(0,t;s(\tau),\tau)v(\tau)d\tau \\ &+\int_0^t \left[N(0,t;s(\tau),\tau) - N(0,t;0,\tau)\right]F(V(\tau))d\tau\ , \end{aligned} \tag{2.5}$$

where G, N *are the Green and Neumann functions and* K *is the fundamental solution of the heat equation, defined respectively by*

$$G(x,t,\xi,\tau) = K(x,t,\xi,\tau) - K(-x,t,\xi,\tau) \tag{2.6}$$

$$N(x,t,\xi,\tau) = K(x,t,\xi,\tau) + K(-x,t,\xi,\tau) \tag{2.7}$$

$$K(x,t,\xi,\tau) = \begin{cases} \frac{1}{2\sqrt{\pi(t-\tau)}}\exp\left(-\frac{(x-\xi)^2}{4(t-\tau)}\right) & t > \tau \\ 0 & t \le \tau \end{cases} \tag{2.8}$$

where $s(t)$ *is given by* (2.2).

In order to prove the local existence and uniqueness of solution v, $V \in C^0[0,\sigma]$ (σ is a positive small number) to the system of two Volterra integral equations (2.4)–(2.5) we will use the Banach fixed point theorem. Let us define the Banach space:

$$C_{M,\sigma} = \left\{ \overrightarrow{w} = \begin{pmatrix} v \\ V \end{pmatrix} / \ v, V : [0,\sigma] \to \mathbb{R}, \text{continuous, with } \left\|\overrightarrow{w}\right\|_\sigma \le M \right\}$$

with the norm

$$\left\|\overrightarrow{w}\right\|_\sigma := \|v\|_\sigma + \|V\|_\sigma := \max_{t\in[0,\sigma]} |v(t)| + \max_{t\in[0,\sigma]} |V(t)| . \tag{2.9}$$

We define the map $A : C_{M,\sigma} \longrightarrow C_{M,\sigma}$, such that

$$\overrightarrow{\widetilde{w}}\ (t) = A\left(\overrightarrow{w}\ (t)\right) = \begin{pmatrix} A_1(v(t), V(t)) \\ A_2(v(t), V(t)) \end{pmatrix} \tag{2.10}$$

where

$$A_1(v(t), V(t)) = F_1(v(t)) + 2\int_0^t \left[N(s(t), t, s(\tau), \tau) - N(s(t), t, 0, \tau)\right] F(V(\tau))d\tau \tag{2.11}$$

with

$$F_1(v(t)) = 2\int_0^b N(s(t), t, \xi, 0)h^{'}(\xi)d\xi - 2\int_0^t N(s(t), t, 0, \tau)\ \dot{f}\ (\tau)d\tau$$
$$+2\int_0^t G_x(s(t), t, s(\tau), \tau)v(\tau)d\tau$$

and

$$A_2(v(t), V(t)) = F_2(v(t)) + \int_0^t \left[N(0, t, s(\tau), \tau) - N(0, t, 0, \tau)\right] F(V(\tau))d\tau. \tag{2.12}$$

with

$$F_2(v(t)) = \int_0^b N(0, t, \xi, 0)h^{'}(\xi)d\xi - \int_0^t N(0, t, 0, \tau)\ \dot{f}\ (\tau)d\tau \tag{2.13}$$
$$+\int_0^t G_x(0, t, s(\tau), \tau)v(\tau)d\tau$$

Then we have the following property:

Theorem 2.2. *If $f \in C^1[0, T]$, $h \in C^1[0, b]$, $f(0) = h(0)$, $h(b) = 0$ and F is a Lipschitz function over $C^0[0, T]$, then the map $A : C_{M,\sigma} \longrightarrow C_{M,\sigma}$ is well defined and is a contraction map if $\sigma > 0$ is small enough. Then there exists an unique solution on $C_{M,\sigma}$ to the system of integral equations* (2.4), (2.5).

3. Explicit solution of a one-phase Stefan problem for a non-classical heat equation

Now, we consider a free boundary problem which consists in determining the temperature $u = u(x, t)$ and the free boundary $x = s(t)$ with a control function F which depends on time and the evolution of the heat flux at the boundary $x = 0$, satisfying the following conditions

$$\rho c u_t - k u_{xx} = -\gamma F(u_x(0, t), t)\ ,\ 0 < x < s(t)\ ,\ t > 0, \tag{3.1}$$

$$u(0, t) = f = \text{Const.} > 0,\ t > 0, \tag{3.2}$$

$$u(s(t), t) = 0,\ \ k u_x(s(t), t) = -\rho l\ \dot{s}\ (t)\ ,\ \ t > 0, \tag{3.3}$$

$$s(0) = 0, \tag{3.4}$$

where the thermal coefficients $k,\ \rho,\ c,\ l,\ \gamma > 0$ and the control function F is given by the expression

$$F(V,t) = \frac{\lambda_0}{\sqrt{t}} V \ (\lambda_0 > 0)\,. \tag{3.5}$$

In order to obtain an explicit solution of a similarity type, we define

$$\Phi(\eta) = u(x,t),\ \eta = \frac{x}{2a\sqrt{t}} \tag{3.6}$$

where $a^2 = k/\rho c$ is the diffusion coefficient.

After some elementary computations we obtain

$$\Phi(\eta) = f\left[1 - \frac{E(\eta)}{E(\eta_0)}\right]\ ,\ 0 < \eta < \eta_0, \tag{3.7}$$

where

$$E(x) = erf\,(x) + \frac{4\lambda}{\sqrt{\pi}} \int_0^x f_1(r)dr,\ \ \lambda = \frac{\gamma\lambda_0}{\rho c a} > 0\ ,\ [\lambda] = 1 \tag{3.8}$$

and

$$f_1(x) = \exp(-x^2) \int_0^x \exp(r^2)dr \tag{3.9}$$

is Dawson's integral [1] and η_0 is an unknown positive parameter to be determined which characterizes the free boundary given by

$$s(t) = 2a\eta_0\sqrt{t}. \tag{3.10}$$

We remark that Dawson's integral also appears in the explicit solution for the supercooled one-phase Stefan problem with a constant temperature boundary condition on the fixed face [16].

Taking into account the Stefan condition we have that $\eta_0 = \eta_0(\lambda, Ste)$ must be the solution of the following equation

$$\frac{Ste}{\sqrt{\pi}}[\exp(-x^2) + 2\lambda f_1(x)] = x[erf(x) + \frac{4\lambda}{\sqrt{\pi}} \int_0^x f_1(z)dz]\ \ ,\ \ x > 0 \tag{3.11}$$

where $Ste = \dfrac{fc}{l} > 0$ is the Stefan number and

$$erf(x) = \frac{2}{\sqrt{\pi}} \int_0^x \exp(-z^2)dz. \tag{3.12}$$

The equation (3.11) is equivalent to the equation

$$W_1(x) = 2\lambda W_2(x)\ ,\ x > 0 \tag{3.13}$$

where functions W_1 and W_2 are defined by

$$W_1(x) = Ste\ \exp\left(-x^2\right) - \sqrt{\pi} erf(x)x \tag{3.14}$$

$$W_2(x) = 2x \int_0^x f_1(r)dr - Ste\ f_1(x). \tag{3.15}$$

Remark 3.1. If $\lambda = 0$ (that is $\gamma = 0$) in the free boundary problem (3.1)–(3.4) we obtain the classical Lamé-Clapeyron [11] solution and there exists a unique solution η_{00} of the equation (3.11) which is given now by

$$F_0(x) = \frac{Ste}{\sqrt{\pi}} \ , \ x > 0 \tag{3.16}$$

where

$$F_0(x) = erf(x)\exp(x^2)x \ . \tag{3.17}$$

Theorem 3.2. *For each $\lambda > 0$ there exists a unique solution η_0 of Eq.* (3.13). *This solution $\eta_0 = \eta_0(\lambda)$ has the following properties*

(i) $\eta_0(0^+) = \eta_{00} > 0$

(ii) $\eta_0(+\infty) = x_4 < +\infty$ (3.18)

(iii) $\eta_0 = \eta_0(\lambda)$ *is an increasing function on* λ

where η_{00} is the unique solution of Equation (3.16) *and $x_4 > 0$ is the unique positive zero of W_2.*

Theorem 3.3. *For each $\lambda > 0$ the free boundary problem* (3.1)–(3.4) *has a unique similarity solution of the type*

$$u(x,t,\lambda) = f\left[1 - \frac{E(n,\lambda)}{E(\eta_0(\lambda),\lambda)}\right] \ , \quad 0 < \eta = \frac{x}{2a\sqrt{t}} < \eta_0(\lambda) \tag{3.19}$$

$$s(t,\lambda) = 2a\,\eta_0(\lambda)\sqrt{t} \tag{3.20}$$

where

$$E(\eta,\lambda) = erf(\eta) + \frac{4\lambda}{\sqrt{\pi}}\int_0^\eta f_1(r)dr \tag{3.21}$$

and $\eta_0 = \eta_0(\lambda)$ is the unique solution of Eq. (3.13), *with $\eta_{00} < \eta_0(\lambda) < x_4$.*

Theorem 3.4. *The explicit solution* (3.19), (3.20) *of the problem* (3.1)–(3.4) *has the following properties:*

(i) $u_x(0,t,\lambda) = \dfrac{-f}{aE(\eta_0(\lambda),\lambda)}\dfrac{1}{\sqrt{\pi t}} < 0 \ , \ \forall t > 0$

(ii) $u(x,t,\lambda) \geq u_0(x,t) \ , \ \forall\, 0 \leq x \leq s_0(t) \ , \ t > 0$

(iii) $s(t,\lambda) \geq s_0(t) \ , \ \forall\, t > 0$

where $u_0(x,t) = f\left[1 - \dfrac{erf(\eta)}{erf(\eta_{00})}\right] \ , \ 0 < \eta = \dfrac{x}{2a\sqrt{t}} < \eta_{00} \ , \ t > 0$

$s_0(t) = s(t,0) = 2a\eta_{00}\sqrt{t}$

(iv) $1 \leq \dfrac{u(x,t,\lambda)}{u_0(x,t)} \leq \dfrac{1}{1 - \dfrac{\eta(x,t)}{\eta_{00}}}\left[1 - \dfrac{2}{Ste}\dfrac{\eta_0(\lambda)\left(1 + 2\lambda\left\|f_1\right\|_\infty\right)}{\exp\left(-\eta_0^2(\lambda)\right) + 2\lambda f_1\left(\eta_0(\lambda)\right)}\,\eta(x,t)\right]$

(v) $\lim\limits_{t\to+\infty}\dfrac{u(x,t,\lambda)}{u_0(x,t)} = 1$ *uniformly* $\forall\, x \in$ *compact sets* $\subset [0, s_0(t))$.

References

[1] M. Abramowitz, I.E. Stegun (Eds.), *Handbook of Mathematical Functions with Formulas, Graphs, and Mathematical Tables,* National Bureau of Standards, Washington (1972).

[2] V. Alexiades, A.D. Solomon, *Mathematical modeling of melting and freezing processes,* Hemisphere – Taylor & Francis, Washington (1983).

[3] L.R. Berrone, D.A. Tarzia and L.T. Villa, *Asymptotic behavior of a non-classical heat conduction problem for a semi-infinite material*, Math. Meth. Appl. Sci., 23 (2000), 1161–1177.

[4] J.R. Cannon, *The one-dimensional heat equation*, Addison-Wesley, Menlo Park (1984).

[5] J.R. Cannon, H.M. Yin, *A class of non-linear non-classical parabolic equations*, J. Diff. Eq., 79 (1989), 266–288.

[6] J. Crank, *Free and moving boundary problems,* Clarendon Press Oxford (1984).

[7] A. Friedman, *Free boundary problems for parabolic equations I. Melting of solids*, J. Math. Mech., 8 (1959), 499–517.

[8] G. Gripenberg, S.O. Londen and O. Staffans, *Volterra integral and functional equations*, Cambridge Univ. Press, Cambridge (1990).

[9] S.C. Gupta, *The classical Stefan problem. Basic concepts, modelling and analysis*, Elsevier, Amsterdam (2003).

[10] N. Kenmochi, M. Primicerio, *One-dimensional heat conduction with a class of automatic heat source controls*, IMA J. Appl. Math. 40 (1988), 205–216.

[11] G. Lamé, B.P. Clapeyron, *Mémoire sur la solidification par refroidissement d'un globe liquide*, Annales chimie Physique 47 (1831), 250–256.

[12] V.J. Lunardini, *Heat transfer with freezing and thawing,* Elsevier, Amsterdam (1991).

[13] W.R. Mann, F. Wolf, *Heat transfer between solid and gases under nonlinear boundary conditions*, Quart. Appl. Math., 9 (1951), 163–184.

[14] R.K. Miller, *Nonlinear Volterra integral equations*, W. A. Benjamin, Menlo Park (1971).

[15] W.E. Olmstead, R.A. Handelsman, *Diffusion in a semi-infinite region with nonlinear surface dissipation*, SIAM Rev., 18 (1976), 275–291.

[16] A. Petrova, D.A. Tarzia and C.V. Turner, *The one-phase supercooled Stefan problem with temperature boundary condition*, Adv. Math. Sci. and Appl., 4 (1994), 35–50.

[17] J.H. Roberts, W.R. Mann, *A certain nonlinear integral equation of the Volterra type*, Pacific J. Math., 1 (1951), 431–445.

[18] L.I. Rubinstein, *The Stefan problem*, Trans. Math. Monographs # 27, Amer. Math. Soc., Providence (1971).

[19] B. Sherman, *A free boundary problem for the heat equation with prescribed flux at both fixed face and melting interface*, Quart. Appl. Math., 25 (1967), 53–63.

[20] D.A. Tarzia, *A bibliography on moving-free boundary problems for the heat-diffusion equation. The Stefan and related problems*, MAT-Serie A, Rosario, # 2 (2000), (with 5869 titles on the subject, 300 pages).
See www.austral.edu.ar/MAT-SerieA/2(2000)/

[21] D.A. Tarzia, *A Stefan problem for a non-classical heat equation,* MAT-Serie A, Rosario, # 3 (2001), 21–26.

[22] D.A. Tarzia, L.T. Villa, *Some nonlinear heat conduction problems for a semi-infinite strip with a non-uniform heat source*, Rev. Un. Mat. Argentina, 41 (2000), 99–114.

[23] L.T. Villa, *Problemas de control para una ecuación unidimensional del calor*, Rev. Un. Mat. Argentina, 32 (1986), 163–169.

Adriana C. Briozzo
Depto. Matemática
FCE, Univ. Austral
Paraguay 1950, S2000FZF Rosario, Argentina
e-mail: `Adriana.Briozzo@fce.austral.edu.ar`

Domingo A. Tarzia
Depto. Matemática – CONICET
FCE, Univ. Austral
Paraguay 1950, S2000FZF Rosario, Argentina
e-mail: `Domingo.Tarzia@fce.austral.edu.ar`

International Series of Numerical Mathematics, Vol. 154, 125–135

Dislocation Dynamics: a Non-local Moving Boundary

P. Cardaliaguet, F. Da Lio, N. Forcadel and R. Monneau

Abstract. In this article, we present briefly the mathematical study of the dynamics of line defects called dislocations, in crystals. The mathematical model is an eikonal equation describing the motion of the dislocation line with a velocity which is a non-local function of the whole shape of the dislocation. We present some partial existence and uniqueness results. Finally we also show that the self-dynamics of a dislocation line at large scale is asymptotically described by an anisotropic mean curvature motion.

Mathematics Subject Classification (2000). Primary 35F25; Secondary 35Q99.

Keywords. Dislocations dynamics, non-local equations, viscosity solutions.

1. Introduction

1.1. What are dislocations?

The crystal defects called dislocations are lines whose typical length in metallic alloys is of the order of $10^{-6}m$, with thickness of the order of $10^{-9}m$ (see Figure 1 for an example of observations of dislocations by electron microscopy).

In the face centered cubic structure, dislocations move at low temperature in well defined crystallographic planes (the slip planes), at velocities of the order of 10 ms^{-1}. We refer for instance to Hirth and Lothe [17] for a description at the atomic level of these dislocations.

The concept of dislocations has been introduced and developed in the XXth century, as the main microscopic explanation of the macroscopic plastic behavior of metallic crystals (see for instance the physical monographs Nabarro [20], Hirth and Lothe [17], or Lardner [19] for a mathematical presentation). Since the beginning of the 90's, the research field of dislocations has enjoyed a new boom based on the increasing power of computers, allowing simulations with a large number of dislocations (see for instance Kubin et al. [18]). This simultaneously motivated new theoretical developments for the modelling of dislocations. Recently Rodney,

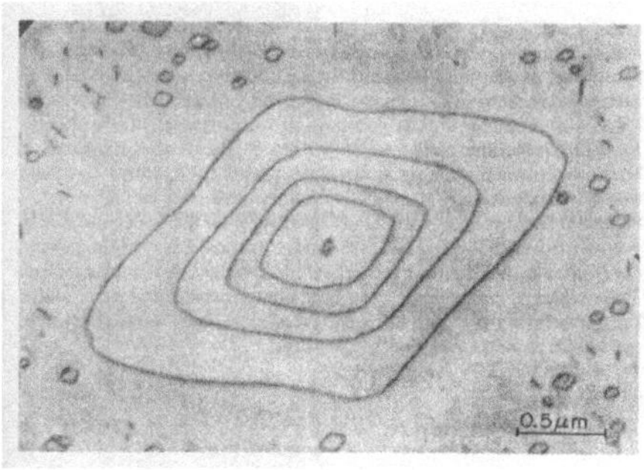

FIGURE 1. Dislocations in a Al-Mg alloy (from [23])

Le Bouar and Finel introduced in [21] a new model that we present and study mathematically in this paper. We also refer the reader to [6] and the references therein for a more detailed introduction to dislocations dynamics. This model has also been numerically studied by Alvarez, Carlini, Monneau and Rouy in [3] and [4]; see also Alvarez, Carlini, Hoch, Le Bouar and Monneau [2]

1.2. Mathematical modelling of dislocations dynamics

An idealization consists in assuming that the thickness of these lines is zero, and in the case of a single line, in assuming that this line is contained and moves in the $x = (x_1, x_2)$ plane. The motion of the line Γ_t (where the subscript t denotes the time) is simply given by the normal velocity c (see Figure 2).

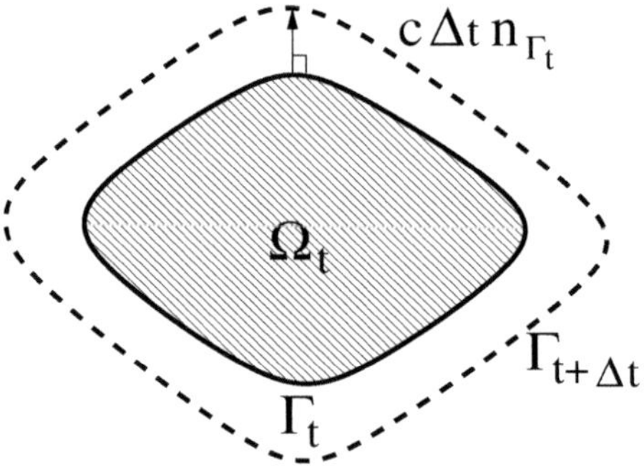

FIGURE 2. Schematic evolution of a dislocation line Γ_t by normal velocity c between the times t and $t + \Delta t$ with unit normal n_{Γ_t}.

The velocity c is proportional to the shear stress in the material. This stress can be computed solving the equations of linearized elasticity where the shape of the dislocation line appears as a source term. This gives a coupled system where the dislocation line evolution is a function of the velocity c, and the velocity c is a function of the dislocation line Γ_t itself. In the case of a single dislocation line it is possible to write the velocity c as a non-local quantity depending on the whole shape of the dislocation line (see Alvarez et al. [6]):

$$c(x,t) = \left(c_0 \star \rho(\cdot,t)\right)(x) + c_1(x,t)$$

where ρ is the characteristic function of an open set $\Omega_t \subset \mathbb{R}^2$ whose the boundary is the dislocation line $\Gamma_t = \partial\Omega_t$:

$$\rho(x,t) = 1_{\Omega_t} := \begin{cases} 1 & \text{if} \quad x \in \Omega_t \\ 0 & \text{if} \quad x \in \mathbb{R}^2 \backslash \Omega_t, \end{cases}$$

and $c_0(x)$ is a given kernel depending on the material. Here the convolution is only done in space on $\mathbb{R}^2$.

It can be easily checked (at least formally), that the evolution on the time interval $(0,T)$ of the dislocation line Γ_t is described by the equation of dislocations dynamics:

$$\begin{cases} \dfrac{\partial \rho}{\partial t} = (c_0 \star \rho + c_1)\ |D\rho| \quad \text{on} \quad \mathbb{R}^2 \times (0,T) \\ \rho(\cdot,0) = \rho^0(\cdot) := 1_{\Omega_0} \quad \text{on} \quad \mathbb{R}^2 \end{cases} \tag{1.1}$$

where Ω_0 is an open set whose boundary $\Gamma_0 = \partial\Omega_0$ is the position of the dislocation line at initial time $t = 0$.

In what follows, we will study this equation in the framework of discontinuous viscosity solutions (see Barles [7] for an introduction to this notion). To simplify the presentation we will state results in dimension $n = 2$, assuming smooth (C^∞) regularity of the initial position Γ_0 of the dislocation line, of the kernel c_0, and of the velocity c_1. We also assume the following behavior of the kernel at infinity (for some function g)

$$c_0(x) = \frac{1}{|x|^3}\, g\left(\frac{x}{|x|}\right) \quad \text{for} \quad |x| \geq 1 \tag{1.2}$$

which is a natural assumption for dislocations.

For considerably weakened assumptions and in any dimensions n, we refer the reader to the original articles cited in the references.

1.3. Organization of the paper

Although equation (1.1) seems very simple, general results of existence and uniqueness are unknown up to our knowledge. Technically, the main difficulty comes from the fact that we have no sign conditions on the kernel c_0, and then that there is no inclusion principle for this evolution.

In this paper we present some partial results. In Section 2, we give a short time existence (and uniqueness) result for a smooth initial dislocation loop. In Section 3, we give a long time existence (and uniqueness) result for a smooth initial curve with non-negative velocity. Finally in Section 4, we consider the "monotone case" where the kernel satisfies $c_0 \geq 0$. In this particular case, a Slepčev "level sets" formulation of equation (1.1) is available. In this framework, we show that at large scales, the dislocation dynamics is asymptotically described by an (anisotropic) mean curvature motion related to the behavior of the kernel $c_0(x)$ as $|x| \to +\infty$.

2. Short time existence results in the general case

We will make the following global assumptions on the smooth velocity $c_1(x,t)$ and the smooth kernel $c_0(x,t) := c_0(x)$, for $i = 0, 1$ and some constants M, L_0, L_1:

$$\begin{cases} \text{i)} & |c_i(y,t)| \le M & \forall (y,t) \in \mathbb{R}^2 \times [0,+\infty) \\ \text{ii)} & |c_i(y_2,t) - c_i(y_1,t)| \le L_0|y_2 - y_1| & \forall (y_1,y_2,t) \in \mathbb{R}^2 \times \mathbb{R}^2 \times [0,+\infty) \\ \text{iii)} & |Dc_i(y_2,t) - Dc_i(y_1,t)| \le L_1|y_2 - y_1| & \forall (y_1,y_2,t) \in \mathbb{R}^2 \times \mathbb{R}^2 \times [0,+\infty). \end{cases} \tag{2.1}$$

To state our results, we first need to recall the notion of discontinuous viscosity solution. We recall that for a function ρ locally bounded on $\mathbb{R}^2 \times [0,T)$, the function ρ^* designates its upper-semicontinuous envelope (i.e., the smallest u.s.c. function $\ge \rho$), and the function ρ_* its lower semi-continuous envelope.

Definition 2.1. i) We say that a function $\rho \in C\left([0,T); L^1(\mathbb{R}^2)\right) \cap L^\infty\left(\mathbb{R}^2 \times (0,T)\right)$ is a discontinuous viscosity subsolution (resp. supersolution) of (1.1), if

$$\rho^*(\cdot,0) \le (\rho^0)^* \quad (\text{resp.} \quad \rho_*(\cdot,0) \ge (\rho^0)_*)$$

and for every point $(\overline{x},\overline{t}) \in \mathbb{R}^2 \times (0,T)$ and every test function $\phi \in C^1(\mathbb{R}^2 \times (0,T))$ satisfying

$$\rho^* \le \phi \quad (\text{resp.} \quad \rho_* \ge \phi) \quad \text{in} \quad \mathbb{R}^2 \times (0,T) \quad \text{and} \quad \rho^*(\overline{x},\overline{t}) = \phi(\overline{x},\overline{t}),$$

we have with $c = c_0 \star \rho + c_1$:

$$\frac{\partial \phi}{\partial t}(\overline{x},\overline{t}) \le c(\overline{x},\overline{t})|D\phi(\overline{x},\overline{t})| \quad \left(\text{resp.} \ \frac{\partial \phi}{\partial t}(\overline{x},\overline{t}) \le c(\overline{x},\overline{t})|D\phi(\overline{x},\overline{t})|\right).$$

ii) We say that ρ is a discontinuous viscosity solution of (1.1), if it is a discontinuous viscosity subsolution and a discontinuous viscosity supersolution.

We are now able to state the first result

Theorem 2.2 (Short time existence and uniqueness, [5], [6]). *Let us assume* (1.2)–(2.1), *and that* Ω_0 *is a smooth bounded open set in* $\mathbb{R}^2$. *Then there exists a time* $T^* > 0$ *and let us consider functions* $\rho \in C\left([0,T^*); L^1(\mathbb{R}^2)\right)$ *with* $0 \le \rho \le 1$, *solutions of equation* (1.1) *on the interval of time* $(0,T^*)$ *with initial data* $\rho(\cdot,0) = 1_{\Omega_0}$. *Then*

i) (existence): *There exists such a solution* ρ.

ii) (uniqueness): *The solution is unique, where the uniqueness has the following meaning: if* ρ_1 *and* ρ_2 *are two such solutions, then* $(\rho_1)^* = (\rho_2)^*$, $(\rho_1)_* = (\rho_2)_*$ *and for every* $t \in [0,T^*)$, $\rho_1(\cdot,t) = \rho_2(\cdot,t)$ *a.e. on* $\mathbb{R}^2$.

Let us remark that on the time interval $(0,T^*)$ where the theorem is proved to hold, the solution can be written $\rho(\cdot,t) = 1_{\Omega_t}$ where Ω_t is a Lipschitz open set. Theorem 2.2 says nothing when Ω_t ceases to be a Lipschitz open set. This is for instance the case when the topology of Ω_t changes.

The proof of Theorem 2.2 is based on the application of a fixed point theorem in the framework of viscosity solutions.

Up to our knowledge, existence and uniqueness for all times (excepted in the case of non-negative velocities (see Theorem 3.1 below)) is still an open problem in general.

3. Long time existence for non-negative velocities

In this section we make the following assumption

$$c_1(y,t) \geq ||c_0||_{L^1(\mathbb{R}^2)} \quad \forall (y,t) \in \mathbb{R}^2 \times [0,+\infty). \tag{3.1}$$

Because we are interested in solutions ρ satisfying $0 \leq \rho \leq 1$, we see that condition (3.1) implies that $c = c_0 \star \rho + c_1 \geq 0$.

Theorem 3.1. [**existence and uniqueness for all time for non-negative velocity,** [1]] *Let us assume* (1.2)–(2.1)–(3.1), *and that* Ω_0 *is a smooth bounded open set in* $\mathbb{R}^2$. *Then there exists a unique function* $\rho \in C\big([0,+\infty); L^1(\mathbb{R}^2)\big)$ *with* $0 \leq \rho \leq 1$, *solution of equation* (1.1) *on the interval of time* $(0,+\infty)$ *with initial data* $\rho(\cdot,0) = 1_{\Omega_0}$.

In [1], Alvarez *et al.* used a geometrical proof. A similar result was also proved by Barles and Ley [8] using a level sets approach and arguments based on a nice L^1 estimate on the level sets of the solution. We also refer to Cardaliaguet, Marchi [11] for a geometrical study of a similar problem on a bounded set in the plane with Neumann boundary conditions. The proof of Theorem 3.1 in [1] uses strongly the following monotonicity formula that we state in any dimension N:

Theorem 3.2 (Monotonicity formula, [1]**).** *Let* K *be a compact subset of* $\mathbb{R}^N$, *and* d_K *the distance to the set* K. *Then for any* $t_2 > t_1 > 0$, *we have*

$$\frac{1}{t_2^{N-1}} \mathcal{H}^{N-1}\left(\{d_K(x) = t_2\}\right) \leq \frac{1}{t_1^{N-1}} \mathcal{H}^{N-1}\left(\{d_K(x) = t_1\}\right).$$

Here $\mathcal{H}^{N-1}$ *stands for the* $(N-1)$*-dimensional Hausdorff measure.*

Main formal arguments in the proof of Theorem 3.1

Argument 1: interior ball condition: Let us call $R(t) > 0$ the radius of the largest ball included in Ω_t and tangent at any point of the boundary $\partial\Omega_t$. Then we can easily check (at least formally) that this radius satisfies the following ODE:

$$\dot{R} = c - R\,(n \cdot Dc) + R^2\left(D^2_{\tau\tau} c\right)$$

where n is the outward unit normal to Ω_t and τ is a tangent unit vector to $\Gamma_t = \partial\Omega_t$. Using the fact that $c \geq 0$, we deduce that

$$R(t) \geq C_1 e^{-\gamma t}$$

for some constants $C_1, \gamma > 0$.

Argument 2: length of the dislocation: We denote by $|\Gamma_t|$ the length of Γ_t. Then using the fact that the curvature K of Γ_t satisfies $K \le 1/R(t)$, we deduce

$$\frac{d}{dt}|\Gamma_t| = \int_{\Gamma_t} cK \le \int_{\Gamma_t} \frac{c}{R(t)} \le \frac{||c||_{L^\infty}}{R(t)}|\Gamma_t|$$

which gives an estimate $|\Gamma_t| \le l(t) < +\infty$.

Argument 3: error estimate: based on the monotonicity formula Theorem 3.2, this is possible to show that if ρ_i satisfy for $i = 1, 2$

$$\begin{cases} \dfrac{\partial \rho_i}{\partial t} = c_i \, |D\rho_i| & \text{on} \quad \mathbb{R}^2 \times (0,T) \\ \rho_i(\cdot, 0) = 1_{\Omega_0} & \text{on} \quad \mathbb{R}^2 \end{cases} \tag{3.2}$$

then we have for any t small enough and some constant $C_2 > 0$:

$$||\rho_2(\cdot,t) - \rho_2(\cdot,t)||_{L^\infty(\mathbb{R}^2)} \le C_2 \; l(t) ||c_2 - c_1||_{L^\infty(\mathbb{R}^2 \times (0,T))} \left(\frac{e^{L_0 t} - 1}{L_0} \right).$$

Combined with the fact that for dislocation dynamics $c_i = c_0 \star \rho_i + c_1$, we get

$$||\rho_2(\cdot,t) - \rho_2(\cdot,t)||_{L^1(\mathbb{R}^2)} \le \alpha(t) ||\rho_2 - \rho_1||_{L^\infty((0,T);L^1(\mathbb{R}^2))}$$

with $\alpha(t) = C_2 \; l(t) ||c_0||_{L^\infty(\mathbb{R}^2)} \left(\frac{e^{L_0 t} - 1}{L_0} \right)$. This shows in particular the uniqueness of the solution for small time, which can also be used as a contraction argument for a fixed point theorem.

4. Convergence to the mean curvature motion at large scale for nonnegative kernels

In this section we assume that the kernel c_0 satisfies the following condition

$$c_0(-x) = c_0(x) \ge 0 \quad \forall x \in \mathbb{R}^2 \tag{4.1}$$

and consider solutions ρ of (1.1) with $c_1 = -\frac{1}{2} \int_{\mathbb{R}^2} c_0$. This particular choice of c_1 insures the equilibrium of straight dislocations lines and is physically relevant for the description at large scales of isolated dislocations lines without exterior stress.

In this section, we are interested in the dynamics of dislocations lines of large diameter of the order of $1/\varepsilon$ and in the limit as $\varepsilon \to 0$. To this end, we define for $\varepsilon > 0$ the rescaled characteristic function

$$\rho^\varepsilon(x,t) = \rho \left(\frac{x}{\varepsilon}, \frac{t}{\varepsilon^2 |\ln \varepsilon|} \right)$$

which satisfies the following equation

$$\frac{\partial \rho^\varepsilon}{\partial t} = \left(c_0^\varepsilon \star \rho^\varepsilon - \frac{1}{2} \int_{\mathbb{R}^2} c_0^\varepsilon \right) |D\rho^\varepsilon| \tag{4.2}$$

with the rescaled kernel

$$c_0^\varepsilon(x) = \frac{1}{\varepsilon^3 |\ln \varepsilon|} \; c_0 \left(\frac{x}{\varepsilon} \right).$$

From the fact that $c_0^\varepsilon \geq 0$, it can be seen (at least formally) that equation (4.2) preserves the inclusion principle. In this section we do not study directly equation (4.2), but prefer to use the following Slepčev "level sets" formulation for a continuous function u^ε:

$$\begin{cases} \dfrac{\partial u^\varepsilon}{\partial t} = \left(\left(c_0^\varepsilon \star 1_{\{u^\varepsilon(\cdot,t) > u^\varepsilon(x,t)\}} \right)(x) - \dfrac{1}{2}\displaystyle\int_{\mathbb{R}^2} c_0^\varepsilon \right) |Du^\varepsilon| \quad \text{on} \quad \mathbb{R}^2 \times (0,T) \\ u^\varepsilon(\cdot,0) = u_0 \quad \text{on} \quad \mathbb{R}^2. \end{cases} \tag{4.3}$$

In this new formulation each level set $\{u^\varepsilon = \lambda\}$ represents a dislocation line associated to a function $\rho_\lambda^\varepsilon = 1_{\{u^\varepsilon > \lambda\}}$ which satisfies (4.2) (at least formally).

In the limit $\varepsilon \to 0$, this dynamics is well approximated by the following anisotropic mean curvature motion (see for instance Crandall, Ishii, Lions [12] for a definition of viscosity solutions of the second order equation (4.4)):

$$\begin{cases} \dfrac{\partial u^0}{\partial t} + F(D^2u^0, Du^0) = 0 \quad \text{on} \quad \mathbb{R}^2 \times (0,T) \\ u^0(\cdot,0) = u_0 \quad \text{on} \quad \mathbb{R}^2 \end{cases} \tag{4.4}$$

with

$$F(M,p) = -g\left(\frac{p^\perp}{|p|}\right) \text{ trace } \left(M \cdot \left(Id - \frac{p}{|p|} \otimes \frac{p}{|p|} \right) \right)$$

where g is introduced in (1.2). In particular we see that equation (4.4) describes the anisotropic mean curvature motion with velocity

$$g(\tau)\ \kappa$$

where κ is the curvature of the level line of u^0 and τ is a unit tangent vector to the level line of u^0.

Before to state our convergence result as $\varepsilon \to 0$, we need to give the precise definition of viscosity solutions we use for the non-local equation (4.3) which is less standard. This definition has been introduced by Slepčev [22] (see also Da Lio, Kim, Slepčev [13]).

Definition 4.1. (Viscosity sub/super/solution for the non-local eikonal equation) A locally bounded upper semicontinuous (usc) function u^ε is a viscosity subsolution of (4.3) if it satisfies:

(i) $u^\varepsilon(x,t=0) \leq u_0(x)$ in $\mathbb{R}^2$,

(ii) for every $(x_0,t_0) \in \mathbb{R}^2 \times (0,\infty)$ and for every test function $\Phi \in C^\infty(\mathbb{R}^2 \times [0,\infty))$, that is tangent from above to u^ε at (x_0,t_0), the following holds:

$$\Phi_t^\varepsilon(x_0,t_0) \leq \left((c_0^\varepsilon \star 1_{\{u^\varepsilon(\cdot,t_0) \geq u^\varepsilon(x_0,t_0)\}})(x_0) - \frac{1}{2}\int_{\mathbb{R}^2} c_0^\varepsilon \right) |D\Phi^\varepsilon(x_0,t_0)|. \tag{4.5}$$

A locally bounded lower semicontinuous (lsc) function u^ε is a viscosity supersolution of (4.3) if it satisfies:

(i) $u^\varepsilon(x,t=0) \geq u_0(x)$ in $\mathbb{R}^2$,

(ii) for every $(x_0,t_0)\in\mathbb{R}^2\times(0,\infty)$ and for every test function $\Phi\in C^\infty\left(\mathbb{R}^2\times[0,\infty)\right)$, that is tangent from below to u^ε at (x_0,t_0), the following holds:

$$\Phi_t^\varepsilon(x_0,t_0)\geq\left((c_0^\varepsilon\star 1_{\{u^\varepsilon(\cdot,t_0)>u^\varepsilon(x_0,t_0)\}})(x_0)-\frac{1}{2}\int_{\mathbb{R}^2}c_0^\varepsilon\right)|D\Phi^\varepsilon(x_0,t_0)|. \tag{4.6}$$

A locally bounded continuous function u^ε is a viscosity solution of (4.3) if, and only if, it is a sub- and a supersolution of (4.3).

Then the main result of this section is

Theorem 4.2. [**Convergence of dislocations dynamics to mean curvature motion,** [14]] *There exists a constant $C_0>0$ only depending on $||c_0||_{L^\infty(\mathbb{R}^2)}$. Given $\varepsilon\in(0,1)$ and a bounded and globally Lipschitz continuous function u_0, there exists a unique viscosity solution $u^\varepsilon\in L^\infty_{loc}\left(\mathbb{R}^2\times[0,+\infty)\right)$ of problem (4.3). The function u^ε satisfies*

$$||Du^\varepsilon||_{L^\infty(\mathbb{R}^2\times[0,+\infty))}\leq||Du_0||_{L^\infty(\mathbb{R}^2)}$$

and for every $\varepsilon\in(0,1/2)$:

$$|u^\varepsilon(x,t+s)-u^\varepsilon(x,s)|\leq C_0||Du_0||_{L^\infty(\mathbb{R}^2)}\sqrt{t},\quad\forall(x,s,t)\in\mathbb{R}^2\times[0,+\infty)\times[0,1)$$

Moreover, the solution u^ε converges locally uniformly in compact sets of $\mathbb{R}^2\times[0,+\infty)$ to the unique solution u^0 of (4.4) with the same initial condition u_0.

Remark 4.3. In a future work, we will apply this result to propose a numerical scheme for anisotropic mean curvature motion or crystalline motion.

While the proof of this convergent result is quite simple in the case where the gradient of the limit function u^0 is non-zero, the case where the gradient of u^0 vanishes is quite delicate and requires more attention.

We will now present a further property of the limit mean curvature motion. To this end, we need the following:

Definition 4.4. Let $g\in C^\infty(\mathbb{R}^2\setminus\{0\})$ satisfying

$$g(\lambda p)=\frac{g(p)}{|\lambda|^3},\ \forall\lambda\in\mathbb{R}\setminus\{0\},\ \forall p\in\mathbb{R}^2.$$

We then associate to g a temperate distribution L_g defined by

$$\langle L_g,\varphi\rangle=\int_{\mathbb{R}^2}dx\ \frac{g\left(\frac{x}{|x|}\right)}{|x|^3}\left(\varphi(x)-\varphi(0)-x\cdot D\varphi(0)1_{B_1(0)}(x)\right)$$

for $\varphi\in\mathcal{S}(\mathbb{R}^2)$, where $\mathcal{S}(\mathbb{R}^2)$ is the Schwarz space of test functions on $\mathbb{R}^2$, and $B_1(0)$ denotes the unit ball centered in zero.

We define the Fourier transform

$$\hat{\varphi}(\xi)=\int_{\mathbb{R}^2}dx\ \varphi(x)e^{-i\xi\cdot x}.$$

Then we have

Theorem 4.5. [**Variational origin of the anisotropic mean curvature motion,** [14]]
Let $g \in C^\infty(\mathbb{R}^2 \setminus \{0\})$ *satisfying* $g(\lambda p) = \dfrac{g(p)}{|\lambda|^3}$, $\forall \lambda \in \mathbb{R}\setminus\{0\}$, $\forall p \in \mathbb{R}^2$. *Let*

$$G := -\frac{1}{2\pi}\hat{L_g} \tag{4.7}$$

where $\hat{L_g}$ *is the Fourier transform of* L_g. *Then* $G(\lambda p) = |\lambda| G(p)$, $\forall \lambda \in \mathbb{R}\setminus\{0\}$, $\forall p \in \mathbb{R}^2$, *and*

$$g\left(\frac{p^\perp}{|p|}\right)\frac{p^\perp}{|p|} \otimes \frac{p^\perp}{|p|} = D^2 G\left(\frac{p}{|p|}\right). \tag{4.8}$$

In particular, we see that G is convex if and only if $g \geq 0$. Moreover (4.8) means that in (4.4), we have

$$-F(D^2u^0, Du^0) = \operatorname{div}\left(\nabla G\left(\frac{Du^0}{|Du^0|}\right)\right)|Du^0|,$$

i.e., this anisotropic mean curvature motion derives from a convex energy $\int G\left(Du^0\right)\big)$.

Remark 4.6. Physically the quantity $\hat{L_g}$ is naturally given, and then the function g can be computed using (4.7)–(4.8) where we can check if g is non-negative or not.
In the simplest case of applications for dislocation dynamics, the crystal is described by isotropic elasticity. When the Burgers vector is along the x_1 direction, we have

$$G(p) = \frac{p_2^2 + \frac{1}{1-\nu}\,p_1^2}{|p|} \quad \text{with} \quad \nu \in (-1, \frac{1}{2})$$

where ν is the Poisson ratio of the material, and

$$g(\theta) = \frac{(2\gamma - 1)(\theta_1)^2 + (2-\gamma)(\theta_2)^2}{|\theta|^5} \geq 0 \quad \text{with} \quad \gamma = \frac{1}{1-\nu} \in (\frac{1}{2}, 2).$$

Our result is very natural for dislocation dynamics. Indeed, in many references in physics, the authors describe dislocations dynamics by line tension terms deriving from an energy associated to the dislocation line. See for instance Brown [10], Barnet, Gavazza [9] for physical references and Garroni, Müller [16] for a variational approach. We also refer to Forcadel [15] for the study of dislocation dynamics with a mean curvature term. As far as we know, Theorem 3.2, completed by Theorem 4.2, is the first rigorous proof for the convergence of dislocations dynamics to mean curvature motion.

Acknowledgment

The authors would like to thanks G.Barles, B.Devincre, A.Finel and Y.Le Bouar for fruitful discussions in the preparation of this paper. The two last authors were supported by the contract JC 1025 called "ACI jeunes chercheuses et jeunes chercheurs" of the French Ministry of Research (2003–2005).

References

[1] O. Alvarez, P. Cardaliaguet, R. Monneau, *Existence and uniqueness for dislocation dynamics with nonnegative velocity.* To appear in Interfaces and Free Boundaries.

[2] O. Alvarez, E. Carlini, P. Hoch, Y. Le Bouar, R. Monneau, *Dislocation dynamics described by non-local Hamilton-Jacobi equations.* To appear in Materials Science and Engineering: A (special issue, proceedings).

[3] O. Alvarez, E. Carlini, E. Rouy, R. Monneau, *Convergence of a first order scheme for a non local eikonal equation*, preprint.

[4] O. Alvarez, E. Carlini, E. Rouy, R. Monneau, *A convergent scheme for a non local Hamilton Jacobi equation modelling dislocation dynamics*, preprint.

[5] O. Alvarez, P. Hoch, Y. Le Bouar, R. Monneau, *Résolution en temps court d'une équation de Hamilton-Jacobi non locale décrivant la dynamique d'une dislocation.* C. R. Acad. Sci. Paris, Sér. I **338** (2004), 679–684.

[6] O. Alvarez, P. Hoch, Y. Le Bouar, R. Monneau, *Dislocation dynamics: short time existence and uniqueness of the solution.* To appear in Archive for Rational Mechanics and Analysis.

[7] G. Barles, *Solutions de Viscosité des Equations de Hamilton-Jacobi.* Springer-Verlag, Berlin, 1994.

[8] G. Barles, O. Ley, *Nonlocal first-order Hamilton-Jacobi equations modelling dislocations dynamics.* To appear in Comm. Partial Differential Equations.

[9] D.M. Barnett, S.D. Gavazza, *The self-force on a planar dislocation loop in an anisotropic linear-elastic medium,* Phy. stat. sol. (a), **38** (1976), p. 637.

[10] L. Brown, *The self-stress of dislocations and the shape of extended nodes,* Phil. Mag., **10** (1964), p. 441.

[11] P. Cardaliaguet, C. Marchi, *Regularity of the eikonal equation with Neumann boundary conditions in the plane: application to fronts with nonlocal terms,* preprint.

[12] M.G. Crandall, H. Ishii, P.-L. Lions: *User's guide to viscosity solutions of second order partial differential equations*, Bull. Amer. Math. Soc. (N.S.), **27** (1992), 1–67.

[13] F. Da Lio, C.I. Kim, D. Slepčev, *Nonlocal front propagation problems in bounded domains with Ncmann-type boundary conditions and applications,* Asymptot. Anal., **37** (2004), pp. 257–292.

[14] F. Da Lio, N. Forcadel, R. Monneau, *Convergence of a non-local eikonal equation to anisotropic mean curvature motion. Application to dislocation dynamics*, preprint.

[15] N. Forcadel, *Dislocations dynamics with a mean curvature term: short time existence and uniqueness*, preprint.

[16] A. Garroni, S. Müller, *A variational model for dislocations in the line tension limit,* preprint.

[17] J.R. Hirth, L. Lothe, *Theory of dislocations*, Second Edition. Malabar, Florida: Krieger, 1992.

[18] L. P. Kubin, G. Canova, M. Condat, B. Devincre, V. Pontikis, Y. Bréchet, *Dislocation Microstructures and Plastic Flow: A 3D Simulation.* Nonlinear Phenomena in Materials Science. Solid State Phenomena **23 & 24** (1992), 455–472.

[19] R.W. Lardner, *Mathematical theory of dislocations and fracture,* Mathematical Expositions 17, University of Toronto Press, 1974.

[20] F.R.N. Nabarro, *Theory of crystal dislocations*, Oxford, Clarendon Press, 1969.

[21] D. Rodney, Y. Le Bouar, A. Finel, *Phase field methods and dislocations.* Acta Mater. **51** (2003), 17–30.

[22] D. Slepčev, *Approximation schemes for propagation of fronts with nonlocal velocities and Neumann boundary conditions*, Nonlinear Anal., **52** (2003), pp. 79–115.

[23] Westmacott, Barnes and Smallman, Phil. Mag. **7** (1962).

P. Cardaliaguet
Département de Mathématiques
Université de Brest
6, Avenue Victor Le Gorgeu
B.P. 809
F-29285 Brest Cedex, France
e-mail: `Pierre.Cardaliaguet@univ-brest.fr`

F. Da Lio
Università degli Studi di Padova
Via Belzoni
I-735131 Padova, Italy
e-mail: `dalio@math.unipd.it`

N. Forcadel
CERMICS – ENPC
6 et 8 avenue Blaise Pascal
Cité Descartes – Champs sur Marne
F-77455 Marne la Vallée Cedex 2, France
e-mail: `forcadel@cermics.enpc.fr`

R. Monneau
CERMICS – ENPC
6 et 8 avenue Blaise Pascal
Cité Descartes – Champs sur Marne
F-77455 Marne la Vallée Cedex 2, France
e-mail: `monneau@cermics.enpc.fr`

International Series of Numerical Mathematics, Vol. 154, 137–147

Bermudean Approximation of the Free Boundary Associated with an American Option

E. Chevalier

Abstract. American options valuation leads to solve an optimal stopping problem or a variational inequality. These two approaches involve the knowledge of a free boundary, boundary of the so-called exercise region. As we are not able to get a closed formula for the American option value function, we will approximate the free boundary by this of a Bermudean option. Indeed a Bermudean option value function is the solution of an optimal stopping problem which can be viewed as a free boundary problem. Thanks to a maximum principle, we evaluate the difference between Bermudean and American boundaries.

Mathematics Subject Classification (2000). Primary 91B24; Secondary 60G40.

Keywords. Free boundary, optimal stopping, variational inequality.

1. American options

An American option is a financial product which gives to its owner the right to earn a specific amount of money at any time he wishes between the initial date 0 and the maturity T. This amount of money, so-called the option pay-off, is very often based on the values of one or several underlying assets. The two main problems of the American option theory are to give a price to this product and to determine the optimal strategy for the owner: the optimal time to exercise his right is assumed to be the time for which his gain is greater as possible.

The first step to solve these two linked problems is to make assumptions on the market. We will assume that the market is composed by d risky assets and denote by S_t^i their respective value at time t. We assume that $(S_t)_{0\leq t\leq T}$ is solution of the following stochastic differential equation:

$$dS_t = \operatorname{diag}(S_t)((r\mathbb{1} - \delta)dt + \sigma dW_t) \tag{1.1}$$

where $\mathbb{1} = (1)_{1\leq i\leq d} \in \mathbb{R}^d$, $r > 0$ is the interest rate of the market, $\delta \in [0, +\infty)^d$ is such that δ_i is the dividend rate of the asset i, $\sigma \in \mathbb{R}^d \times \mathbb{R}^d$ is called the market volatility, and $(W_t)_{0\leq t\leq T}$ is a standard Brownian motion on $\mathbb{R}^d$. Moreover, we assume that σ satisfies the following hypothesis which insures the non-degeneracy for the infinitesimal generator of the diffusion S:

$$H_1: \quad \exists M > m > 0,\ \forall x \in \mathbb{R}^d,\ m\|x\|^2 \leq x^*\sigma\sigma^* x \leq M\|x\|^2.$$

We denote by $\mathbb{F}$ the filtration associated to W, and for $x \in [0,+\infty)^d$, $(S_t^x)_{0\leq t\leq T}$ is the solution of the stochastic differential equation (1.1) such that $S_0^x = x$.

Our goal here is to study a specific class of options, called basket options. These options offer a pay-off which is the positive part of the difference between a positive constant (the strike price) and a linear combination of several assets. We define the pay-off function f such that:

$$\forall x \in [0,+\infty)^d,\ f(x) = (K - \langle \alpha, x\rangle)^+,$$

where $K > 0$ is the strike price, $\alpha \in \mathbb{R}^d$, and $\langle .,.\rangle$ is the usual scalary product on $\mathbb{R}^d$.

In this setting, the option theory (see [2] and [11]) asserts that at time $t \in [0,T]$, the price of the American option associated with the pay-off f is $P(T-t, S_t)$ where:

$$P(t,x) = \sup_{\tau\in\mathcal{T}_{0,t}} \mathbb{E}[e^{-r\tau} f(S_\tau^x)], \quad \forall x \in [0,+\infty)^d,$$

where $\mathcal{T}_{0,t}$ is the set of $\mathbb{F}$-stopping times with values in $[0,t]$.

At this point two approaches enable us to get information on the value function P. First the optimal stopping theory (see [9]) asserts that the supremum is attained and more precisely, we have:

$$P(t,x) = \mathbb{E}\left[e^{-r\tau^*} f(S_{\tau^*}^x)\right],$$

$$\text{where } \tau^* = \inf\{t \geq 0 : P(T-t, S_t^x) = f(S_t^x)\} \wedge T.$$

A second point of view gives a variational characterisation for P. We know (see [3] and [10]) that P is the solution of the following variational inequality

$$\begin{cases} (\mathcal{M}P - rP) \leq 0, \quad f \leq P, \quad (\mathcal{M}P - rP)(P-f) = 0 \quad \text{a.s.} \\ P(0,x) = f(x) \quad \text{on} \quad \mathbb{R}^+, \end{cases}$$

where we set:

$$\mathcal{M}h(t,x) = -\frac{\partial h}{\partial t} + \frac{1}{2}\sum_{i,j=1}^d (\sigma\sigma^*)_{i,j} x_i x_j \frac{\partial^2 h}{\partial x_i x_j} + \sum_{i=1}^d (r-\delta_i) x_i \frac{\partial h}{\partial x_i}.$$

A specific region of $(0,+\infty)\times[0,+\infty)^d$ appears in these two approaches, it is called the exercise region:

$$\mathcal{E} = \left\{(t,x) \in (0,T] \times [0,+\infty)^d : \ P(T-t,x) = f(x)\right\}.$$

In fact if we know this region we would be able on one hand to compute τ^*, on the other hand to compute P as a solution of a partial differential equation. From a financial point of view, this region is very interesting because it determines the optimal strategy the option owner has to follow. Our goal here is to give an estimation of this region and more specifically of its temporal sections:

$$\forall t \in (0,T], \quad \mathcal{E}_t = \{x \in [0,+\infty)^d : P(t,x) = f(x)\}.$$

Indeed, we are not able to get a closed formula for the price of an American option and to determine its exercise region so a lot of numerical methods have been developed to compute American options prices. The first idea is to solve the variational inequality satisfied by the value function thanks to a finite differences method. However, for problems with high dimension this approach become very difficult to implement. In this case, we solve the optimal stopping problem with Monte-Carlo methods (see [5], [13], and [1]). For that, we consider a Bermudean option, this is an American one which can be exercised only at a finite number of dates. From a financial point of view, it gives less rights to its owner than an American option, then its price is lower than the price of the corresponding American option. However, if the number of exercise opportunities goes to infinity, it is well known that the Bermudean option price tends to the American ! one.

Some estimations of the convergence rate have been found (see [1]). Since, the Bermudean option value function is the solution of an optimal stopping problem which can be seen as a free boundary problem, our goal is the evaluation of the convergence rate of the Bermudean free boundary to the American one when the number of exercise opportunities goes to infinity.

2. American and Bermudean exercise regions

Let $n \in \mathbb{N}^*$. We define the Bermudean option value function for an option offering n exercise opportunities. For $k \in \{0,\dots,n\}$ and $x \in [0,+\infty)^d$, we set

$$P^n(kh,x) = \sup_{\tau \in \mathcal{T}^n_{0,kh}} \mathbb{E}[e^{-r\tau} f(S^x_\tau)],$$

where $h = T/n$ and $\mathcal{T}^n_{0,kh}$ is the set of $\mathbb{F}$-stopping times with values in $\{ph : p \in \mathbb{N} \text{ et } 0 \le p \le k\}$.

This function P^n is easy to compute thanks to Monte-Carlo methods because it satisfies the so-called dynamic programming principle:

$$\begin{cases} P^n(0,x) &= f(x) \\ P^n((k+1)h,x) &= \max\left(f(x);\ e^{-rh}\mathbb{E}[P^n(kh,S^x_h)]\right), \end{cases}$$

for all $k \in \{0,\dots,n-1\}$. Hence it just remains to evaluate the expectancies thanks to Monte-Carlo method.

Now we have to control the error made by approximating the American option by the Bermudean one. It has already been proved that the error on value functions is lower than a constant multiplied by h (see [6], [12], and [1]).

In this paper, we deduce from this error estimate on the value functions an approximation of the American exercise region by the Bermudean one. Indeed, we can define an exercise region for the Bermudean option in the same way than before. Especially, we will be interested in its temporal sections:

$$\forall 1 \leq k \leq n, \quad \mathcal{E}^n_{kh} = \{x \in [0,+\infty)^d : P^n(kh,x) = f(x)\}.$$

Now we have to give a sense to the convergence of a region to another. For that we notice that the temporal sections of the American and Bermudean exercise regions satisfy some useful properties:

$\forall t \in (0,T], \quad \mathcal{E}_t$ and $\mathcal{E}^n_t$ contain 0 and are convex and closed subsets in $[0,+\infty)^d$.

Hence, for $\varepsilon \in [0,+\infty)^d$ such that $\|\varepsilon\| = 1$, we can define:

$$\begin{aligned} s(t,\epsilon) &= \inf\{\lambda \in \mathbb{R}^+ : P(t,\lambda\epsilon) > f(\lambda\epsilon)\} \\ s^n(t,\epsilon) &= \inf\{\lambda \in \mathbb{R}^+ : P^n(t,\lambda\epsilon) > f(\lambda\epsilon)\}, \end{aligned}$$

and with these notations, we get

$$\begin{aligned} \mathcal{E}_t &= \{x \in [0,+\infty)^d : \quad \|x\| \leq s(t, \frac{x}{\|x\|})\} \\ \mathcal{E}^n_{kh} &= \{x \in [0,+\infty)^d : \quad \|x\| \leq s^n(kh, \frac{x}{\|x\|})\}. \end{aligned}$$

Finally, our convergence rate in a direction $\varepsilon \in [0,+\infty)^d$ such that $\|\varepsilon\| = 1$, will be the positive quantity $s^n - s$. Our method consists in applying a parabolic maximum principle to deduce from the error estimate on the value functions that the difference between s^n and s is lower than a constant multiplied by $\sqrt{h}$.

We conclude this paper by studying the particular case of a put option on a single asset. We prove that the error $P - P^n$ admits a lower bound: a constant multiplied by h for h small enough. We will see that this estimation lies on the previous result: the upper bound for the error on the free boundaries.

3. Upper bound for the error on the free boundaries

In this section we establish our main result: the following Theorem.

Theorem 3.1. *Let $\varepsilon \in [0,+\infty)^d$ such that $\|\varepsilon\| = 1$. There exists a constant $C_{T,\varepsilon} > 0$ such that*

$$0 \leq s^n(T,\varepsilon) - s(T,\varepsilon) \leq C_{T,\varepsilon}\sqrt{h},$$

when $h = T/n$ is small enough.

This result is quite easy to prove in the case $d = 1$ because we are able to control $\frac{\partial^2 P}{\partial x^2}$ thanks to the variational inequality satisfied by P. However, for $d > 1$, we can not get this control. To prove the result for $\varepsilon \in (0,+\infty)^d$ such that $\|\varepsilon\| = 1$, we will use a maximum principle. If there exists $i \in \{1,\ldots,d\}$ such that $\varepsilon_i = 0$, then we come back to the case with $d-1$ assets.

3.1. The case $d = 1$

Throughout this section, we assume that $d = 1$ and that $\alpha = 1$. We will establish Theorem 3.1 in this particular case.

We recall the definition of the critical price for an American option with pay-off f. For $t \in (0, T]$, we set

$$s(t) = \inf\{x \in \mathbb{R}^+ : P(t, x) > f(x)\}.$$

As the value function of the corresponding Bermudean option, P^n, is convex, it is possible to define a critical price for the Bermudean option. For $k \in \{1, \ldots, n\}$, we set

$$s^n(kh) = \inf\{x \in \mathbb{R}^+ : P^n(kh, x) > f(x)\}.$$

As $P^n \leq P$ for all $n \in \mathbb{N}^*$ and $\lim_{n \to +\infty} P^n(T, \cdot) = P(T, \cdot)$, it is easy to see that $s^n \geq s$ for all $n \in \mathbb{N}^*$ and $\lim_{n \to +\infty} s^n(T) = s(T)$. We recall that we know that there exists a constant $C_1 > 0$ such that

$$\sup_{x \in [0, +\infty)} [P - P^n](T, x) \leq C_1 h.$$

We deduce from this an upper bound for the difference between s and s^n.

On the open set $(s(T), s^n(T))$, we have:

$$\begin{aligned}
\frac{\sigma^2 \xi^2}{2} \frac{\partial^2 P}{\partial x^2}(T, \xi) &= rP(T, \xi) - (r - \delta)\xi \frac{\partial P}{\partial x}(T, \xi) + \frac{\partial P}{\partial t}(T, \xi) \\
&\geq rP(T, \xi) - (r - \delta)\xi \frac{\partial P}{\partial x}(T, \xi) \\
&\geq r(K - \xi) - (\delta - r)^+ \xi \\
&\geq rK - \min\{\delta, r\} s^n(T),
\end{aligned}$$

then there exists $C_1 > 0$ such that $\frac{\partial^2 P}{\partial x^2}(T, \xi) \geq C_1$.

As we know that $s(T) < K$, we assume that n is great enough to have $s^n(T) < K$ and we integrate this inequality between $s(T)$ and $x \in (s(T), s^n(T))$. We get:

$$\begin{aligned}
\frac{\sigma^2 s^n(T)^2}{2} \left(\frac{\partial P}{\partial x}(T, x) + 1 \right) &\geq (rK - \delta s^n(T))\,(x - s(T)) \\
&\quad -(r - \delta)^+ s^n(T)\,(P(T, x) - f(x)) \\
&\geq (rK - \delta s^n(T))\,(x - s(T)) \\
&\quad -(r - \delta)^+ s^n(T)\,(P(T, s^n(T)) - f(s^n(T))),
\end{aligned}$$

because the function $x \to P(T, x) - f(x)$ is nondecreasing on $[0, K]$. Integrating a second time between $s(T)$ and $s^n(T)$, we obtain:

$$\begin{aligned}
\frac{\sigma^2 s^n(T)^2}{2} (P(T, s^n(T)) - f(s^n(T))) &\geq \frac{1}{2} (rK - \delta s^n(T))\,(s^n(T) - s(T))^2 \\
-(r - \delta)^+ s^n(T)\,(P(T, s^n(T)) &- f(s^n(T)))\,(s^n(T) - s(T)).
\end{aligned}$$

As $\lim_{n\to+\infty} s^n(T) = s(T)$, for $\eta > 0$ and n great enough, we have

$$(1+\eta)\sigma^2 s^n(T)^2 \left(P(T, s^n(T)) - f(s^n(T))\right) \geq \left(rK - \delta s^n(T)\right)\left(s^n(T) - s(T)\right)^2.$$

It follows that

$$\begin{aligned}
\left(s^n(T) - s(T)\right)^2 &\leq (1+\eta)\frac{\sigma^2 s^n(T)^2}{(rK - \delta s^n(T))}\left(P(T, s^n(T)) - f(s^n(T))\right)\\
&\leq (1+\eta)\frac{\sigma^2 s(T)^2}{(rK - \delta s(T))} C_1 h + o(h).
\end{aligned}$$

To conclude, we have:

$$\limsup_{n\to+\infty} \frac{s^n(T) - s(T)}{s(T)\sqrt{h}} \leq \sigma\sqrt{\frac{C_1(1+\eta)}{rK - \delta s(T)}},$$

and if we let η going to 0, we get the result. $\diamondsuit$

3.2. The case $d > 1$

To prove Theorem 3.1, we need much information on the regularity of the function $x \to s^n(T, x)$. The following lemma will provide us with this regularity result.

Lemma 3.2. *Let $\varepsilon \in (0, +\infty)^d$ such that $\|\varepsilon\| = 1$. For $\eta > 0$, we set*

$$V_\eta^\varepsilon = \left\{x \in]0, +\infty[^d \colon \ \|x - s(T, \varepsilon)\varepsilon\| < \eta\right\}.$$

Let $\eta > 0$ such that the set V_η^ε is included in a compact subset of $]0, +\infty)^d$. There exists a constant $s_\eta^\varepsilon > 0$ such that, for $n \in \mathbb{N}$ great enough, we have

$$\sup_{x \in V_\eta^\varepsilon} \frac{\left| s^n\left(T, \frac{x}{\|x\|}\right) - s^n(T, \varepsilon)\right|}{\|x - s(T, \varepsilon)\varepsilon\|} \leq s_\eta^\varepsilon.$$

Proof of Lemma 3.2. Let $n \in \mathbb{N}^*$. As 0 belongs to $\mathcal{E}_T \cap \mathcal{E}_T^n$ and as these two sets are convex subsets of $\mathbb{R}^d$, we can define the functions g and g_n such that $\forall x \in (0, +\infty)^d$,

$$g(x) = \inf\{\mu > 0, \ x/\mu \in \mathcal{E}_T\} \quad \text{and} \quad g_n(x) = \inf\{\mu > 0, \ x/\mu \in \mathcal{E}_T^n\}.$$

They are convex and homogeneous functions (see [4], Lemma I.2, p. 5). On the other hand, for $x \in (0, +\infty)^d$, it is easy to see that

$$g(x) = \frac{\|x\|}{s\left(T, \frac{x}{\|x\|}\right)} \quad \text{and} \quad g_n(x) = \frac{\|x\|}{s^n\left(T, \frac{x}{\|x\|}\right)}.$$

For $x \in V_\eta^\varepsilon$, we have

$$\frac{\left| s^n\left(T, \frac{x}{\|x\|}\right) - s^n(T, \varepsilon)\right|}{\|x - s(T, \varepsilon)\varepsilon\|} = \frac{1}{\|x\| g_n(x) g_n(\varepsilon)} \frac{\left| g_n(\|x\|\varepsilon) - g_n(x)\right|}{\|x - s(T, \varepsilon)\varepsilon\|}.$$

It follows from the definition of s^n that $s^1 \geq s^n \geq s$ so $g_1 \leq g_n \leq g$. Since g_1 and g are continuous with positive values on $(0,+\infty)^d$, there exists two constants C_1, $C_2 > 0$ such that $C_1 < g_n < C_2$. Hence, there exists a constant $C > 0$ such that

$$\sup_{x\in V_\eta^\varepsilon} \frac{\mid s^n\left(T,\frac{x}{\|x\|}\right) - s^n\left(T,\varepsilon\right)\mid}{\|x - s(T,\varepsilon)\varepsilon\|} \leq C \sup_{x\in V_\eta^\varepsilon} \frac{\mid g_n(\|x\|\varepsilon) - g_n\left(x\right)\mid}{\|x - s(T,\varepsilon)\varepsilon\|}.$$

Writing $g_n(x) = g_n\left(s(T,\varepsilon)\varepsilon\right) + \left(g_n(x) - g_n\left(s(T,\varepsilon)\varepsilon\right)\right)$ and using the homogeneity of g and g_n, we have

$$\begin{aligned}\sup_{x\in V_\eta^\varepsilon} \frac{\mid s^n\left(T,\frac{x}{\|x\|}\right) - s^n\left(T,\varepsilon\right)\mid}{\|x - s(T,\varepsilon)\varepsilon\|} \quad &\leq \quad C g_n(\varepsilon) \sup_{x\in V_\eta^\varepsilon} \frac{\mid \|x\| - s(T,\varepsilon)\mid}{\|x - s(T,\varepsilon)\varepsilon\|} \\ &\quad + C \sup_{x\in V_\eta^\varepsilon} \frac{\mid g_n\left(s(T,\varepsilon)\varepsilon\right) - g_n\left(x\right)\mid}{\|x - s(T,\varepsilon)\varepsilon\|}.\end{aligned}$$

As $(g_n(\varepsilon))_{n\in\mathbb{N}^*}$ is bounded, there exists $\tilde{C} > 0$ such that

$$C g_n(\varepsilon) \sup_{x\in V_\eta^\varepsilon} \frac{\mid \|x\| - s(T,\varepsilon)\mid}{\|x - s(T,\varepsilon)\varepsilon\|} \leq \tilde{C}.$$

Moreover, g_n is convex, denoting by $\delta V_\eta^\varepsilon$ the boundary of V_η^ε, then we obtain

$$\begin{aligned}\sup_{x\in V_\eta^\varepsilon} \frac{\mid g_n\left(s(T,\varepsilon)\varepsilon\right) - g_n\left(x\right)\mid}{\|x - s(T,\varepsilon)\varepsilon\|} \quad &\leq \quad \sup_{x\in \delta V_\eta^\varepsilon} \frac{\mid g_n\left(s(T,\varepsilon)\varepsilon\right) - g_n\left(x\right)\mid}{\|x - s(T,\varepsilon)\varepsilon\|} \\ &\leq \quad \frac{2}{\eta} \sup_{x\in \delta V_\eta^\varepsilon} \mid g_n\left(x\right)\mid \\ &\leq \quad \frac{2C_2}{\eta}. \quad \diamond\end{aligned}$$

Moreover, we recall a parabolic maximum principle on which lies the proof of Theorem 3.1. It appears in FRIEDMAN A. (1975).

Let D a bounded domain of $(0,T)\times\mathbb{R}^d$. We define the parabolic boundary of D by $\delta_p D = \delta D - \{(t,x) \in \delta D :\ t = T\}$ where δD is the boundary of D and introduce the operator $\tilde{\mathcal{M}}$ such that $\tilde{\mathcal{M}}h = \mathcal{M}h - rh$.

Let u a function defined on $[0,T]\times\mathbb{R}^d$, continuous on $\bar{D}$, and such that

$$u \in \mathcal{C}^{1,2}(D), \quad \tilde{\mathcal{M}}u \geq 0 \text{ on } D \quad \text{and} \quad u \leq 0 \text{ on } \delta_p D.$$

Then we have $u \leq 0$ on D.

Now we are able to prove Theorem 3.1.

Proof of Theorem 3.1. Assume that there exists a constant $b > 0$ such that $s(T,\varepsilon) < s^n(T,\varepsilon) - b\sqrt{h}$. We will prove that it leads to a contradiction by proving that it implies that there exists $\lambda \in [s(T,\varepsilon), s^n(T,\varepsilon) - b\sqrt{h}]$ such that

$$0 \geq [P - f]\,(T, \lambda\varepsilon)\,.$$

For that we will apply the maximum principle on the following domain:

$$D = \Big\{(t,x) \in (0,T)\times(0,+\infty)^d : \ \|x - s(T,\varepsilon)\varepsilon\| < \eta\sqrt{h}$$
$$\text{and } s\left(t, \frac{x}{\|x\|}\right) < \|x\| < s^n\left(T, \frac{x}{\|x\|}\right)\Big\},$$

where $\eta > 0$ is a constant which will be determined later.

Since for all $x \in (0,+\infty)^d$, the function $t \to s\left(t, \frac{x}{\|x\|}\right)$ is non-increasing and since $s\left(T, \frac{x}{\|x\|}\right) < s^n\left(T, \frac{x}{\|x\|}\right)$, we can assert that D is a bounded domain in $(0,T) \times (0,+\infty)^d$.
For $t \in (0,T)$, we set $\bar{t} = h \min\{k \in \{0,\dots,n\} : \ t \le kh\}$. First we notice that $P^n = f$ on $\bar{D}$ because for all $t \in (0,T)$, we have $s^n\left(T, \frac{x}{\|x\|}\right) \le s^n\left(\bar{t}, \frac{x}{\|x\|}\right)$.

Hence, it follows from the estimation of the value functions error and the fact that P is non-increasing with respect to time that we have:

$$[P - f]\,(t,x) \le P(\bar{t},x) - P^n(\bar{t},x) \le C_d h \quad \text{on } D.$$

Notice that on $\delta_p D$, we have

$$[P-f](t,x) \le \begin{cases} 0 & \text{if } \ \|x\| = s\left(t, \frac{x}{\|x\|}\right) \ \text{ or } t = 0 \\ Ch & \text{if } \ \|x\| = s^n\left(T, \frac{x}{\|x\|}\right) \ \text{ or } \|x - s(T,\varepsilon)\varepsilon\| = \eta\sqrt{h}. \end{cases}$$

Hence we will introduce a function which will kill the positive part of $P - f$ on $\delta_p D$. On $(0,T) \times (0,+\infty)^d$, we define the function $\beta(t,x) = \beta_1(x) + \beta_2(x)$, with

$$\beta_1(x) = \frac{a}{\sqrt{h}}\left(\left(\|x\| - s^n(T,\varepsilon) + b\sqrt{h}\right)^+\right)^3$$
$$\beta_2(x) = \frac{c}{\sqrt{h}}\left(\left(\|x - s(T,\varepsilon)\varepsilon\| - \frac{\eta}{2}\sqrt{h}\right)^+\right)^3,$$

where a, b, and c are positive constants which will be determined later.

Now we want to prove that $P - f - \beta \le 0$ on D, so we just have to prove that the function $P - f - \beta$ satisfies the assumptions of the maximum principle.

Indeed β is a $\mathcal{C}^{1,2}$ function on D and D is included in a compact subset of $(0,T) \times (0,+\infty)^d$ so we can apply the maximum principle on D to the function $P - f - \beta$.

First step: We prove that $P - f - \beta \le 0$ on $\delta_p D$.

Let $(t,x) \in \delta_p D$. We have four cases to study, corresponding to four part of $\delta_p D$.

- *First case*: Assume that $\|x\| = s\left(t, \frac{x}{\|x\|}\right)$.

In this case, we have

$$[P - f - \beta]\,(t,x) = -\beta(t,x) \le 0.$$

- *Second case*: Assume that $\|x - s(T,\varepsilon)\varepsilon\| = \eta\sqrt{h}$.

From the estimation of the error on value functions (see [1]), we have

$$\begin{aligned}[P-f-\beta](t,x) &\le C_d h - \beta_2(x)\\ &\le \left(C_d - c\frac{\eta^3}{8}\right)h.\end{aligned}$$

Hence, $[P-f-\beta](t,x) \le 0$ if we choose c and η such that $C_d - c\frac{\eta^3}{8} < 0$.

- *Third case*: Assume that $\|x\| = s^n\left(T, \frac{x}{\|x\|}\right)$.

In this case, we have

$$\begin{aligned}[P-f-\beta](t,x) &\le C_d h - \beta_1(t,x)\\ &\le C_d h - \frac{a}{\sqrt{h}}\left(\left(s^n\left(T,\frac{x}{\|x\|}\right) - s^n(T,\varepsilon) + b\sqrt{h}\right)^+\right)^3.\end{aligned}$$

Since there exists $R > 0$ such that $D \subset V_R^\varepsilon$ and V_R^ε is included in a compact subset of $(0,+\infty)^d$, we can apply Lemma 3.2, to prove that there exists $s_L > 0$ such that

$$[P-f-\beta](t,x) \le \left(C_d - a(b-s_L\eta)^3\right)h.$$

We conclude by asserting that $[P-f-\beta](t,x) \le 0$ as soon as we choose a, b, and η such that $C_d - a(b-s_L\eta)^3 < 0$.

- *Fourth case*: We assume that $t = 0$. We have

$$[P-f-\beta](0,x) = -\beta(t,x) \le 0.$$

In conclusion, $P - f - \beta \le 0$ on $\delta_p D$ if the two following conditions are satisfied:

$$C_d < c\frac{\eta^3}{8} \quad \text{and} \quad C_d < a(b-s_L\eta)^3.$$

Second step: We prove that $\tilde{\mathcal{M}}[P-f-\beta] \ge 0$ on D.

We begin with evaluating $\tilde{\mathcal{M}}\beta(t,x)$ when h goes to 0. Computing the derivatives of β on D and using Lemma 3.2, we get the following upper bound for h going to 0.

$$\tilde{\mathcal{M}}\beta(t,x) \le 3aMs(T,\varepsilon)^2(b+s_L\eta) + 3c\eta Ms\,(T,\varepsilon)^2 + o(1),$$

where the $o(1)$ does not depend on x. As D is included in the continuation region of the American option, we have

$$\tilde{\mathcal{M}}P(t,x) = 0 \quad \text{and} \quad \tilde{\mathcal{M}}f(t,x) = -rK + \langle \alpha\delta, x\rangle \quad \text{on } D.$$

We obtain:

$$\begin{aligned}\tilde{\mathcal{M}}\left[P-f-\beta\right](t,x) &= rK - \langle\alpha\delta,x\rangle - \tilde{\mathcal{M}}\beta(t,x)\\ &\le rK - \langle\alpha\delta,x\rangle\\ &\quad -3aMs(T,\varepsilon)^2(b+s_L\eta - 3c\eta Ms\,(T,\varepsilon)^2 + o(1)\\ &\le rK - s(T,\varepsilon)\langle\alpha\delta,\varepsilon\rangle\\ &\quad -3aMs(T,\varepsilon)^2(b+s_L\eta - 3c\eta Ms\,(T,\varepsilon)^2 + o(1).\end{aligned}$$

We have given some conditions on the constants a, b, c, and η such that if there are satisfied, the assumptions of the maximum principle are too. Indeed, we have showed that for h small enough, $P-f-\beta \leq 0$ on $\delta_p D$ and $\tilde{\mathcal{M}}\left[P-f-\beta\right](t,x) \geq 0$ on D if the constants a, b, c, and η satisfy $C_d < c\frac{\eta^3}{8}$, $C_d < a(b-s_L\eta)^3$ and $3aMs(T,\varepsilon)^2(b+s_L\eta) + 3c\eta Ms\left(T,\varepsilon\right)^2 < rK - s(T,\varepsilon)\langle\alpha\delta,\varepsilon\rangle$. It is quite easy to find some constants a, b, c, and η satisfying these conditions, then with these constants, we can apply the maximum principle on D and prove that $P-f-\beta \leq 0$ on D. However, if we assume that $s(T,\varepsilon) < s^n(T,\varepsilon) - b\sqrt{h}$, then there exists $\lambda \in (s(T,\varepsilon), s^n(T,\varepsilon)-b\sqrt{h})$ such that $(T,\lambda\varepsilon)$ owns to $\bar{D}$. That leads to a contradiction because the continuity of the function $P-f-\beta$ should imply that

$$0 \geq \left[P-f-\beta\right](T,\lambda\varepsilon) = \left[P-f\right](T,\lambda\varepsilon) > 0.$$

In conclusion, we have proved that $s(T,\varepsilon) \geq s^n(T,\varepsilon) - b\sqrt{h}$. □

4. Conclusion

A first consequence of Theorem 3.1 is the following result:

Proposition 4.1. *Let $\eta \in (0,T)$. If $d=1$ and $\alpha=1$, there exists $0<a\leq 1$ such that for $x \in (s(T),+\infty)$, we have*

$$P(T,x) - P^n(T,x) \geq ah\mathbb{E}\left[e^{-r\tau^*}\left(rK - \delta s(\tau^*)\right)1\!\!1_{\{\tau^*<T-\eta\}}\right],$$

with $\tau^* = \inf\left\{u\geq 0:\ S^x_u \leq s\left(u\right)\right\}\wedge T$.

The complete the proof of Proposition 4.1 is in [7]. The main idea for the proof is to construct a new boundary, greater than the Bermudean one but close enough to the American boundary. For that, we deduce from Theorem 3.1 that there exists a constant $C_\eta > 0$ such that $\tilde{s}^n(t) = s(t)(1+C_\eta\sqrt{h}) \geq s^n(t)$. Then we define

$$\tilde{\tau}^h = \inf\left\{u\geq 0: S^x_u \leq \tilde{s}^n\left(u\right)\right\}\wedge(T-\eta) \quad \text{and} \quad \bar{\tau}^h = h\inf\{k\in\mathbb{N}: kh\geq\tilde{\tau}^h\},$$

and notice that $\bar{\tau}^h$ is lower than the optimal stopping time for the Bermudean option: $\tau^h = h\inf\{k\in\mathbb{N}: S^x_{kh}\leq s^n(kh)\}$. Now we use the fact that $P-f\geq 0$ and then apply Itô's formula in the second inequality to get:

$$\begin{aligned}
P(T,x)-P^n(T,x) &\geq \mathbb{E}\left[P(T,x) - e^{-r\tau^h}P(T-\tau^h, S^x_{\tau^h})\right]\\
&\geq \mathbb{E}\left[\int_0^{\tau^h} e^{-ru}\left(rK-\delta S^x_u\right)1\!\!1_{\{S^x_u\leq s(u)\}}\,du\right]\\
&\geq \mathbb{E}\left[\int_{\tilde{\tau}^h}^{\bar{\tau}^h} e^{-ru}\left(rK-\delta S^x_u\right)1\!\!1_{\{S^x_u\leq s(u)\}}\,du 1\!\!1_{\{\tilde{\tau}^h<T-\eta\}}\right].
\end{aligned}$$

The result follows from the fact that

$$\mathbb{P}\left(\bar{\tau}^h - \tilde{\tau}^h \geq \frac{h}{2}\right) \geq \frac{1}{4}.$$

Proposition 4.1 shows that the convergence rate of P^n to P is proportional to $1/n$, but we do not get an equivalent for $n\,(P^n - P)$. Considering a call option with infinite maturity, Dupuis and Wang achieved to give a first-order expansion of the error (see [8]). However, when maturity is finite, the value functions and the free boundaries are time-dependent and finding a first-order expansion of the errors is still an open problem.

References

[1] Bally V., Pagès G., *Error Analysis of the quantization algorithm for obstacle problems*, Stochastic Processes and their Applications **106** (2003), 1–40.

[2] Benssoussan A., *On the theory of option pricing*, Acta Applicandae Mathematicae **2** (1984), 139–158.

[3] Benssoussan A., Lions J.L., *Applications of Variational Inequalities in Stochastic Control*, North-Holland, 1982.

[4] Brezis H., *Analyse fonctionnelle. Théorie et applications*, Masson, 1983.

[5] Broadie M., Glasserman P., *Pricing American-style securities using simulation*, Journal of Economic Dynamics and Control **21** (1997), 1323–1352.

[6] Carverhill A.P., Webber N., *American options: theory and numerical analysis*, Options: Recent Advances in Theory and Practice, Manchester Univ. Press, 1990.

[7] Chevalier E., *American options and free boundaries*, Ph.D. dissertation, University of Marne-la-Vallée, available on *http://perso-math.univ-mlv.fr/users/chevalier.etienne*, 2004.

[8] Dupuis P., Wang H., *On the Convergence from Discrete to Continuous Time in an Optimal Stopping Problem*, available on *www.cfm.brown.edu/people/dupuis*, 2004.

[9] El Karoui N., *Les aspects probabilistes du contrôle stochastique*, Lecture Notes in Mathematics **876** (1981), 72–238. Springer-Verlag.

[10] Jaillet P., Lamberton D., Lapeyre B., *Variational inequalities and the pricing of American options*, Acta Applicandae Mathematicae, **21** (1990), 263–289.

[11] Karatzas I., *On the pricing of American options*, Applied Math. Optimization **17** (1988), 37–60.

[12] Lamberton D., *Error estimates for the binomial approximation of American put options*, Annals of Applied Probability **8** (1998), 206–233.

[13] Longstaff F.A., Schwartz E.S., *Valuing American Options by Simulations: a Simple Least Squares Approach*, Review of Financial Studies **14** (2001), 113–147.

E. Chevalier
Université d'Evry Val d'Essonne
Laboratoire d'Analyse et Probabilité
Boulevard François Mitterrand
F-91025 Evry Cedex, France
e-mail: `echevali@univ-evry.fr`

International Series of Numerical Mathematics, Vol. 154, 149–157

Steady-state Bingham Flow with Temperature Dependent Nonlocal Parameters and Friction

L. Consiglieri and J.F. Rodrigues

Abstract. We consider Bingham incompressible flows with temperature dependent viscosity and plasticity threshold and with mixed boundary conditions, including a friction type boundary condition. The coupled system of motion and energy steady-state equations may be formulated through a variational inequality for the velocity and variational methods provide a weak solution to the model. In the asymptotic limit case of a high thermal conductivity, the temperature becomes a constant solving an implicit total energy equation involving the viscosity function, the plasticity threshold and the friction yield coefficient. The limit model corresponds to a steady-state Bingham flow with nonlocal parameters, which has therefore at least one solution.

Mathematics Subject Classification (2000). Primary 35J85; Secondary 76D03, 80A20.

Keywords. Bingham fluid, friction law, thermal conductivity, variational methods.

1. Introduction

In the sixties, Ladyzhenskaya [8] proposed a modified Navier-Stokes system with nonlocal viscosity. In [5], the authors proved that the nonlocal model, as well as other nonlocal non-Newtonian models, can be obtained as an asymptotic limit case of a very large thermal conductivity when the viscosities depend on temperature. In the present work, we extend some of those models for the nonlocal Bingham flow when the friction behavior on a part of the boundary is also taken into account. The principal difficulty is that the quadratic term due to the energy dissipation arising in the right-hand side of the heat equation leads to the L^1-analysis of the partial differential equation. The new feature in the limit model is due to a Fourier type boundary condition, and consists in the appearance of a nonlocal energy term on the boundary part where friction is taken into account.

The Bingham viscoplastic fluid does not flow as a fluid unless the stress tensor achieves at least some critical shear stress η (the plasticity threshold):

$$D(\mathbf{u}) = 0 \quad \text{if} \quad |\tau| \le \eta \tag{1.1}$$

$$D(\mathbf{u}) = \frac{|\tau| - \eta}{\mu|\tau|}\tau \quad \text{if} \quad |\tau| > \eta \tag{1.2}$$

where $\mathbf{u}$ is the velocity vector, $D(\mathbf{u}) = (\nabla\mathbf{u} + (\nabla\mathbf{u})^T)/2$ the symmetric part of the gradient of the velocity vector, μ the viscosity and τ the deviator tensor of the Cauchy stress tensor σ, that is, $\sigma = -pI + \tau$ where p denotes the pressure and I is the identity matrix. The law (1.1)–(1.2) is an inverse form of the constitutive law [6]

$$\tau = \mu(\theta)D(\mathbf{u}) + \eta(\theta)\frac{D(\mathbf{u})}{|D(\mathbf{u})|} \text{ if } |D(\mathbf{u})| \neq 0$$

$$|\tau| \le \eta(\theta) \text{ if and only if } |D(\mathbf{u})| = 0$$

considering the viscosity and the plasticity threshold dependent on the temperature θ, and $|D(\mathbf{u})| = (D_{ij}(\mathbf{u})D_{ij}(\mathbf{u}))^{1/2}$, with the convention on implicit summation over repeated indices.

Here, let Ω be a bounded open subset of $\mathbb{R}^n (n = 2, 3)$ with Lipschitz continuous boundary $\partial\Omega$, which is assumed to consist of two disjoint parts Γ_0 and Γ such that $\partial\Omega = \bar{\Gamma}_0 \cup \bar{\Gamma}$ and $\text{meas}(\Gamma_0) > 0$. The governing equations to the Bingham incompressible thermal flow at steady-state are given by

$$(\mathbf{u} \cdot \nabla)\mathbf{u} - \nabla \cdot \tau = -\nabla p + \mathbf{f} \text{ in } \Omega; \tag{1.3}$$

$$\nabla \cdot \mathbf{u} = \sum_{i=1}^{n} \frac{\partial u_i}{\partial x_i} = 0 \text{ in } \Omega; \tag{1.4}$$

$$\mathbf{u} \cdot \nabla\theta - \kappa\Delta\theta = \tau : D(\mathbf{u}) - \alpha\theta \text{ in } \Omega, \qquad (\alpha \ge 0), \tag{1.5}$$

where the density and the specific heat are assumed equal to one, $\mathbf{f}$ denotes the external forces, and κ is the thermal conductivity. Note that we admit a possible external heat source proportional to the temperature if $\alpha > 0$, in addition to the dissipation energy factor $\tau : D(\mathbf{u})$.

We introduce a thermal friction law on the part Γ of the boundary, keeping the no-slip condition on the other part Γ_0:

$$\text{on } \Gamma_0 : \quad \mathbf{u} = 0 \tag{1.6}$$

$$\text{on } \Gamma : \quad u_N = 0 \text{ and} \tag{1.7}$$

$$|\sigma_T| < \nu(\theta) \Rightarrow \mathbf{u}_T = 0 \tag{1.8}$$

$$|\sigma_T| = \nu(\theta) \Rightarrow \exists\lambda \ge 0, \ \mathbf{u}_T = -\lambda\sigma_T. \tag{1.9}$$

Here the tangential and normal velocities and the components of the tangential stress tensor are given, respectively, by

$$\mathbf{u}_T = \mathbf{u} - u_N\mathbf{n}, \quad u_N = u_i n_i, \quad \sigma_{Ti} = \sigma_{ij}n_j - \sigma_N n_i$$

where $\mathbf{n} = (n_i)$ denotes the unit outward normal to $\partial\Omega$. In (1.8)–(1.9), we assume a temperature dependent function ν, $\nu \geq 0$, to represent the friction yield coefficient (see, for instance, [6] for solid-solid interface or [2, 3, 4] for liquid-solid interface).

Finally we consider a homogeneous Neumann boundary condition

$$\frac{\partial\theta}{\partial n} = 0 \text{ on } \Gamma_0, \tag{1.10}$$

and the Fourier boundary condition

$$\kappa\frac{\partial\theta}{\partial n} + \beta\theta = \nu(\theta)|\mathbf{u}_T| \text{ on } \Gamma, \quad (\beta \geq 0). \tag{1.11}$$

In the framework of Lebesgue and Sobolev spaces with $W^{1,2}(\Omega) = H^1(\Omega)$, we introduce

$$\begin{aligned}
\mathcal{V} &= \{\mathbf{v} \in (C^\infty(\Omega))^n : \nabla\cdot\mathbf{v} = 0 \text{ in } \Omega\};\\
H_s &= \{\mathbf{v} \in (L^s(\Omega))^n : \nabla\cdot\mathbf{v} = 0 \text{ in } \Omega,\ v_N = 0 \text{ on } \partial\Omega\}, \qquad (s > 1);\\
V &= \{\mathbf{v} \in (H^1(\Omega))^n : \nabla\cdot\mathbf{v} = 0 \text{ in } \Omega,\ \mathbf{v} = 0 \text{ on } \Gamma_0,\ v_N = 0 \text{ on } \Gamma\},
\end{aligned}$$

endowed with the standard norm

$$\|\mathbf{v}\|_V = \|D(\mathbf{v})\|_{2,\Omega} = \|D(\mathbf{v})\|_{L^2(\Omega)}.$$

For fixed $\kappa > 0$, we formulate the problem (1.1)–(1.11) in variational form [6]: find a weak solution $(\mathbf{u}, \theta) \in V \times W^{1,q}(\Omega)$, for $1 < q < n/(n-1)$, such that,

$$\begin{aligned}
\int_\Omega \{\mu(\theta)D(\mathbf{u}) - \mathbf{u}\otimes\mathbf{u}\} : D(\mathbf{v}-\mathbf{u})dx + J(\theta,\mathbf{v}) - J(\theta,\mathbf{u}) \qquad\qquad (1.12)\\
\geq \int_\Omega \mathbf{f}\cdot(\mathbf{v}-\mathbf{u})dx, \quad \forall\mathbf{v}\in V;
\end{aligned}$$

$$\begin{aligned}
\kappa\int_\Omega \nabla\theta\cdot\nabla\phi dx - \int_\Omega \theta\mathbf{u}\cdot\nabla\phi dx + \alpha\int_\Omega \theta\phi dx + \beta\int_\Gamma \theta\phi ds \quad (1.13)\\
= \int_\Omega \{\mu(\theta)|D(\mathbf{u})|^2 + \eta(\theta)|D(\mathbf{u})|\}\phi dx + \int_\Gamma \nu(\theta)|\mathbf{u}_T|\phi ds, \quad \forall\phi\in W^{1,q'}(\Omega);
\end{aligned}$$

where $J : W^{1,1}(\Omega) \times V \to \mathbb{R}_0^+$ is defined by

$$J(\theta,\mathbf{v}) = \int_\Omega \eta(\theta)|D(\mathbf{v})|dx + \int_\Gamma \nu(\theta)|\mathbf{v}_T|ds.$$

The main idea is to pass to the limit on κ ($\kappa \to +\infty$) in order to reformulate the local system (1.12)–(1.13) into a nonlocal problem with constant parameters for the viscosity, the plasticity threshold and the friction yield coefficient calculated at the constant homogenized temperature, which is implicitly given through a scalar equation. We notice that the argument used in this work is applicable to the Newtonian as well as non-Newtonian fluids, as shown in [5].

2. The main result

Let us state the weak nonlocal formulation to the problem (1.1)–(1.4) and (1.6)–(1.9) corresponding formally to the limit model $\kappa = \infty$.

PROBLEM. Find $(\mathbf{u}, \Theta) \in V \times \mathbb{R}$ satisfying

$$\mu(\Theta) \int_\Omega D(\mathbf{u}) : D(\mathbf{v}-\mathbf{u})dx - \int_\Omega \mathbf{u} \otimes \mathbf{u} : D(\mathbf{v}-\mathbf{u})dx \tag{2.1}$$
$$+\eta(\Theta) \int_\Omega \{|D(\mathbf{v})| - |D(\mathbf{u})|\}dx + \nu(\Theta) \int_\Gamma \{|\mathbf{v}_T| - |\mathbf{u}_T|\}ds$$
$$\geq \int_\Omega \mathbf{f} \cdot (\mathbf{v}-\mathbf{u})dx, \quad \forall \mathbf{v} \in V,$$

where Θ is a solution to the implicit scalar equation

$$(\alpha|\Omega| + \beta|\Gamma|)\,\Theta = \mu(\Theta) \int_\Omega |D(\mathbf{u})|^2 dx + \eta(\Theta) \int_\Omega |D(\mathbf{u})|dx + \nu(\Theta) \int_\Gamma |\mathbf{u}_T|ds. \tag{2.2}$$

Remark 2.1. Notice that the antisymmetry of the convective term $\int_\Omega (\mathbf{u} \cdot \nabla)\mathbf{u} \cdot \mathbf{v}$ is valid by the incompressibility property (1.4) and the boundary condition $u_N = 0$ on $\partial\Omega$ given by (1.6)–(1.7).

We assume

$$\mu \in C^0(\mathbb{R}): \quad \exists \mu_*, \mu^* > 0, \quad \mu_* \leq \mu(s) \leq \mu^*, \quad \forall s \in \mathbb{R}; \tag{2.3}$$
$$\eta \in C^0(\mathbb{R}): \quad \exists \eta^* > 0, \quad 0 \leq \eta(s) \leq \eta^*, \quad \forall s \in \mathbb{R}; \tag{2.4}$$
$$\nu \in C^0(\mathbb{R}): \quad \exists \nu^* > 0, \quad 0 \leq \nu(s) \leq \nu^*, \quad \forall s \in \mathbb{R}; \tag{2.5}$$
$$\alpha, \beta \geq 0: \quad \alpha + \beta > 0; \tag{2.6}$$
$$\mathbf{f} \in V'. \tag{2.7}$$

The main result of this work is the following theorem.

Theorem 2.2. *Under the assumptions (2.3)–(2.7), there exists $(\mathbf{u}, \Theta) \in V \times \mathbb{R}$ a solution to the problem (2.1)–(2.2), which can be obtained as a limit in $V \times W^{1,q}(\Omega)$, $1 < q < n/(n-1)$, as $\kappa \to \infty$ of solutions $(\mathbf{u}_\kappa, \theta_\kappa)$ of (1.12)–(1.13).*

3. Auxiliary existence results

The following propositions are essential in the proof of Theorem 2.2.

Proposition 3.1. *For every $\mathbf{w} \in H_s$, $s \geq n$, and $\xi \in W^{1,1}(\Omega)$ there exists a unique solution $\mathbf{u} = \mathbf{u}(\mathbf{w}, \xi) \in V$ to the variational inequality*

$$\int_\Omega \{\mu(\xi)D(\mathbf{u}) - \mathbf{w} \otimes \mathbf{u}\} : D(\mathbf{v}-\mathbf{u})dx + J(\xi, \mathbf{v}) - J(\xi, \mathbf{u}) \tag{3.1}$$
$$\geq \int_\Omega \mathbf{f} \cdot (\mathbf{v}-\mathbf{u})dx, \quad \forall \mathbf{v} \in V,$$

and it satisfies the estimate

$$\|\mathbf{u}\|_V \leq \frac{\|\mathbf{f}\|_{V'}}{\mu_*}. \tag{3.2}$$

Moreover, if $\mathbf{w}_m$ *and* ξ_m *are sequences in* H_s *and* $W^{1,1}(\Omega)$, *respectively, such that* $\mathbf{w}_m \to \mathbf{w}$ *in* H_s, $\xi_m \to \xi$ *in* $L^1(\Omega)$ *and* $\xi_m \to \xi$ *in* $L^1(\Gamma)$, *and* $\mathbf{u}_m = \mathbf{u}(\mathbf{w}_m, \xi_m)$ *are the corresponding solutions satisfying* (3.1), *then there exists* $\mathbf{u} = \mathbf{u}(\mathbf{w}, \xi) \in V$ *the solution to* (3.1) *such that* $\mathbf{u}_m \to \mathbf{u}$ *in* V.

Proof. The existence and uniqueness of the solution are consequences of classical results (for instance, see [9]) on variational inequalities with convex continuous functionals. The estimate (3.2) follows by choosing $\mathbf{v} = 0$ as a test function in (3.1).

Let $\mathbf{w}_m, \xi_m, \mathbf{u}_m = \mathbf{u}(\mathbf{w}_m, \xi_m)$ be sequences in the conditions of the Proposition. From estimate (3.2) we have $\mathbf{u}_m \rightharpoonup \mathbf{u}$ in V for a subsequence of $\mathbf{u}_m$, still denoted by $\mathbf{u}_m$, and consequently

$$\mathbf{u}_m \to \mathbf{u} \text{ in } H_s, \qquad \text{for } s < 2n/(n-2) \tag{3.3}$$

$$\text{and in } L^r(\Gamma), \qquad \text{for } r < 2(n-1)/(n-2). \tag{3.4}$$

The convective term $\mathbf{w}_m \otimes \mathbf{u}_m : D(\mathbf{v})$ easily passes to the limit in m. Since $\xi_m \to \xi$ a.e. in Ω and on Γ, the functions μ, η and ν are continuous, and due to the sequential weak lower semicontinuity of the continuous and convex functional J, we obtain as in [7]

$$\int_\Omega \{\mu(\xi)D(\mathbf{u}) - \mathbf{w} \otimes \mathbf{u}\} : D(\mathbf{v})dx + J(\xi, \mathbf{v}) - \int_\Omega \mathbf{f} \cdot (\mathbf{v} - \mathbf{u})dx$$

$$\geq \liminf_{m \to +\infty} \int_\Omega \mu(\xi_m)|D(\mathbf{u}_m)|^2 dx + \liminf_{m \to +\infty} J(\xi_m, \mathbf{u}_m) \geq \int_\Omega \mu(\xi)|D(\mathbf{u})|^2 dx + J(\xi, \mathbf{u})$$

So $\mathbf{u}$ is a solution to (3.1), and its uniqueness is due to the standard variational argument.

Choosing $\mathbf{v} = (\mathbf{u}_m + \mathbf{u})/2$ as a test function in (3.1) for the solutions $\mathbf{u}_m$ and $\mathbf{u}$, and subtracting the obtained inequalities, it results

$$\mu_* \int_\Omega |D(\mathbf{u}_m - \mathbf{u})|^2 dx + \int_\Omega \{\eta(\xi_m) - \eta(\xi)\}|D(\mathbf{u}_m)|dx + \int_\Gamma \{\nu(\xi_m) - \nu(\xi)\}|\mathbf{u}_m|ds$$

$$\leq \int_\Omega (\mathbf{w} - \mathbf{w}_m) \otimes \mathbf{u}_m : D(\mathbf{u})dx + \int_\Omega \{\mu(\xi) - \mu(\xi_m)\}D(\mathbf{u}) : D(\mathbf{u}_m - \mathbf{u})dx$$

$$+ \int_\Omega \{\eta(\xi_m) - \eta(\xi)\}|D(\mathbf{u})|dx + \int_\Gamma \{\nu(\xi_m) - \nu(\xi)\}|\mathbf{u}_T|ds.$$

Applying Fatou lemma to the second and third terms on the left-hand side of the above inequality and using Lebesgue theorem to the convergences on the right-hand side, the required strong convergence holds. □

Proposition 3.2. *Let* $\mathbf{u} = \mathbf{u}(\mathbf{w}, \xi)$ *be the solution given by Proposition* 3.1. *Then there exists* $\theta = \theta(\mathbf{u}, \xi) \in W^{1,q}(\Omega)$ *a solution to the variational problem*

$$\int_\Omega (\kappa \nabla\theta - \theta \mathbf{u}) \cdot \nabla\phi dx + \alpha \int_\Omega \theta\phi dx + \beta \int_\Gamma \theta\phi ds \tag{3.5}$$
$$= \int_\Omega \{\mu(\xi)|D(\mathbf{u})|^2 + \eta(\xi)|D(\mathbf{u})|\}\phi dx + \int_\Gamma \nu(\xi)|\mathbf{u}_T|\phi ds, \quad \forall\phi \in W^{1,q'}(\Omega),$$

that satisfies the estimate

$$\alpha\|\theta\|_{q,\Omega} + \beta\|\theta\|_{q,\Gamma} + \sqrt{\kappa}\|\nabla\theta\|_{q,\Omega} \le \mathcal{F}\left(\|\mathbf{f}\|_{V'}, \frac{\mu^*}{\mu_*}, \eta^*, \nu^*\right) \tag{3.6}$$

for an arbitrary $1 < q < n/(n-1)$, *and* $\mathcal{F}$ *is a positive function. Moreover, let* $\mathbf{w}_m$ *and* ξ_m *be sequences in* H_s *and* $W^{1,1}(\Omega)$, *respectively, such that* $\mathbf{w}_m \to \mathbf{w}$ *in* H_s, $\xi_m \to \xi$ *in* $L^1(\Omega)$ *and* $\xi_m \to \xi$ *in* $L^1(\Gamma)$, *and* $\mathbf{u}_m = \mathbf{u}(\mathbf{w}_m, \xi_m)$ *be the corresponding solutions given by Proposition* 3.1. *If* $\theta_m = \theta(\mathbf{u}_m, \xi_m)$ *are solutions satisfying* (3.5), *then there exists* $\theta = \theta(\mathbf{u}, \xi)$ *a solution to* (3.5) *such that* $\theta_m \to \theta$ *in* $W^{1,q}(\Omega)$*-weak,* $L^1(\Omega)$*-strong and* $L^1(\Gamma)$*-strong.*

Remark 3.3. In (3.5), the terms on the right-hand side have sense, since $\phi \in W^{1,q'}(\Omega)) \hookrightarrow C(\bar{\Omega})$ for $q' > n$, that is, $q < n/(n-1)$, and the term $\int_\Omega \theta\mathbf{u} \cdot \nabla\phi$ has meaning for $\theta \in W^{1,q}(\Omega)$, $\mathbf{u} \in H_s$ with $s \ge n$, and $\phi \in W^{1,q'}(\Omega)$.

Proof. Let us define $F = \mu(\xi)|D(\mathbf{u})|^2 + \eta(\xi)|D(\mathbf{u})|$ and $G = \nu(\xi)|\mathbf{u}_T| \in L^r(\Omega)$ for r as in (3.4), and, for each $m \in \mathbb{N}$, take

$$F_m = \frac{mF}{m + |F|} \in L^\infty(\Omega).$$

From the Lax-Milgram theorem, there exists a unique solution $\theta_m \in H^1(\Omega)$ to the following variational problem

$$\int_\Omega (\kappa\nabla\theta_m - \theta_m\mathbf{u}) \cdot \nabla\phi dx + \alpha \int_\Omega \theta_m\phi dx + \beta \int_\Gamma \theta_m\phi ds \tag{3.7}$$
$$= \int_\Omega F_m\phi dx + \int_\Gamma G\phi ds, \quad \forall\phi \in H^1(\Omega).$$

From L^1–data theory (see, for instance, [5] or [10]), the estimate (3.6) follows for θ_m. Indeed, choosing

$$\phi = \operatorname{sign}(\theta_m)[1 - 1/(1 + |\theta_m|)^\varsigma] \in W^{1,2}(\Omega) \cap L^\infty(\Omega), \quad \text{for } \varsigma > 0,$$

as a test function in (3.7) it follows

$$\kappa \int_\Omega \frac{\varsigma|\nabla\theta_m|^2}{(1 + |\theta_m|)^{\varsigma+1}} dx + \beta C(\varsigma) \int_\Gamma |\theta_m| ds \le \|F\|_{1,\Omega} + \|G\|_{r,\Gamma}.$$

Arguing as in [10] and [5] we conclude, for $q < n/(n-1)$, that

$$\int_\Omega |\nabla\theta_m|^q dx \le \left(\frac{\|F\|_{1,\Omega} + \|G\|_{r,\Gamma}}{\kappa\varsigma}\right)^{q/2} \left(\varepsilon \left(\int_\Omega |\theta_m|^{qn/(n-q)}\right)^{(2-q)/2} + C(\varepsilon)\right)$$

for arbitrary $\varepsilon > 0$. If $\beta > 0$, using a Poincaré-Sobolev type inequality we obtain

$$\|\theta_m\|_{qn/(n-q),\Omega} \leq C\left(\|\nabla\theta_m\|_{q,\Omega} + \beta\int_\Gamma |\theta_m|ds\right) \tag{3.8}$$

$$\leq C\sqrt{\frac{\|F\|_{1,\Omega} + \|G\|_{r,\Gamma}}{\varsigma}}(\varepsilon^{1/q}\|\theta_m\|_{qn/(n-q),\Omega}^{n(2-q)/[2(n-q)]} + C'(\varepsilon)) \quad \text{for } \kappa > 1.$$

If $\beta = 0$, the assumption (2.6) implies that $\alpha > 0$. Choosing $\phi \equiv 1$ as a test function in (3.7) we get

$$0 \leq \int_\Omega \theta_m = \int_\Omega F_m dx + \int_\Gamma G ds \leq \|F\|_{1,\Omega} + \|G\|_{1,\Gamma};$$

and instead of (3.8) we obtain

$$\|\theta_m\|_{qn/(n-q),\Omega} \leq C\|\nabla\theta_m\|_{q,\Omega} + |\Omega|^{(n-q)/(nq)}|\fint_\Omega \theta_m|$$

$$\leq C\sqrt{\frac{\|F\|_{1,\Omega} + \|G\|_{r,\Gamma}}{\varsigma}}(\varepsilon^{1/q}\|\theta_m\|_{qn/(n-q),\Omega}^{n(2-q)/[2(n-q)]} + C'(\varepsilon)) \quad \text{for } \kappa > 1,$$

where C is a constant depending on Ω, and $\fint_\Omega$ denotes $\frac{1}{|\Omega|}\int_\Omega$.

Consequently, for ε sufficiently small it follows

$$\|\theta_m\|_{qn/(n-q),\Omega} \leq C \text{ (independent of } m \text{ and } \kappa),$$

and then θ_m satisfies the estimate (3.6). Thus, we can extract a subsequence of θ_m, still denoted by θ_m, such that it weakly converges to θ in $W^{1,q}(\Omega)$, where θ solves the limit problem (3.5).

Let $\mathbf{w}_m, \xi_m$ and $\mathbf{u}_m = \mathbf{u}(\mathbf{w}_m, \xi_m)$ be sequences in the conditions of Proposition 3.1, that is, $\mathbf{u}_m$ is such that $\mathbf{u}_m \to \mathbf{u}$ in V. In order to pass to the limit in (3.5) for solutions $\theta_m = \theta(\mathbf{u}_m, \xi_m)$ when m tends to infinity, from estimate (3.6), we can extract a subsequence of θ_m, still denoted by θ_m, such that it converges to θ, which is the solution to (3.5). Note that by (3.3) and $\theta_m \rightharpoonup \theta$ in $L^{qn/(n-q)}(\Omega)$ we obtain $\theta_m\mathbf{u}_m \rightharpoonup \theta\mathbf{u}$ in $L^{sqn/[qn+s(n-q)]}(\Omega) \hookrightarrow L^q(\Omega)$ for $n = 2, 3$. □

4. Proof of Theorem 2.2

This proof is divided in two parts.

4.1. Existence for the coupled system (1.12)–(1.13)

Consider the multi-valued mapping $\mathcal{L}$ defined on

$$K := \{(\mathbf{w}, \xi) \in V \times W^{1,q}(\Omega) : \|\mathbf{w}\|_V \leq R_1 \text{ and } \|\xi\|_{W^{1,q}(\Omega)} \leq R_2\},$$

taking $R_1 \geq \|\mathbf{f}\|_{V'}/\mu_*$ and R_2 conveniently chosen from estimate (3.6), such that $\mathcal{L}$ applies $(\mathbf{w}, \xi)$ into the nonempty convex set $\{(\mathbf{u}, \theta)\} \subset K$, where $\mathbf{u}$ and θ are the solutions given at Propositions 3.1 and 3.2, respectively. Thus the Tychonof-Kakutani-Glicksberg fixed point theorem (see [1, pages 218–220]) guarantees a solution, $(\mathbf{u}, \theta) \in \mathcal{L}(\mathbf{u}, \theta)$, to (1.12)–(1.13) still satisfying the estimates (3.2) and

(3.6), provided $\mathcal{L}(\mathbf{w},\xi)$ is a closed set and $\mathcal{L}$ is upper semicontinuous for the weak topology in $V \times W^{1,q}(\Omega)$, for $1 < q < n/(n-1)$. From the closed graph theorem [1, page 413], it remains therefore to prove that if $(\mathbf{w}_m, \xi_m) \rightharpoonup (\mathbf{w}, \xi)$ in $V \times W^{1,q}(\Omega)$ and $(\mathbf{u}_m, \theta_m) \in \mathcal{L}(\mathbf{w}_m, \xi_m)$ then

$$(\mathbf{u}_m, \theta_m) \rightharpoonup (\mathbf{u}, \theta) \in \mathcal{L}(\mathbf{w}, \xi). \tag{4.1}$$

By the Rellich-Kondrachof embedding

$$\begin{array}{lll} V \hookrightarrow\hookrightarrow H_s, & \text{for } n \leq s < 2n/(n-2); & \\ W^{1,q}(\Omega) \hookrightarrow\hookrightarrow L^1(\Omega), & \text{and} & W^{1,q}(\Omega) \hookrightarrow\hookrightarrow L^1(\Gamma), \end{array}$$

the final assertion (4.1) derives from Propositions 3.1 and 3.2.

4.2. Passage to the limit on κ

Let $(\mathbf{u}_\kappa, \theta_\kappa)$ be a solution to (1.12)–(1.13), corresponding to each $\kappa > 0$ and let $\kappa \to +\infty$. From the estimates (3.2) and (3.6), we can extract a subsequence of $(\mathbf{u}_\kappa, \theta_\kappa)$, still denoted by $(\mathbf{u}_\kappa, \theta_\kappa)$, satisfying

$$\begin{aligned} &\nabla\theta_\kappa \to 0 \text{ in } L^q(\Omega), \\ &\theta_\kappa \to \Theta = \ \text{constant in } W^{1,q}(\Omega). \end{aligned}$$

We can proceed as in the proof of Proposition 3.1 to get a strong convergence of $\mathbf{u}_\kappa$ to $\mathbf{u}$ in $H^1(\Omega)$. Then, we can pass to the limit (1.13) on κ $(\kappa \to +\infty)$, taking $\phi \equiv 1$ to obtain (2.2). Now, taking the limit $\kappa \to +\infty$ in (1.12), it follows that the limit $\mathbf{u}$ solves the nonlocal problem (2.1). □

Acknowledgment

The authors were partially supported by FCT research program POCTI (Portugal/FEDER-EU).

References

[1] C. Baiocchi and A. Capelo, *Variational and quasivariational inequalities: Applications to free boundary problems.* Wiley-Interscience, Chichester-New York, 1984.

[2] D. Cioranescu, *Sur quelques équations aux derivées partielles posées par la mécanique des milieux continus.* Thèse de Doctorat, Université Pierre et Marie Curie, Paris **6**, 1977.

[3] L. Consiglieri, Stationary solutions for a Bingham flow with nonlocal friction. In *Mathematical topics in fluid mechanics*, J.F. Rodrigues and A. Sequeira (eds.), Pitman Res. Notes in Math., Longman (1992), 237–243.

[4] L. Consiglieri, A nonlocal friction problem for a class of non-Newtonian flows, *Portugaliae Mathematica* **60** :2 (2003), 237–252.

[5] L. Consiglieri and J.F. Rodrigues, *On stationary flows with energy dependent nonlocal viscosities.* Zapiski Nauchn. Seminarov POMI **295** (2003), 99–117.

[6] G. Duvaut and J.L. Lions, *Les inéquations en mécanique et en physique.* Dunod, 1972.

[7] G. Duvaut and J.L. Lions, *Transfert de chaleur dans un fluide de Bingham dont la viscosité dépend de la température.* Journal of Functional Analysis **11** (1972), 93–110.

[8] O.A. Ladyzhenskaya, *On new equations for the description of the motion of viscous incompressible fluids and on the solvability "in large" of some boundary-value problems for them.* Proceedings of Steklov Math. Inst. **102** (1967), 85–104.

[9] J.L. Lions, *Quelques méthodes de résolution des problèmes aux limites non linéaires.* Dunod et Gauthier-Villars, 1969.

[10] A. Prignet, *Conditions aux limites non homogènes pour des problèmes elliptiques avec second membre mesure.* Ann. Fac. Sci. Toulouse Math. **6** (1997), 297–318.

L. Consiglieri
Department of Mathematics and CMAF
Faculdade de Ciências da Universidade de Lisboa
1749-016 Lisboa, Portugal
e-mail: `lcconsiglieri@fc.ul.pt`

J.F. Rodrigues
CMUC/ University of Coimbra and University of Lisbon
Av. Prof. Gama Pinto 2
1649-003 Lisboa, Portugal
e-mail: `rodrigue@fc.ul.pt`

International Series of Numerical Mathematics, Vol. 154, 159–168

Some P.D.E.s with Hysteresis

Michela Eleuteri

Abstract. We present some results concerning two classes of P.D.E.s containing a continuous hysteresis operator. We introduce a weak formulation in Sobolev spaces for a Cauchy problem; under suitable assumptions on the hysteresis operator, we state some existence results. The presentation of the paper is quite general, as we avoid to describe all the details of the proof of the theorems involved.

Introduction

Hysteresis is a phenomenon that occurs in several and quite different settings, for example in plasticity, in ferromagnetism, in phase transitions.

We can certainly fix in the important monograph of Krasnosel'skiĭ and Pokrovskiĭ [8] the starting point of the mathematical research on hysteresis, which, from that moment onwards, has been considerably increasing, providing many interesting results (at this purpose we can certainly quote the recent monographes devoted to this topic, see Brokate and Sprekels [4], Krejčí [9], Mayergoyz [10] and Visintin [11], together with the references therein). In particular, a great number of contributions has been obtained for classes of P.D.E.s containing hysteresis nonlinearities, involving quasilinear and semilinear, parabolic and hyperbolic equations.

The aim of this paper is to present an overview on some results obtained by the author concerning two new classes of P.D.E.s containing a continuous hysteresis operator; all these results can be found in [5]. More in detail, we deal with the following two model equations

$$\frac{\partial^2 u}{\partial t^2} + \frac{\partial u}{\partial t} - \triangle \left(\frac{\partial u}{\partial t} + \overline{\mathcal{G}}(u) \right) = f \qquad \text{in } \Omega \times (0,T) \tag{0.1}$$

$$\frac{\partial}{\partial t}(u + \overline{\mathcal{F}}(u)) + \vec{v} \cdot \nabla (u + \overline{\mathcal{F}}(u)) - \triangle u = f \qquad \text{in } \Omega \times (0,T), \tag{0.2}$$

where Ω is an open bounded set of $\mathbb{R}^N$, $N \geq 1$, $\triangle$ is the Laplace operator, with suitable boundary conditions, $\vec{v} : \Omega \times (0,T) \to \mathbb{R}^N$ is known, $\overline{\mathcal{G}}$ and $\overline{\mathcal{F}}$ are hysteresis operators and f in both cases is a given function. Actually our results turn to be valid for the more general class of memory operators, i.e., operators fulfilling only (1.3), but in the most frequent applications $\overline{\mathcal{G}}$ and $\overline{\mathcal{F}}$ are instead also rate independent (i.e., hysteresis operators).

We focus our attention on the well-posedness of our model problems; our treatment is quite general because we do not propose all the details of the proof of our results. We instead spend some time describing the physical interpretation of our model equations and the choice of the functional setting in which we work.

The plan of the paper is the following: in Section 1 we recall the concept of hysteresis operator together with its most common properties which will be used later. In Section 2 we study the model equation (0.1): we first explain its physical meaning; then after a small presentation of the functional setting in which the problem will be framed (which is a bit unusual), we introduce a weak formulation in Sobolev spaces of the Cauchy problem related to equation (0.1). Under suitable assumptions on the data and on the operator $\overline{\mathcal{G}}$, we state an existence and uniqueness theorem for the solutions of our model equation. Section 3 is devoted instead to the analysis of the model equation (0.2): also in this case we briefly explain in which physical context the equation arises; then we still introduce a weak formulation in the Sobolev spaces of the initial and boundary value problem (with Dirichlet boundary conditions) associated to equation (0.2). Under suitable assumptions on the operator $\overline{\mathcal{F}}$ and on the data we state an existence result for the solutions of this model equation. We moreover include in some remarks further results which can be obtained for our model equations, without entering into details.

1. Some basic facts concerning hysteresis and hysteresis operators

We can illustrate the basic concept of hysteresis by means of a very simple example. Let us consider a sort of black box which transforms a continuous time-dependent scalar variable u - which is often called *input* – into a time-dependent variable w – which plays the role of the *output*. Just to do a couple of examples, in ferromagnetism u may correspond to the magnetic field H and w to the magnetic induction B; otherwise in plasticity u can be identified to the strain ε and w can represent instead the stress σ and so on.

At any instant t, $w(t)$ depends on the evolution of the input u and on the initial state of the system, or, more in general, on a variable η^0 which contains all the information about the initial state. So we have

$$\mathcal{F} : C^0([0,T]) \times X \to C^0([0,T]) \qquad w(t) = [\mathcal{F}(u,\eta^0)](t) \qquad \forall t \in [0,T], \qquad (1.1)$$

where X is a suitable metric space.

We have the *memory effect* when at any instant t the output $w(t)$ is not simply determined by the value $u(t)$ of the input at the same instant, but it depends also on the previous evolution of u. We have the *rate independence property* when the path of the couple $(u(t), v(t))$ is invariant with respect to any increasing time homeomorphism, i.e., there is no dependence on the derivative of u. It is this fact that allows us to draw the characteristic pictures of hysteresis in the (u, w)-plane. The operators showing the memory effect are called *memory (or Volterra) operators*, while those fulfilling both the memory and the rate independence property are called *hysteresis operators*.

The notion of hysteresis operator just introduced, (which is essentially due to Krasnosel'skiĭ, see [7]), can be used to model phenomena in which time is the only independent variable, as in O.D.E.s but it cannot be directly applied in situations like in problems involving P.D.E.s, where also space variables appear. In order to overcome this difficulty, it is enough to extend the operator like $\mathcal{F}$ in (1.1), acting on time-dependent functions, to some operator $\overline{\mathcal{F}}$ acting also on space-dependent functions.

The choice we make in this paper is to work with a more general class of memory operators of the following type

$$\overline{\mathcal{F}} : \mathcal{M}(\Omega; \mathcal{C}^0([0,T])) \to \mathcal{M}(\Omega; \mathcal{C}^0([0,T])), \tag{1.2}$$

where Ω is an open bounded set of $\mathbb{R}^N$, $N \geq 1$ and $\mathcal{M}(\Omega; \mathcal{C}^0([0,T]))$ is the Fréchet space of strongly measurable functions $\Omega \to \mathcal{C}^0([0,T])$, i.e., $\mathcal{M}(\Omega; \mathcal{C}^0([0,T]))$ is the set of functions $v : \Omega \to \mathcal{C}^0([0,T])$ such that there exists a sequence v_n of simple functions with $v_n \to v$ strongly in $\mathcal{C}^0([0,T])$, a.e. in Ω, endowed with the quasi-norm

$$||v||_{\mathcal{M}(\Omega;\mathcal{C}^0([0,T]))} := \int_\Omega \frac{||v||_{\mathcal{C}^0([0,T])}}{1 + ||v||_{\mathcal{C}^0([0,T])}} \, dx.$$

Nevertheless, we shall have always in mind the case in which the operator $\mathcal{F}$ is rate independent (and so when $\mathcal{F}$ is a hysteresis operator), which is the most frequent situation one can encounter in the applications.

We conclude the section by recalling some properties, useful in the following, that an operator like $\overline{\mathcal{F}}$ in (1.2) can satisfy. In particular $\overline{\mathcal{F}}$ can be:

• CAUSAL if

$$\begin{cases} \forall\, v_1, v_2 \in \mathcal{M}(\Omega; \mathcal{C}^0([0,T])), \ \ \forall\, t \in [0,T], \text{ if } v_1 = v_2 \text{ in } [0,t], \text{ a.e. in } \Omega, \\ \text{then } [\overline{\mathcal{F}}(v_1)](\cdot, t) = [\overline{\mathcal{F}}(v_2)](\cdot, t) \text{ a.e. in } \Omega; \end{cases} \tag{1.3}$$

• LIPSCHITZ CONTINUOUS from $L^2(\Omega; \mathcal{C}^0([0,T]))$ into $L^2(\Omega; \mathcal{C}^0([0,T])$ itself, if there exists a constant L such that, for any $u_1, u_2 \in L^2(\Omega; \mathcal{C}^0([0,T]))$

$$||\overline{\mathcal{F}}(u_1) - \overline{\mathcal{F}}(u_2)||_{L^2(\Omega;\mathcal{C}^0([0,T]))} \leq L \, ||u_1 - u_2||_{L^2(\Omega;\mathcal{C}^0([0,T]))}; \tag{1.4}$$

• STRONGLY CONTINUOUS, if

$$\begin{cases} \forall \{v_n \in \mathcal{M}(\Omega; \mathcal{C}^0([0,T]))\}_{n \in \mathbb{N}}, \ \text{ if } v_n \to v \text{ uniformly in } [0,T] \\ \text{a.e. in } \Omega, \text{ then } \overline{\mathcal{F}}(v_n) \to \overline{\mathcal{F}}(v) \text{ uniformly in } [0,T], \text{ a.e. in } \Omega; \end{cases} \tag{1.5}$$

• AFFINELY BOUNDED, if

$$\begin{cases} \exists\, \bar{L}, \ \exists\, \tau \in L^2(\Omega) : \forall v \in \mathcal{M}(\Omega; \mathcal{C}^0([0,T])), \\ ||[\overline{\mathcal{F}}(v)](x, \cdot)||_{\mathcal{C}^0([0,T])} \leq \bar{L} \, ||v(x, \cdot)||_{\mathcal{C}^0([0,T])} + \tau(x) \ \text{ a.e. in } \Omega; \end{cases} \tag{1.6}$$

- PIECEWISE MONOTONE, if

$$\begin{cases} \forall v \in \mathcal{M}(\Omega; \mathcal{C}^0([0,T])), \ \forall [t_1, t_2] \subset [0,T], \\ \text{if } v(x,\cdot) \text{ is affine in } [t_1,t_2] \text{ a.e. in } \Omega, \text{ then} \\ \{[\overline{\mathcal{F}}(v)](x,t_2) - [\overline{\mathcal{F}}(v)](x,t_1)\} \cdot [v(x,t_2) - v(x,t_1)] \geq 0 \text{ a.e. in } \Omega; \end{cases} \tag{1.7}$$

- PIECEWISE LIPSCHITZ CONTINUOUS, if

$$\begin{cases} \exists\, \hat{L} > 0 : \ \forall v \in \mathcal{M}(\Omega; \mathcal{C}^0([0,T])), \ \forall [t_1, t_2] \subset [0,T] \\ \text{if } v(x,\cdot) \text{ is affine in } [t_1,t_2] \text{ a.e. in } \Omega, \text{ then} \\ |[\overline{\mathcal{F}}(v)](x,t_2) - [\overline{\mathcal{F}}(v)](x,t_1)| \leq \hat{L}\, |v(x,t_1) - v(x,t_2)| \qquad \text{a.e. in } \Omega. \end{cases} \tag{1.8}$$

It is not restrictive to assume that $L = \bar{L} = \hat{L}$.

2. First model equation

2.1. Physical interpretation of the model equation (0.1)

If we couple in a suitable way the Maxwell equations with the Ohm law, considered in a domain $\mathcal{D} \subset \mathbb{R}^3$ which represents our electromagnetic material, (for more details on the topic see for example [6]), we get the following equation

$$\epsilon \frac{\partial^2 \vec{B}}{\partial t^2} + 4\pi\sigma \frac{\partial \vec{B}}{\partial t} + c^2 \nabla \times \nabla \times \vec{H} = 4\pi\, c\, \sigma \nabla \times \vec{g} \qquad \text{in } \mathcal{D}_T := \mathcal{D} \times (0,T),$$

where c is the speed of light in vacuum, ϵ the dielectric permittivity, $\vec{H}$ is the magnetic field, $\vec{B}$ is the magnetic induction, σ is the electric conductivity and finally $\vec{g}$ is an applied electromotive force.

We can simplify this equation by imposing severe restrictions on the geometry of the system; more precisely, let Ω be a domain of $\mathbb{R}^2$, we can assume that $\vec{H}$ is parallel to the z-axis and depends on the first two cartesian coordinates, x, y, i.e., $\vec{H} = (0, 0, H(x,y))$. This implies that

$$\nabla \times \nabla \times \vec{H} = (0, 0, -\triangle_{x,y} H) \qquad \text{where} \qquad \triangle_{x,y} = \frac{\partial^2}{\partial x^2} + \frac{\partial^2}{\partial y^2}.$$

Dealing with a strongly anisotropic material we can also assume $\vec{B} = (0, 0, B(x,y))$ and the same can be done for the known term. Choosing to not display the coefficients for the sake of simplicity, the previous equation becomes

$$\frac{\partial^2 B}{\partial t} + \frac{\partial B}{\partial t} - \triangle_{x,y} H = \tilde{f} \qquad \text{in } \Omega_T := \Omega \times (0,T). \tag{2.1}$$

At this point it is necessary to introduce a constitutive relation between B and H; we choose the following one

$$H = \mathcal{G}(B) + \lambda\, \dot{B}, \tag{2.2}$$

where $\mathcal{G}$ is a suitable hysteresis operator and $\lambda > 0$ is a given constant, depending on the geometry of the system. This relation can be for example obtained combining in series a ferromagnetic element with hysteresis (which is a rate independent element) and a conducting solenoid with a paramagnetic core (which is instead a rate dependent element). Coupling (2.1) and (2.2) we get exactly (0.1).

2.2. Statement of the main results

We fix an open bounded set $\Omega \subset \mathbb{R}^2$ of Lipschitz class with boundary Γ and set $Q := \Omega \times (0,T)$. We took $\mathbb{R}^2$ in consideration of the physical meaning of our model equation, even if actually our results are true also for the more general case of $\mathbb{R}^N$, $N \geq 1$. We consider the following operator

$$\overline{\mathcal{G}} : \mathcal{M}(\Omega; \mathcal{C}^0([0,T])) \to \mathcal{M}(\Omega; \mathcal{C}^0([0,T]))$$

which is assumed to satisfy (1.3) and (1.4).

We denote by $\gamma_0 : H^1(\Omega) \to H^{1/2}(\Gamma)$ the unique linear continuous trace operator such that $\gamma_0 v = v_{|\Gamma}$ for all $v \in \mathcal{C}^\infty(\bar{\Omega}) \cap H^1(\Omega)$.

Now we introduce the operator $A : H_0^1(\Omega) \to H^{-1}(\Omega)$ defined as follows

$$_{H^{-1}(\Omega)}\langle Au, v\rangle_{H_0^1(\Omega)} := \int_\Omega \nabla u \cdot \nabla v \, dx \qquad \forall u, v \in H_0^1(\Omega); \tag{2.3}$$

so it is clear that $Au = -\Delta_{x,y}\, u \left(:= -\dfrac{\partial^2 u}{\partial x^2} - \dfrac{\partial^2 u}{\partial y^2} \right)$ in the sense of distributions. Moreover we also define the operator A^{-1} which can be interpreted as the inverse of the operator $-\Delta_{x,y}$ associated with the homogeneous Dirichlet boundary conditions, i.e., for any $v \in H^{-1}(\Omega)$, $u = A^{-1}v$ if and only if $u \in H^1(\Omega)$, $-\Delta_{x,y} u = v$ in the sense of distributions and $\gamma_0 u = 0$ on $\partial\Omega = \Gamma$.

Before presenting the main result, we discuss a bit the setting of our model problem, as the choice of the right functional spaces to work with plays a fundamental role in order to have some positive results. In our case the choice we make is a bit unusual in the sense that we consider the following Hilbert triplet

$$L^2(\Omega) \subset H^{-1}(\Omega) \equiv (H^{-1}(\Omega))' \subset (L^2(\Omega))'$$

where the role of the pivot space is played by $H^{-1}(\Omega)$ and the injection (continuous and dense) of the space $L^2(\Omega)$ into $H^{-1}(\Omega)$ is defined in this way

$$_{H^{-1}(\Omega)}\langle j(f), \varphi\rangle_{H_0^1(\Omega)} := \int_\Omega f\, \varphi \, dx \qquad \forall\, f \in L^2(\Omega),\ \forall\, \varphi \in H_0^1(\Omega).$$

The Sobolev space $H^{-1}(\Omega)$ is endowed with the scalar product

$$(u, v)_{H^{-1}(\Omega)} :=_{H^{-1}(\Omega)}< u, A^{-1}v >_{H_0^1(\Omega)};$$

therefore, if we identify $H^{-1}(\Omega)$ to its dual $(H^{-1}(\Omega))'$ by means of the Riesz theorem, we have

$$_{(L^2(\Omega))'}\langle\psi, f\rangle_{L^2(\Omega)} =(\psi, f)_{H^{-1}(\Omega)} = (f,\psi)_{H^{-1}(\Omega)} =_{H^{-1}(\Omega)} \langle f, A^{-1}\psi\rangle_{H_0^1(\Omega)}$$

$$= \int_\Omega f\, A^{-1}\psi\, dx \qquad \forall f \in L^2(\Omega),\ \forall \psi \in H^{-1}(\Omega),$$

where we also used the fact that the scalar product is symmetric.

For the sake of simplicity, from now on we set $L^2(\Omega) := V$, $H^{-1}(\Omega) := H$ and $(L^2(\Omega))' := V'$. We assume that $u^0 \in V$, $v^0 \in H$ and $f \in L^2(0,T;H)$.

We want to solve the following problem

Problem 2.1. *To find two functions $u \in \mathcal{M}(\Omega;\mathcal{C}^0([0,T])) \cap L^2(Q)$ and $v \in L^2(Q)$ such that $\overline{\mathcal{G}}(u) \in L^2(Q)$ and for any $\psi \in H^1(0,T;V)$ with $\psi(\cdot,T) = 0$ a.e. in Ω*

$$\begin{aligned}&\int_0^T -{}_{V'}\langle v+u, \frac{\partial\psi}{\partial t}\rangle_V\, dt + \int_0^T\int_\Omega (v+\overline{\mathcal{G}}(u))\,\psi\, dx\, dt\\ &= \int_0^T {}_{V'}\langle f,\psi\rangle_V\, dt + {}_{V'}\langle (v^0+u^0)(\cdot),\psi(\cdot,0)\rangle_V\end{aligned} \tag{2.4}$$

$$-\int_0^T\int_\Omega u\frac{\partial\psi}{\partial t}\, dx\, dt = \int_0^T\int_\Omega v\,\psi\, dx\, dt + \int_\Omega u^0(\cdot)\,\psi(\cdot,0)\, dx. \tag{2.5}$$

Interpretation. It is not difficult to show that (2.4) and (2.5) yield

$$\begin{cases} A^{-1}\dfrac{\partial v}{\partial t} + A^{-1}\dfrac{\partial u}{\partial t} + v + \overline{\mathcal{G}}(u) = A^{-1}f \\ \dfrac{\partial u}{\partial t} = v \end{cases} \qquad \text{in } L^2(\Omega), \text{ a.e. in } (0,T)$$

and $u_{|_{t=0}} = u^0$, $v_{|_{t=0}} = v^0$ in the sense of the traces. If in addition the solution (u,v) is more regular in space, (as indeed happens at the end, see Remark 2.3) then (2.4) and (2.5) yield directly (0.1) in $L^2(0,T;V')$.

The main theorem which we obtain is the following

Theorem 2.2. (*Existence and uniqueness*) *Let us assume that the operator $\overline{\mathcal{G}} : \mathcal{M}(\Omega;\mathcal{C}^0([0,T])) \to \mathcal{M}(\Omega;\mathcal{C}^0([0,T]))$ fulfills* (1.3) *and* (1.4). *Suppose moreover that $u^0 \in V$, $v^0 \in H$ and $f \in L^2(0,T;H)$. Then Problem* 2.1 *has a unique solution $u \in H^1(0,T;V)$ and $v \in L^2(Q)$ such that $\overline{\mathcal{G}}(u) \in L^2(\Omega;\mathcal{C}^0([0,T]))$. If moreover $\overline{\mathcal{G}}$ fulfills* (1.8), *then we get that $\overline{\mathcal{G}}(u) \in H^1(0,T;L^2(\Omega))$.*

The proof of this result is based on the contraction mapping principle. We first consider any given function $z \in H^1(0,T;V)$ and solve the counterpart of Problem 2.1 with $\overline{\mathcal{G}}(u)$ replaced by $\overline{\mathcal{G}}(z)$. This procedure allows us to get rid of the nonlinearity in the hysteresis term. In a second step we construct an operator $J : H^1(0,T;V) \to H^1(0,T;V)$ which associates to any z the first element u of the pair (u,v) solution of the modified problem just considered. We consequently prove that J has a fixed point, which will be the desired unique solution of our model equation.

Remark 2.3. Once that the solution of our problem is also unique, it is not hard to obtain the Lipschitz continuous dependence of u from the data of our model problem. Moreover, using a standard characterization of the spaces $H_0^1(\Omega)$ and $H^1(\Omega)$ (which can be for example found in [3], Sect. IX.1, IX.4) we achieve a higher regularity result in space for the solution of our model problem.

3. Second model problem

Consider an open bounded set of Lipschitz class $\Omega \subset \mathbb{R}^N$, $N \geq 1$ with boundary Γ and set $Q := \Omega \times (0,T)$. Assume that the operator $\overline{\mathcal{F}} : \mathcal{M}(\Omega; \mathcal{C}^0([0,T])) \to \mathcal{M}(\Omega; \mathcal{C}^0([0,T]))$ satisfies (1.3), (1.5), (1.6), (1.7) and (1.8).

The causality property entails that $[\overline{\mathcal{F}}(v)](\cdot,0)$ depends just on $\overline{\mathcal{F}}$ and $v(\cdot,0)$; so we can set $\mathcal{H}_{\overline{\mathcal{F}}}(v(\cdot,0)) := [\overline{\mathcal{F}}(v)](\cdot,0)$ for all $v \in \mathcal{M}(\Omega; \mathcal{C}^0([0,T]))$.

For the sake of simplicity we set this time $V := H_0^1(\Omega)$, $H := L^2(\Omega)$ and $V' := H^{-1}(\Omega)$ and we consider V endowed with the norm $||u||_V := ||\nabla u||_{L^2(\Omega)^N}$. We then identify the space $L^2(\Omega)$ to its topological dual $(L^2(\Omega))'$; as the injection of V into $L^2(\Omega)$ is continuous and dense, $(L^2(\Omega))'$ can be identified to a subspace of V'. This yields the Hilbert triplet $V \subset H \equiv H' \subset V'$ with dense and continuous injections.

Now we denote by ${}_{V'}\langle \cdot, \cdot \rangle_V$ the duality pairing between V' and V and we then define the linear and continuous operator $A : V \to V'$ as in (2.3).

We assume that $u^0, w^0 = \mathcal{H}_{\overline{\mathcal{F}}}(u^0) \in L^2(\Omega)$ are given initial data.

The problem we want to solve is the following

Problem 3.1. *Let us consider a known function $\vec{v}$ such that $\vec{v}$, $\frac{\partial \vec{v}}{\partial t} \in L^\infty(Q)^N$ and $\nabla \cdot \vec{v} = 0$ a.e. in Q. We search for a function $u \in \mathcal{M}(\Omega; \mathcal{C}^0([0,T])) \cap L^2(0,T;V)$ such that $\overline{\mathcal{F}}(u) \in \mathcal{M}(\Omega; \mathcal{C}^0([0,T])) \cap L^2(Q)$ and for any $\psi \in H^1(0,T;L^2(\Omega)) \cap L^2(0,T;V)$ with $\psi(\cdot,T) = 0$ a.e. in Ω*

$$\begin{aligned}
\int_0^T \int_\Omega -(u + \overline{\mathcal{F}}(u)) \frac{\partial \psi}{\partial t}\, dx\, dt - \int_0^T \int_\Omega [\vec{v} \cdot \nabla \psi]\, (u + \overline{\mathcal{F}}(u))\, dx\, dt \\
+ \int_0^T \int_\Omega \nabla u \cdot \nabla \psi\, dx\, dt \\
= \int_0^T {}_{V'}\langle f, \psi \rangle_V\, dt + \int_\Omega [u^0(x) + w^0(x)]\, \psi(x,0)\, dx. \qquad (3.1)
\end{aligned}$$

Interpretation. The variational equation (3.1) can be interpreted as

$$\begin{cases} \dfrac{\partial w}{\partial t} + \vec{v} \cdot \nabla w - \triangle u = f & \text{in } V', \text{ a.e. in } (0,T) \\ w = (I + \overline{\mathcal{F}})(u) \end{cases} \qquad (3.2)$$

hence, integrating by parts in time in (3.1) we get $[u+\overline{\mathcal{F}}(u)]_{|t=0} = u^0 + w^0$ in V', in the sense of the traces.

Physical interpretation. Let $D \subset \mathbb{R}^3$ represent the region occupied by a porous medium. We consider the equation of continuity

$$\frac{\partial \theta}{\partial t} + \nabla \cdot \vec{q} = 0 \qquad \text{in } D_T = D \times (0,T),$$

where θ is the water content of the medium and $\vec{q}$ is the flux. We have $\theta = \varphi\, s$ where $\varphi : D \to [0,1]$ is the porosity of the medium and s is the saturation.

We couple this equation with Darcy's law

$$\vec{q} = -k\,(\nabla u + \rho\, g\, \vec{z})$$

where k is the hydraulic conductivity, u is the pressure, ρ is the density of the fluid, g is the gravity acceleration and $\vec{z}$ is the upward vertical unit vector. The saturation s and the pressure u are unknown.

Therefore the system we find is the following

$$\begin{cases} \varphi \dfrac{\partial s}{\partial t} - \nabla \cdot k\,(\nabla u + \rho\, g\, \vec{z}) = 0 \\ s = \mathcal{F}(u), \end{cases} \qquad \text{in } D_T, \tag{3.3}$$

where the dependence of s upon u is formally represented by the operator $\mathcal{F}$. Experimental results show the occurrence of a quite relevant hysteresis effect which has occasionally been represented by Preisach models in engineering literature. Some interesting results on this topic have been obtained for example in [1], [2].

Now, first of all we make the following strong assumption on the hydraulic conductivity: we suppose that k does not depend on the saturation s (as indeed happens) but it is a constant; then we suppose that in (3.3) the derivative in time is not a Eulerian derivative but a *material* derivative. At this point, if we want to express this resulting equation in terms of the Eulerian derivative, the system we get is included in our model system (3.2), which may then represent a model with saturation versus pressure constitutive relation with hysteresis and with a term of transport.

The existence result we are able to state is the following

Theorem 3.2. (*Existence*) *Let us assume that the operator* $\overline{\mathcal{F}} : \mathcal{M}(\Omega; \mathcal{C}^0([0,T])) \to \mathcal{M}(\Omega; \mathcal{C}^0([0,T]))$ *fulfills* (1.3), (1.5) (1.6), (1.7) *and* (1.8). *Moreover let*

$$f \in L^2(Q), \;\; u^0 \in V, \;\; w^0 \in H.$$

Then Problem 3.1 *admits at least one solution* $u \in H^1(0,T;H) \cap L^\infty(0,T;V)$ *such that* $\overline{\mathcal{F}}(u) \in H^1(0,T;H)$.

The technique we use for proving this existence result is based on approximation by implicit time discretization, a priori estimates and passage to the limit by compactness. This approximation procedure is quite convenient in the analysis of equations that include a hysteresis operator, as in any time-step we have to solve a stationary problem in which the hysteresis operator is reduced to the superposition with a nonlinear function.

Remark 3.3. As the equation is quasilinear, one cannot expect to have uniqueness for all the choices of hysteresis operators $\overline{\mathcal{F}}$. However when $\overline{\mathcal{F}}$ is a Prandtl-Ishlinskiĭ operator of play type, the solution of Problem 3.1 is also unique. The proof does not relay on the Hilpert's inequality as indeed happens in [11], Sect. IX.2, but exploits instead the properties of $\overline{\mathcal{F}}$. Another interesting result which can be established is the dependence of the solutions from the data. Assume to have a sequence of memory operators $\overline{\mathcal{G}}_n$ converging pointwise in $\mathcal{C}^0([0,T])$, a.e. in Ω to some operator $\overline{\mathcal{G}}$; if u_n are the solutions of Problem 3.1 with $\overline{\mathcal{F}}$ replaced by $\overline{\mathcal{G}}_n$, then it turns out that there exists u such that $u_n \to u$ in some suitable topology and $\overline{\mathcal{G}}_n(u_n) \to \overline{\mathcal{G}}(u)$ strongly in $L^2(Q)$. Moreover u is a solution of Problem 3.1 with $\overline{\mathcal{F}}$ replaced by $\overline{\mathcal{G}}$. The idea contained in the proof is new and also the assumptions we take on $\overline{\mathcal{G}}_n$ are weaker than the ones usually employed in results of this kind. The proof exploits the properties of the operators $\overline{\mathcal{G}}_n$ and the uniform convergence in time of the sequence of our approximate solutions (pointwise convergence would not be enough for our purposes).

Remark 3.4. We develop the results of this section working with Dirichlet boundary conditions, other alternative choices of boundary conditions are possible; the discussion of these different situations is still work in progress.

Acknowledgment

The author wishes to thank Prof. A. Visintin for having proposed the subject developed in the paper and for many useful discussions. This work is partially supported by INdAM.

References

[1] F. Bagagiolo, A. Visintin: *Hysteresis in filtration through porous media*, Z. Anal. Anwendungen **19** (4) (2000) 977–998.

[2] F. Bagagiolo, A. Visintin: *Porous media filtration with hysteresis*, Adv. Math. Sci. Appl. **14** (2004) 379–403.

[3] H. Brezis: *Analyse fonctionnelle*, Théorie et applications, Masson, Paris (1983).

[4] M. Brokate, J. Sprekels: *Hysteresis and phase transitions*, Applied Mathematical Sciences, **121**, Springer-Verlag, New York (1996).

[5] M. Eleuteri: *On some P.D.E.s with hysteresis*, Ph.D. Thesis, UTM PhDTS n. 47, Trento, February 2006.

[6] J.D. Jackson: *Classical electrodynamics*, New York, Wiley, 1975.

[7] M.A. Krasnosel'skiĭ, B.M. Darinskiĭ, I.V. Emelin, P.P. Zabreĭko, E.A. Lifsic, A.V. Pokrovskiĭ: *Hysterant operator*, Soviet Math. Dokl., **11** (1970) 29–33.

[8] M.A. Krasnosel'skiĭ, A.V. Pokrovskiĭ: *Systems with hysteresis*, Springer, Berlin (1989), Russian edition: Nauka, Moscow (1983).

[9] P. Krejčí: *Hysteresis, convexity and dissipation in hyperbolic equations*, Tokyo: Gakkotosho, (1996).

[10] I.D. Mayergoyz: *Mathematical models of hysteresis*, Springer, New York (1991).

[11] A. Visintin: *Differential Models of Hysteresis*, Springer, (1994).

Michela Eleuteri
Department of Mathematics of Trento
Via Sommarive, 14
I-38050 Povo, Trento, Italy
e-mail: `eleuteri@science.unitn.it`

International Series of Numerical Mathematics, Vol. 154, 169–178

Embedding Theorem for Phase Field Equation with Convection

Takesi Fukao

Abstract. In this paper, we shall prove the existence of solutions for the system of second order partial differential equations. This system is constructed by the phase field equations with a convection described by the Navier-Stokes equations in a liquid region. In our setting, this liquid region is also unknown, which is defined by the solution of the phase field equations. In order to determine the liquid region by the unknown parameter, which is called order parameter, we need to get the continuity. From the L^2 framework, we shall obtain the smoothness of the order parameter by the compactness theorem of Aubin's type.

Mathematics Subject Classification (2000). Primary 35K55; Secondary 76D05.

Keywords. Phase field system, Navier-Stokes equations, compactness method.

1. Introduction

We consider a material which has two physical phases, liquid and solid. We are interested in the dynamics of the interface, taking account of a convective flow in the liquid region. Our idea is to describe this phenomenon by the weak formulation of the phase field equations with a convection. And this convection is governed by the Navier-Stokes equations in the unknown liquid region. Our purpose is to show the existence of some solutions. The phase field system is one of the famous model which describes the solid-liquid phase transition phenomena by the partial differential equations. This model has the strong relationship between the Stefan problem which is also well known as the effective model for the phase transition. About the Stefan problem we have a result [6] from the stand point of the practical situation. But the weak solution of the Stefan problem is discontinuous in general. So it is not easy to guarantee the fact that the liquid region is the exact open set. Actually the result of [6] is the existence theory for an approximate problem. On the other hand, in the case of the prototype phase filed equations we can gain the regularity of the order parameter if we regularize the convection. So we apply the same methods in [6] and find the exact solution. Considering the problem as the weak formulation and applying the penalty method, we settle the difficulty of the unknown region. And then we shall obtain the existence result.

2. System of phase field and Navier-Stokes equations

In this section we introduce a simplified system. After define our solution we shall give the main theorem. Let $0 < T < +\infty$, $t \in [0,T]$ and $\Omega_m(t) \subset \mathbb{R}^3$ be the bounded time dependent domain with smooth boundary $\Gamma_m(t) := \partial\Omega_m(t)$. Moreover we define a non-cylindrical domain and its lateral boundary by

$$Q_m := \bigcup_{t\in(0,T)} \{t\} \times \Omega_m(t), \qquad \Sigma_m := \bigcup_{t\in(0,T)} \{t\} \times \Gamma_m(t).$$

$\Omega_m(t)$ is occupied by a material having two phases, liquid and solid. Our purpose is to find the distribution of these regions. For any $(t,x) \in Q_m$, let $\chi := \chi(t,x)$ be the order parameter which stands for the state of the material. These regions are separated by an unknown interface. From the stand point of the Stefan problem it is a natural setting that the sharp interface is defined by the 0-level curve of χ. On the other hand in the case of the phase field equations the set $\{(t,x) \in Q_m; \chi(t,x) = 0\}$ has the measure in general. So in our phase field equations we image the virtual solid-liquid interface namely we call the set

$$\Omega_\ell(t) := \Omega_m(t) \setminus \overline{\{x \in \Omega_m(t); \chi(t,x) < 0\}}$$

by the liquid region, the set $\Omega_s(t) := \Omega_m(t) \setminus \overline{\Omega_\ell(t)}$ by the solid region and $S(t) := \Omega_m(t) \setminus \{\Omega_\ell(t) \cup \Omega_s(t)\}$ by the virtual interface. If χ is continuous in Q_m, then $\Omega_s(t)$ and $\Omega_\ell(t)$ are open sets and $\Omega_m(t) = \Omega_\ell(t) \cup S(t) \cup \Omega_s(t)$ for each $t \in [0,T]$. And then we define

$$S(\chi) := \bigcup_{t\in(0,T)} \{t\} \times S(t), \quad Q_i(\chi) := \bigcup_{t\in(0,T)} \{t\} \times \Omega_i(t) \quad \text{for } i = s, \ell.$$

We consider the following system of a couple of the phase field equations and the Navier-Stokes equations: $\theta := \theta(t,x)$ be the temperature, $\mathbf{v} = \mathbf{v}(t,x)$ be the convective vector and $p_\ell := p_\ell(t,x)$ be the pressure,

$$D_t^\varepsilon(\mathbf{v})\theta + D_t^\varepsilon(\mathbf{v})\chi - \Delta\theta = f \quad \text{in } Q_m, \tag{2.1}$$

$$D_t^\varepsilon(\mathbf{v})\chi - \Delta\chi + \chi^3 - \chi = \theta \quad \text{in } Q_m, \tag{2.2}$$

$$\frac{\partial \mathbf{v}}{\partial t} + (\mathbf{v}\cdot\nabla)\mathbf{v} - \Delta\mathbf{v} = \boldsymbol{g}(\theta) - \nabla p_\ell \quad \text{in } Q_\ell(\chi), \tag{2.3}$$

$$\mathrm{div}\mathbf{v} = 0 \quad \text{in } Q_\ell(\chi), \tag{2.4}$$

$$\mathbf{v} = \mathbf{v}_D \quad \text{in } Q_s(\chi) \cup S(\chi), \tag{2.5}$$

$$\frac{\partial\theta}{\partial\mathbf{n}} = 0, \quad \frac{\partial\chi}{\partial\mathbf{n}} = 0, \quad \mathbf{v} = \mathbf{v}_D \quad \text{on } \Sigma_m, \tag{2.6}$$

$$\theta(0,\cdot) = \theta_0, \quad \chi(0,\cdot) = \chi_0, \quad \mathbf{v}(0,\cdot) = \mathbf{v}_0 \quad \text{in} \quad \Omega_{m0} := \Omega_m(0), \tag{2.7}$$

where for each $\varepsilon > 0$, the operator $D_t^\varepsilon(\mathbf{v})$ stands for the regularization of the material derivative $\partial/\partial t + (\rho_\varepsilon * \mathbf{v}) \cdot \nabla$ with the unknown $\mathbf{v}$, where the convolution $(\rho_\varepsilon * \mathbf{v})(t,x) = (\rho_\varepsilon(t) * v_1(t,x), \rho_\varepsilon(t) * v_2(t,x), \rho_\varepsilon(t) * v_3(t,x))$ is the mollification of $\mathbf{v}$ and $\rho_\varepsilon(t)$ is the usual mollifier with respect to t; $f, \boldsymbol{g}, \mathbf{v}_D, \theta_0, \chi_0, \mathbf{v}_0$ are given functions; $\mathbf{n} = \mathbf{n}(t,x) := (n_1(t,x),\ n_2(t,x),\ n_3(t,x))$ is the 3-dimensional unit vector outward normal to $\Gamma_m(t)$ at $x \in \Gamma_m(t)$. Throughout this paper we assume that the shape of domain $\Omega_m(t)$ smoothly changes in time in the following sense:

(A1) There exists a bounded domain $\Omega \subset \mathbb{R}^3$ with smooth boundary $\Gamma := \partial\Omega$ such that $\Omega_m(t) \subset \Omega$ for all $t \in [0,T]$. Moreover there exists a transformation $\mathbf{y} \in \mathbf{C}^3(\overline{Q}) := C^3(\overline{Q})^3$ with $Q := (0,T) \times \Omega$ which gives a C^3-diffeomorphism $\mathbf{y}(t,\cdot) := (y_1(t,\cdot),\ y_2(t,\cdot),\ y_3(t,\cdot))$ from $\overline{\Omega}$ onto itself for all $t \in [0,T]$ such that

$$\mathbf{y}(t, \overline{\Omega_m(t)}) = \overline{\Omega_{m0}} \quad \text{for all } t \in [0,T], \quad \mathbf{y}(0,\cdot) = \mathbf{I} \text{ (identity) on } \overline{\Omega_{m0}}.$$

We can say that in the liquid region $Q_\ell(\chi)$ the vector $\mathbf{v}$ coincides with the convection described by the Navier-Stokes equations (2.3) and (2.4). In the solid region $Q_s(\chi)$ the vector $\mathbf{v}$ coincides with the given vector $\mathbf{v}_D := \mathbf{v}_D(t,x)$, for example we interpret it as the deformation speed. It is given by the above assumption (A1) by $\partial\mathbf{x}/\partial t(t, \mathbf{y}(t,x))$ where $\mathbf{x} := \mathbf{y}^{-1}$. Anyway we assume that the given function $\mathbf{v}_D$ satisfies the following compatibility condition:

(A2) A vector function $\mathbf{v}_D \in \mathbf{C}^2(\overline{Q})$ so that

$$\mathrm{div}\mathbf{v}_D(t,\cdot) = 0 \quad \text{in } \Omega_m(t) \quad \text{for all } t \in [0,T],$$

$$\mathbf{v}_D \cdot \mathbf{n} = v_{\mathbf{n}} \quad \text{on } \Sigma_m,$$

where $v_{\mathbf{n}}(t,\cdot)$ is the normal speed of $\Gamma_m(t)$ defined by $\partial\mathbf{x}/\partial t(t,\mathbf{y}(t,\cdot)) \cdot \mathbf{n}(t,\cdot)$.

Let $\Sigma := (0,T) \times \Gamma$ and $Q_T := (0,T) \times \Omega_{m0}$. We use the following notations:

$$H := L^2(\Omega_{m0}), \quad V := H^1(\Omega_{m0}) \ (:= W^{1,2}(\Omega_{m0})),$$

with the usual norms. H is a Hilbert space with standard inner product $(\cdot,\cdot)_H$. We see that the following relations hold:

$$V \hookrightarrow H \hookrightarrow V^*,$$

where $\hookrightarrow$ means that the embedding is compact. Moreover we use the following notations for vector valued function spaces:

$$\boldsymbol{\mathcal{D}}_\sigma(\Omega) := \left\{ \mathbf{z} \in \mathbf{C}_0^\infty(\Omega); \mathrm{div}\mathbf{z} = 0 \text{ in } \Omega \right\},$$

$$\mathbf{H} := \mathbf{L}_\sigma^2(\Omega), \quad \mathbf{Y} := \mathbf{L}_\sigma^4(\Omega), \quad \mathbf{V} := \mathbf{H}_\sigma^1(\Omega), \quad \mathbf{X} := \mathbf{W}_\sigma^{1,4}(\Omega),$$

where $\mathbf{L}_\sigma^2(\Omega)$, $\mathbf{L}_\sigma^4(\Omega)$, $\mathbf{H}_\sigma^1(\Omega)$ and $\mathbf{W}_\sigma^{1,4}(\Omega)$ are the closures of $\boldsymbol{\mathcal{D}}_\sigma(\Omega)$ in spaces $\mathbf{L}^2(\Omega)$, $\mathbf{L}^4(\Omega)$, $\mathbf{H}^1(\Omega)$ and $\mathbf{W}^{1,4}(\Omega)$, respectively. They are equipped with the usual product norms. We see that $\mathbf{H}$ is a Hilbert space with the usual inner product $(\cdot,\cdot)_{\mathbf{H}}$. Now we formulate the Navier-Stokes equations as a homogeneous Dirichlet boundary value problem. We put $\mathbf{w} := \mathbf{v} - \mathbf{v}_D$ on Q_m and $\mathbf{w}_0 := \mathbf{v}_0 - \mathbf{v}_D(0)$ on Ω_{m0}. Then

$$\mathbf{w} = 0 \quad \text{on } Q_s(\chi) \cup S(\chi) \quad \text{and on } \Sigma_m. \tag{2.8}$$

For simplicity, we denote by $\mathbf{w}$ and $\mathbf{w}_0$ again the 0-extensions of them onto Q and Ω, respectively. A weak variational formulation of Navier-Stokes equations is described in terms of $\mathbf{w}$ and $\mathbf{w}_0$ as follows:

$$-\int_0^T (\mathbf{w},\boldsymbol{\eta}')_{\mathbf{H}}dt + \int_0^T a(\mathbf{w},\boldsymbol{\eta})dt + \int_0^T b(t;\mathbf{w},\mathbf{w},\boldsymbol{\eta})dt + \int_0^T c(t;\mathbf{w},\boldsymbol{\eta})dt \quad (2.9)$$
$$= \int_0^T (\boldsymbol{g}_L(\theta),\boldsymbol{\eta})_{\mathbf{H}}dt + (\mathbf{w}_0,\boldsymbol{\eta}(0))_{\mathbf{H}} \quad \text{for all } \boldsymbol{\eta} \in \mathbf{W}(\chi),$$

subject to the constraint $\mathbf{w} = 0$ a.e. on $Q_s(\chi)$, where

$$\mathbf{W}(\chi) := \left\{ \boldsymbol{\eta} \in L^4(0,T;\mathbf{X}); \begin{array}{l} \boldsymbol{\eta}' \in L^2(0,T;\mathbf{H}), \quad \boldsymbol{\eta}(T,\cdot) = 0 \text{ a.e. on } \Omega, \\ \boldsymbol{\eta} = 0 \quad \text{a.e. on } Q \setminus Q_\ell(\chi) \end{array} \right\};$$

here $\boldsymbol{\eta}' := \partial\boldsymbol{\eta}/\partial t$. In order to emphasize the dependence of the class of test functions upon χ we denote it by $\mathbf{W}(\chi)$. Moreover $a(\cdot,\cdot) : \mathbf{V}\times\mathbf{V} \to \mathbb{R}$ and for each $t \in [0,T]$, $b(t;\cdot,\cdot,\cdot) : \mathbf{V}\times\mathbf{V}\times\mathbf{V} \to \mathbb{R}$, $c(t;\cdot,\cdot) : \mathbf{H}\times\mathbf{H} \to \mathbb{R}$ are defined by

$$a(\mathbf{z},\boldsymbol{\eta}) := \sum_{i=1}^3 \int_\Omega \nabla z_i \cdot \nabla \eta_i dx \quad \text{for all } \mathbf{z},\boldsymbol{\eta} \in \mathbf{V},$$

$$b(t;\mathbf{z},\bar{\mathbf{z}},\boldsymbol{\eta}) := \sum_{i=1}^3 \int_\Omega ((\mathbf{z}+\mathbf{v}_D(t))\cdot\nabla \bar{z}_i)\eta_i dx \quad \text{for all } \mathbf{z},\bar{\mathbf{z}},\boldsymbol{\eta} \in \mathbf{V},$$

$$c(t;\mathbf{z},\boldsymbol{\eta}) := \sum_{i=1}^3 \int_\Omega (\mathbf{z}\cdot\nabla(v_D)_i(t))\eta_i dx \quad \text{for all } \mathbf{z},\boldsymbol{\eta} \in \mathbf{H},$$

and for each $z \in L^2(Q)$, $\boldsymbol{g}_L(z) \in L^2(0,T;\mathbf{H})$ is defined by

$$\boldsymbol{g}_L(z) := \begin{cases} P_L\left[\boldsymbol{g}(z) - \dfrac{\partial \mathbf{v}_D}{\partial t} - (\mathbf{v}_D\cdot\nabla)\mathbf{v}_D + \nu_\ell \Delta \mathbf{v}_D\right] & \text{on } Q_m, \\ 0 \quad \text{otherwise,} \end{cases}$$

where $P_L : \mathbf{L}^2(\Omega) \to \mathbf{H}$ is the Leray projector.

Definition 2.1. The triplet $\{\theta,\chi,\mathbf{v}\} \in L^\infty(Q_m)\times C(\overline{Q_m})\times \mathbf{L}^2(Q_m)$ is called a weak solution of the system if (D1)–(D3) are satisfied:

(D1) θ and χ satisfy the following estimates

$$\sup_{t\in(0,T)} |\theta(t)|_{H^1(\Omega_m(t))} < +\infty, \qquad \int_0^T |\theta(t)|^2_{H^2(\Omega_m(t))}dt < +\infty,$$

$$\int_0^T \left|\frac{\partial\chi}{\partial t}(t)\right|^2_{H^1(\Omega)} dt < +\infty, \qquad \sup_{t\in(0,T)} |\chi(t)|_{H^2(\Omega_m(t))} < +\infty;$$

(D2) $\mathbf{w} := \mathbf{v} - \mathbf{v}_D \in L^\infty(0,T;\mathbf{H})\cap L^2(0,T;\mathbf{V})$, $\mathbf{w}$ is weakly continuous from $[0,T]$ into $\mathbf{H}$ and $\mathbf{w} = 0$ a.e. on $Q_s(\chi)$;

(D3) $\{\theta,\chi,\mathbf{v}\}$ satisfy (2.1)–(2.7) in the variational sense.

For each $\varepsilon > 0$, our main theorem is now stated as follows:

Theorem 2.2 (Main Theorem). *Assume that* (A1) *and* (A2) *hold. Moreover* $f \in L^\infty(Q_m)$, $\boldsymbol{g} \in \mathbf{C}^{0,1}(\mathbb{R})$, $\theta_0 \in H^1(\Omega_{m0})$, $\chi_0 \in H^2(\Omega_{m0})$ *and* $\mathbf{v}_0 \in \mathbf{L}^2(\Omega_{m0})$ *with* $\mathrm{div}\mathbf{v}_0 = 0$ *in* Ω_{m0}. *Then there exists at least one solution* $\{\theta, \chi, \mathbf{v}\}$ *of our system.*

Remark 2.3. In the 2-dimensional case, Planas and Boldrini [10] obtained the existence result for the same kind of the problem without the time regularization for the convection by applying the L^p-theory of parabolic equations.

3. Phase field equations with given convection

In this section we discuss about the solvability of the phase field equations with given convection, and recall the known result for the Navier-Stokes equation in the non-cylindrical domain. Finally we shall note the key of the proof. Firstly, we assume that the convective vector $\tilde{\mathbf{v}} := (\tilde{v}_1, \tilde{v}_2, \tilde{v}_3)$ is given. Now for each $s_0, s \in [0, T]$ with $0 \le s_0 < s \le T$, we use the following notations:

$$Q(s_0, s) := (s_0, s) \times \Omega, \quad Q_m(s_0, s) := \bigcup_{t \in (s_0, s)} \{t\} \times \Omega_m(t).$$

Moreover we consider the following auxiliary system:

$$-\int_{Q_m(s_0,s)} \theta \tilde{D}_t^\varepsilon \eta dxdt - \int_{Q_m(s_0,s)} \chi \tilde{D}_t^\varepsilon \eta dxdt + \int_{Q_m(s_0,s)} \nabla\theta \cdot \nabla\eta dxdt \tag{3.1}$$
$$= \int_{Q_m(s_0,s)} f\eta dxdt + \int_{\Omega_m(s_0)} \theta_{s_0}\eta(s_0)dx + \int_{\Omega_m(s_0)} \chi_{s_0}\eta(s_0)dx,$$

$$-\int_{Q_m(s_0,s)} \chi \tilde{D}_t^\varepsilon \eta dxdt + \int_{Q_m(s_0,s)} \nabla\chi \cdot \nabla\eta dxdt + \int_{Q_m(s_0,s)} (\chi^3 - \chi)\eta dxdt \tag{3.2}$$
$$= \int_{Q_m(s_0,s)} \theta\eta dxdt + \int_{\Omega_m(s_0)} \chi_{s_0}\eta(s_0)dx,$$

for all $\eta \in H^1(Q_m(s_0, s))$ with $\eta(s, \cdot) = 0$ a.e. on $\Omega_m(s)$, where $\tilde{D}_t^\varepsilon := D_t^\varepsilon(\tilde{\mathbf{v}})$. Assume that $\theta_{s_0} \in H^1(\Omega_m(s_0)), \chi_{s_0} \in H^2(\Omega_m(s_0))$. Moreover $\tilde{\mathbf{v}} - \mathbf{v}_D \in L^2(0, T; \mathbf{V}) \cap L^\infty(0, T; \mathbf{H})$ and $\tilde{\mathbf{v}}$ satisfies the following compatibility condition

$$\tilde{\mathbf{v}} \cdot \mathbf{n} = v_{\mathbf{n}} \quad \text{on } \Sigma_m. \tag{3.3}$$

Then there exists uniquely $\{\tilde{\theta}, \tilde{\chi}\} \in H^1(Q_m(s_0, s)) \times H^1(Q_m(s_0, s))$ such that

$$\sup_{t \in (s_0,s)} |\tilde{\theta}(t)|_{H^1(\Omega_m(t))} < +\infty, \quad \int_{s_0}^{s} |\tilde{\theta}(t)|^2_{H^2(\Omega_m(t))} dt < +\infty,$$

$$\sup_{t \in (s_0,s)} |\tilde{\chi}(t)|_{H^1(\Omega_m(t))} < +\infty, \quad \int_{s_0}^{s} |\tilde{\chi}(t)|^2_{H^2(\Omega_m(t))} dt < +\infty,$$

and $\{\tilde{\theta}, \tilde{\chi}\}$ satisfy the weak formulations (3.1) and (3.2). See Fukao [5], or more general approach by Schimperna [13]. Here we recall an important result of the embedding theorem for spaces $L^2(0,T;H^1(\Omega)) \cap L^\infty(0,T;L^2(\Omega))$. For example, Chapter 3, Section 2 in the book of Ladyženskaja, Solonnikov and Ural'ceva [9]

$$|u|_{L^r(0,T;L^q(\Omega))} \le c_1 |\nabla u|^{1-2/r}_{L^2(0,T;\mathbf{L}^2(\Omega))} |u|^{2/r}_{L^\infty(0,T;L^2(\Omega))},$$

where q and r are arbitrary positive numbers satisfying the condition

$$\frac{1}{r} + \frac{3}{2q} = \frac{3}{4} \quad \text{with } q \in [2,6], \quad r \in [2,+\infty], \tag{3.4}$$

and c_1 is a positive constant. We have the following estimate especially the key point is the independence of $\tilde{\mathbf{v}}$.

Lemma 3.1. *For any $s_0, s \in [0,T]$ with $0 \le s_0 < s \le T$, there exists a positive constant M_1 depend on $|\theta_{s_0}|_{L^2(\Omega_m(s_0))}$, $|\chi_{s_0}|_{L^2(\Omega_m(s_0))}$ and $|f|_{L^2(Q_m(s_0,s))}$, independent of $\tilde{\mathbf{v}}$ such that*

$$\sup_{t\in(s_0,s)} |\tilde{\theta}(t)|_{L^2(\Omega_m(t))} + \int_{s_0}^{s} |\tilde{\theta}(t)|^2_{H^1(\Omega_m(t))} dt \le M_1. \tag{3.5}$$

$$\sup_{t\in(s_0,s)} |\tilde{\chi}(t)|_{L^2(\Omega_m(t))} + \int_{s_0}^{s} |\tilde{\chi}(t)|^2_{H^1(\Omega_m(t))} dt + |\tilde{\chi}|_{L^4(Q_m(s_0,s))} \le M_1. \tag{3.6}$$

Proof. Using Green-Stokes' formula with the help of the divergence freeness, the compatibility condition (3.3) and Gronwall's inequality we get the conclusion. □

Using the same method of Theorem 7.1 in Chapter 3, Section 7 of the book by Ladyženskaja, Solonnikov and Ural'ceva [9], we obtain the following global boundedness:

Lemma 3.2. *For any $s_0, s \in [0,T]$ with $0 \le s_0 < s \le T$, there exists a positive constant M_2 depend on $|\theta_{s_0}|_{L^2(\Omega_m(s_0))}$ and $|\chi_{s_0}|_{L^2(\Omega_m(s_0))}$ independent of $\tilde{\mathbf{v}}$ such that*

$$|\tilde{\chi}|_{L^\infty(Q_m(s_0,s))} \le M_2. \tag{3.7}$$

Proof. From the independence of $\tilde{\mathbf{v}}$ in the estimate of Lemma 3.1, we take $\eta = [\tilde{\chi} - M]^+$ in (3.2) with some large positive constant M. And then $\tilde{\chi} - \tilde{\chi}^3 = \tilde{\chi}(1-\tilde{\chi}^2) \le \tilde{\chi}$ on $\{(t,x) \in Q_m(s_0,s); \tilde{\chi}(t,x) \ge M\}$. So thanks to the result of [9], it is enough to show that $\tilde{\theta}$ is bounded with respect to the norm of $L^{r^*}(s_0,s)$ as the $L^{q^*}(\Omega_m(t))$ valued function, where q^* and r^* are arbitrary positive numbers satisfying the condition

$$\frac{1}{r^*} + \frac{3}{2q^*} = 1 - \kappa, \tag{3.8}$$

with

$$q^* \in \left[\frac{3}{2(1-\kappa)}, +\infty\right], \quad r^* \in \left[\frac{1}{1-\kappa}, +\infty\right], \quad 0 < \kappa < 1.$$

By virtue of (3.4) and Lemma 3.1 with $\kappa = 1/4$ we get the conclusion. □

Lemma 3.3. *For any $s_0, s \in [0,T]$ with $0 \le s_0 < s \le T$, there exists a positive constant M_3 depend on $|\theta_{s_0}|_{L^2(\Omega_m(s_0))}$, $|\chi_{s_0}|_{L^2(\Omega_m(s_0))}$, $|f|_{L^\infty(Q_m(s_0,s))}$ and $|\tilde{\mathbf{v}}|_{L^2(s_0,s;\mathbf{V})}$ such that*

$$\int_{s_0}^{s} |\tilde{\chi}(t)|^2_{H^2(\Omega_m(t))} dt + \sup_{t\in(s_0,s)} |\tilde{\chi}(t)|_{H^1(\Omega_m(t))} \le M_3. \tag{3.9}$$

Proof. Consider the strong formulation of (3.2). For any $\tau \in [s_0, s]$, multiplying the function $-\Delta\tilde{\chi}$ and integrating over $Q_m(s_0,\tau)$ with respect to t and x. Recall the Gagliardo-Nirenberg inequality

$$|\nabla\tilde{\chi}|^2_{\mathbf{L}^4(\Omega_m(t))} \le c_2 |\tilde{\chi}|_{H^2(\Omega_m(t))} |\tilde{\chi}|_{L^\infty(\Omega_m(t))},$$

where c_2 is a positive constant. Then by using Lemma 3.2 and Young's inequality for $\sigma_1 > 0$ the following estimate holds

$$\begin{aligned} &|\nabla\tilde{\chi}(\tau)|^2_{\mathbf{L}^2(\Omega_m(\tau))} + 2(1-2\sigma_1)\int_{s_0}^{\tau} |\tilde{\chi}|^2_{H^2(\Omega_m(t))} dt \\ \le\ & \frac{(c_2 M_2)^2 |\tilde{\mathbf{v}}|^2_{L^2(s_0,s;\mathbf{V})}}{2\sigma_1} + M_1^2 + \frac{M_1}{4\sigma_1} + 2M_1 + |\chi_{s_0}|^2_{H^1(\Omega_m(s_0))}, \end{aligned}$$

for all $\tau \in [s_0, s]$. Thus we get the conclusion. □

Lemma 3.4. *For any $s_0, s \in [0,T]$ with $0 \le s_0 < s \le T$, there exists a positive constant M_4 depend on $|\theta_{s_0}|_{H^1(\Omega_m(s_0))}$, $|\chi_{s_0}|_{H^2(\Omega_m(s_0))}$ and $|\tilde{\mathbf{v}}|_{L^2(s_0,s;\mathbf{V})}$ such that*

$$\sup_{t\in(s_0,s)} |\tilde{\chi}(t)|_{H^2(\Omega_m(t))} \le M_4. \tag{3.10}$$

Proof. We consider the following auxiliary equation with $U = \partial\tilde{\chi}/\partial t$.

$$\tilde{D}_t^\varepsilon U - \Delta U + 3\chi^2 U - U = G \quad \text{in } Q_m(s_0,s),$$

$$\frac{\partial U}{\partial \mathbf{n}} = 0 \quad \text{on } \Gamma_m(s_0,s),$$

$$U(s_0) = U_{s_0} := -D_t^\varepsilon(\tilde{\mathbf{v}}(s_0))\chi(s_0) + \Delta\chi_{s_0} - \chi_{s_0}^3 + \chi_{s_0} + \theta_{s_0} \quad \text{on } \Omega_m(s_0),$$

where $G := \partial\tilde{\theta}/\partial t - (\partial(\rho_\varepsilon * \tilde{\mathbf{v}})/\partial t)\cdot\nabla\tilde{\chi})$. Now $\theta_{s_0} \in H^1(\Omega_m(s_0))$, $\chi_{s_0} \in H^2(\Omega_m(s_0))$ and $\mathbf{v}_D \in \mathbf{C}^2(\overline{Q})$, so the above equation of the initial and boundary value problem with given coefficient can be solved. Then U satisfies

$$\sup_{t\in(s_0,s)} |U(t)|_{L^2(\Omega_m(t))} + \int_{s_0}^{s} |U(t)|^2_{H^1(\Omega_m(t))} dt \le M_4', \tag{3.11}$$

where M_4' is a positive constant depend on $|\theta_{s_0}|_{L^2(\Omega_m(s_0))}$, $|\chi_{s_0}|_{H^2(\Omega_m(s_0))}$ and $|\tilde{\mathbf{v}}|_{L^2(s_0,s;\mathbf{V})}$. Finally thanks to Lemma 3.1, 3.2, 3.3 and 3.4 with the equation $\Delta\tilde{\chi} = \tilde{D}_t^\varepsilon\tilde{\chi} - \tilde{\chi}^3 + \tilde{\chi} + \tilde{\theta}$ we get the conclusion. □

Lemma 3.5. *For any $s_0, s \in [0,T]$ with $0 \le s_0 < s \le T$, there exists a positive constant M_5 depend on $|\theta_{s_0}|_{H^1(\Omega_m(s_0))}$, $|\chi_{s_0}|_{H^2(\Omega_m(s_0))}$, $|f|_{L^\infty(Q_m(s_0,s))}$ and $|\tilde{\mathbf{v}}|_{L^2(s_0,s;\mathbf{V})}$ such that*

$$|\tilde{\theta}|_{L^\infty(Q_m(s_0,s))} + \int_{s_0}^{s} |\tilde{\theta}(t)|^2_{H^2(\Omega_m(t))} dt + \sup_{t\in(s_0,s)} |\nabla\tilde{\theta}(t)|_{\mathbf{L}^2(\Omega_m(t))} \le M_5. \tag{3.12}$$

Proof. Thanks to the estimate (3.10), the same argument of Lemma 3.2 and 3.3 works to the equation (3.1) of $\tilde{\theta}$. □

In order to show the main theorem, especially to obtain the uniformly convergence of approximation for χ, we prepare the compactness theorem of Aubin's type, see the paper of Simon [14]. Let $\tilde{D}_t u := \partial u/\partial t + \mathbf{v}\cdot\nabla u$ and $\tilde{\mathbf{v}} - \mathbf{v}_D \in L^2(0,T;\mathbf{V})\cap L^\infty(0,T;\mathbf{H})$ and it satisfies (3.3) then the following proposition holds:

Proposition 3.6. *Let $\bar{F}$ be a bounded set in $L^\infty(0,T;H^2(\Omega_{m0}))$ and*

$$\int_0^T |\tilde{D}_t u(t)|^2_{L^2(\Omega_m(t))} dt < M_6 \quad \text{for all } u(t,x) := \bar{u}(t,\mathbf{y}(t,x)) \text{ with } \bar{u}\in\bar{F},$$

where M_6 is a positive constant. Then $\bar{F}$ is relatively compact in $C([0,T]\times\overline{\Omega_{m0}})$.

Proof. In our setting the domain is time dependent, but we have the enough estimate for $\tilde{\mathbf{v}}$. So the boundedness of the time derivative is coming from the one of $\tilde{D}_t u$. Thus we get the conclusion.

We can find the related topics in Fukao [5]. □

Proof of Theorem 2.1. The proof is the same way in Fukao and Kenmochi [6] with Proposition 3.6. The essential idea is due to Fujita and Sauer [4]. We denote by $((\mathrm{PF});\tilde{\mathbf{v}},\theta_{s_0},\chi_{s_0})$ on $[s_0,s]$ the variational problem associated with the phase field equations on $Q_m(s_0,s)$ with given convection $\tilde{\mathbf{v}}$. And any functions $\{\tilde{\theta},\tilde{\chi}\}$ satisfying the above lemmas are called solutions of $((\mathrm{PF});\tilde{\mathbf{v}},\theta_{s_0},\chi_{s_0})$ on $[s_0,s]$. On the other hand the solvability for the Navier-Stokes equations in non-cylindrical domain was discussed by many authors, for example Fujita and Sauer [4]. Here we apply the result of Kenmochi [7, 8]. In the existence proofs of [7, 8], one of main points is an extensive use of a compactness theorem of Aubin's type and its extension. We denote $((\mathrm{NS})_\delta;\tilde{\theta},\tilde{\chi},\mathbf{v}_{s_0})$ on $[s_0,s]$ the following variational problem associated with the penalized Navier-Stokes equations on $Q_m(s_0,s)$:

$$-\int_{s_0}^{s} (\boldsymbol{\eta}',\tilde{\mathbf{w}})_{\mathbf{H}} d\tau + \int_{s_0}^{s} a(\tilde{\mathbf{w}},\boldsymbol{\eta}) d\tau + \int_{s_0}^{s} b(\tau;\tilde{\mathbf{w}},\tilde{\mathbf{w}},\boldsymbol{\eta}) d\tau$$
$$+\int_{s_0}^{s} c(\tau;\tilde{\mathbf{w}},\boldsymbol{\eta}) d\tau + \frac{1}{\delta}\int_{s_0}^{s} (P_L([\tilde{\chi}]^-\tilde{\mathbf{w}}),\boldsymbol{\eta})_{\mathbf{H}} d\tau = \int_{s_0}^{s} (\boldsymbol{g}_L(\tilde{\theta}),\boldsymbol{\eta})_{\mathbf{H}} d\tau + (\mathbf{w}_{s_0},\boldsymbol{\eta}(0))_{\mathbf{H}}$$
$$\text{for all } \boldsymbol{\eta}\in\mathbf{W}_0(s_0,s),$$

where $[\tilde{\chi}]^-$ is the negative part of $\tilde{\chi}$, $\mathbf{w}_{s_0} := \mathbf{v}_{s_0} - \mathbf{v}_D(s_0)$ and

$$\mathbf{W}_0(s_0,s) := \left\{ \boldsymbol{\eta}\in L^4(s_0,s;\mathbf{X});\ \begin{array}{l} \boldsymbol{\eta}'\in L^2(s_0,s;\mathbf{H}),\ \boldsymbol{\eta}(s)=0 \text{ a.e. on } \Omega, \\ \boldsymbol{\eta}=0 \text{ a.e. on } Q(s_0,s)\setminus Q_m(s_0,s) \end{array} \right\}.$$

We know from the result of [7] there exist functions $\tilde{\mathbf{w}}_\delta \in L^\infty(s_0, s; \mathbf{H}) \cap L^2(s_0, s; \mathbf{V})$ with $\tilde{\mathbf{w}} = 0$ a.e. on $Q(s_0, s) \backslash Q_m(s_0, s)$ and $\tilde{\mathbf{w}}_\delta$ is weakly continuous from $[s_0, s]$ into $\mathbf{H}$ such that $\tilde{\mathbf{w}}_\delta$ satisfies the above variational formulation. Moreover the following inequality holds:

$$\begin{aligned}
&\frac{1}{2}|\tilde{\mathbf{w}}_\delta(t)|^2_{\mathbf{H}} + \int_{s_0}^t |\tilde{\mathbf{w}}_\delta(\tau)|^2_{\mathbf{V}} d\tau + \frac{1}{\delta}\int_{Q(s_0,t)} [\tilde{\chi}]^- |\tilde{\mathbf{w}}_\delta|^2 dx d\tau \\
\leq\ & \frac{1}{2}|\tilde{\mathbf{w}}_{s_0}|^2_{\mathbf{H}} + \int_{s_0}^t (\boldsymbol{g}_L(\tilde{\theta}), \tilde{\mathbf{w}}_\delta)_{\mathbf{H}} d\tau \quad \text{for all } t \in [s_0, s].
\end{aligned}$$

Any functions $\tilde{\mathbf{v}}_\delta := \tilde{\mathbf{w}}_\delta + \mathbf{v}_D$ satisfying the above estimate are called solutions of $((\mathrm{NS})_\delta; \tilde{\theta}, \tilde{\chi}, \mathbf{v}_{s_0})$ on $[s_0, s]$. Let $0 = t_0^N < t_1^N < t_2^N < \cdots < t_N^N = T$, be the partition of $[0, T]$ given by $t_k^N = kh_N$ for $k = 0, 1, \ldots, N$ with $h_N = T/N$. We are now going to construct a sequence of approximate solutions. For each $s, t \in [0, T]$, $\Theta_{t,s}(\cdot)$ be the C^3-diffeomorphism in Ω given by $\Theta_{t,s}(x) = \mathbf{x}(s, \mathbf{y}(t, x))$ for all $x \in \Omega$, note that $\Theta_{t,s}$ maps $\Omega_m(t)$ onto $\Omega_m(s)$ for each $s, t \in [0, T]$. Now, for fixed positive parameters $\delta \in (0, 1]$, let us define a set of functions $\theta_\delta^N, \chi_\delta^N$ on Q_m and $\mathbf{v}_\delta^N$ on Q by

$$\begin{aligned}
\theta_\delta^N(t, x) &:= \theta_{\delta,k}^N(t, x), \quad \text{if } t \in [t_{k-1}^N, t_k^N) \text{ and } x \in \Omega_m(t), \\
\chi_\delta^N(t, x) &:= \chi_{\delta,k}^N(t, x), \quad \text{if } t \in [t_{k-1}^N, t_k^N) \text{ and } x \in \Omega_m(t), \\
\mathbf{v}_\delta^N(t, x) &:= \mathbf{v}_{\delta,k}^N(t, x) \quad \text{if } t \in [t_{k-1}^N, t_k^N) \text{ and } x \in \Omega,
\end{aligned}$$

where $\theta_{\delta,k}^N, \chi_{\delta,k}^N$ and $\mathbf{v}_{\delta,k}^N$ are solutions of the Navier-Stokes equations $((\mathrm{NS})_\delta; \tilde{\theta}, \tilde{\chi}, \mathbf{v}_{\delta,k-1}^N(t_{k-1}))$ and phase field equations $((\mathrm{PF}); \mathbf{v}_{\delta,k}^N, \theta_{\delta,k-1}^N(t_{k-1}), \chi_{\delta,k-1}^N(t_{k-1}))$ on $[t_{k-1}^N, t_k^N]$ where

$$\begin{aligned}
\tilde{\theta}(t, x) &= \theta_{\delta,k-1}^N(t - h_N, \Theta_{t,t-h_N}(x)) \quad \text{for } (t, x) \in Q_m(t_{k-1}^N, t_k^N), \\
\tilde{\chi}(t, x) &= \chi_{\delta,k-1}^N(t - h_N, \Theta_{t,t-h_N}(x)) \quad \text{for } (t, x) \in Q_m(t_{k-1}^N, t_k^N).
\end{aligned}$$

By virtue of the Proposition 3.6 with Lemma 3.4, the compact embedding $\{\chi_\delta^N\}$ to $C(\overline{Q_m})$ is ensured. Thus in order to discuss convergences as $N \to 0$ and $\delta \to 0$ we can use the standard compactness argument. Finally in order to show that the convective vector coincides with $\mathbf{v}_D$ in the solid region, the idea of the compact cylinder by Fujita and Sauer [4] can be applied, because our solid region is exact open set. □

References

[1] G. Caginalp, An analysis of a phase field model of a free boundary, Arch. Rat. Mech. Anal., **92**(1986), 205–245.

[2] E. Casella and M. Giangi, An analytical and numerical study of the Stefan problem with convection by means of an enthalpy method, Math. Methods Appl. Sci., **24**(2001), 623–639.

[3] G.J. Fix, Phase field methods for free boundary problems, pp. 580–589 in *Free Boundary Problems: Theory and Applications*, Pitman Rese. Notes Math. Ser., **Vol. 79**, Longman, London, 1983.

[4] H. Fujita and N. Sauer, On existence of weak solutions of the Navier-Stokes equations in regions with moving boundarics, J. Fac. Sci., Univ. Tokyo., Sec. IA. Math., **17**(1970), 403–420.

[5] T. Fukao, Phase field equations with convections in non-cylindrical domains, pp. 42–54 in *Mathematical Approach to Nonlinear Phenomena; Modelling, Analysis and Simulations*, GAKUTO Internat. Ser. Math. Sci. Appl., **Vol. 23**, Gakkōtosho, Tokyo.

[6] T. Fukao and N. Kenmochi, Stefan problems with convection governed by Navier-Stokes equations, Adv. Math. Sci. Appl., **15**(2005), 29–48.

[7] N. Kenmochi, Résolution de problèmes variationels paraboliques non linéaires par les méthodes de compacité et monotonie, Theses, Universite Pierre et Marie Curie, Paris 6, (1979).

[8] N. Kenmochi, Résultats de compacité dans des espaces de Banach dépendant du temps, Séminaire d'analyse convexe, Montpellier, Exposé 1, (1979), 1–26.

[9] O.A. Ladyženskaja, V.A. Solonnikov and N.N. Ural'ceva, *Linear and quasilinear equations of parabolic type*, Translations of Mathematical Monographs, **Vol. 23**, Amer. Math. Soc., 1968.

[10] G. Planas and J.L. Boldrini, A bidimensional phase-field model with convection for change phase of an alloy, J. Math. Anal. Appl. **303**(2005), 669–687,

[11] J.F. Rodrigues, Variational methods in the Stefan problem, pp.147–212 in *Phase Transitions and Hysteresis*, Lecture Notes Math., **Vol. 1584**, Springer-Verlag, 1994.

[12] J.F. Rodrigues and F. Yi, On a two-phase continuous casting Stefan problem with nonlinear flux, European J. Appl. Math., **1**(1990), 259–278.

[13] G. Schimperna, Abstract approach to evolution equations of phase-field type and applications, J. Differential Equations **164**(2000), 395–430,

[14] J. Simon, Compact sets in the spaces $L^p(0,T;B)$, Ann. Mate. Pura. Appl., **146** (1987), 65–96.

[15] V.N. Strarovoitov, On the Stefan problem with different phase densities, Z. Angew. Math. Mech., **80**(2000), 103–111.

[16] A. Visintin, *Models of phase transitions*, PNLDE, Birkhäuser, Boston, 1996.

Takesi Fukao
General Education
Gifu National College of Technology
2236-2 Kamimakuwa, Motosu-shi,
Gifu 501-0495 Japan
e-mail: `fukao@gifu-nct.ac.jp`

International Series of Numerical Mathematics, Vol. 154, 179–188

A Dynamic Boundary Value Problem Arising in the Ecology of Mangroves

Gonzalo Galiano and Julián Velasco

Abstract. We consider an evolution model describing the vertical movement of water and salt in a domain split in two parts: a water reservoir and a saturated porous medium below it, in which a continuous extraction of fresh water takes place (by the roots of mangroves). The problem is formulated in terms of a coupled system of partial differential equations for the salt concentration and the water flow in the porous medium, with a dynamic boundary condition which connects both subdomains.

We study the existence and uniqueness of solutions, the stability of the trivial steady state solution, and the conditions for the root zone to reach, in finite time, the threshold value of salt concentration under which mangroves may live.

Keywords. Dynamic boundary condition, system of partial differential equations, existence, uniqueness, stability, dead core.

1. Introduction

Mangrove forests or swamps can be found on low, muddy, tropical coastal areas around the world. Mangroves are woody plants that form the dominant vegetation of mangrove forests. They are characterized by their ability to tolerate regular inundation by tidal water with salt concentration c_w close to that of sea water (see, for example, [19]). The mangrove roots take up fresh water from the saline soil and leave behind most of the salt, resulting in a net flow of water downward from the soil surface, which carries salt with it. As pointed out by Passioura *et al.* [26], in the absence of lateral flow, the steady state salinity profile in the root zone must be such that the salinity around the roots is higher than c_w, and that the concentration gradient is large enough so that the advective downward flow of salt is balanced by the diffusive flow of salt back up to the surface. In [26] the authors presented steady state equations governing the flow of salt and uptake of water in

---

Supported by the Spanish DGI Project MTM2004-05417 and by the European RTN Contract HPRN-CT-2002-00274.

the root zone, assuming that there is an upper limit c_c to the salt concentration at which roots can take up water, and that the rate of uptake of water is proportional to the difference between the local concentration c and the assumed upper limit c_c. They also assumed that the root zone is unbounded, and that the constant of proportionality for root water uptake is independent of depth through the soil. In [12], the model was extended in two important ways. First, considering more general root water uptake functions and second, limiting the root zone to a bounded domain. The authors proved mathematical properties such as the existence and uniqueness of solutions of the evolution and steady state problems, the conditions under which the threshold level of salt concentration is attained, and others. In [12], it is assumed that tides, or other sources of fresh or not too saline water, renew the water on the soil-water interface allowing to prescribe the salt concentration at this boundary (Dirichlet boundary data). Although this is the usual situation in which mangroves live, in this article we shall focus in the situation in which the inflow of fresh or sea water is impeded. In this situation, the continuous extraction of fresh water by the roots of mangroves drives the ecosystem to a complete salinization and, henceforth, to death. This work is motivated by the occurrences observed at Ciénaga Grande de Santa Marta, Colombia. As reported by Botero [8] (see also [29]), the construction of a highway along the shore in the 1950s obstructed the natural circulation of water between both parts of the road (Caribbean sea and lagoon). In addition, in the 1970s, inflow of fresh water from the river Magdalena was reduced due to the construction of smaller roads and flooding control dikes. These changes caused a hypersalinization of water and soil, which resulted in approximately 70% mangrove mortality (about 360 Km^2 of mangrove forests), see [8], [18]. Although other causes, like evaporation or sedimentation, may have had an important contribution to the salinization of the Ciénaga, we shall keep our attention in the mechanisms of mangroves and their influence in this process.

The main mathematical difficulty of this model when compared with that studied in [12] is that the closure of the natural system, the lagoon, implies a new type of boundary condition in the water-soil interface, which is no longer of Dirichlet type. Balance equations for salt and water content lead to a *dynamical boundary condition* at such interface, i.e., a boundary condition involving the time derivative of the solution. Although not too widely considered in the literature, dynamic boundary conditions date back at least to 1901 in the context of heat transfer [27]. Since then, they have been studied in many applied investigations in several disciplines like Stefan problems [30, 33], fluid dynamics [16], diffusion in porous medium [28, 15], mathematical biology [14] or semiconductor devices [31]. From a more abstract point of view the reader is referenced to, among others, [10, 24, 20, 11, 13, 1, 2, 7].

Apart from the mathematical technical details, one of the main features of the dynamic boundary condition when compared to the Dirichlet boundary condition is the elimination of the boundary layer the latter creates in a neighborhood of the water-soil interface, layer in which the salt concentration keeps well below the threshold salinity level. Thus, this new model allows us to describe the situation in which a continuous increase of fresh water uptake by the roots of mangroves drives the ecosystem to a complete salinization.

The outline of the paper is the following: in Section 2 we formulate the mathematical model. We assume that mangroves roots are situated in a porous medium in the top of which a water reservoir keeps the soil saturated. As in [12], coupled partial differential equations for salt concentration and water discharge are considered in the porous medium. Above it, in the water reservoir, balance laws for salt and water are formulated. The assumption of homogeneous salt concentration in the water reservoir leads to a dynamic boundary condition in the water-soil interface. In Section 3 we state our hypothesis and formulate our main results on existence and uniqueness of solutions of the evolution problem, as well as the convergence of this solution to the steady state solution. We also study the conditions under which the complete salinization of the root zone is attained in finite time (dead core). The proofs of these results will appear elsewhere [17].

2. The mathematical model

In this section we formulate the mathematical model which describes the salt and water movement in the water-soil system. We consider the case where the mangroves are present in the horizontal x, y plane, with a homogeneous porous medium located below this plane and a water reservoir above it. The porous medium is characterized by a constant porosity θ, indicating that we are assuming the mangroves roots to be homogenized throughout the porous medium, without affecting its properties. Assuming further that the hydrodynamic dispersion tensor, $\mathbf{D}$, is constant and isotropic, i.e., neglecting the velocity dependence in the mechanical dispersion, we find for the salt concentration the equation, see [6],

$$\theta \frac{\partial c}{\partial t} + \operatorname{div}\left(c\mathbf{q} - \theta \mathbf{D} \nabla c\right) = 0, \tag{2.1}$$

where the vector $\mathbf{q}$ denotes the specific discharge of the fluid, $\mathbf{D} = D\mathbf{I}$, $\mathbf{I}$ is the identity matrix and t denotes time. We also have a fluid balance in the porous medium. Disregarding density variations in the mass balance equation of the fluid, we obtain a fluid volume balance expressed by

$$\operatorname{div} \mathbf{q} + S = 0, \tag{2.2}$$

where S is the volume of water taken up by the roots per unit volume of porous material per unit time. If the mangroves are uniformly distributed throughout the x, y-plane and there is no lateral fluid flow, we may consider the problem as one-dimensional in the vertical direction. If the z-axis is positive when pointing downwards, the flow domain is characterized by the interval $0 < z < H < \infty$. In the one-dimensional setting equations (2.1) and (2.2) become

$$\theta c_t + (cq - \theta D c_z)_z = 0, \tag{2.3}$$

$$q_z + S = 0, \qquad \text{in } (0, H) \times (0, T) \tag{2.4}$$

For S, we assume to have the form

$$S := \begin{cases} s(z)\left(1 - \dfrac{c}{c_c}\right)^p & \text{for } 0 \leq c \leq c_c, \\ 0 & \text{for } c > c_c, \end{cases} \tag{2.5}$$

where c_c is the upper limit of salt concentration at which mangroves may uptake water, $p > 0$ and $s(z)$ is determined by the root distribution as a function of the depth z below the soil surface. This root distribution function will be non-negative, and non-increasing with z. We shall keep in mind the following characteristic example: we assume that the function s is a positive constant, s_0/z_*, above a certain depth z_*, and zero below that depth, i.e.,

$$s(z) = s_0/z_* \quad \text{if } 0 \leq z \leq z_* \quad \text{and} \quad s(z) = 0 \quad \text{if } z_* < z \leq H. \tag{2.6}$$

The quantity s_0 is the total amount of root water uptake in the profile with no salt present, in volume per unit surface per unit time, i.e., the *transpiration rate* of the mangrove plants in the absence of salinity. On the bottom of the porous medium domain, we assume no flux boundary conditions, resulting in

$$q(H,t) = c_z(H,t) = 0 \quad \text{for } t \in (0,T). \tag{2.7}$$

On the water-soil interface we prescribe a boundary condition which is deduced from conservation laws for salt and water in the whole system water-soil. We assume that salt concentration in the water domain, C, remains uniformly distributed in space. This approximation is justified when assuming a much faster mixing of the salt in the reservoir than in the porous medium. Then, the average height level of the water reservoir, W, and C are functions that only depend on time. We further consider, based on a continuity assumption

$$C(t) = c(0,t) \quad \text{for } t \in (0,T). \tag{2.8}$$

Then we have:

- The salt balance. Assuming that the total amount of salt in the system water-soil remains constant, we have

  $$\frac{d}{dt}\left(CW + \int_0^H \theta c\right) = 0 \quad \text{in } (0,T).$$

  Therefore, from equation (2.3) and the boundary condition (2.7),

  $$\frac{d(CW)}{dt} = c(0,\cdot)q(0,\cdot) - \theta D c_z(0,\cdot) \quad \text{in } (0,T). \tag{2.9}$$

- The fluid balance, which asserts that the amount of water taken up from the soil by the roots of mangroves is replaced by water from the reservoir:

  $$\frac{dW}{dt} = -q(0,\cdot) \quad \text{in } (0,T). \tag{2.10}$$

Combining (2.8)–(2.10) we deduce

$$W(t)c_t(0,t) = \theta D c_z(0,t) \quad \text{for } t \in (0,T), \tag{2.11}$$

which is the dynamic boundary condition for the soil-water interface. Finally, we add to this formulation given initial distributions of salt concentration, $c(\cdot,0) = c_0$ in $(0,H)$, and of water reservoir height level, $W(0) = W_0$.

We recast the above formulation in an appropriate dimensionless form introducing the following variables, unknowns and parameters:

$$\tilde{t} := Dt/z_*^2, \quad x := z/z_*, \quad u := c/c_c, \quad \tilde{q} = qz_*/D\theta,$$

$$w = W/\theta z_*, \quad \tilde{s}(x) := z_* s(Hx)/s_0, \quad d := H/z_*, \quad m := s_0 z_*/D\theta,$$

and we define $f(x,u) := S(Hx, c_c u)$, with $f : [0,d] \times [0,1] \to \mathbb{R}_+$ given by

$$f(x,\sigma) := \tilde{s}(x)(1-\sigma)_+^p, \tag{2.12}$$

with $p > 0$ and

$$\tilde{s}(x) = 1 \quad \text{if } 0 \le x \le 1 \quad \text{and } \tilde{s}(x) = 0 \quad \text{if } 1 < x \le d. \tag{2.13}$$

With the above changes we are led to the following problem (omitting tildes): find $u : \bar{Q}_T \to [0,1]$, $q : \bar{Q}_T \to \mathbb{R}$ and $W : [0,T] \to \mathbb{R}$ such that

$$u_t + (uq - u_x)_x = 0, \tag{2.14}$$

$$q_x + mf(\cdot,u) = 0 \quad \text{in } Q_T = I \times (0,T), \text{ with } I = (0,d), \tag{2.15}$$

$$w'(t) + q(0,t) = 0 \quad \text{for } t \in (0,T),$$

subject to the boundary and initial conditions

$$w(t)u_t(0,t) = u_x(0,t), \tag{2.16}$$

$$u_x(d,t) = q(d,t) = 0 \qquad \text{for } t \in (0,T), \tag{2.17}$$

$$u(\cdot,0) = u_0 \quad \text{in } I, \qquad w(0) = w_0. \tag{2.18}$$

Remark 1. In the recasting of our model there appeared a constant capturing all the important physical parameters, the *mangrove's number*:

$$m := s_0 z_*/D\theta. \tag{2.19}$$

Using [26] and [25] as a reference we find the following values for the physical constants: $D = 7 \cdot 10^{-5}\,\text{m}^2/\text{day}$, $\theta = 0.5$, and $s_0 = 1\ \ell\text{m}^{-2}\text{day}^{-1}$. Taking z^* in the range 0.2–0.5 m, this implies a time scale in the range 2–10 yr and $m \in (6,15)$.

3. Main results

We shall refer to problem (2.14)–(2.18), as to Problem P, for which we assume the following hypothesis:

H_1. The function $f : \bar{I} \times [0,1] \to \mathbb{R}$, with $I = (0,d)$ and $d \geq 1$, satisfies

$$f \in L^\infty(I; C([0,1])), \ |f| \leq 1,$$
$$f(\cdot, s) \text{ is non-increasing in } \bar{I} \text{ and } f(d,s) \geq 0 \text{ for all } s \in [0,1],$$
$$f(x, \cdot) \text{ is non-increasing in } [0,1] \text{ and } f(x,1) = 0 \text{ for a.e. } x \in I.$$

Note that, in particular, $f \geq 0$ in $\bar{I} \times [0,1]$.

H_2. The initial data posses the regularity

$$u_0 \in H^1(I) \quad \text{with } 0 \leq u_0 \leq 1 \quad \text{in } I.$$

H_3. The function w is a positive constant. The number m is positive. We set $w = m = 1$.

Remark 2. The assumption w (or the dimensional W) constant in H_3 has a reasonable range of validity. From (2.4), (2.5), (2.10) and the mean value theorem we infer

$$W(t) = W_0 - \int_0^t q(0,\tau)d\tau = W_0 - ts_0(1 - \frac{\bar{c}}{c_c})^p, \quad \text{for some } \bar{c} \in (0, c_c).$$

Set s_0 as in Remark 1 and $p = 1$. A lower limit for $\bar{c}$ is sea water salt concentration $c_w \sim 0.5c_c$. Then W_0 must be much greater than the 15 cm that the lagoon will decrease per year while keeping the sea water salt concentration. For a value of $c = 0.9c_c$ the decrease of the height level is of about 3 cm per year.

Remark 3. Since the numbers m and w do not play any essential role in the results we prove in this work, we set $m = w = 1$ for clarity.

Under Hypothesis H_1–H_3 we cannot expect the existence of classical solutions. We then introduce the notion of solution we shall work with.

Definition 1. We say that (u, q) is a *strong solution* of Problem P if $u : \bar{Q}_T \to [0,1]$ and $q : \bar{Q}_T \to \mathbb{R}$ satisfy the following properties:

1. For any $r \in (0, \infty)$,
$$u \in W^{1,r}(0,T; L^r(I)) \cap L^r(0,T; W^{2,r}(I)) \cap C((0,T]; C(\bar{I})),$$
$$q \in C((0,T]; \mathcal{W})$$
with $\mathcal{W} := \{\varphi \in W^{1,\infty}(I) : \varphi(d) = 0\}$.
2. The differential equations (2.14) and (2.15) and the boundary conditions (2.16) and (2.17) are satisfied almost everywhere. The initial distribution is satisfied in the sense
$$\lim_{t \to 0} \|u(\cdot, t) - u_0\|_{L^2(I)} = 0.$$

We prove the following result on existence and regularity of solutions.

Theorem 1. *Assume H_1–H_3. Then there exists a strong solution of Problem P satisfying*

$$u \geq u_m := \min_{\bar{I}} u_0 \quad a.e. \text{ in } Q_T. \tag{3.1}$$

In addition, if for some $p > 0$

$$f \in C^p(\bar{I} \times [0,1]) \quad and \quad u_0 \in C^{2+p}(\bar{I}), \tag{3.2}$$

and if u_0 *satisfies the following compatibility condition*

$$u_0'(0) + u_0'(0)\int_0^1 f(x, u_0(x))dx - u_0''(0) = f(0, u_0(0))u_0(0), \tag{3.3}$$

then $u \in C^{1+p,2+p}(\bar{Q}_T)$ *and* $q \in C^{1+p,1+p}(\bar{Q}_T)$.

We prove uniqueness of solution for $f(x,\cdot)$ being Lipschitz continuous in $[0,1]$. For more general functions, we show that uniqueness of solution holds true under an additional condition on the component u. In Proposition 1 we give an example in which solutions of Problem P satisfy such condition.

Theorem 2. *Let* (u_1, q_1) *and* (u_2, q_2) *be two strong solutions of Problem P and let* H_1*-*H_3 *be satisfied. If either*

$$f(x,\cdot) \quad is\ Lipschitz\ continuous\ in\ [0,1] \quad for\ almost\ all\ x \in \Omega, \tag{3.4}$$

or anyone of the solutions satisfies

$$u(x,t) > \int_0^x |u_x(y,t)|\, dy \quad a.e.\ in\ Q_T, \tag{3.5}$$

then $(u_1, q_1) = (u_2, q_2)$ *a.e. in* Q_T.

Proposition 1. *Assume* H_1*–*H_3 *and* (3.2)*–*(3.3)*, and let* (u,q) *be a solution of Problem P. Suppose that* u_0 *satisfies* $u_{0x} \le L$ *in* I *and*

$$f(\cdot, u_m) \le L < u_m - \frac{1}{2}, \tag{3.6}$$

for some positive constant, L*, with* u_m *given by* (3.1)*. Assume*

$$f(\cdot,u) + u f_u(\cdot,u) < 0 \quad in\ Q_T. \tag{3.7}$$

Then condition (3.5) *is satisfied.*

Remark 4. In particular, if $f(x,\sigma) = s(x)(1-\sigma)^p$, with s smooth, and $u_0 \in C^{2+p}(\bar{I})$ satisfies $u_{0x} \le (1-u_m)^p < u_m - \frac{1}{2}$, then condition (3.5) is satisfied. Actually, the smoothness requirement on s may be dropped by using an approximation argument.

One important effect of the dynamic boundary condition when compared to the Dirichlet boundary condition at the boundary water-soil is the elimination of the boundary layer the latter creates. It is straightforward to prove that the unique solution of the steady state problem corresponding to Problem P, i.e., functions $U \in H^1(I)$ and $Q \in \mathcal{W}$ satisfying

$$\begin{aligned} &(QU - U_x)_x = Q_x + f(\cdot, U) = 0 \quad \text{in } I, \\ &U_x(0) = U_x(d) = 0, \end{aligned}$$

is the trivial solution $(U, Q) = (1, 0)$. Regarding the asymptotic convergence of solutions of Problem P to this trivial solution when $t \to \infty$, we have the following result.

Theorem 3. *Assume H_1-H_3 and $u_m > 0$, and let (u, q) be a strong solution of Problem P. Then*

$$(u, q) \to (1, 0) \quad \text{in } L^2(I) \quad \text{and} \quad u(0, t) \to 1 \quad \text{pointwise} \quad \text{as } t \to \infty.$$

We finally state a result on the existence of a dead core for solutions of Problem P, i.e., sets where the threshold salinization $u = 1$ is attained in finite time. The proof of this result, which is of local nature, i.e., independent of the boundary data, can be found in [12]. First, we introduce some notation. For any $t \in (0, T)$ we consider the parabola of vertex (x_0, t),

$$\mathcal{P}(t) := \{(x, \tau) : |x - x_0| < (\tau - t)^\nu, \quad \tau \in (t, T)\},$$

with $0 < \nu < 1$ and $x_0 \in I$ such that $T^\nu < x_0 < 1 - T^\nu$, implying $\mathcal{P}(t) \subset Q_T$ for all $t \in (0, T)$. We define the *local energy functions*

$$E(t) := \int_{\mathcal{P}(t)} |u_x|^2 \, dx \, d\tau \quad \text{and} \quad C(t) := \int_{\mathcal{P}(t)} (1 - u)^{p+1} \, dx \, d\tau. \tag{3.8}$$

In [12] we proved the following theorem using the techniques introduced in [3, 4].

Theorem 4. *Suppose there exist constants s_0 and s_1 such that*

$$0 < s_0 \sigma^{p+1} \leq \sigma f(\cdot, 1 - \sigma) \leq s_1 \sigma^{p+1} \quad \text{for } \sigma \in [0, 1], \tag{3.9}$$

in $\mathcal{P}(t)$ for a.e. $t \in (0, T)$, with $p \in (0, 1)$ and $s_0 > s_1/2$, and let (u, q) be a strong solution of Problem P. Then there exists a positive constant M such that if $E(0) + C(0) \leq M$ then $u \equiv 1$ in $\mathcal{P}(t^*)$, for some $t^* \in (0, T)$.

Let us finish this section with a remark on the assumptions of Theorem 4. First, if function f is given by $f(x, \sigma) = s(x)(1 - \sigma)^p$, with s given by (2.13) then (3.9) is trivially satisfied in the region where $s > 0$ (root zone). Regarding the bound of the initial energy, we have that testing the first equation of (2.14) with $1 - u$ and using the equation (2.15) we obtain

$$2E(0) + C(0) \leq \int_I (1 - u_0)^2 + (1 - u_0(0))^2 (1 + \int_{Q_T} f(\cdot, u)). \tag{3.10}$$

Therefore, if the initial datum is close enough to one then the initial energy bound is satisfied. Combining Theorems 3 and 4 we deduce the following corollary.

Corollary 1. *Let (u, q) be a strong solution of Problem P in Q_T, for T large enough. Under the conditions of Theorems 3 and 4 there exist $T_0, t^* > 0$ such that $u \equiv 1$ in $\mathcal{P}(t^*)$, for some $t^* \in (T_0, T)$.*

Or, in other words, the threshold value of salt concentration is attained in any compact set contained in the root zone in finite time.

Acknowledgements

The essentials of the modelling which gave rise to this article were worked out during a stay of the first author at the University of Eindhoven. We thank Professors Hans van Duijn and Peter Raats for their valuable help.

References

[1] Amann, H., Escher, J., Strongly continuous dual semigroups, 1996, Ann. Mat. Pura et Appl., CLXXI, 41–62.

[2] Arrieta, J.M., Quittner, P., Rodríguez-Bernal, A., Parabolic problems with nonlinear dynamical boundary conditions and singular initial data, 2001, Differential Integral Equations, 14, 1487–1510.

[3] Antontsev, S., Díaz, J.I., Shmarev, S.I., The support shrinking properties for local solutions of quasilinear parabolic equations with strong adsorption terms, 1995, Ann. Fac. Sciences de Toulouse 4(1), 5–30.

[4] Antontsev, S., Díaz, J.I., Shmarev, S.I., *Energy methods for free boundary problems*, 2002, Birkhäuser, Boston.

[5] Arino, O., Gauthier, S., Penot, J.P., A fixed point theorem for sequentially continuous mappings with application to ordinary differential equations, 1987, Funkcial Ekvac. 24, 273–279.

[6] Bear, J., *Dynamics of fluids in porous media*, 1972, New York, American Elsevier.

[7] Bejenerau, I., Díaz, J.I., Vrabie, I.I., An abstract approximate controllability result and applications to elliptic and parabolic systems with dynamic boundary conditions, 2001, Elec. J. Diff. Eq., 50, 1–19.

[8] Botero, L., Massive mangrove mortality of the Caribbean coast of Colombia, 1990, Vida Silve. Neot., 2, 77–78.

[9] Brezis, H., *Analise Fonctionnelle. Théorie et applications*, 1987, Masson, Paris.

[10] Courant, R., Hilbert, D., *Methoden der Mathematischen Physik II*, 1965, Springer, Berlin.

[11] Díaz, J.I., Jiménez, R., Aplicación de la teoría no lineal de semigrupos a un operador pseudodiferencial, 1984, Actas VII CEDYA, 137–182.

[12] van Duijn, C.J., Galiano, G., Peletier, M.A., A diffusion-convection problem with drainage arising in the ecology of mangroves, 2001, Interfaces and Free Boundaries, 3, 15–44.

[13] Escher, J., Quasilinear parabolic systems with dynamical boundary conditions, 1993, Commun in Partial Differential Equations, 18, 1309–1364.

[14] Feller, W., The parabolic differential equations and the associated semi-groups of transforms, 1952, Ann. of Math., 55, 468–519.

[15] Filo, J., Luckhaus, S., Modelling surface runoff and infiltration of rain by an elliptic-parabolic equation coupled with a first order equation on the boundary, 1999, Arch. Ration. Mech. Anal., 146, 157–182.

[16] Friedman, A., Shinbrot, M., The initial value problem for the linearized equations of water waves, 1968, J. Math. Mech., 17, 107–180.

[17] Galiano, G., Velasco, J., A dynamic boundary value problem arising in the ecology of mangroves. Submitted.

[18] González, E., El manglar de la Ciénaga Grande de Santa Marta, ecosistema en peligro de extinción, 1991, Colombia, sus gentes y regiones, 21, 2–21.

[19] Hutchings, P., Saenger, P., *Ecology of mangroves*, 1987, Queensland, University of Queensland Press.

[20] Kačur, J., Nonlinear parabolic equations with the mixed nonlinear and nonstationary boundary conditions, 1980, Math. Slovaca, 30, 213–237.

[21] Ladyzenskaya, O.A., Solonnikov, V.A. and Ural'ceva, N.N., *Quasilinear Equations of Parabolic Type*, 1968, Translations of Mathematical Monographs, 23, American Mathematical Society, Providence.

[22] Langlais, M., Phillips, D., Stabilization of solutions of nonlinear and degenerate evolution equations, 1985, Nonlinear Analysis, Th., Meth. and App. 9, No. 4, 321–333.

[23] Lieberman, G.M., *Second-order parabolic differential equations*, 1996, World Scientific, Singapore.

[24] Lions, J.L., *Quelques méthodes de résolution des problèmes aux limites non linéaires*, 1969, Dunod, Gauthiers-Villars, Paris.

[25] Paliyavuth, C., Clough, B., Patanaponpaiboon, P., Salt uptake and shoot water relations in mangroves, 2004, Aquatic Botany 78, 349–360.

[26] Passioura, J.B., Ball, M.C. and Knight, J.H., Mangroves may salinize the soil and in so doing limit their transpiration rate, 1992, Funct. Ecol. 6, 476–481.

[27] Peddie, W., Note on the cooling of a sphere in a mass of well-stirred liquid, 1901, Proc. Edinburgh Math. Soc., 19, 34–35.

[28] Peek, R.L., Solutions to a problem in diffusion employing a non-orthogonal sine series, 1929, Ann. of Math., 30, 265–269.

[29] Perdomo, L., Ensminger, I., Espinosa, L.F., Elster, C., Wallner-Kersanach, M., Schnetter, M.L., The mangrove ecosystem of the Ciénaga Grande de Santa Marta (Colombia): Observations on regeneration and trace metals in sediment, 1998, Marine Pollution Bull., 37, 393–403.

[30] Primicerio, M., Rodrigues, J.F., The Hele-Shaw problem with nonlocal injection condition, in *Nonlinear Mathematical Problems in Industry*, Gatuko, Tokyo, 1993.

[31] Selberher, S., *Analysis and simulation of semiconductor devices*, 1984, Springer, Wien.

[32] Simon, J., Compact sets in the space $L^p(0,T;B)$, 1987, Ann. Math. Pures et Appl. 146, 65–96.

[33] Solonnikov, V.A., Frolova, E.V., L^p-theory for the Stefan problem, 2000, J. Math. Sci. 989–1006.

Gonzalo Galiano and Julián Velasco
Departamento de Matemáticas
Universidad de Oviedo
c/ Calvo Sotelo s/n
E-33007 Oviedo, Spain
e-mail: `galiano@uniovi.es`
e-mail: `julian@orion.ciencias.uniovi.es`

International Series of Numerical Mathematics, Vol. 154, 189–198

Wave Breaking over Sloping Beaches Using a Coupled Boundary Integral-Level Set Method

M. Garzon and J.A. Sethian

Abstract. We present a numerical method for tracking breaking waves over sloping beaches. We use a fully non-linear potential model for incompressible, irrotational and inviscid flow, and consider the effects of beach topography on breaking waves. The algorithm uses a Boundary Element Method (BEM) to compute the velocity at the interface, coupled to a Narrow Band Level Set Method to track the evolving air/water interface, and an associated extension equation to update the velocity potential both on and off the interface. The formulation of the algorithm is applicable to two- and three-dimensional breaking waves; in this paper, we concentrate on two-dimensional results showing wave breaking and rollup, and perform numerical convergence studies and comparison with previous techniques.

Keywords. Level set methods, Boundary integral methods, Wave breaking .

1. Introduction and overview

The coupling of Level Set Methods (LSM) and Boundary Integral Methods (BIM) is very adequate for solving certain class of free boundary problems, mainly for two reasons. First, the robustness and topological properties of LSM to move the front, and second, the accuracy, sharpness and single fluid approach of BIM to obtain the front velocity.

In this work we are interested in problems where the boundary condition for the BIM has to be obtained from the solution of a partial differential equation posed on the moving front. Examples of that problems are: The Helle-Shaw problem, sprays and electrosprays, and wave breaking among others.

This work was supported by U.S. Department of Energy, Applied mathematical Sciences, and the Division of Mathematical Sciences, National Sciences Foundation and the Spanish DGI project BFM 00-1324.

Wave breaking is a highly non-linear phenomena involving two phase flow and turbulent flow. It has been confirmed by experimental measurements ([7]) and full Navier Stokes simulations ([13]), that the water flow is almost irrotational up to the impact of the jet onto the flat surface of the water. Therefore a fully non-linear potential flow model (FNPM) can be used [4].

Previous works on wave breaking are very extensive. Simulations up to the impact can be found in ([4], [5], [6]). For breaking and postbreaking see, for example, ([13],([8]). Laboratory measurements can be found in ([7]).

The idea of the present work is, as mentioned above, to couple LSM and BIM for the numerical solution of the FNPM for two-dimensional waves shoaling over a constant and sloping bottom. This approach provides a simple and direct way to solve the model equations by reformulating the problem in a complete Eulerian framework, and straightforward upwind numerical schemes give sufficiently accurate wave profiles while shoaling and breaking. The formulation is unchanged in three dimensions, offering the possibility of computing complex breaking wave motions.

2. The governing equations

Let $\Omega(t)$ be the $2D$ fluid domain in the vertical plane (x, z) at time t, with z the vertical upward direction (and $z = 0$ at the undisturbed free surface), and $\Gamma_t(s) = (x(s,t), z(s,t))$ a parametrization of the free boundary at time t (see Figure 1).

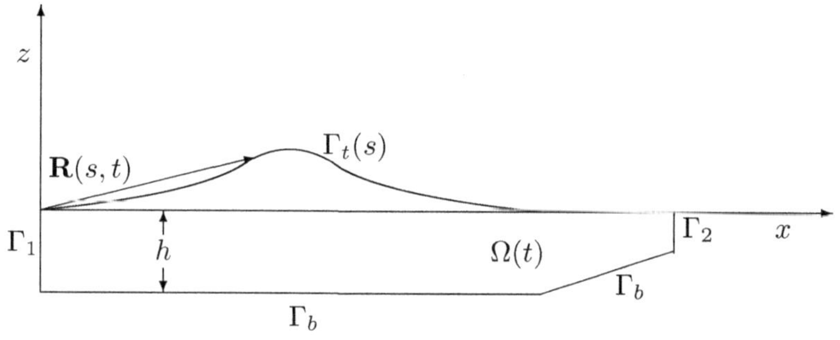

FIGURE 1. The domain

Under the previous above mentioned assumptions, the mass and momentum conservation equations are given by

$$\nabla \mathbf{u} = 0 \text{ in } \Omega(t) \tag{2.1}$$

$$\mathbf{u}_t + \mathbf{u} \cdot \nabla \mathbf{u} = \frac{-\nabla p}{\rho} + \mathbf{b} \text{ in } \Omega(t) \tag{2.2}$$

where $\mathbf{u}(x, z, t)$ is the fluid velocity, $p(x, z, t)$ the pressure field, $\mathbf{b}(x, z, t)$ the body forces (per unit mass), ρ is the fluid density.

Since the motion is irrotational, fluid particles do not rotate and vorticity vanishes everywhere in the field of flow. In this case, this means that the velocity field can be represented as the gradient of a scalar function referred to as the velocity potential $\phi(x, y, t)$. If $\mathbf{u} = \nabla\phi$, and $\mathbf{b} = -g\mathbf{z}$, being $\mathbf{z}$ a unit vector in the vertical direction ($\mathbf{z} = \nabla z$), the momentum equation (2.2) reduces to the so-called Bernoulli's equation:

$$\phi_t + \frac{1}{2}(\nabla\phi \cdot \nabla\phi) + \frac{p - p_a}{\rho} + gz = 0. \tag{2.3}$$

This gives the pressure field once ϕ is known (here p_a denotes the atmospheric pressure).

On the free boundary, the following boundary conditions are imposed:

1. Continuity of stress tensor between water and air leads to $p = p_a$, and thus we have
$$\phi_t + \frac{1}{2}(\nabla\phi \cdot \nabla\phi) + gz = 0 \text{ on } \Gamma_t(s).$$
2. If $\mathbf{R}(s, t) = (X(s, t), Y(s, t))$ is the position vector of a fluid particle on the free surface, we have the kinematic boundary condition
$$\mathbf{R}_t(s, t) = \mathbf{u}(\mathbf{R}(s, t), t) \text{ on } \Gamma_t(s)$$
where s identifies the fluid particle that is in $x = X(s, t), z = Z(s, t)$ at time t.

Therefore, the model equations are:

$$\begin{aligned}
\mathbf{u} &= \nabla\phi \text{ in } \Omega(t) && (2.4)\\
\Delta\phi &= 0 \text{ in } \Omega(t) && (2.5)\\
\mathbf{R}_t &= \mathbf{u} \text{ on } \Gamma_t(s) && (2.6)\\
\frac{D\phi}{Dt} &= -gz + \frac{1}{2}(\nabla\phi \cdot \nabla\phi) \text{ on } \Gamma_t(s) && (2.7)\\
\phi_n &= 0 \text{ on } \Gamma_b \cup \Gamma_1 \cup \Gamma_2. && (2.8)
\end{aligned}$$

3. Embedding the equations of motion in a level set framework

Briefly, the main idea of the LSM ([9], [10], [11]) is to embed the initial position of the front as the zero level set of a higher-dimensional function $\Psi(x, z, t)$. One then links the evolution of this function Ψ to the propagation of the front itself through a time-dependent initial value problem. At any time, the front is given by the zero level set of the time-dependent level set function Ψ. An equation for the motion for this level set function Ψ which matches the zero level set of Ψ with the evolving front comes from observing that the level set value of a particle on the front with path $\mathbf{R}(s, t)$ must always be zero:

$$\Psi(\mathbf{R}(s, t), t) = 0$$

Hence, we have that

$$\Psi_t + \nabla\Psi(\mathbf{R}(s, t), t) \cdot \mathbf{u} = 0. \tag{3.1}$$

For our wave problem, let Ω_1 be a fictitious fixed squared domain that contains the free boundary at any time t. Equation (2.6), which states that the front moves with velocity $\mathbf{u}$ can be replaced by the level set equation (3.1) posed on Ω_1.

To embed equation (2.7) in the level set framework we do the following: On the free boundary $\Gamma_t(s)$ we define

$$\Phi(s,t) = \phi(x,z,t)|_{\Gamma_t(s)} = \phi(\mathbf{R}(s,t),t),$$

and thus by fixing s and moving t, we are constrained to a fluid particle, which means that $\Phi_t(s,t)$ is a total derivative and hence

$$\Phi_t = \phi_t + \mathbf{u}\cdot\nabla\phi = \frac{1}{2}(\nabla\phi\cdot\nabla\phi) - gz.$$

Next, let be $G(x,z,t)$ a function defined on Ω_1 with the following property:

$$G(X(s,t),Z(s,t),t) = \Phi(s,t) \ \text{ on } \Gamma_t(s)$$

Applying the chain rule, we have

$$G_t + \mathbf{u}\cdot\nabla G = \frac{1}{2}(\nabla\phi\cdot\nabla\phi) - gz, \tag{3.2}$$

which holds on $\Gamma_t(s)$. Note that $\mathbf{u}$ and right-hand side of equation (3.2) are only defined on $\Gamma_t(s)$. In order to be able to solve equation (3.2) embedded in the whole domain Ω_1, we need to extend these variables off the front.

The model equations, written in a complete Eulerian framework, are

$$\begin{aligned} \mathbf{u} &= \nabla\phi \ \text{ in } \Omega(t) && (3.3)\\ \Delta\phi &= 0 \ \text{ in } \Omega(t) && (3.4)\\ \Psi_t + \nabla\Psi\cdot\mathbf{u}_{\text{ext}} &= 0 \ \text{ in } \Omega_1 && (3.5)\\ G_t + \mathbf{u}_{\text{ext}}\cdot\nabla G &= f_{\text{ext}} \ \text{ in } \Omega_1 && (3.6)\\ \phi_n &= 0 \ \text{ on } \Gamma_b\cup\Gamma_1\cup\Gamma_2 && (3.7) \end{aligned}$$

being $f = \frac{1}{2}(\nabla\phi\cdot\nabla\phi) - gz$ and f_{ext} the extension of f onto Ω_1.

4. Numerical approximations and algorithms

Integral formulation of equation (3.4), with boundary conditions given by (3.6) and (3.7) is solved using a BEM, that uses a Galerkin approximation of the BI, linear shape functions and a special treatment of the corners. The interface velocity is obtained postprocessing the solution ϕ_n on the interface using a Galerkin technique, see ([3]).

To approximate equations (3.5) and (3.6) in Ω_1 homogeneous boundary conditions are imposed on $\partial\Omega_1$. A second-order upwind finite differences in space and first-order in time is used for equation (3.5), while for equation (3.6) we used a first-order upwind finite differences in space and first-order in time.

The velocity and the velocity potential are both initially defined only on the interface. In order to create values throughout the narrow band, which are required to update the fixed grid Eulerian partial differential equations, we use the extension methodology developed by Adalsteinsson and Sethian in [1] to construct appropriate extensions.

The basic algorithm can be summarized as follows:

1. Compute initial front position and velocity potential $\Phi(s,0)$ on $\Gamma_0(s)$.
2. Extend $\Phi(s,0)$ onto the grid points of Ω_1 to initialize G.
3. Generate $\Omega(t)$ and solve (3.4), using the Boundary Element Method. This yields the velocity $\mathbf{u}$ and source term f on the front nodes.
4. Extend $\mathbf{u}$ and f off the front onto Ω_1.
5. Update G using (3.6) in Ω_1.
6. Move the front with velocity $\mathbf{u}$ using (3.5) in Ω_1
7. Interpolate (bi-cubic interpolation) G from grid points of Ω_1 to the front nodes to obtain new boundary conditions for (3.4). Go back to step 3 and repeat forward in time.

A more complete algorithm with regridding can be found in [2].

5. Numerical results

A common procedure to study the accuracy and convergence properties of the discretized equations with respect the mesh sizes and the time step is by means of an analytical solution. A solitary wave propagating over a constant depth is a travelling wave that moves in the x direction with speed equal to the celerity of the wave (c). The velocity potential and the velocity on the front as functions of x are also translated with the same speed c. Therefore, in this case, by calculating initial wave data with Tanaka's method ([12]) and translating it, we are able to compute the L2 norms of the errors for the various magnitudes. See these results in [2].

For the case of a solitary wave shoaling over a sloping bottom, the accuracy can only be checked looking at the mass and energy conservation properties and comparing breaking wave characteristic obtained here with those reported elsewhere, for example in [5].

5.1. Sloping bottom test

A solitary wave propagating over a sloping bed changes its shape gradually, slightly increasing maximum height and front steepness, till a point where a vertical front tangent is reached. This is usually called the breaking point BP=(t_{bp}, x_{bp}, z_{bp}), where x_{bp} represents the x coordinate, z_{bp} the height at x_{bp} and t_{bp} the time of occurrence. Beyond the BP the wave tip develops, with velocities much bigger than the wave celerity, causing the wave overturning and the subsequent falling of the jet toward the flat water surface. Denote this endpoint as EP=(t_{ep}, x_{ep}, z_{ep}). Total wave mass and total energy should be theoretically, conserved until EP. However beyond the BP a lost in potential energy and the corresponding gain in kinetic energy is expected, due to the large velocities on the wave jet.

Wave breaking characteristics change, mainly according to initial wave amplitude (H_0) and bottom topography. To study how our numerical method predicts wave breaking we run the following test cases:

- (a) $H_0 = 0.6$, $L = 25$, slope=$1:22$, $x_c = 6.05$, $x_s = 6$
- (b) $H_0 = 0.6$, $L = 18$, slope=$1:15$, $x_c = 5.55$, $x_s = 5.4$

and compare the results obtained here for case (b) with those reported in ([4]). Here x_c denotes the x coordinate at the crest for the initial wave and x_s the x coordinate where the bottom slope starts.

A series of numerical experiments have been made, and optimal discretization parameters found are: $\Delta x = 0.01$, $\Delta t = 0.0001$ and $d_0 = 0.005$, $f_0 = 10$ (approximately 193 BEM nodes) for all cases. Front regridding has been made according to maximum height before the BP and according to maximum velocity modulus beyond BP. Beyond the BP, and due to the complex topography of the wave front, reinitialization of Ψ and new $\Phi(s,t)$ extension has been performed every 1000 time steps.

Table 1 shows the breaking characteristics for the test cases. Grilli et all reported in ([4]) for test (b) values of $t_{bp} = 2.41$, $x_{bp} = 15.64$ and $z_{bp} = 0.67$. The discrepancies can be attributed to the slightly different position of the initial wave ($x_c = 5.5$) and the higher-order approximations used in their Lagrangian-Eulerian formulation.

Test	t_{bp}	x_{bp}	z_{bp}	t_{ep}	x_{ep}
(a)	2.76	17.39	0.674	3.36	20.2
(b)	2.34	15.20	0.662	2.90	17.8

TABLE 1. Breaking characteristics

Figure 2 shows wave shape for case (a) at $t = 0, 1, 2, 2.76, 2.94, 2.14, 3.34$ and Figure 3 shows wave shape for case (b) at $t = 0, 1, 2, 2.34, 2, 48, 2.68, 2.90$. In Figures 4 and 5 we show in more detail the wave profiles from the BP to the EP for cases (a) and (b) respectively.

Finally, to see how wave shape and breaking characteristics change with bottom topography, we consider two more tests, this time with a sinusoidal shape bottom. As can be seen in Figures 6 and 7 the breaking characteristics are considerably different, and the wave breaks as a spilling breaker rather than the plunging breaker of case (a) and (b).

From these numerical experiments we conclude that the numerical method presented here is capable of reproducing wave shoaling and breaking till the touchdown of the wave jet. Considering that we use only first-order approximations of the model equations, a piecewise linear approximation of the free boundary, and a first-order linear BEM, the results are quite accurate. The absolute errors in mass and energy seem to be higher than those reported in ([4]). This is not surprising due to the fact that in ([4]) a higher-order BEM is used (both higher-order elements to define local interpolation between nodes and spline approximation of the free boundary geometry) and time integration for the free boundary conditions is at least second-order in time.

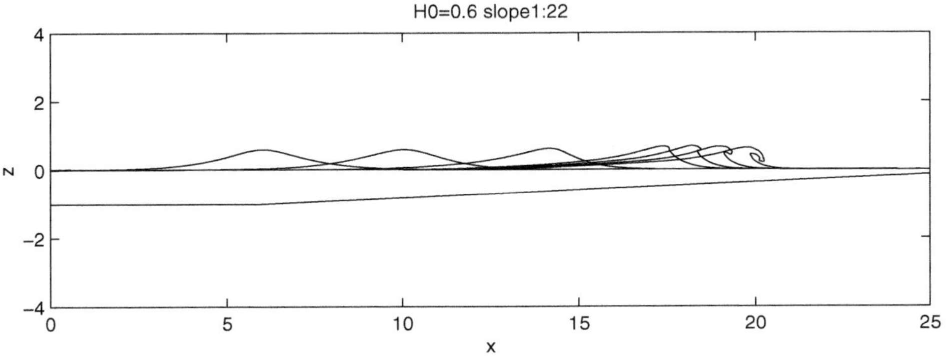

FIGURE 2. Wave shape at various times. Case (a)

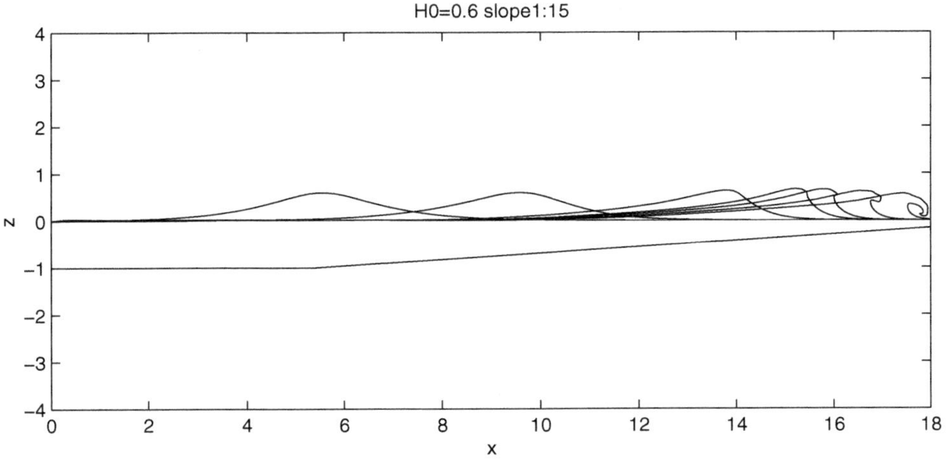

FIGURE 3. Wave shape at various times. Case (b)

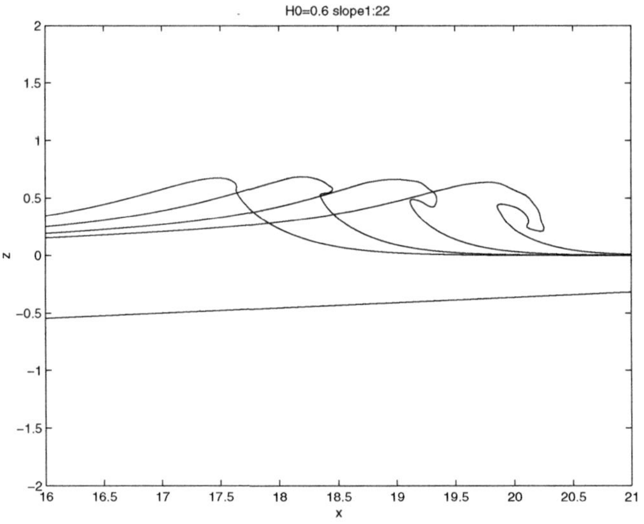

FIGURE 4. Wave shape at various times. Case (a)

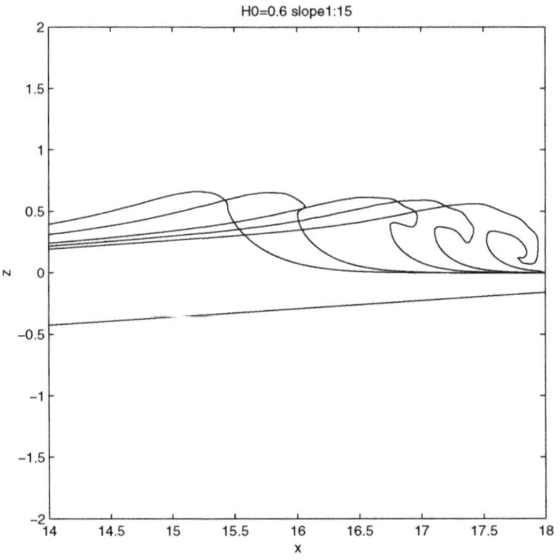

FIGURE 5. Wave shape at various times. Case (b)

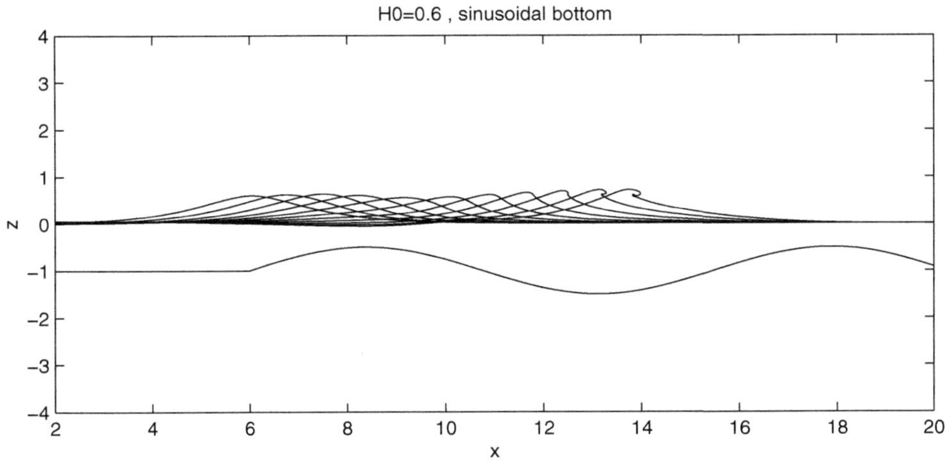

FIGURE 6. Wave shape at various times. Case (c)

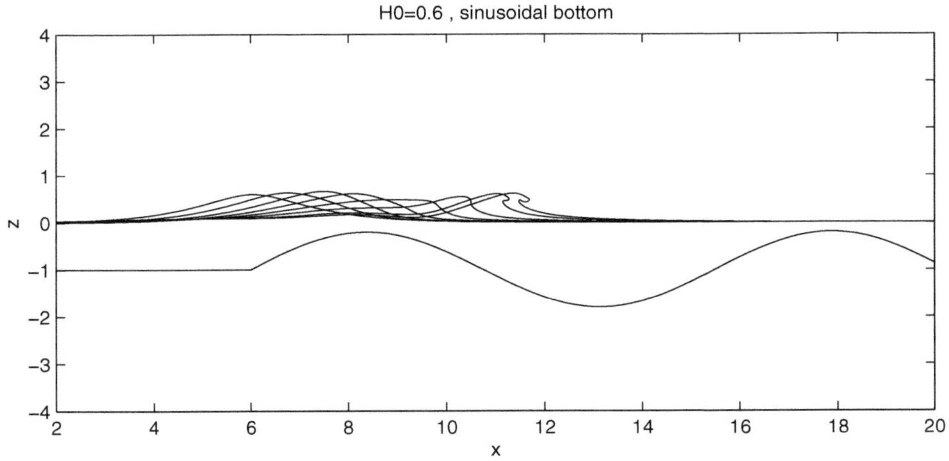

FIGURE 7. Wave shape at various times. Case (d)

Acknowledgment

L. Gray, Computer Science and Mathematics Division, Oak Ridge national laboratory; and D. Adalsteinsson, Dept. of Mathematics, University of North Carolina are also coauthors of this work.

References

[1] Adalsteinsson, D., and Sethian, J.A., *The Fast Construction of Extension Velocities in Level Set Methods*, 148, J. Comp. Phys., 1999, pp. 2–22.

[2] Garzon M., Adalsteinsson D.,Gray L., Sethian J.A. *A coupled level set-boundary integral method for moving boundary simulations* Interfaces and Free Boundaries 7 (2005), 1–26.

[3] Gray L.J., Phan A.-V. and Kaplan T., *Boundary Integral Evaluation of Surface Derivatives*, SIAM J. Sci. Comput., submitted, 2004.

[4] Grilli, S.T., Guyenne, P., and Dias, F., *A Fully Non-linear Model for Three-dimensional Overturning Waves Over an Arbitrary Bottom.* International Journal for Numerical Methods in Fluids 35:829–867 pp. (2001).

[5] Grilli, S.T., Svendsen, I.A., and Subramanya, R., *Breaking Criterion and Characteristics For Solitary Waves On Slopes.* Journal Of Waterway, Port, Coastal, and Ocean Engineering (June 1997).

[6] Grilli, S.T., Subramanya, R., *Numerical Modeling of Wave Breaking Induced by Fixed or Moving Boundaries.* Computational Mechanics 1996; 17:374–391.

[7] Melville W.K., Veron F., White C.J. *The velocity fields under breaking waves: coherent structures and turbulence* to appear

[8] Lin, P., Chang, K., and Liu, P.L., *Runup and Rundown of Solitary Waves on Sloping Beaches.* Journal Of Waterway, Port, Coastal, and Ocean Engineering (Sept./Oct. 1999).

[9] Osher, S., and Sethian, J.A., *Fronts Propagating with Curvature-Dependent Speed: Algorithms Based on Hamilton–Jacobi Formulations*, Journal of Computational Physics, 79, pp. 12–49, 1988.

[10] Sethian, J.A., *A Fast Marching Level Set Method for Monotonically Advancing Fronts*, Proc. Nat. Acad. Sci., 93, 4, pp. 1591–1595, 1996.

[11] Sethian, J.A., *Level Set Methods and Fast Marching Methods.* Cambridge Monographs on Applied and Computational Mathematics. Cambridge University Press (1999).

[12] Tanaka, M., *The stability of solitary waves* , Phys. Fluids, 29 (3), pp. 650–655, 1986.

[13] Chen G., Kharif C., Zaleski S. and Li, J. *Two-dimensional Navier-Stokes simulation of breaking waves* Physics of fluids, Vol. 11, number 1, pp. 121–133, 1999.

M. Garzon
Dept. of Applied Mathematics
Univ. of Oviedo
Calvo Sotelo s/n
33007 Oviedo, Spain
e-mail: maria@orion.ciencias.uniovi.es

J.A. Sethian
Dept. of Mathematics
University of California
Berkeley, USA
e-mail: sethian@math.berkeley.edu

International Series of Numerical Mathematics, Vol. 154, 199–208

Finite Difference Schemes for Incompressible Flows on Fully Adaptive Grids

Frédéric Gibou, Chohong Min and Hector Ceniceros

Abstract. We describe a finite difference scheme for simulating incompressible flows on nonuniform meshes using quadtree/octree data structure. A semi-Lagrangian method is used to update the intermediate fluid velocity in a standard projection framework. Two Poisson solvers on fully adaptive grids are also described. The first one is cell-centered and yields first-order accurate solutions, while producing symmetric linear systems (see Losasso, Gibou and Fedkiw [15]). The second is node-based and yields second-order accurate solutions, while producing nonsymmetric linear systems (see Min, Gibou and Ceniceros [17]). A distinguishing feature of the node-based algorithm is that gradients are found to second-order accuracy as well. The schemes are fully adaptive, i.e., the difference of level between two adjacent cells can be arbitrary. Numerical results are presented in two and three spatial dimensions.

Mathematics Subject Classification (2000). Primary 99Z99; Secondary 00A00.

1. Introduction

Incompressible flows are at the center of countless applications in physical sciences. Uniform Cartesian grids used in numerical simulations are limited in their ability to resolve small scale details and as a consequence nonuniform meshes are often used in practice. For example, see the approach of Almgren *et al.* [1] (and the references therein) for the Navier-Stokes equations on block structured grids. Since the work of [2] on compressible flows, adaptive mesh refinement techniques have been widely used. In the case of incompressible flows, adaptive mesh strategies are quite common (see, e.g., [7]), but implementations based on the optimal quadtree/octree data structure is less common.

In the case of a standard projection method (see, e.g., [4, 3]), the most computationally expensive part comes from solving a Poisson equation for the pressure. This is also the limiting part in terms of accuracy, since high-order accurate (and unconditionally stable) semi-Lagrangian methods exist for the convective part. In

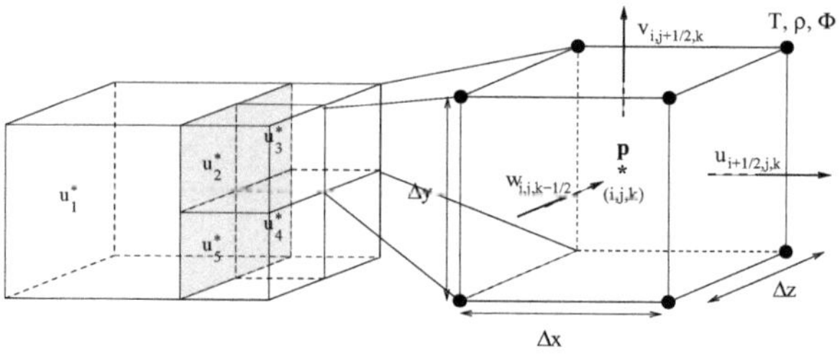

FIGURE 1. Left: the domain is tiled with cells of sizes varying according to the refinement criterion. Right: Zoom of one computational cell. The velocity components u, v and w are defined on the cell faces while the pressure p is defined at the center of the cell. The density ρ, the temperature T and the level set function ϕ are stored at the nodes.

[18], Popinet proposed second-order nonsymmetric numerical method to study the incompressible Navier-Stokes equations using an octree data structure. In [15], Losasso *et al.* extended the nonsymmetric discretization of [18] proposing a symmetric solution of the Poisson equation. This work relies on the observation that, in the case of the Poisson equation, first-order perturbations in the location of the solution yield consistent schemes (see [6]). Losasso *et al.* [14] recently extended the work of Lipnikov *et al.* [12] to the case of arbitrary grids to propose a second-order accurate symmetric discretization of the Poisson equation. In [17], Min *et al.* proposed a second-order accurate scheme that also yields second-order accurate gradients. In this case the linear system is nonsymmetric, but diagonally dominant.

2. The octree data structure

In [15], Losasso *et al.* proposed a solver for the incompressible Euler equations on fully adaptive grids. The domain is tiled with cells as depicted in Figure 1 and the mesh is refined automatically in order to capture the local details critical to realistic simulations and coarsened elsewhere. An octree data structure is used (see [19]) for efficient processing and the different variables are stored as depicted in Figure 1: The velocity components u, v and w are stored at the cell faces while the pressure is stored at the center of the cell. This is the standard MAC grid arrangement used in previous works (see, e.g., [8]). However, in the case of nonuniform meshes it is more convenient to store the other quantities such as the density ρ, the temperature T and the level set function ϕ at the *nodes* of each cell. This stems from the fact that interpolations of ρ, ϕ and T are more difficult with cell-centered data as discussed in [22].

3. Navier-Stokes equations on octrees

The motion of fluids is described by the incompressible Navier-Stokes equations for the conservation of momentum and mass:

$$\mathbf{u}_t + \mathbf{u} \cdot \nabla \mathbf{u} = -\nabla p + \mathbf{f}, \tag{3.1}$$

$$\nabla \cdot \mathbf{u} = 0, \tag{3.2}$$

where $\mathbf{u} = (u, v, w)$ is the velocity field, $\mathbf{f}$ accounts for the external forces and where the spatially constant density of the mixture has been absorbed in the pressure p. Viscous effects are ignored.

A projection method [4] (see also [3]) is used to solve equations 3.1 and 3.2: First an intermediate velocity $\mathbf{u}^*$ is computed over a time step $\triangle t$, ignoring the pressure term

$$\frac{\mathbf{u}^* - \mathbf{u}}{\triangle t} + \mathbf{u} \cdot \nabla \mathbf{u} = \mathbf{f}. \tag{3.3}$$

This step, accounting for the convection and the external forces, is followed by a projection step to account for incompressibility and boils down to solving the Poisson equation

$$\nabla^2 p = \frac{1}{\triangle t} \nabla \cdot \mathbf{u}^*. \tag{3.4}$$

Finally, the pressure correction is added to define the new velocity field $\mathbf{u}$:

$$\mathbf{u} = \mathbf{u}^* - \triangle t \nabla p. \tag{3.5}$$

The reader is referred to [15] and the references therein for details in the simulations of free surface flows.

3.1. Finding the intermediate velocity

The intermediate velocity u^* is found by solving equation 3.3 using a first-order accurate semi-Lagrangian method. In the case of nonuniform grids, the standard high-order accurate upwind methods (see, e.g., [9, 20, 13]) traditionally used in the case of uniform grids are not well suited due to their stringent time step restrictions and the complexity of their implementations. On the other hand, semi-Lagrangian methods (see, e.g., [21]) are unconditionally stable and are straightforward to implement.

3.2. The divergence operator

Equation 3.4 is solved by first evaluating the right-hand side at every grid point in the domain. Then, a linear system for the pressure is constructed and inverted. Consider the discretization of equation 3.4 for a large cell with dimensions $\triangle x$, $\triangle y$ and $\triangle z$ neighboring small cells as depicted in Figure 1 (left). Since the discretization is closely related to the second vector form of Green's theorem that relates a volume integral to a surface integral, we first rescale equation 3.4 by the volume of the large cell to obtain

$$\mathrm{V}_{\text{cell}} \triangle t \nabla^2 p = \mathrm{V}_{\text{cell}} \nabla \cdot \mathbf{u}^*. \tag{3.6}$$

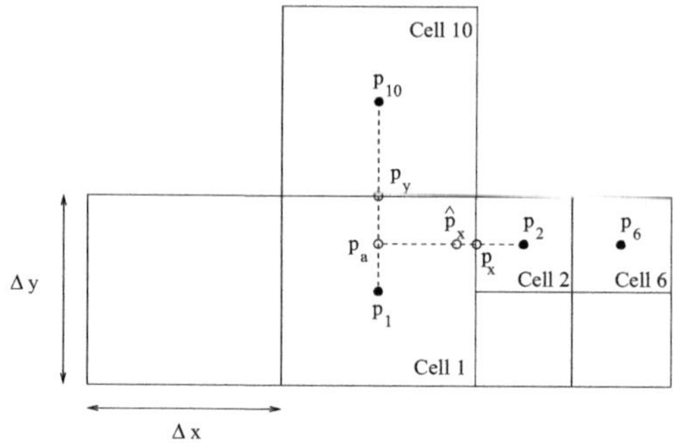

FIGURE 2. Discretization of the pressure gradient. The pressure values p_1, p_2, p_6, and p_{10} are defined at the center of the cells. p_a represents a weighted average pressure value. p_y defines the y component of the pressure gradient between Cell 1 and Cell 10 defined by standard central differencing. $\hat{p}_x$ represents the discretization of the x component of the pressure gradient between Cell 1 and Cell 2, whereas p_x is a $O(\triangle x)$ perturbation of $\hat{p}_x$.

The right-hand side of equation 3.6 now represents the quantity of mass flowing in and out of the large cell within a time step $\triangle t$ in $m^3 s^{-1}$. This can be further rewritten as

$$\mathrm{V}_{\text{cell}} \nabla \cdot (\mathbf{u}^* - \triangle t \nabla p) = 0. \tag{3.7}$$

This equation implies that the term ∇p is most naturally evaluated at the same location as $\mathbf{u}^*$, namely at the cell faces, and that there is a direct correspondence between the components of the vectors ∇p and $\mathbf{u}^*$. That is, there is a direct correspondence between p_x and u, p_y and v, p_z and w, which live on the right and left faces, top and bottom faces, front and back faces, respectively. Moreover, substituting equation 3.5 into equation 3.7 implies $\mathrm{V}_{\text{cell}} \nabla \cdot u = 0$ or $\nabla \cdot u = 0$ as desired.

Invoking the second vector form of Green's theorem, one can write

$$\mathrm{V}_{\text{cell}} \nabla \cdot \mathbf{u}^* = \sum_{\text{faces}} (\mathbf{u}^*{}_{\text{face}} \cdot \mathbf{n})\, \mathrm{A}_{\text{face}}, \tag{3.8}$$

where $\mathbf{n}$ is the *outward* unit normal of the large cell and where A_{face} represents the area of a cell face. In the case of Figure 1 (left), the discretization of the x component $\partial \mathbf{u}^* / \partial x$ of the divergence reads

$$\triangle x \triangle y \triangle z \frac{\partial \mathbf{u}^*}{\partial x} = u_2^* \mathrm{A}_2 + u_3^* \mathrm{A}_3 + u_4^* \mathrm{A}_4 + u_5^* \mathrm{A}_5 - u_1^* \mathrm{A}_1, \tag{3.9}$$

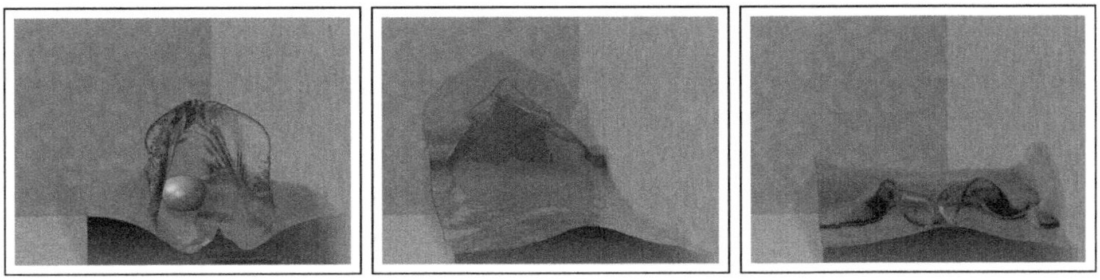

FIGURE 3. High-resolution simulation illustrating the motion of a solid object through shallow water (left), the subsequent flow that reflects off the wall (center), and the eventual cresting and overturning of waves generated by this process (right). Reprinted from [15]

where the *minus* sign in front of $u_1^* \mathrm{A}_1$ accounts for the fact that the unit normal points to the left. In this example, the discretization of $\partial \mathbf{u}^*/\partial x$ amounts to

$$\frac{\partial \mathbf{u}^*}{\partial x} = \frac{1}{\triangle x} \left(\frac{u_2^* + u_3^* + u_4^* + u_5^*}{4} - u_1^* \right). \tag{3.10}$$

The y and z directions are treated similarly.

3.3. Defining the pressure derivative to obtain a symmetric linear system

Once, the divergence is computed at the grid node, equation 3.4 is used to construct a linear system of equations for the pressure. Invoking again the second vector form of Green's theorem, one can write

$$\mathrm{V}_{\text{cell}} \nabla \cdot (\triangle t \nabla p) = \sum_{\text{faces}} \left((\triangle t \nabla p)_{\text{face}} \cdot \mathbf{n} \right) \mathrm{A}_{\text{face}}. \tag{3.11}$$

Therefore, once the pressure gradient is computed at every face, we can carry out the computation in a similar manner as for the divergence of the velocity described above.

In Gibou *et al.* [6], we showed that $O(\triangle x)$ perturbations in the location of the solution sampling still yield consistent approximations. This was then exploited in [15] to define ∇p in order to construct a symmetric linear system. We simply define

$$p_x = \frac{p_2 - p_1}{\triangle},$$

where $\triangle$ can be defined as $\triangle = \triangle x$, which is the size of the large cell or $\triangle = \frac{1}{2}\triangle x$, which is the size of the small cell, or as the Euclidean distance between the locations of p_1 and p_2 or as the distance along the x direction between the locations of p_1 and p_2 etc. In fact, in light of Theorem 4.1, we understand that there is some leeway in defining $\triangle$. Numerical tests against analytical solutions equation were presented in [15] to demonstrate that this scheme is convergent.

FIGURE 4. High-resolution simulation of the formation of a milk crown illustrating the capability of emulating surface tension. Reprinted from [15].

4. Second-order schemes for Poisson

The method in [15] is first-order accurate on fully adaptive grid and yields a symmetric linear system. This work was extended to second-order accuracy in Losasso *et al.* [14] using ideas from Lipnikov *et al.* [12]. We have presented in Min *et al.* [17] a Poisson solver on fully adaptive grids that produces second-order accurate solutions with second-order accurate gradients. This scheme yields diagonally dominant linear systems and is straightforward to implement. In particular, the discretization associated with one grid nodes involves only two (2D) or three (3D) adjacent cells, producing a scheme straightforward to implement.

4.1. Analysis

Theorem 4.1 (from [17]). *Consider a finite difference for $\triangle u = f$ that is mth-order accurate at locally uniform cell and nth-order accurate at locally nonuniform cell with $m, n \geq 1$. Then, it is globally $\min(m, n+1)$th-order accurate in L^∞ norm.*

This means that second-order accuracy in the maximum norm can be achieved with discretizations that are only first-order accurate at locally nonuniform points, but that reduce to second-order accurate at locally uniform points. Consider a Cartesian domain $\Omega \in \mathbb{R}^n$ with boundary $\partial\Omega$ and the variable Poisson equation $\nabla \cdot (\rho \nabla u) = f$ on Ω with Dirichlet boundary condition $u_{|\partial\Omega} = g$. We assume that the variable coefficient ρ is bounded from below by a positive constant.

In one spatial dimension, standard central differencing formulas read:

$$\left(\frac{u_{i-1} - u_i}{s_{i-\frac{1}{2}}} \cdot \frac{\rho_{i-1} + \rho_i}{2} + \frac{u_{i+1} - u_i}{s_{i+\frac{1}{2}}} \cdot \frac{\rho_{i+1} + \rho_i}{2} \right) \cdot \frac{2}{s_{i-\frac{1}{2}} + s_{i+\frac{1}{2}}} = f_i,$$

where $s_{i-1/2}$ is the distance between nodes $i-1$ and i. Authors have often been mislead by Taylor analysis and concluded that such schemes are only first-order accurate. However, they are second-order accurate. This has been observed analytically (see, e.g., [16, 11] and numerically [5], [10]).

This discretization can be applied in a dimension by dimension framework. However, special care needs to be taken when vertices are no longer aligned (see,

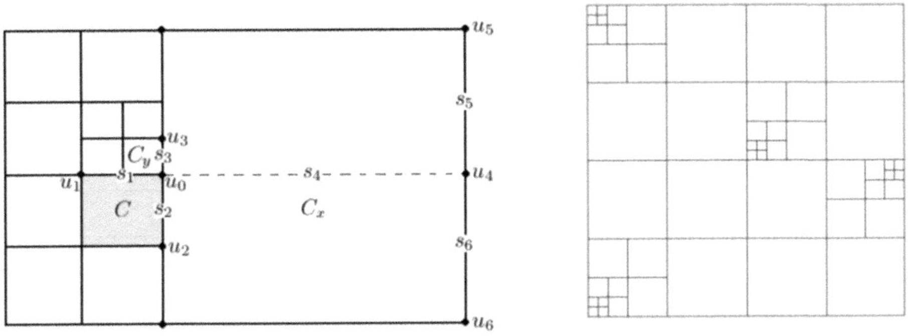

FIGURE 5. Left: Discretization at u_0 in the case of a nonuniform mesh. Right: Domain $\Omega = [0,\pi]^2$ and original mesh used in example 4.2.1.

e.g., Figure 5 [left]). In this case, [17] propose to use the truncation error in linear interpolation in the transverse direction as part of the stencil for the derivative in the other direction, leading to a more compact stencil, and an M-matrix. For example, referring to Figure 5 (left) the discretizations for $(\rho u_x)_x$ and $(\rho u_y)_y$ along with their Taylor analysis are given respectively by

$$\left(\frac{u_1-u_0}{s_1}\cdot\frac{\rho_1+\rho_0}{2}+\frac{s_6D_5+s_5D_6}{s_5+s_6}\right)\cdot\frac{2}{s_1+s_4} = (\rho u_x)_x + \frac{s_5s_6}{(s_1+s_4)s_4}(\rho u_y)_y + O(h), \tag{4.1}$$

and

$$\left(\frac{u_2-u_0}{s_2}\cdot\frac{\rho_2+\rho_0}{2}+\frac{u_3-u_0}{s_3}\cdot\frac{\rho_3+\rho_0}{2}\right)\cdot\frac{2}{s_2+s_3} = (\rho u_y)_y + O(h), \tag{4.2}$$

with

$$\begin{aligned} D_5 &= \frac{u_5-u_0}{s_4}\cdot\frac{\rho_5+\rho_0}{2}, \\ D_6 &= \frac{u_6-u_0}{s_4}\cdot\frac{\rho_6+\rho_0}{2}. \end{aligned}$$

The spurious term $\frac{s_5s_6}{(s_1+s_4)s_4}(\rho u_y)_y$ is cancelled by weighting appropriately equations (4.1) and (4.2) as

$$\left(\frac{u_1-u_0}{s_1}\cdot\frac{\rho_1+\rho_0}{2}+\frac{s_6a_5+s_5a_6}{s_5+s_6}\right)\cdot\frac{2}{s_1+s_4} + \left(\frac{u_2-u_0}{s_2}\cdot\frac{\rho_2+\rho_0}{2}+\frac{u_3-u_0}{s_3}\cdot\frac{\rho_3+\rho_0}{2}\right)\cdot\frac{2}{s_2+s_3}\cdot\left(1-\frac{s_5s_6}{(s_1+s_4)s_4}\right) = f_0 + O(h).$$

The discretization obtained is now first-order accurate at locally nonuniform points and second-order accurate at locally uniform points, hence yields a globally second-order accurate scheme in the maximum norm in light of Theorem 4.1.

Effective Resolution	Error in the L^∞ norm	Order	Error in the L^1 norm	Order
32^2	4.18×10^{-1}		9.25×10^{-2}	
64^2	9.11×10^{-2}	2.20	2.09×10^{-2}	2.15
128^2	2.31×10^{-2}	1.98	5.29×10^{-3}	1.99
256^2	5.89×10^{-3}	1.97	1.33×10^{-3}	1.99
512^2	1.49×10^{-3}	1.98	3.35×10^{-4}	1.99

TABLE 1. Second-order accuracy in the solution of example 4.2.1.

4.2. Numerical experiments

We report numerical evidences that confirm the schemes described above yield second-order accuracy in the L^1 and L^∞ norms for both the solution *and* its gradients, on highly irregular grids. In particular the difference of level between cells can be greater that one, illustrating that the method preserves its second-order accuracy on *fully* adaptive meshes. The linear systems of equations are solved using a bi-conjugate gradient method with an incomplete Cholesky preconditioner.

4.2.1. Accuracy on solution. Consider a domain and a grid depicted in Figure 5 [right] and $\nabla(\rho\nabla u) = f$ with an exact solution of $u(x, y) = \sin(x) + \sin(y)$ and density $\rho = \sin(x)\sin(y) + 2$. Dirichlet boundary conditions are imposed on the boundary. Table 1 demonstrates second-order accuracy in the L^1 and L^∞ norms.

4.2.2. Accuracy on gradient. One distinguishing feature of this algorithm is that it yields second-order accuracy in the maximum norm for the solution's *gradients* as well. This is achieved by removing spurious error with procedures related to those presented above, i.e., referring to the notations in Figure 5 (left), the gradient at u_0 is calculated as:

$$\begin{aligned} u_x &= \tfrac{u_4-u_0}{s_4} \cdot \tfrac{s_1}{s_1+s_4} + \tfrac{u_0-u_1}{s_1} \cdot \tfrac{s_4}{s_1+s_4} - \tfrac{s_5 s_6 s_1}{2s_4(s_1+s_4)} u_{yy}, \\ u_y &= \tfrac{u_3-u_0}{s_3} \cdot \tfrac{s_2}{s_2+s_3} + \tfrac{u_0-u_2}{s_2} \cdot \tfrac{s_3}{s_2+s_3}, \end{aligned}$$

where

$$u_{yy} = \left(\frac{u_3 - u_0}{s_3} + \frac{u_2 - u_0}{s_2} \right) \cdot \frac{2}{s_2 + s_3}.$$

Consider a domain and a grid depicted in Figure 5 (right) and $\nabla(\rho\nabla u) = f$ with an exact solution of $u(x, y) = \sin(x) + \sin(y)$ and density $\rho = \sin(x)\sin(y) + 2$. Dirichlet boundary conditions are imposed on the boundary. Table 2 demonstrates second-order accuracy of the gradient in the L^1 and L^∞ norms.

Effective Resolution	Error in the L^∞ norm	Order	Error in the L^1 norm	Order
64^2	4.25×10^{-2}		9.48×10^{-3}	
128^2	1.66×10^{-2}	1.36	2.41×10^{-3}	1.98
256^2	4.42×10^{-3}	1.91	5.81×10^{-4}	2.05
512^2	1.12×10^{-3}	1.98	1.41×10^{-4}	2.04
1024^2	2.78×10^{-4}	2.01	3.46×10^{-5}	2.03

TABLE 2. Convergence rate for example 4.2.2.

5. Conclusion

We have described finite difference schemes for simulating incompressible flows on nonuniform meshes. The quadtree (2D) and octree (3D) data structures allowed for an efficient representation of the mesh. In particular, we have described two different schemes for solving the Poisson equation. The first one is cell-centered and yields first-order accurate solutions, while producing symmetric linear systems [15]. The second is node-based and yields second-order accuracy for the solution and its gradients, while producing nonsymmetric linear systems [17]. The schemes are fully adaptive, i.e., the difference of level between two adjacent cells can be arbitrary.

References

[1] A. Almgren, J. Bell, P. Colella, L. Howell, and M. Welcome. A conservative adaptive projection method for the variable density incompressible Navier-Stokes equations. *J. Comput. Phys.*, 142:1–46, 1998.

[2] M. Berger and J. Oliger. Adaptive mesh refinement for hyperbolic partial differential equations. *J. Comput. Phys.*, 53:484–512, 1984.

[3] D. Brown, R. Cortez, and M. Minion. Accurate projection methods for the incompressible Navier-Stokes equations. *J. Comput. Phys.*, 168:464–499, 2001.

[4] A. Chorin. A Numerical Method for Solving Incompressible Viscous Flow Problems. *J. Comput. Phys.*, 2:12–26, 1967.

[5] F. Gibou and R. Fedkiw. A fourth-order accurate discretization for the laplace and heat equations on arbitrary domains, with applications to the Stefan problem. *J, Comput. Phys.*, 202:577–601, 2005.

[6] F. Gibou, R. Fedkiw, L.-T. Cheng, and M. Kang. A second-order accurate symmetric discretization of the Poisson equation on irregular domains. *J. Comput. Phys.*, 176:205–227, 2002.

[7] F. Ham, F. Lien, and A. Strong. A fully conservative second-order finite difference scheme for incompressible flow on nonuniform grids. *J. Comput. Phys.*, 117:117–133, 2002.

[8] F. Harlow and J. Welch. Numerical calculation of time-dependent viscous incompressible flow of fluid with free surface. *Phys. Fluids*, 8:2182–2189, 1965.

[9] A. Harten, B. Enquist, S. Osher, and S Chakravarthy. Uniformly high-order accurate essentially non-oscillatory schemes III. *J. Comput. Phys.*, 71:231–303, 1987.

[10] H. Johansen and P. Colella. A Cartesian grid embedded boundary method for Poisson's equation on irregular domains. *J. Comput. Phys.*, 147:60–85, 1998.

[11] H.O. Kreiss, H.-O. Manteuffel, T.A. Schwartz, B. Wendroff, and A.B. White Jr. Supra-convergent schemes on irregular grids. *Math. Comp.*, 47:537–554, 1986.

[12] K. Lipnikov, J. Morel, and M. Shashkov. Mimetic finite difference methods for diffusion equations on non-orthogonal non-conformal meshes. *J. Comput. Phys.*, 199:589–597, 2004.

[13] X.-D. Liu, S. Osher, and T. Chan. Weighted essentially non-oscillatory schemes. *J. Comput. Phys.*, 126:202–212, 1996.

[14] F. Losasso, R. Fedkiw, and S. Osher. Spatially adaptive techniques for level set methods and incompressible flow. *Computers and Fluids (in press)*.

[15] F. Losasso, F. Gibou, and R. Fedkiw. Simulating water and smoke with an octree data structure. *ACM Trans. Graph. (SIGGRAPH Proc.)*, pages 457–462, 2004.

[16] T. Manteuffel and A. White. The numerical solution of second-order boundary value problems on nonuniform meshes. *Math. Comput.*, 47 (176):511–535, 1986.

[17] C. Min, F. Gibou, and H. Ceniceros. A simple second-order accurate finite difference scheme for the variable coefficient poisson equation on fully adaptive grids. *CAM report* 05-29, *Submitted to J. Comput. Phys.*

[18] S. Popinet. Gerris: A tree-based adaptive solver for the incompressible euler equations in complex geometries. *J. Comput. Phys.*, 190:572–600, 2003.

[19] H. Samet. *The Design and Analysis of Spatial Data Structures.* Addison-Wesley, New York, 1989.

[20] C.-W. Shu and S. Osher. Efficient implementation of essentially non-oscillatory shock capturing schemes II (two). *J. Comput. Phys.*, 83:32–78, 1989.

[21] A. Staniforth and J. Cote. Semi-Lagrangian Integration Schemes for Atmospheric Models: A Review. *Monthly Weather Review*, 119:2206–2223, 1991.

[22] J. Strain. Fast tree-based redistancing for level set computations. *J. Comput. Phys.*, 152:664–686, 1999.

Frédéric Gibou
Department of Mechanical Engineering & Department of Computer Science
Bldg II, room 2334, UCSB, CA 93111, USA
e-mail: `fgibou@engineering.ucsb.edu`

Chohong Min
Department of Mathematics, UCSB, CA 93106, USA
e-mail: `chohong@math.ucsb.edu`

Hector Ceniceros
Department of Mathematics, UCSB, CA 93106, USA
e-mail: `hdc@math.ucsb.edu`

International Series of Numerical Mathematics, Vol. 154, 209–218

Global Solvability of Constrained Singular Diffusion Equation Associated with Essential Variation

Yoshikazu Giga, Hirotoshi Kuroda and Noriaki Yamazaki

Abstract. We consider a gradient flow system of total variation with constraint. Our system is often used in the color image processing to remove a noise from picture. In particular, we want to preserve the sharp edges of picture and color chromaticity. Therefore, the values of solutions to our model is constrained in some fixed compact Riemannian manifold. By this reason, it is very difficult to analyze such a problem, mathematically. The main object of this paper is to show the global solvability of a constrained singular diffusion equation associated with total variation. In fact, by using abstract convergence theory of convex functions, we shall prove the existence of solutions to our models with piecewise constant initial and boundary data.

Mathematics Subject Classification (2000). Primary 35K55, 68U10; Secondary 47J35.

Keywords. Singular diffusion, total variation, color image processing, subdifferential.

1. Introduction

We consider a constrained singular diffusion equation associated with total variation as follows:

$$u' = -\pi_u \left(-\mathrm{div} \left(\frac{\nabla u}{|\nabla u|} \right) \right) \qquad \text{in } (0,T) \times \Omega, \tag{1.1}$$

where $u'(t) := \dfrac{d}{dt}u(t)$, $0 < T < +\infty$ and Ω is a bounded domain in $\mathbb{R}^2$ with boundary Γ. Let S^{n-1} be the unit sphere in $\mathbb{R}^n$ $(n \geq 1)$, i.e.,

$$S^{n-1} := \{w \in \mathbb{R}^n; |w| = 1\}.$$

For each element $u \in S^{n-1}$, let $\pi_u : \mathbb{R}^n \to T_u S^{n-1}$ be an orthogonal projection from $\mathbb{R}^n = T_u\mathbb{R}^n$ to tangent space $T_u S^{n-1}$ of S^{n-1} at u.

The motivation of this paper is the color image processing. The constrained singular diffusion equation (1.1) was proposed by Tang-Sapiro-Caselles [22, 23] in order to remove a noise from the chromaticity of the initial image preserving the sharp edges of picture and color chromaticity.

For the gray image processing, there is a vast literature. For instance, we refer to [1, 2, 6, 19, 20]. In the simplest model of the gray image processing, the Gaussian filter was used for a grey-level function. Namely, for a given initial grey-level function u_0, we solve the heat equation

$$u' = \Delta u \quad \text{in } (0,T) \times \Omega$$

to get a denoised grey-level function $u(t,\cdot)$ at scale t. However, this way has a drawback since all characteristic function is mollified and a sharp contrast become ambiguous. In order to keep the sharp edges, one use a (unconstrained) singular diffusion equation governed by total variation flow

$$\begin{cases} u' = \operatorname{div}\left(\dfrac{\nabla u}{|\nabla u|}\right) & \text{in} \quad (0,T)\times\Omega, \\ u = g(x) & \text{on} \quad (0,T)\times\Gamma, \\ u(0,\cdot) = u_0 & \text{in} \quad \Omega, \end{cases} \tag{1.2}$$

where g and u_0 are given data. Then, the grey-level function is not mollified, and a Heaviside type function is a stationary solution to (1.2). Since (1.2) is the gradient system, we easily get the results on existence and uniqueness of solutions. In fact, we can define the energy functional ψ on $L^2(\Omega)$ by

$$\psi(u) = \begin{cases} \displaystyle\int_\Omega |\nabla \widetilde{u}| & \text{if } \widetilde{u} \in BV(\Omega), \\ +\infty & \text{otherwise.} \end{cases} \tag{1.3}$$

Here $\widetilde{u}$ is the extension of $u \in L^2(\Omega)$ to $\mathbb{R}^2$ such that $\widetilde{u}(x) = \widetilde{g}(x)$ for $x \in \mathbb{R}^2 \setminus \Omega$, where $\widetilde{g}$ is a Lipschitz extension of the boundary data g to $\mathbb{R}^2$. Then, ψ is proper, lower semi-continuous and convex on $L^2(\Omega)$. Moreover, the (unconstrained) gradient system (1.2) can be reformulated as in the abstract form:

$$u'(t) \in -\partial\psi(u(t)) \qquad \text{in } L^2(\Omega), \qquad \text{for } t > 0. \tag{1.4}$$

Thus, by applying the general theory established by Brézis [5] and Kōmura [16], we can get the solution to (1.2). For another detailed analysis, we refer to [1, 2], for instance. The (unconstrained) singular diffusion equation is also important to describe nonlinear physical phenomena (cf. [8, 10, 14, 15, 21, 24]). For instance, Shirakawa-Kimura [21] studied Allen-Cahn type equation with the total variation functional as the interfacial energy.

In this paper we discuss the global existence of solution $u : [0,T) \times \Omega \to S^{n-1} \subset \mathbb{R}^n$ to the following Dirichlet problem

$$\begin{cases} u' = -\pi_u \left(-\mathrm{div} \left(\dfrac{\nabla u}{|\nabla u|} \right) \right) & \text{in} \quad (0,T) \times \Omega, \\ u = g(x) & \text{on} \quad (0,T) \times \Gamma, \\ u(0,\cdot) = u_0 & \text{in} \quad \Omega, \end{cases} \tag{1.5}$$

where g and u_0 are given data which are maps from Ω to S^{n-1}. In 2003, Giga-Kobayashi [10] considered the problem (1.5) in the one-dimensional case. Then, they showed that for each piecewise constant initial data u_0 on Ω, there is a unique global solution u on $[0,\infty)$ such that $u(t)$ is a piecewise constant on Ω. Moreover, they studied the stationary problem in the case when the manifold is the unit circle S^1 in $\mathbb{R}^2$. In 2004, Giga-Kashima-Yamazaki [9] studied (1.1) in the n-dimensional torus domain $\Omega := \mathbb{T}^n = \prod_{i=1}^n (\mathbb{R}/\omega_i\mathbb{Z})$ for given $\omega_i > 0$ $(i = 1,2,\ldots,n)$. Assuming that the initial data u_0 is (sufficiently) small in some sense, they [9] constructed the local solution to (1.1) in the torus domain $\mathbb{T}^n$ as the limit of solutions to p-harmonic map flow equations with $p > 1$

$$u' = -\pi_u \left(-\mathrm{div} \left(|\nabla u|^{p-2} \nabla u \right) \right) \qquad \text{in} \quad (0,T) \times \mathbb{T}^n \tag{1.6}$$

by passing to the limit of $p \to 1$. In 2005, Giga-Kuroda-Yamazaki [12] proved the global existence of solution to a discretized version of (1.1) with Neumann boundary condition by restricting a class of mappings into that of piecewise constant mappings.

The main object of this paper is to show a global solvability of (a discretized version of) Dirichlet problem (1.5) by using the idea of [9, 10, 12]. Namely, for each piecewise constant initial and boundary data we find the piecewise constant solution $u(t)$ to (1.5) on Ω. Then, the problem is reduced to a system of ordinary differential equations unless two different values merges. This is the key point and idea in order to construct the global solution to the discretized Dirichlet problem (1.5). Of course, merging may occur, so, it is very difficult to study the detailed dynamics in 2-dimensional case. Different from one dimensional problem, our approach may not correspond to a solution of an original problem with a piecewise constant initial data. Such a difficulty is also observed in the unconstrained problem of crystalline flow [4] and [8], for instance.

The plan of this paper is as follows. In Section 2, we reformulate the problem (1.5) as in the evolution equation in some real Hilbert space by using subdifferential of convex functional. Then, we mention main result (Theorem 2.3) in this paper, which is concerned with the global existence of solution. In Section 3, we recall the convergence theorem established in [9]. In Section 4, we consider the approximating problem to (1.5). In the final Section 5 we give the proof of Theorem 2.3.

Notation

Throughout this paper, let Ω be a bounded domain in $\mathbb{R}^2$ with boundary Γ. We denote by $L^2(\Omega;\mathbb{R}^n)$ the space of $\mathbb{R}^n$-valued square integrable functions. For the unit sphere S^{n-1} in $\mathbb{R}^n$ $(1 \leq n < +\infty)$, let $L^2(\Omega; S^{n-1})$ be the closed subset of $L^2(\Omega; \mathbb{R}^n)$ of the form

$$L^2(\Omega; S^{n-1}) := \{v \in L^2(\Omega; \mathbb{R}^n) \ ; \ v(x) \in S^{n-1} \text{ a.e. } x \in \Omega\}.$$

Let H be a real Hilbert space with the inner product $\langle \cdot, \cdot \rangle$, and $\varphi : H \to (-\infty, +\infty]$ be a proper (i.e., not identically equal to infinity), l.s.c. (lower semi-continuous) and convex function on H. Then, we denote by $\partial\varphi$ the subdifferential of φ, which is defined by the set

$$\partial\varphi(u) = \{f \in H \mid \varphi(u+h) - \varphi(u) \geq \langle f, h \rangle \text{ for any } h \in H\}.$$

For basic properties of subdifferential, we refer to the monograph by Brézis [5].

2. Subdifferential formulation and main theorem

We begin with the definition of rectangular decompositions of Ω.

Definition 2.1 (Rectangular decomposition). For the bounded domain Ω in $\mathbb{R}^2$, let $\mathcal{C}$ be a rectangular decomposition of $\mathbb{R}^2$ so that $\mathcal{C} := \{R_j\}_{j \in \Lambda}$ is a disjoint family of open rectangles $R_j = (a_j, b_j) \times (c_j, d_j)$ which covers $\mathbb{R}^2$ expect a Lebesgue measure zero set. Then, we define a decomposition Δ of Ω associated with $\mathcal{C}$ by

$$\Delta := \{\Omega_i\}_{i \in I} \text{ with } \Omega_i = R_i \cap \Omega, \ \ I = \{i \in \Lambda \ ; \ \Omega_i \neq \emptyset\}.$$

Note that I is a finite index set, since Ω is a bounded domain.

Throughout this paper we fix the family $\Delta = \{\Omega_i\}_{i \in I}$. Then, let H_Δ be the set of all $\mathbb{R}^n$-valued step functions on $\bigcup_{i \in I} \Omega_i$, i.e.,

$$H_\Delta = \left\{ \sum_{i \in I} a_i \chi_{\Omega_i} \ ; \ a_i \in \mathbb{R}^n \right\},$$

where χ_{Ω_i} is the characteristic function on Ω_i. We easily see that H_Δ is the subset of $L^2(\Omega; \mathbb{R}^n)$, and the total variation of $u \in H_\Delta$ is given by this form

$$\int_\Omega |\nabla u| = \sum_{i<j} c_{ij} |a_i - a_j| \quad \text{ if } u \in H_\Delta,$$

which is also called a essential variation of u. Here, we set $c_{ij} = \mathcal{H}^1(\partial\Omega_i \cap \partial\Omega_j)$, where $\mathcal{H}^1$ is the Hausdorff measure and $\partial\Omega_i$ is the boundary of Ω_i. More precisely, c_{ij} implies a length of $\partial\Omega_i \cap \partial\Omega_j$. For the precise definition and basic properties of total variation, see monographs by Evans-Gariepy [7] or Giusti [13], for instance.

Now, by the similar argument in the gray image processing (1.2)–(1.4), we reformulate the problem (1.5) as in some evolution equation. To do so, let us define two functions on real Hilbert spaces. For given boundary data $g \in H_\Delta$, we put

$$\varphi_\Delta(u) = \begin{cases} \displaystyle\int_\Omega |\nabla u| & \text{if } u \in H_\Delta \text{ with } u(x) = g(x) \text{ for } x \in \Gamma, \\ +\infty & \text{otherwise.} \end{cases} \tag{2.1}$$

Then, from [7] or [13] it follows that φ_Δ is the proper, l.s.c. and convex function on $L^2(\Omega;\mathbb{R}^n)$. Also, we can define the proper, l.s.c. and convex function Φ_Δ^T on $L^2(0,T;L^2(\Omega;\mathbb{R}^n))$ by the form (cf. [9, Proposition 2.1])

$$\Phi_\Delta^T(u) = \int_0^T \varphi_\Delta(u(t))\,dt \qquad \text{for all } u \in L^2(0,T;L^2(\Omega;\mathbb{R}^n)). \tag{2.2}$$

Next, for each $h \in L^2(0,T;L^2(\Omega;S^{n-1}))$ we define a map $P_h(\cdot)$ from $L^2(0,T;L^2(\Omega;\mathbb{R}^n))$ to $L^2(0,T;L^2(\Omega;\mathbb{R}^n))$ by

$$P_h(f)(t,x) := \pi_{h(t,x)}(f(t,x)) \quad \text{for a.e. } (t,x) \in [0,T]\times\Omega \tag{2.3}$$

for any $f \in L^2(0,T;L^2(\Omega;\mathbb{R}^n))$.

By using these notations as above, we easily see that the problem (1.5) can be reformulated as in the following form:

$$\begin{cases} u' \in -P_u(\partial\Phi_\Delta^T(u)) & \text{in } L^2(0,T;L^2(\Omega;\mathbb{R}^n)), \\ u|_{t=0} = u_0 & \text{in } \Omega. \end{cases} \tag{2.4}$$

Now, let us give the definition of a solution to (2.4) (i.e., (1.5)).

Definition 2.2. Let $0 < T < +\infty$. For given data g, $u_0 \in H_\Delta$, a function $u : \Omega \times [0,T] \to \mathbb{R}^n$ is called a solution of (2.4) (i.e., (1.5)), if $u \in L^2(0,T;L^2(\Omega;S^{n-1})) \cap C([0,T];L^2(\Omega;\mathbb{R}^n))$, $u_t \in L^2(0,T;L^2(\Omega;\mathbb{R}^n))$ and (2.4) holds.

Now, let us mention our main result in this paper, which is concerned with the global existence of a solution to (2.4) (i.e., (1.5)).

Theorem 2.3. *Suppose the initial and boundary data u_0, $g \in H_\Delta$ with u_0, $g \in L^2(\Omega;S^{n-1})$. Then, for any time $T > 0$ there exists at least one solution u on $[0,T]$ to the problem* (2.4), *i.e.,* (1.5).

Note that we cannot apply the general theory (cf. [5, 16]) to the problem (2.4), because of the projection P_u. Hence, in order to prove Theorem 2.3, we consider the approximating problem of (1.5), and apply the abstract convergence theorem established in [9].

3. Abstract convergence theory

In this section, we recall the abstract convergence theory in [9]. We begin with the notion of Graph-convergence for multi-valued operators on a real Hilbert space H.

Definition 3.1 (e.g. [3]). For (multi-valued) operators A_n $(n = 1,2,\dots)$ and A on a real Hilbert space H, we say that A_n converges to A in the sense of Graph as $n \to +\infty$, if for any $(u,v) \in Graph(A)$ there exists $(u_n,v_n) \in Graph(A_n)$ such that $u_n \to u$ and $v_n \to v$ strongly in H as $n \to +\infty$.

Example. (cf. [3] or [9, Appendix]). Let ψ, ψ_n $(n = 1,2,\dots)$ be proper, l.s.c. and convex functions on H. Assume that ψ_n converges to ψ on H as $n \to +\infty$ in the sense of Mosco [18], namely, the following two conditions are satisfied:

(i) For any subsequence $\{\psi_{n_k}\} \subset \{\psi_n\}$, if $z_k \to z$ weakly in H as $k \to +\infty$, then, $\liminf_{k\to+\infty} \psi_{n_k}(z_k) \geq \psi(z)$.

(ii) For any $z \in D(\psi) = \{z \in H \mid \psi(z) < +\infty\}$, there is a sequence $\{z_n\}$ in H such that $z_n \to z$ in H and $\psi_n(z_n) \to \psi(z)$ as $n \to +\infty$.

Then, $\partial\psi_n$ converges to $\partial\psi$ on H in the sense of Graph as $n \to +\infty$.

Next, let us introduce the class $\mathcal{L}(K)$ of the operator $B(\cdot)(\cdot) : L^2(0,T;G) \times L^2(0,T;H) \to L^2(0,T;H)$, where G is a non-empty closed subset of H and $L^2(0,T;G)$ is a closed subset of $L^2(0,T;H)$ of the form

$$L^2(0,T;G) := \{u \in L^2(0,T;H) \ ; \ u(t) \in G \text{ a.e. } t \in [0,T]\}.$$

Definition 3.2 (cf. [9, Section 3]**).** We denote by $B \in \mathcal{L}(K)$ the set of all operator $B(\cdot)(\cdot) : L^2(0,T;G) \times L^2(0,T;H) \to L^2(0,T;H)$ satisfying the following three conditions:

(i) For any $u \in L^2(0,T;G)$, $B(u)(\cdot)$ is a bounded linear operator on $L^2(0,T;H)$.

(ii) There exists a constant $K > 0$ such that $\sup_{u\in L^2(0,T;G)} \|B(u)(\cdot)\|_{\mathcal{L}} \leq K$, where
$$\|B(u)(\cdot)\|_{\mathcal{L}} = \sup_{v\in L^2(0,T;H),\|v\|=1} \|B(u)(v)\|_{L^2(0,T;H)}.$$

(iii) If a sequence $\{u_k\}_{k=1}^{+\infty} \subset L^2(0,T;G)$ strongly converges to some u in $L^2(0,T;G)$, then, there is a subsequence $\{u_{k(l)}\}_{l=1}^{+\infty} \subset \{u_k\}_{k=1}^{+\infty}$ such that
$$B(u_{k(l)})^*(v) \longrightarrow B(u)^*(v) \text{ strongly in } L^2(0,T;H)$$
for any $v \in L^2(0,T;H)$, where $B(u)^*(\cdot)$ is the adjoint operator of $B(u)(\cdot)$.

Example. The projection operator $P_h(\cdot)$ defined in (2.3) is contained in the class $\mathcal{L}(K)$ in Definition 3.2.

Now, let us recall the abstract convergence theory established in [9].

Proposition 3.3 (Abstract convergence theorem) (cf. [9, Theorem 3.1] **).** *Let Ψ_n $(n = 1,2,\dots)$ and Ψ be proper, convex, l.s.c. functionals on $L^2(0,T;H)$. Let $B \in \mathcal{L}(K)$. Assume that $\partial\Psi_n$ converges to $\partial\Psi$ in the sense of Graph. Assume that there is a constant $R > 0$ so that $u_n \in L^2(0,T;H)$ $(n = 1,2,\dots)$ satisfies following conditions;*

$$\begin{cases} u_n' \in -B(u_n)(\partial\Psi_n(u_n) \cap B_R) & \text{in } L^2(0,T;H), \\ u_n \in L^2(0,T;G), \\ u_n|_{t=0} = u_{0,n} \in G, \end{cases}$$

where $B_R := \{u \in L^2(0,T;H) \ ; \ \|u\|_{L^2(0,T;H)} \leq R\}$. If $u_{0,n} \to u_0$ strongly in H and $u_n \to u$ in $C([0,T],H)$ as $n \to +\infty$, then, the function u is the solution of

$$\begin{cases} u' \in -B(u)(\partial\Psi(u)) & \text{in } L^2(0,T;H), \\ u \in L^2(0,T;G), \\ u|_{t=0} = u_0 \in G. \end{cases}$$

4. Approximating problem

In this section we consider the approximating problem of (1.5). At first we shall define the approximating energy function to (2.1).

For each $\varepsilon > 0$, let us define the function $\varphi_\Delta^\varepsilon$ by the form

$$\varphi_\Delta^\varepsilon(u) = \begin{cases} \sum_{j<k} c_{jk}\sqrt{|a_j - a_k|^2 + \varepsilon^2} & \text{if } u \in H_\Delta \\ & \quad \text{with } u(x) = g(x) \text{ for } x \in \Gamma, \\ +\infty & \text{otherwise.} \end{cases} \tag{4.1}$$

Clearly, $\varphi_\Delta^\varepsilon$ is proper, l.s.c. and convex on $L^2(\Omega;\mathbb{R}^n)$ such that $\partial\varphi_\Delta^\varepsilon(\cdot)$ is single-valued for any i with $\overline{\Omega}_i \cap \Gamma = \emptyset$. More precisely, we have

$$\partial\varphi_\Delta^\varepsilon(u) = \sum_{i,j} \frac{c_{ij}}{|\Omega_i|} \frac{a_i - a_j}{\sqrt{|a_i - a_j|^2 + \varepsilon^2}} \chi_{\Omega_i} \quad \text{for } \Omega_i \text{ with } \overline{\Omega}_i \cap \Gamma = \emptyset \tag{4.2}$$

for all $u = \sum_{i \in I} a_i \chi_{\Omega_i} \in H_\Delta$, where $|\Omega_i|$ is the volume of Ω_i.

Since $\varphi_\Delta^\varepsilon$ is the approximating function of our energy φ_Δ defined by (2.1), the approximating problem to (2.4) is given by the following form

$$\begin{cases} u_\varepsilon' = -P_{u_\varepsilon}(\partial\varphi_\Delta^\varepsilon(u_\varepsilon)) & \text{in } L^2(\Omega;\mathbb{R}^n), \quad \text{a.e. } t \in (0,T), \\ u_\varepsilon|_{t=0} = u_0 & \text{in } \Omega. \end{cases} \tag{4.3}$$

Here let us mention the result on existence-uniqueness of solutions to (4.3).

Proposition 4.1. *Suppose the same condition in Theorem* 2.3. *Then, for any* $\varepsilon > 0$ *and* $T > 0$, *there exists at most one solution* u_ε *on* $[0, T]$ *to the approximating problem* (4.3).

Proof. For each given data $u_0, g \in H_\Delta$ we can prove this Proposition by the slight modification in [10, Subsection 4.3] or [12, Proposition 3.1].

In fact, let $g = \sum_{i \in I} g_i \chi_{\Omega_i} \in H_\Delta$. Then, by taking account of u_0, $g \in H_\Delta$, (2.3) and (4.2), we observe that the approximating problem (4.3) is reduced to the ODE (ordinary differential equation) system:

(ODE): Find a unique function $u_\varepsilon(t) = \sum_{i \in I} a_i(t)\chi_{\Omega_i}$ on $[0,\infty)$ such that $a_i(t)$ is Lipschitz continuous from $[0,\infty)$ to S^{n-1} satisfying

$$a_i(t) \equiv g_i \quad \text{on } \Omega_i \text{ with } \overline{\Omega}_i \cap \Gamma \neq \emptyset,$$

$$\frac{da_i(t)}{dt} = -\pi_{a_i(t)}\left(\sum_j \frac{c_{ij}}{|\Omega_i|} \frac{a_i(t) - a_j(t)}{\sqrt{|a_i(t) - a_j(t)|^2 + \varepsilon^2}} \right) \tag{4.4}$$
$$\text{on } \Omega_i \text{ with } \overline{\Omega}_i \cap \Gamma = \emptyset,$$

for each $i \in I$.

Since $a_i(t) \in S^{n-1}$ and the projection $\pi_{a_i(t)} : \mathbb{R}^n \to T_{a_i(t)}S^{n-1}$, we observe that the right hand side of (4.4) is bounded independent of t. Hence, by applying the classical theory of ODE (e.g. Cauchy-Lipschitz Theorem), we can get the unique global solution u_ε on $[0,\infty)$ to (ODE), i.e., to our approximating problem (4.3). □

5. Proof of Theorem 2.3

In this section, we prove Theorem 2.3 by applying the abstract convergence theory [Proposition 3.3]. We begin with the key lemma to show Theorem 2.3.

Lemma 5.1. *Let φ_Δ and $\varphi_\Delta^\varepsilon$ be proper, l.s.c. and convex functions on $L^2(\Omega;\mathbb{R}^n)$ defined in* (2.1) *and* (4.1), *respectively. Then, we have:*

(i) *$\varphi_\Delta^\varepsilon$ converges to φ_Δ on $L^2(\Omega;\mathbb{R}^n)$ in the sense of Mosco* [18] *as $\varepsilon \to 0$.*

(ii) *$\Phi_\Delta^{T,\varepsilon}$ converges to Φ_Δ^T on $L^2\big(0,T;L^2(\Omega;\mathbb{R}^n)\big)$ in the sense of Mosco* [18] *as $\varepsilon \to 0$, where $\Phi_\Delta^{T,\varepsilon}$ is proper, l.s.c. and convex on $L^2\big(0,T;L^2(\Omega;\mathbb{R}^n)\big)$ defined by*

$$\Phi_\Delta^{T,\varepsilon}(u) = \int_0^T \varphi_\Delta^\varepsilon\big(u(t)\big)\,dt \qquad \text{for all } u \in L^2\big(0,T;L^2(\Omega;\mathbb{R}^n)\big).$$

Proof. By the general theory of convex analysis and the lower semi-continuity of the total variation, we can easily show (i). The assertion (ii) is the direct consequence of (i). □

Proof of Theorem 2.3. By applying the abstract convergence theory [Proposition 3.3], we can get the solution of our problem (2.4) as the limit of the function u_ε of (4.3) when $\varepsilon \to 0$.

Note that the function u_ε is also a solution to the approximating problem

$$\begin{cases} u_\varepsilon' = -P_{u_\varepsilon}\big(\partial\Phi_\Delta^{T,\varepsilon}(u_\varepsilon)\big) & \text{in } \ L^2\big(0,T;L^2(\Omega;\mathbb{R}^n)\big), \\ u_\varepsilon|_{t=0} = u_0 & \text{in } \ \Omega, \end{cases} \tag{5.1}$$

since we observe that $f \in \partial\Phi_\Delta^{T,\varepsilon}(u_\varepsilon)$ in $L^2\big(0,T;L^2(\Omega;\mathbb{R}^n)\big)$ if and only if $f(t) \in \partial\varphi_\Delta^\varepsilon(u_\varepsilon(t))$ for a.e. $t \in [0,T]$ (for instance, we refer to Brézis [5]).

Now, we take $L^2(\Omega;\mathbb{R}^n)$ as a real Hilbert space H, and choose $L^2(\Omega;S^{n-1})$ as a non-empty closed subset G in Proposition 3.3. Moreover, from Examples in Section 3 and Lemma 5.1 we observe that the projection operator $P_h(\cdot) \in \mathcal{L}(K)$, and $\partial\Phi_\Delta^{T,\varepsilon}$ converges to $\partial\Phi_\Delta^T$ on $L^2\big(0,T;L^2(\Omega;\mathbb{R}^n)\big)$ in the sense of Graph as $\varepsilon \to 0$.

By the expression (4.2) of $\partial\varphi_\Delta^\varepsilon(u_\varepsilon)$, we see that the subdifferential $\partial\varphi_\Delta^\varepsilon(u_\varepsilon)$ is bounded in $L^2(\Omega;\mathbb{R}^n)$ uniformly in ε. Therefore, the subdifferential $\partial\Phi_\Delta^{T,\varepsilon}(u_\varepsilon)$ is also bounded in $L^2\big(0,T;L^2(\Omega;\mathbb{R}^n)\big)$ uniformly in ε for each $T>0$, hence, there is a closed ball B_R of $L^2\big(0,T;L^2(\Omega;\mathbb{R}^n)\big)$ such that

$$\partial\Phi_\Delta^{T,\varepsilon}(u_\varepsilon) \subset B_R \quad \text{ uniformly in } \varepsilon > 0 \text{ for each } T > 0.$$

Since u_ε is the solution to (4.3) on $(0,T)$, there is an element $u_\varepsilon^* \in \partial\varphi_\Delta^\varepsilon(u_\varepsilon)$ such that $u_\varepsilon'(\tau,x) = -\pi_{u_\varepsilon(\tau,x)}(u_\varepsilon^*(\tau,x))$ for a.e. $(\tau,x) \in (0,T)\times\Omega$. By the definition of $\pi_{u_\varepsilon(\tau,x)}(\cdot)$, we see that $u'_\varepsilon(\tau,x) \in T_{u_\varepsilon(\tau,x)}S^{n-1}$ for a.e. $(\tau,x) \in (0,T)\times\Omega$. Thus, we have

$$\begin{aligned}\int_\Omega |u'_\varepsilon(\tau,x)|^2 dx &= (u'_\varepsilon(\tau), -\pi_{u_\varepsilon(\tau)}(u^*_\varepsilon)(\tau))_{L^2(\Omega;\mathbb{R}^n)} \\ &= -(u'_\varepsilon(\tau), u^*_\varepsilon(\tau))_{L^2(\Omega;\mathbb{R}^n)} = -\frac{d}{d\tau}\varphi^\varepsilon_\Delta(u_\varepsilon(\tau))\end{aligned} \tag{5.2}$$

for a.e. $\tau \in (0,T)$. By integrating (5.2) over $(0,T)$, we get the energy equation

$$\int_0^t \int_\Omega |u'_\varepsilon(\tau,x)|^2 dx d\tau + \varphi^\varepsilon_\Delta(u_\varepsilon(t)) = \varphi^\varepsilon_\Delta(u_0) \quad \text{for any } t \in [0,T]. \tag{5.3}$$

From (5.3) and the compactness theory (cf. [13, Theorem 1.19]) it follows that $\{u_\varepsilon(t)\}$ is relatively compact in $L^2(\Omega;\mathbb{R}^n)$ for any $t \in [0,T]$. Thus, Ascoli-Arzela's theorem implies that there exist a subsequence $\{u_{\varepsilon_m}\}_{m=1}^{+\infty} \subset \{u_\varepsilon\}$ and a function $u \in C([0,T];L^2(\Omega;\mathbb{R}^n))$ such that $\varepsilon_m \to 0$ and

$$u_{\varepsilon_m} \longrightarrow u \text{ strongly in } C([0,T];L^2(\Omega;\mathbb{R}^n)) \quad \text{as } m \to \infty.$$

Therefore, since assumptions of the abstract convergence theory [Proposition 3.3] are satisfied, we can apply Proposition 3.3 to our problem. Thus, we conclude that u is the solution on $[0,T]$ to (2.4) (i.e., (1.5)) for each $T > 0$. □

Acknowledgment. This work was studied during the stay of the third author at caesar (Bonn) in 2005, and was supported by Ministry of Education, Culture, Sports, Science and Technology, Japan. The third author wishes to thank Professor Karl-Heinz Hoffmann for his valuable comments and kind invitation to caesar in Germany.

References

[1] F. Andreu, C. Ballester, V. Caselles and J.M. Mazón, *Minimizing total variation flow,* Differential and Integral Equations, **14**(2001), 321–360.

[2] F. Andreu, C. Ballester, V. Caselles and J.M. Mazón, *The Dirichlet problem for the total variation flow,* J. Funct. Anal., **180**(2001), 347–403.

[3] H. Attouch, *Variational Convergence for Functions and Operators*, Pitman Advanced Publishing Program, Boston-London-Melbourne, 1984.

[4] G. Bellettini, M. Novaga and M. Paolini, *Facet-breaking for three-dimensional crystals evolving by mean curvature,* Interfaces Free Bound., **1**(1999), 39–55.

[5] H. Brézis, *Opérateurs Maximaux Monotones et Semi-Groupes de Contractions dans les Espaces de Hilbert*, North-Holland, Amsterdam, 1973.

[6] F. Catté, P.-L. Lions, J.-M. Morel and T. Coll, *Image selective smoothing and edge detection by nonlinear diffusion*, SIAM J. Numer. Anal., **29**(1992), 182–193.

[7] L.C. Evans and R.F. Gariepy, *Measure Theory and Fine Properties of Functions,* Studies in Advanced Mathematics, CRC Press, Boca Raton, FL, 1992.

[8] M.-H. Giga, Y. Giga and R. Kobayashi, *Very singular diffusion equations,* Proc. Taniguchi Conf. on Math., Advanced Studies in Pure Math., **31**(2001), 93–125.

[9] Y. Giga, Y. Kashima and N. Yamazaki, *Local solvability of a constrained gradient system of total variation,* Abstract and Applied Analysis, **2004**(2004), 651–682.

[10] Y. Giga and R. Kobayashi, *On constrained equations with singular diffusivity,* Methods and Applications of Analysis, **10**(2003), 253–278.
[11] Y. Giga and H. Kuroda, *On breakdown of solutions of a constrained gradient system of total variation*, Bol. Soc. Parana. Mat.(3), **22**(2004), 9–20.
[12] Y. Giga, H. Kuroda and N. Yamazaki, *An existence result for a discretized constrained gradient system of total variation flow in color image processing*, Interdisciplinary Information Sciences **11**(2005), 199–204.
[13] E. Giusti, *Minimal surfaces and functions of bounded variation*, Monographs in Mathematics, **80**, Birkhäuser Verlag, Basel, 1984.
[14] N. Kenmochi and K. Shirakawa, *Stability for a phase field model with the total variation functional as the interfacial energy,* Nonlinear Anal., **53**(2003), 425–440.
[15] R. Kobayashi and Y. Giga, *Equations with singular diffusivity,* J. Statist. Phys., **95**(1999), 1187–1220.
[16] Y. Kōmura, *Nonlinear semi-groups in Hilbert space,* J. Math. Soc. Japan, **19**(1967), 493–507.
[17] F. Mémoli, G. Sapiro, and S. Osher, *Solving variational problems and partial differential equations mapping into general target manifolds,* Journal of Computational Physics, **195**(2004), 263–292.
[18] U. Mosco, *Convergence of convex sets and of solutions variational inequalities,* Advances Math., **3**(1969), 510–585.
[19] S. Osher, A. Solé and L. Vese, *Image decomposition and restoration using total variation minimization and the H^{-1} norm*, Multiscale Model. Simul., **1**(2003), 349–370.
[20] G. Sapiro, *Geometric Partial Differential Equations and Image Analysis,* Cambridge University Press, Cambridge, 2001.
[21] K. Shirakawa and M. Kimura, *Stability analysis for Allen-Cahn type equation associated with the total variation energy,* Nonlinear Anal., **60**(2005), 257–282.
[22] B. Tang, G. Sapiro and V. Caselles, *Diffusion of general data on non-flat manifolds via harmonic maps theory: The direction diffusion case*, Int. J. Computer Vision, **36(2)**(2000), 149–161.
[23] B. Tang, G. Sapiro and V. Caselles, *Color image enhancement via chromaticity diffusion,* IEEE Transactions on Image Processing, **10**(2001), 701–707.
[24] A. Visintin, *Models of Phase Transitions,* Progress in Nonlinear Differential Equations and their Applications **28**, Birkhäuser, Boston-Basel-Berlin, 1996.

Yoshikazu Giga
Department of Mathematics, Graduate School of Mathematical Sciences
University of Tokyo, 3-8-1 Komaba, Meguro, Tokyo, 153-8914, Japan
e-mail: `labgiga@ms.u-tokyo.ac.jp`

Hirotoshi Kuroda
Department of Mathematics, Graduate School of Science, Hokkaido University
Kita 10, Nishi 8, Kita-Ku, Sapporo, Hokkaido, 060-0810, Japan
e-mail: `s023012@math.sci.hokudai.ac.jp`

Noriaki Yamazaki
Department of Mathematical Science, Common Subject Division
Muroran Institute of Technology, 27-1 Mizumoto-cho, Muroran, 050-8585, Japan
e-mail: `noriaki@mmm.muroran-it.ac.jp`

International Series of Numerical Mathematics, Vol. 154, 219–230

Capillary Mediated Melting of Ellipsoidal Needle Crystals

M.E. Glicksman, A. Lupulescu and M.B. Koss

Abstract. Measurements of video data on melting dendritic crystal fragments in reduced gravity show that a fragment's ellipsoidal axial ratio, C/A, rises initially until it melts down to a pole-to-pole length of $C \approx 5$ mm. At that point we observe a sudden fall in the C/A ratio with time, as the polar regions melt toward each other more rapidly than C/A times the melting speed, dA/dt, of the equatorial region. This accelerated melting allows the C/A ratio to fall from values around 10-20 (needle-like) towards values approaching unity (spheres) just before total extinction occurs. Analytical and numerical modeling will be presented that suggest that the cause of these sudden changes in kinetics and morphology during melting at small length scales is due to a crystallite's extreme shape anisotropy. Shape anisotropy leads to steep gradients in the mean curvature of the solid-melt interface near the ellipsoid's poles. These curvature gradients act through the Gibbs-Thomson effect to induce unusual thermo-capillary heat fluxes *within the crystallite* that account for the observed enhanced polar melting rates. Numerical evaluation of the thermo-capillary heat fluxes shows that they increase rapidly with the C/A ratio, and with decreasing length scale, as melting progresses toward total extinction.

PACS numbers: 66.30.-h, 81.10.-h, 81.30.Fb, 83.10.-y

Keywords. Mushy zones, dendrites, melting, capillarity, microgravity.

1. Introduction

Dendritic (branched) crystals are often encountered whenever metals, semiconductors, ceramics, and some organic materials solidify from supercooled or supersaturated melts and solutions, or when alloys solidify directionally under low thermal gradients [1]. Thermal dendrites are the simplest cases to treat theoretically, as their formation involve only the redistribution of latent heat in a pure molten phase. Two comprehensive reviews summarizing the kinetics of dendritic crystal growth are available [2, 3].

The most precise quantitative experiments that test theories of dendritic growth were accomplished about six years ago as a series of microgravity space flight experiments using transparent plastic crystals crystallizing from their melts at small supercoolings [4, 5]. Dendritic growth measurements made in microgravity advantageously simplify the transport of latent heat during crystal growth by eliminating nearly all sources of melt convection [6]. Specifically, the kinetics of dendritic growth in high-purity succinonitrile (SCN), a BCC crystal, and pivalic acid (PVA), an FCC crystal, were measured under convection-free conditions on three United States Microgravity Payload Missions (USMP-2, 1994, USMP-3, 1996, and USMP-4, 1997). These missions were flown as a coordinated series on the space shuttle *Columbia*. Video data were recorded for the first time on USMP-4, allowing measurements of the kinetics of dendritic fragments undergoing melting in microgravity. Microgravity melting of crystals is unique in avoiding any relative motions or natural convection between solid and melt because of the virtual absence of buoyancy forces.

The initial experimental melting data were analyzed on the basis of conduction-limited quasi-static melting theory. This theory was developed specifically to predict the growth rate or melting speed of needle-like crystallites that were modeled as slender ellipsoids with time-dependent major and minor axes of lengths $C(t)$ and $A(t)$, respectively [7]. Close agreement was found between that theory and the microgravity experiments when the melting process occurred under "shape-preserving" conditions. Shape preservation implies a constant value during melting of the C/A ratios of individual ellipsoidal crystal fragments [7].

2. Mushy Zones

The measurements taken with video recording of PVA crystals during USMP-4 provide the first significant compilation of convection-free melting data over a large variation of length scales, covering a range from about 10^{-2} m down to 2×10^{-4} m. As melting progresses, see the video sequence in figure 1-left, dendritic side-arms detach and shorten. Each fragment eventually melts away to extinction. It is especially important to note that in microgravity the individual crystallites remain motionless, and that melting proceeds by thermal conduction from the heated surrounding melt to the crystallites.

3. Melting and freezing

3.1. Background

Until recently, quantitative studies of melting kinetics comprise a relatively small body of experimental literature, and researchers had concentrated on mathematical analysis of the kinetics of melting in pure materials, such as heat storage media

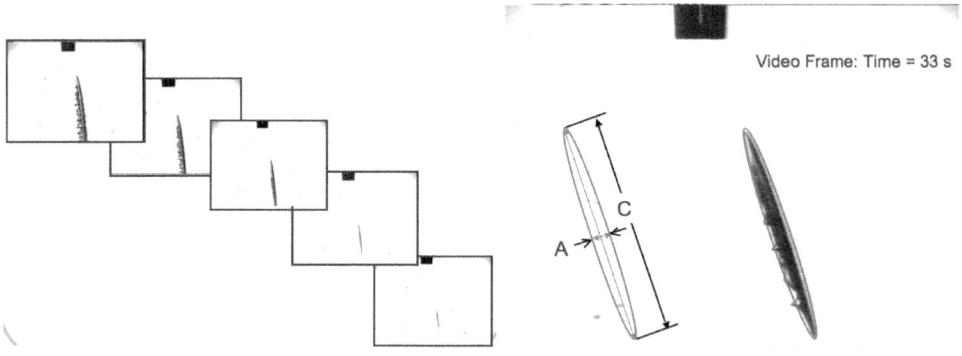

FIGURE 1. **Left**: Sequence of video frames after most crystal fragments in the dendritic mush zone are totally melted. **Right**: Last fragment to melt in the same video field of view. Digital image analysis of the middle frame shown in the video sequence on the left. Automated edge detection is employed to find the elliptical profile of this PVA crystallite to determine its major and minor axes, $\mathcal{C}(t)$ and $\mathcal{A}(t)$, respectively. Tip of glass injector at top of all frames is 1 mm in diameter.

or so-called phase change materials or PCM's [8]–[12]. A mathematical review is available of moving and free boundary approaches applied to the analysis of melting kinetics by Herraiz et al. [13]. Their comprehensive survey includes most of the known analytical solutions for melting. Melting kinetics in superheated alloys was also analyzed recently by Coriell et al., who demonstrated the interesting occurrence of multiple similarity solutions [14, 15]. Rettenmayr et al. published a series of interesting papers on the non-equilibrium melting of alloys that provide experimental observations and theoretical analysis of melting phenomena where solute diffusion plays a key rôle [16]–[19].

Both quasi-static and moving boundary solutions are well known for the *growth* of spherical crystals [20, 21]. Also, a class of moving boundary similarity solutions was developed by Ham that describes the freezing of crystals in the form of ellipsoids and hyperboloids [22]. Ham's solutions for the growth of ellipsoidal crystals, however, based as they are on similarity solutions to the heat conduction equation, are neither morphologically stable for the case of freezing, nor are they applicable via time reversal to the process of melting.

The present authors also recently analyzed a specific class of conduction-limited melting problems related to the present work, namely, the quasi-static self-similar melting of prolate spheroids [7]. Analytic solutions were developed earlier by assuming the presence of small superheating in the melt (quasi-static conditions), ignoring the influences of capillarity at small length scales, and employing the simplification that the crystal's C/A ratio remains constant during melting. Specifically, potential theory predicts the changes with time, t, of a crystallite's semi-major axis, $C(t)$, rescaled with respect to the crystal's initial semi-major axis length, C_0. The melting kinetics derived in [7] may be written as

$$C(t)/C_0 = \sqrt{1 \pm K_{\text{prolate}} \times St \cdot Fo}. \tag{1}$$

In Eq. (1) the kinetic coefficient, K_{prolate}, is defined as

$$K_{\text{prolate}} \equiv \frac{\pm 16}{\Xi \log \frac{\Xi+1}{\Xi-1}} \left(\frac{C}{A}\right)^2,$$

where the parameter $\Xi \geq 1$ is the prolate spheroidal interface coordinate denoting the crystal-melt interface. $K_{\text{prolate}} < 0$ for melting, and positive for freezing, respectively. An interface becomes progressively more needle-like in shape as $\Xi \to 1$, and more spherical as $\Xi \to \infty$. The melting kinetic constant can be expressed solely in terms of the crystal's C/A ratio by using the relationship for prolate spheroids that $\frac{C}{A} = \frac{\Xi}{\sqrt{\Xi^2-1}}$. The dimensionless melting time, or Fourier number, is defined as $Fo \equiv \frac{\alpha_\ell}{C_0^2} t$, where α_ℓ is the thermal diffusivity of the melt. The melting Stefan number, St, is defined as the ratio of the superheating in the surrounding melt, $T_\infty - T_m$, to the material's characteristic temperature, $\mathbb{T} \equiv \frac{\Delta H_f}{C_p}$. Thus, $St \equiv (T_\infty - T_m)/\mathbb{T}$, where T_∞ is the temperature of the melt far from the crystallite; T_m is the equilibrium melting temperature; ΔH_f is the molar enthalpy of fusion, and ΔS_f is the molar entropy of fusion. It is apparent from the form of Eq. (1) that the conduction-limited *freezing* of prolate spheroids at a constant C/A ratio will progress at long times as approximately the square-root of time. In the case of conduction-limited melting, the rate of melting – as measured by the major axis – will accelerate continuously, approaching an infinite rate at the extinction time. This last important kinetic prediction, along with independent estimation of the melting Stefan number, St, were checked quantitatively from the video images derived from the IDGE microgravity experiments [7]. The correspondence between Eq. (1) and experiment was generally excellent, as may be seen in figure 2. The theoretical curve approaching the extinction time tends to under-predict the rate of melting slightly.

4. Dendritic melting experiments

4.1. Background

The Isothermal Dendritic Growth Experiment (IDGE) was designed to test existing mathematical models that predict how melt supercooling affects dendrite growth speed and tip radius [5], [32]–[37]. As mentioned in the introduction, only the final space flight, USMP-4, provided the capability for real-time digital streaming of 30 fps full gray-scale video during crystal growth and melting. During the final minute of melting, just prior to extinction, the thermal fields within the melting chamber stabilize. Individual video frames covering that period were converted to .tiff files and exported to permit the post-flight kinetic melting analysis discussed here.

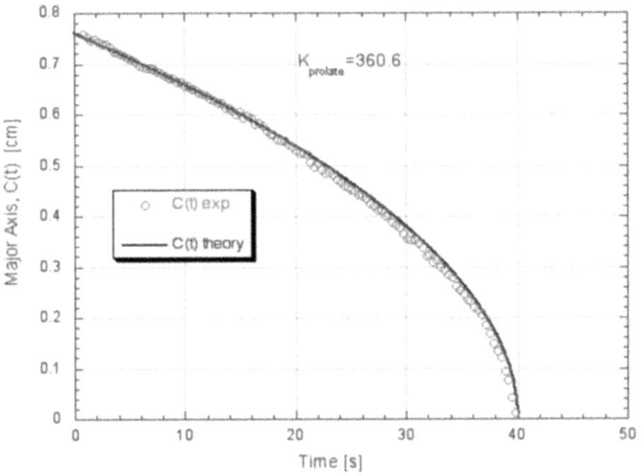

FIGURE 2. Comparison with theory, Eq. (1), of the observed crystallite major axis length, $C(t)$, versus melting time. The extinction time observed for this melting experiment was approximately 40 s. The value of $K_{\text{prolate}} = 360.6$ selected to predict these data corresponds to a steady $\mathcal{C}/\mathcal{A}$ ratio of approximately 12.

4.2. Video data analysis

The size and shapes of the melting dendritic fragments were determined from individual digitized video frames using commercial image analysis software [38]. The equivalent ellipse of the two-dimensional profiles of the crystal-melt interfaces surrounding selected fragments were calculated using the measured lengths of their semi-major axes, $C(t)$, and semi-minor axes, $A(t)$. These crystallites often, but not always, approximated ellipsoidal bodies of revolution, excepting some remnant surface irregularities on the otherwise smooth crystal-melt interface. These projections, moreover, represent the profile of the crystallites viewed normal to their C-axis.

We have already reported details [7] on the kinetic analysis using Eq. (1) to predict the behavior of a needle crystal melting self-similarly with a nearly constant $C/A \approx 12$. See again Figure 2. We now present the analysis of melting crystal fragments close to the extinction time, when the axial ratio $C/\mathcal{A}$ changes significantly with time. Specifically, in the current study of the measured C/A ratios of melting PVA crystal fragments, the axial ratio rose during most of the melting process but then suddenly decreased as the extinction point, $t^{\dagger}$, approached. The extinction point corresponds to the time at which a fragment completely disappears within one additional video frame. Kinetic data for melting, as reported here, usually started at about $t = (t^{\dagger} - 60)$ s. For example, the time-dependent behavior of the C/A ratio is shown for these crystallites in Figure 3-left, and Figure 3-right. In both of these melting cycles the lC/A ratio during the last minute climbs steadily from its initial value of $C/A \approx 7$ to more than about $C/A \approx 17$ – indicating a

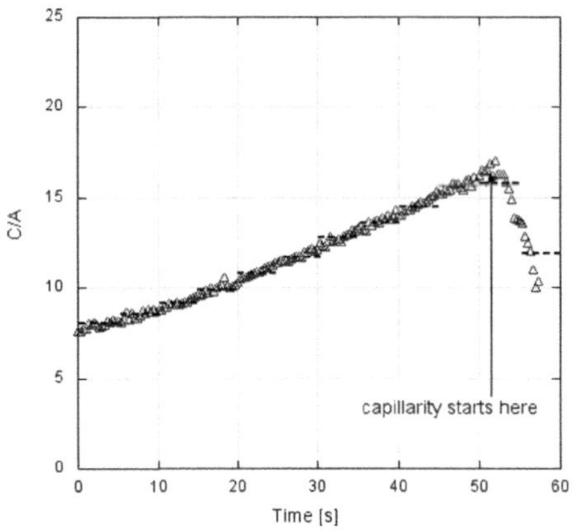

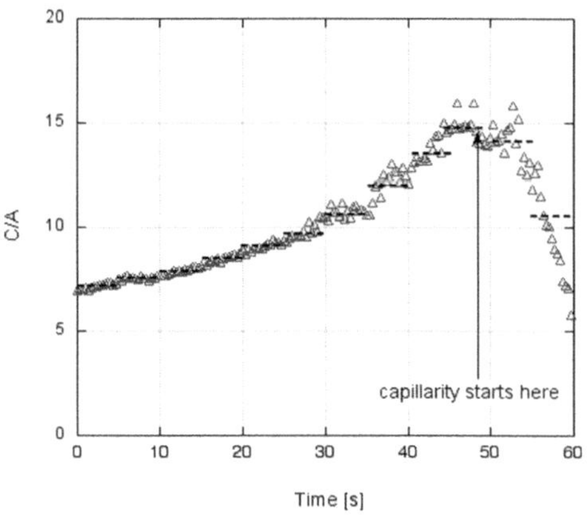

FIGURE 3. **Left:** $\mathcal{C}/\mathcal{A}$ ratio versus time. Data for PVA crystallites formed at an initial supercooling of 0.42K in the IDGE, during USMP-4. The $\mathcal{C}/\mathcal{A}$ values rise steadily until at about 50 s. they suddenly drop. **Right:** The $\mathcal{C}/\mathcal{A}$ ratio versus time. Data for crystallites formed at a similar initial supercooling of 0.46K. The $\mathcal{C}/\mathcal{A}$ almost doubles during the first 50 s. of melting, and then fall precipitously.

strongly increasing anisotropy of the shape as the crystallite melts and becomes increasingly slender and more needle-like. Then, at the point where the total length of the crystallite, $2l \times C(t)$, is reduced by melting to less than about 5 mm, and the corresponding crystallite diameter $2A(t)$ is reduced to 0.03 mm, one consistently observes (see again Figure 3-left, and Figure 3-right), that the rising C/A ratios suddenly reverse their trend and begin to fall rapidly. When the $\mathcal{C}/\mathcal{A}$ ratios reach values between roughly 10 and 5, respectively, the speed of melting becomes so high that the video framing rate cannot keep pace with the changes leading to extinction. This sudden reduction in the crystallite's axial ratio just precedes its extinction from melting at $t^{\dagger} \approx 60$ s. As it will be discussed in more detail later in this paper, we suggest that this unusual behavior, i.e., the sudden decrease during melting in a crystallite's shape anisotropy, or slenderness, reflects that its dimensions have become sufficiently small for the crystallite to begin exhibiting significant capillary or surface tension effects. Such an observation or suggestion for its cause has, to our knowledge, not been reported, heretofore.

More specifically, as melting progresses, the reduction in crystal size increases the mean curvature at all points on the ellipsoidal interface, but especially near its poles. The increased curvature reduces the equilibrium melting point, via the Gibbs-Thomson effect [12, 39], which in turn speeds up the melting process. The presence of highly curved regions near the poles results in enhanced melting rates that force the poles to approach each other *faster* than even the self-similar rate of decrease in the equatorial diameter. In order that the poles on a slender crystallite melt fast enough to cause a reduction in the crystallite's axial ratio, the following inequality must hold:

$$\left|\frac{dC}{dt}\right| > \frac{C}{A} \cdot \left|\frac{dA}{dt}\right|. \tag{2}$$

The enhanced speed of melting at and near the poles suggested in Eq. (2) would result in the crystallite achieving a more spherical shape, consistent with the observed drop in the C/A ratio. If relationship (2) above were an equation rather than the expressed inequality, the ellipsoid would melt away self-similarly with a constant axial ratio.

5. Interfacial capillary

5.1. Curvature of ellipsoids

Quasi-static conduction theory, as mentioned earlier, is able to correlate nearly all the observed kinetic melting data up to the time, $t^\dagger$, at which crystallite extinction via melting occurs. This observation may be ascribed to the fact that although capillarity can drastically alter the shape of a small melting crystal fragment, it has only minor effects on a crystallite's equilibrium melting point over the range of size scales that were observed in our experiments.

The equilibrium melting point of a small crystal with an anisotropic interfacial energy, $\gamma(\theta)$, is given by the anisotropic Gibbs-Thomson-Herring equation [39], namely

$$T_e = T_m - \frac{(\gamma(\varphi) + \gamma_{\varphi\varphi})}{\Delta S_f/\Omega} \cdot \mathcal{H}, \tag{3}$$

where Ω is the molar volume of the crystal. Here we have chosen to parameterize the location along the interface using the orientation angle, φ, of the *local* unit normal on the interface with respect to the C-axis, or $\langle 100 \rangle$ zone axis of the PVA crystallite, rather than the polar projection angle, θ, from the ellipsoid's centroid, $x = y = z = 0$. The second angular derivative of the interfacial energy, $\gamma_{\varphi\varphi}$, that appears in Eq. (3) is also taken with respect to this normal orientation angle.

5.2. Thermo-capillary effects

As discussed earlier in Section 3.1, a crystal surrounded by a large body of hotter melt that is heated to a temperature, $T_\infty \geq T_m$, experiences a rate of melting caused by heat conduction, the intensity of which is controlled by the melting Stefan number, St. A *local* Stefan number for melting, including the effect of curvature on temperature, Eq. (3), may be written as

$$St_{\text{loc}} = (T_\infty - T_m)\frac{C_p}{\Delta H_f} + \frac{(\gamma(\varphi) + \gamma_{\varphi\varphi})}{\Delta S_f/\Omega}\left(\frac{C_p}{\Delta H_f}\right) \cdot \mathcal{H}. \tag{4}$$

The first term on the right-hand side of Eq. (4) is the melting Stefan number applicable to a large crystal, where surface tension effects may be neglected, whereas

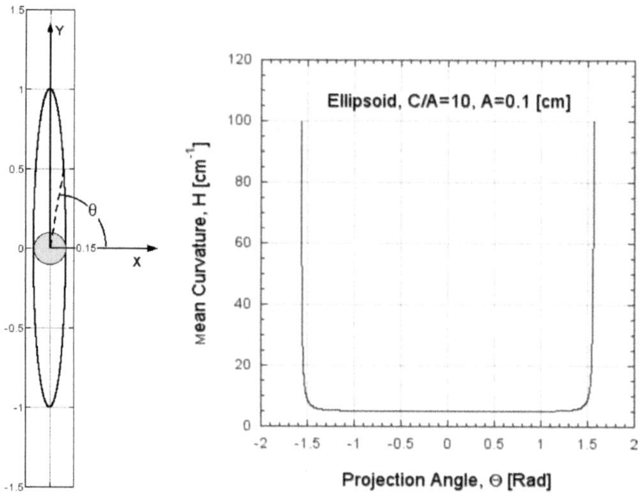

FIGURE 4. **Left:** Principle sections through a needle crystal. The crystal is in the form of an ellipsoid with its $\mathcal{C}$-axis parallel to the y-axis, and is viewed here as a projection onto the plane $x = 0$, and rotated into the plane of the page. Its $\mathcal{A}$-axis is parallel to the x-axis, and is viewed here with its projection onto the plane $y = 0$, and rotated into the plane of the page. This ellipsoid's $\mathcal{C}/\mathcal{A}$ ratio equals 10, which is typical of the axial ratios encountered in the present melting experiments. The polar projection angle θ, defined as $\theta = \arctan(y/x)$, provides a convenient running variable, against which the mean curvature, $\mathcal{H}(\theta)$, of the elliptical cross-section may be plotted. **Right:** Mean curvature, $\mathcal{H}$, around the upper half of the ellipsoid's surface. (Note use of a logarithmic scale on the ordinate.) The angle $\theta = 0$ locates the ellipsoid's equator, shown as the circular projection on the left, around which the curvature is a minimum. The north and south poles of the ellipsoid occur at $\theta = \pm\pi/2$, which designate interfacial locations that exhibit the maximum curvature.

the second term provides the local correction arising from anisotropic capillarity. Equation (4) can be written conveniently as

$$St_{\mathrm{loc}} = St + \Delta St_{cap}, \tag{5}$$

where ΔSt_{cap} is the local capillary correction added at small length scales (large $\mathcal{H}$ values).

During the microgravity melting experiments reported here, capillary effects became significant for crystals melted sufficiently to reduce their semi-major axis lengths to less than $\mathcal{C} \simeq 0.25$ cm. The additional thermo-chemical data for pure PVA needed to evaluate eqs. (4) and (5), include: $T_m = 309$ K, $C_p/\Delta H_f = 0.091$ K^{-1}. The 4-fold anisotropy of the interfacial free energy taken about the

⟨100⟩ crystallographic direction that coincides with the major growth axes of PVA dendritic crystallites may be expressed in polar form as [40],

$$\frac{(\gamma + \gamma_{\varphi\varphi})}{\Delta S_f/\Omega} = 4.5 \times 10^{-6}(1 - 0.75\cos 4\varphi)\ [\mathrm{cm}\cdot\mathrm{K}]. \tag{6}$$

The normal angles at the poles of a prolate spheroid are $\varphi = \pm\pi/2$. When these materials data are substituted back into Eq. (6) to evaluate the capillary constants at the poles one finds that $(\gamma + \gamma_{\varphi\varphi})\Omega/\Delta S_f = 1.1 \times 10^{-6}$ [cm·K]. Inserting all of these terms into Eq. (4) shows that the maximum capillary correction, or contribution, to the apparent melting Stefan number at the poles of the crystallites observed in the experiments reported in figure 1

$$\Delta St_{cap} \simeq 1.0 \times 10^{-7}\ \mathcal{H}_{\mathrm{pole}}. \tag{7}$$

At the poles, $\kappa_1 = \kappa_2$, so the maximum mean curvature, $\mathcal{H}_{\mathrm{pole}}$, for an ellipsoid with a given $\mathcal{C}/\mathcal{A}$ ratio and minor axis length, $\mathcal{A}$, may be shown to be

$$\mathcal{H}_{\mathrm{pole}} = \kappa_{\mathrm{pole}} = \left(\frac{l\cdot C}{A}\right)\frac{1}{A}. \tag{8}$$

Inserting the right-hand side of Eq. (8) back into Eq. (7) yields the maximum[1] capillary correction to the melting Stefan number for these crystallites ($C/A \simeq 17$ and $A \simeq 0.012$ cm), namely

$$\Delta St_{cap} = 1.0 \times 10^{-7}\left(\frac{l\cdot C}{A}\right)\frac{1}{l\cdot A} \simeq 1.4 \times 10^{-4}. \tag{9}$$

5.3. Internal thermo-capillary fluxes

Such a small change in St caused by capillarity would have negligible influence on the heat flux delivered *externally* to the poles from the surrounding hotter melt. As suggested, however, by the behavior of the mean curvature shown in Figure 4 right, needle-like ellipsoidal crystals also exhibit an enormous *gradient* of their curvature near their poles. A steep gradient in the mean curvature of the crystal-melt interface also corresponds, via the Gibbs-Thomson-Herring effect, to a large capillary-induced local temperature gradient: with the polar regions at the coldest location, i.e., largest mean curvature, and the equator and mid-latitudes slightly warmer. The strong gradients of the interfacial curvature near the poles add an additional thermo-capillary flux of heat conducted *within* the crystallite that increases its melting rates near the poles. Figure 2-left shows a sketch suggesting this mechanism of enhanced heat currents that reside internally and externally to the crystallite. The authors calculated the detailed dependence of the internal thermo-capillary flux at the ellipsoid's poles as a function of the C/A ratio. These data, presented in Figure 2-right, are based on finite element numerical solutions of Laplace's equation.

---

[1] For the anisotropic case considered here, the actual maximum capillary correction to the melting point occurs at a latitude slightly below the true pole. This offset in the coldest latitude is, however, small and will be ignored for the purposes of the present analysis.

6. Conclusion

The heat flux data in Figure 2 support our hypothesis that sharp curvature gradients account for internal thermo-capillary heat fluxes that cause the observed onset of a falling C/A ratio, whenever crystallites melt below a critical size. In the specific example drawn from the melting cycle analyzed in figure 3-right, the critical size at which thermo-capillary effects become manifest occurs at $t = 52$ s, corresponding to the time at which the major axis length $C = 0.4$ cm, and the minor axis $A = 0.024$ cm.

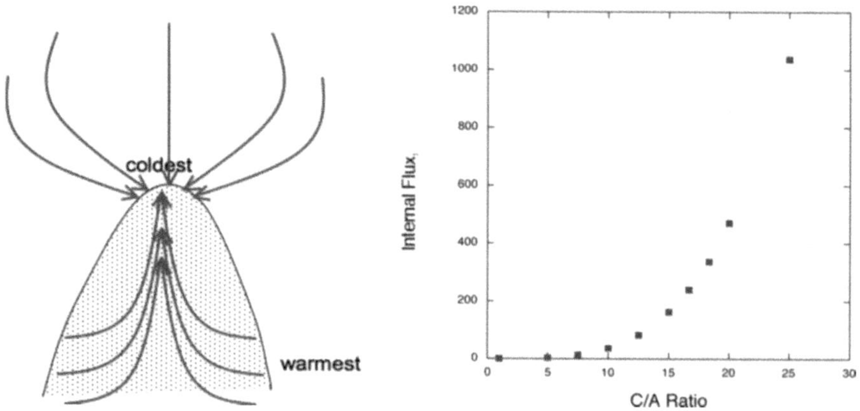

FIGURE 5. **Left:** External fluxes, controlled by the Stefan number of the heated melt, and internal fluxes, controlled by the curvature gradient near a sharp tip. Extra (internal) flux causes the poles to melt more rapidly, which accounts for the observation that the $\mathcal{C}/\mathcal{A}$ ratio drops rapidly as a melting crystallite approaches its extinction point. **Right**: Finite element calculations of capillary induced fluxes within a melting crystallite.

References

[1] M.E. Glicksman, S.P. Marsh, "Handbook of Crystal Growth", Ed. D.T.J. Hurle, Elsevier Science Publishers, Amsterdam, (1993) 1075.

[2] M.E. Glicksman, "Crystal Growth of Electronic Materials". Chapter 5, Ed. Kaldis E., Elsevier Science Publishers, Amsterdam, (1985) 57.

[3] M.E. Glicksman and A. O. Lupulescu, *J. Cryst. Growth*, **264** (2004) 541.

[4] M.E. Glicksman, M.B. Koss, J.C. LaCombe, L.A. Tennenhouse, *J. Cryst. Growth*, 174 (1997) 82.

[5] M.B. Koss, L.A. Tennenhouse, J.C. LaCombe, M.E. Glicksman, E.A. Winsa, *Metall. and Mat. Trans. A*, **30A** (1999) 3177.

[6] D.T.J. Hurle, G. Müller, and R. Nitsche, in "Fluid Sciences and Materials Science in Space", H.U. Walter Ed., Ch.X, Springer-Verlag, Berlin, (1987) 313.

[7] M.E. Glicksman, A. Lupulescu, and M.B. Koss, *J. Thermophys. and Heat Trans.*, **17**, 1 (2003) 69.

[8] A.D. Solomon, *Solar Energy*, **22** (1979) 251.

[9] A.D. Solomon, *Lett. Heat and Mass Trans.*, **7** (1980) 379.

[10] A.D. Solomon, *Lett. Heat and Mass Trans.*, **8** (1981) 237.

[11] J. Crank, "Free and Moving Boundary Problems", Clarendon Press, Oxford, (1984).

[12] V. Alexiades and A.D. Solomon, "Mathematical Modeling of Melting and Freezing Processes", Hemisphere Publishing Co., Washington, (1993).

[13] L.A. Herraiz and M.A. Herrero and J.J.L. Velazquez, *Proc. Roy. Soc. Edinburgh A*, **131** (2001) 371.

[14] S.R. Coriell, G.B. McFadden, and R.F. Sekerka. *J. Cryst. Growth*, **200** (1999) 276.

[15] S.R. Coriell, G.B. McFadden, R.F. Sekerka, and W.J. Boettinger, *J. Crys. Growth*, **191** (1998) 573.

[16] M. Rettenmayr, O. Warkentin, and H.E. Exner, *Zeit. für Metall.*, **88** (1997) 617.

[17] B. Dutta and M. Rettenmayr, *Metall. and Mat. Trans. A*, **31A** (2000) 2713.

[18] M. Rettenmayr, O. Warkentin, M. Rappaz, and H.E. Exner, *Acta Mat.*, **49** (2001) 2499.

[19] M. Hillert and M. Rettenmayr, *Acta Mat.*, **51** (2003) 2803.

[20] R.F. Sekerka, C.L. Jeanfils, R.W Heckel, "Lectures on the Theory of Phase Transformations", Ed. H.I. Aaronson, The Metallurgical Society, AIME, New York, (1982).

[21] M.E Glicksman, "Diffusion in Solids: Field Theory, Solid-state Principles, and Applications". John Wiley & Sons, New York, (2000).

[22] F.S. Ham, *Quart. J. Appl. Math.*, **17** (1959) 137.

[23] W.W. Mullins, R.F. Sekerka, *J. Appl. Phys.*, **34** (1961) 323.

[24] J.S. Langer, H. Muller-Krumbhaar, *Acta Met.*, **26** (1978) 1681.

[25] J.S. Langer, *Princeton Series in Physics*, "Critical Problems in Physics", Ch. 2, Princeton University Press, (1996).

[26] J. Lipton, M.E. Glicksman, W. Kurz, *Metall. Trans. A*, **18A** (1987) 341.

[27] M.E. Glicksman, R.J. Schaefer, J.D. Ayers, *Metall. Trans. A*, **7A** (1976) 1747.

[28] M.E. Glicksman, "Crystal Growth of Electronic Materials", Ch. 5, Ed. E. Kaldis, Elsevier Science Publishers, (1985) 57.

[29] S.C. Huang, M.E. Glicksman, *Acta Met.*, **29** (1981) 701.

[30] R. Trivedi, J.T. Mason, *Metall. Trans. A*, **22A** (1991), 235.

[31] G. Poots, *Int. J. Heat Mass Transfer*, **5** (1963) 339.

[32] G.P. Ivantsov, *Dokl. Akad. Nauk. USSR*, **58**, 4 (1947) 567.

[33] A. Lupulescu, M.E. Glicksman, M.B. Koss, 42nd AIAA Aerospace Sciences Meeting, (2004) 626.

[34] M.E Glicksman, M.B. Koss, L.T. Bushnell, J.C. Lacombe, E.A. Winsa, *Mat. Sci. Forum*, **215-216** (1996) 179.

[35] J.C. LaCombe, M.B. Koss, V.E. Fradkov, M.E. Glicksman, *Phys. Rev. E*, **52**, **3** (1995) 2778.

[36] J.C. LaCombe, M.B. Koss, D.C. Corrigan, A. Lupulescu, L.A. Tennenhouse, *J. Cryst. Growth*, **206** (1999) 331.

[37] A. Lupulescu, J.C. LaCombe, J.E. Frei, M.E. Glicksman, M.B. Koss, Microgravity Symposium: Proc. TMS Meeting, Nashville, CD-ROM, (2000).

[38] A. Lupulescu, M.E. Glicksman, M.B. Koss, *J. Cryst. Growth*, **276** (2005) 549-565.

[39] S.H. Davis, "Theory of Solidification", Cambridge University Press, Cambridge, (2001).

[40] N.B. Singh and M.E. Glicksman, *J. Cryst. Growth*, **98** (1989) 573.

M.E. Glicksman and A. Lupulescu
Materials Science and Engineering Department
Rensselaer Polytechnic Institute
Troy, New York, 12180-3590, USA

M.B. Koss
Department of Physics, College of the Holy Cross
Worcester, MA, 01610-2395 USA

International Series of Numerical Mathematics, Vol. 154, 231–240

Boundary Regularity at $\{t = 0\}$ for a Singular Free Boundary Problem

Eurica Henriques and José Miguel Urbano

Abstract. In this note it is shown that the weak solutions of the Stefan problem for the singular p-Laplacian are continuous up to $\{t = 0\}$. The result is a follow-up to a recent paper of the authors concerning the interior regularity.

Mathematics Subject Classification (2000). 35B65, 35D10, 35K65.

Keywords. Singular PDE, boundary regularity, intrinsic scaling, Stefan problem.

1. The problem and the regularity result

In a recent paper (cf. [5]), the authors obtained interior continuity results for the weak solutions of the singular parabolic PDE

$$\partial_t \eta - \Delta_p \theta = 0 \ , \quad \eta \in \gamma(\theta) \ ; \qquad 1 < p < 2 \ , \tag{1.1}$$

where γ is a maximal monotone graph and $\Delta_p u = \operatorname{div} |\nabla u|^{p-2} \nabla u$ is the p-Laplacian. When γ has a single jump at the origin, this equation generalizes to a nonlinear setting the modelling of the classical Stefan problem that corresponds to the case $p = 2$ and describes a phase transition at constant temperature for a substance obeying Fourier's law. Equation (1.1) is singular both in space and time since $1 < p < 2$ and, roughly speaking, $\gamma'(0) = \infty$.

In this note it is shown that, for continuous initial data, the continuity result holds up to $\{t = 0\}$ so that, in a way, the solution inherits the continuity properties of the boundary data. We consider a regularized approximated problem and show that the sequence of approximate solutions is equicontinuous up to $\{t = 0\}$. Due to the singularities in the equation we need to use intrinsic scaling to uniformly reduce the oscillation of the approximate solutions in a sequence of shrinking cylinders laying at the bottom of the space-time domain. For a modern account of intrinsic scaling and related matters, we suggest the reading of the recent survey [4].

To fix ideas, assume that an incompressible material (say pure water) occupies a bounded domain $\Omega \subset \mathbb{R}^N$, with two phases, a solid phase corresponding to the

---

Research supported by CMUC/FCT, Project POCI/MAT/57546/2004 and PRODEP-FSE.

region $\{\theta < 0\}$ and a liquid phase corresponding to the region $\{\theta > 0\}$, separated by an interface $\Phi = \{\theta = 0\}$, the free boundary. We denote $\Omega_T = \Omega \times (0,T)$ and $\Sigma = \partial\Omega \times (0,T)$, for some $T > 0$. The problem in its strong formulation reads

$$
(\mathbf{P}) \left\{ \begin{array}{rcll} \partial_t \theta & = & \Delta_p \theta & \text{in} \quad \Omega_T \setminus \Phi = \{\theta < 0\} \cup \{\theta > 0\} \\ \left[|\nabla\theta|^{p-2}\nabla\theta\right]_-^+ \cdot \mathbf{n} & = & \lambda\, \mathbf{w}\cdot\mathbf{n} & \text{on} \quad \Phi = \{\theta = 0\} \\ \theta & = & 0 & \text{on} \quad \Sigma \\ \theta(0) & = & \theta_0 & \text{in} \quad \Omega \times \{0\} \end{array} \right.
$$

where $\mathbf{n}$ is the unit normal to Φ, pointing to the solid region, $\mathbf{w}$ the velocity of the free boundary and $\lambda = [e]_-^+ > 0$ the latent heat of phase transition (e is the internal energy), with $[\,.\,]_-^+$ denoting the jump across Φ.

As usual, a weak formulation, in which all explicit references to the free boundary are absent, is obtained considering the maximal monotone graph H associated with the Heaviside function, and introducing a new unknown function, the enthalpy η, such that

$$\eta \in \gamma(\theta) := \theta + \lambda H(\theta) \ .$$

A formal integration by parts against appropriate test functions and the replacement of the initial condition for θ by a more adequate initial condition for η, leads to an integral relation that we adopt as definition of weak solution.

Definition 1.1. We say that (η,θ) is a weak solution of problem **(P)**, if

$$\theta \in L^p\Big(0,T;W_0^{1,p}(\Omega)\Big) \cap L^\infty(\Omega_T) \ ;$$
$$\eta \in L^\infty(\Omega_T) \quad \text{and} \quad \eta \in \gamma(\theta) \ , \ \text{a.e. in } \Omega_T \ ;$$
$$-\int_{\Omega_T} \eta\, \partial_t\xi + \int_{\Omega_T} |\nabla\theta|^{p-2}\nabla\theta\cdot\nabla\xi = \int_\Omega \eta_0\, \xi(0) \ , \quad \forall \xi \in \mathcal{T}(\Omega_T) \ .$$

The space of test functions we are considering is

$$\mathcal{T}(\Omega_T) := \left\{ \xi \in L^p\Big(0,T;W_0^{1,p}(\Omega)\Big) \ : \ \partial_t\xi \in L^2(\Omega_T) \ , \ \xi(T) = 0 \right\} ,$$

and we assume that

$$\eta_0 \in \gamma(\theta_0) \ , \quad \text{and} \quad \exists M > 0 \ : \ |\theta_0(x)| \le M \ , \ \text{a.e. } x \in \Omega \ . \tag{1.2}$$

Let $0 < \epsilon \ll 1$ and consider the bilipschitzian function

$$\gamma_\epsilon(s) = s + \lambda H_\epsilon(s) \ ,$$

where H_ϵ is a $\mathcal{C}^\infty$-approximation of the Heaviside function. Taking also functions $\theta_{0\epsilon} \in W^{1,p}(\Omega)$ such that

$$\theta_{0\epsilon} \to \theta_0 \ , \ \gamma_\epsilon(\theta_{0\epsilon}) \to \eta_0 \quad \text{in} \quad L^p(\Omega) \quad \text{and} \quad |\theta_{0\epsilon}| \le M \ , \quad \text{a.e. in } \Omega$$

we define a sequence of approximated problem as follows:

(P_ϵ): For each $0 < \epsilon \ll 1$, find a function

$$\theta_\epsilon \in H^1(0,T;L^2(\Omega)) \cap L^\infty(0,T;W_0^{1,p}(\Omega)) \cap L^\infty(\Omega_T)$$

such that

$$-\int_{\Omega_T} \gamma_\epsilon(\theta_\epsilon)\, \partial_t \xi + \int_{\Omega_T} |\nabla \theta_\epsilon|^{p-2} \nabla \theta_\epsilon \cdot \nabla \xi = \int_\Omega \gamma_\epsilon(\theta_{0\epsilon})\, \xi(0) \ , \quad \forall \xi \in \mathcal{T}(\Omega_T) \ . \quad (1.3)$$

In the presence of the regularity required, equation (1.3) can be shown to be equivalent to the two conditions: $\theta_\epsilon(0) = \theta_{0\epsilon}$ and, for a.e. $t \in (0,T)$,

$$\int_{\Omega \times \{t\}} \partial_t[\gamma_\epsilon(\theta_\epsilon)]\, \varphi + \int_{\Omega \times \{t\}} |\nabla \theta_\epsilon|^{p-2} \nabla \theta_\epsilon \cdot \nabla \varphi = 0 \ , \quad \forall \varphi \in W_0^{1,p}(\Omega) \ . \quad (1.4)$$

It was shown in [7] that this approximated problem has a unique solution and enough *a priori* estimates were derived to pass to the limit and obtain a solution of the original problem. In particular, the sequence of approximate solutions was shown to be equibounded.

We show here that there exists a uniform, i.e., independent of ϵ, modulus of continuity for θ_ϵ up to $\{t = 0\}$ and this will allow us to obtain a continuous solution up to $\{t = 0\}$ for the original problem as a consequence of Ascoli's theorem. We need to assume, in addition to (1.2), that

$$\theta_0 \in C(\Omega) \quad \text{and} \quad (\theta_{0\epsilon})_\epsilon \text{ is equicontinuous.} \quad (1.5)$$

This means that, over a compact $K \subset \Omega$, each $\theta_{0\epsilon}$ and θ_0 have the same modulus of continuity.

We will prove the following regularity result.

Theorem 1.2. *The sequence $(\theta_\epsilon)_\epsilon$ is equicontinuous up to $\{t = 0\}$. Then the weak solution of problem (P) is continuous up to $\{t = 0\}$. Moreover, for any compact $K \subset \Omega$, there exists a non-decreasing continuous function $\omega_K : \mathbb{R}^+ \to \mathbb{R}^+$, $\omega_K(0) = 0$, depending only upon the data and the modulus of continuity of θ_0, such that*

$$|\theta(x_1,t_1) - \theta(x_2,t_2)| \leq \omega_K\left(|x_1 - x_2| + |t_1 - t_2|^{\frac{1}{p}}\right) \ ,$$

for every pair of points $(x_1,t_1),(x_2,t_2) \in K' \times [0,T]$, and every compact $K' \subset K$.

In face of the recent results of [5], we clearly only need to prove the continuity at $t = 0$.

2. Energy and logarithmic estimates near $\{t = 0\}$

The building blocks of regularity theory leading to continuity results are energy and logarithmic estimates. These are the fundamental tools to proof Proposition 3.1 and will be derived next.

The crucial observation here is that, when deriving estimates for (1.1) in cylinders laying at the bottom of Ω_T, the term involving $(\theta_\epsilon - k)_\pm$ with power one, is absent, unlike in the interior case, which strongly simplifies the analysis. This is due to the choice of an independent of time cutoff function, which suffices for our purposes, and an appropriate selection of levels k, according to the initial data.

Given a point $x_0 \in \mathbb{R}^N$, $K_\rho(x_0)$ denotes the N-dimensional cube with centre at x_0 and wedge 2ρ:

$$K_\rho(x_0) := \left\{ x \in \mathbb{R}^N : \max_{1 \le i \le N} |x_i - x_{0i}| < \rho \right\} .$$

Fix $(x_0, t_0) \in \Omega_T$ and consider the cylinder

$$(x_0, t_0) + Q(\tau, \rho) = K_\rho(x_0) \times (t_0 - \tau, t_0)$$

where τ is such that $t_0 - \tau = 0$ so the cylinder lies at the bottom of Ω_T. Consider a piecewise smooth cutoff function ξ, independent of $t \in (0, t_0)$, satisfying

$$0 \le \xi \le 1 , \quad |\nabla \xi| < \infty \quad \text{and} \quad \xi(x) = 0 , \ x \notin K_\rho(x_0) . \tag{2.1}$$

In the weak formulation (1.4), take $\varphi = \pm (\theta_\epsilon - k)_\pm \, \xi^p \in W_0^{1,p}(K_\rho(x_0))$ and then integrate in time over $(0, t)$, for $t \in (0, t_0)$. Since

$$\pm \partial_t \left(\gamma_\epsilon(\theta_\epsilon) \right) (\theta_\epsilon - k)_\pm = \pm \gamma_\epsilon'(\theta_\epsilon) \, \partial_t \theta_\epsilon \, (\theta_\epsilon - k)_\pm = \partial_t \left(\int_0^{(\theta_\epsilon - k)_\pm} \gamma_\epsilon'(k \pm s) \, s \, ds \right) ,$$

recalling the t-independence of ξ and the definition of γ_ϵ, we obtain the following bound from below for the term involving the time derivative

$$\frac{1}{2} \int_{K_\rho(x_0) \times \{t\}} (\theta_\epsilon - k)_\pm^2 \, \xi^p - 2(M + \lambda) \int_{K_\rho(x_0) \times \{0\}} (\theta_\epsilon - k)_\pm \, \xi^p . \tag{2.2}$$

Observe that, if we choose

$$k \ge \sup_{x \in K_\rho(x_0)} \theta_{0\epsilon}(x) \tag{2.3}$$

when working with $(\theta_\epsilon - k)_+$, and

$$k \le \inf_{x \in K_\rho(x_0)} \theta_{0\epsilon}(x) \tag{2.4}$$

for $(\theta_\epsilon - k)_-$, the second term of (2.2) vanishes. On the other hand, the term concerning the space derivatives is estimated above by

$$\frac{1}{2} \int_0^t \int_{K_\rho(x_0)} |\nabla (\theta_\epsilon - k)_\pm|^p \, \xi^p - C(p) \int_0^t \int_{K_\rho(x_0)} (\theta_\epsilon - k)_\pm^p \, |\nabla \xi|^p$$

using Young's inequality with $\varepsilon = (2(p-1))^{\frac{p-1}{p}}$. We thus obtain

Proposition 2.1. *There exists a constant C, that can be determined a priori in terms of the data and independently of ϵ, such that for every $(x_0, t_0) \in \Omega_T$, for*

every cylinder $(x_0,t_0)+Q(\tau,\rho)$ *such that* $t_0-\tau=0$, *and for every level* k *verifying* (2.3) *or* (2.4),

$$\sup_{0<t<t_0}\int_{K_\rho(x_0)\times\{t\}}(\theta_\epsilon-k)_\pm^2\,\xi^p+\int_0^{t_0}\int_{K_\rho(x_0)}|\nabla(\theta_\epsilon-k)_\pm|^p\,\xi^p \leq \int_0^{t_0}\int_{K_\rho(x_0)}(\theta_\epsilon-k)_-^p\;|\nabla\xi|^p\,. \tag{2.5}$$

Now consider the logarithmic function

$$\Psi^\pm=\Psi\left(H_k^\pm,(\theta_\epsilon-k)_\pm,c\right)=\left(\ln\left(\frac{H_k^\pm}{H_k^\pm+c-(\theta_\epsilon-k)_\pm}\right)\right)_+\,,\qquad 0<c<H_k^\pm$$

where

$$H_k^\pm=\operatorname*{ess\,sup}_{(x_0,t_0)+Q(\tau,\rho)}(\theta_\epsilon-k)_\pm\,.$$

In the weak formulation (1.4) take

$$\varphi=\left[\left(\Psi^\pm\right)^2\right]'\xi^p=2\Psi^\pm\,\left(\Psi^\pm\right)'\,\xi^p\,,$$

where ξ is defined as in (2.1). Observing that

$$\begin{cases}\Psi^+(x,0)=0 \quad \text{for} \quad k\geq \displaystyle\sup_{x\in K_\rho(x_0)}\theta_{0\epsilon}(x)\\[2ex] \Psi^-(x,0)=0 \quad \text{for} \quad k\leq \displaystyle\inf_{x\in K_\rho(x_0)}\theta_{0\epsilon}(x)\end{cases},$$

and using Young's inequality with $\varepsilon=(2(p-1))^{\frac{p-1}{p}}$ we arrive at

Proposition 2.2. *There exists a constant* C, *determined a priori only in terms of the data and independently of* ϵ, *such that for every* $(x_0,t_0)\in\Omega_T$, *for every cylinder* $(x_0,t_0)+Q(\tau,\rho)$ *such that* $t_0-\tau=0$ *and for every level* k *verifying* (2.3) *or* (2.4),

$$\sup_{0<t<t_0}\int_{K_\rho(x_0)\times\{t\}}\left(\Psi^\pm\right)^2\xi^p\leq\int_0^{t_0}\int_{K_\rho(x_0)}\Psi^\pm\left|\left(\Psi^\pm\right)'\right|^{2-p}|\nabla\xi|^p\,. \tag{2.6}$$

3. Reduction of the oscillation in rescaled cylinders

Fix $(x_0,0)\in\Omega\times\{0\}$, and take $R>0$ such that $K_{2R}(x_0)\subset\Omega$. By translation, we may assume that $x_0=0$. Introduce the cylinder

$$Q\left(R^p,2R\right):=K_{2R}\times(0,R^p)$$

and define

$$\mu^+=\operatorname*{ess\,sup}_{Q(R^p,2R)}\theta_\epsilon\,;\quad \mu^-=\operatorname*{ess\,inf}_{Q(R^p,2R)}\theta_\epsilon\,;\quad \omega=\operatorname*{ess\,osc}_{Q(R^p,2R)}\theta_\epsilon=\mu^+-\mu^-\,.$$

Construct the cylinder

$$Q\left(a_0 R^p, R\right) = K_R \times (0, a_0 R^p) \ , \quad a_0 = \left(\frac{\omega}{2^m}\right)^{2-p} \ ,$$

where $m > 1$ is to be chosen. Without loss of generality, we may assume that $\frac{\omega}{2^m} \leq 1$ so that the following relations hold:

$$Q\left(a_0 R^p, R\right) \subset Q\left(R^p, 2R\right) \quad \text{and} \quad \operatorname*{ess\,osc}_{Q(a_0 R^p, R)} \theta_\epsilon \leq \omega \ .$$

The proof of Theorem 1.2 is a well-known consequence of the following iterative argument.

Proposition 3.1. *There exist constants $\sigma \in (0,1)$, and $C, m > 1$, that can be determined a priori only in terms of the data, such that constructing the sequences*

$$\begin{cases} \omega_0 = \omega \\ \omega_{n+1} = \sigma\, \omega_n \end{cases} \quad \text{and} \quad \begin{cases} R_0 = R \\ R_{n+1} = \frac{R}{C^n} \end{cases}$$

and the family of boxes

$$Q_n = \left(a_n R_n^p, R_n\right) \ , \qquad a_n = \left(\frac{\omega_n}{2^m}\right)^{2-p} \ ,$$

we have

$$Q_{n+1} \subset Q_n \quad \text{and} \quad \operatorname*{ess\,osc}_{Q_n} \theta_\epsilon \leq \max\left\{\omega_n, 2 \operatorname*{ess\,osc}_{K_{R_n}} \theta_{0\epsilon}\right\} \ , \tag{3.1}$$

for all $n = 0, 1, 2, \ldots$.

To prove Proposition 3.1, assume first that both inequalities

$$\mu^+ - \frac{\omega}{4} \leq \mu_0^+ := \operatorname*{ess\,sup}_{K_R} \theta_{0\epsilon} \qquad \text{and} \qquad \mu^- + \frac{\omega}{4} \geq \mu_0^- := \operatorname*{ess\,inf}_{K_R} \theta_{0\epsilon} \tag{3.2}$$

hold. Subtracting the second inequality from the first one we get

$$\frac{\omega}{2} \leq \mu_0^+ - \mu_0^- = \operatorname*{ess\,osc}_{K_R} \theta_{0\epsilon} \ .$$

and the proposition is trivially proved.

Without loss of generality, assume that the second inequality in (3.2) fails. Then the levels $k = \mu^- + \frac{\omega}{2^s}$, for $s \geq 2$, verify $k \leq \mu_0^-$ and, consequently, the energy and logarithmic estimates (2.5) and (2.6), respectively, hold for $(\theta_\epsilon - k)_-$. The next result has a double scope: it determines the parameter m that defines the height of the constructed initial cylinder and defines a level such that the subset of $Q\left(a_0 R^p, \frac{R}{2}\right)$ where θ_ϵ is below that level is small.

Lemma 3.2. *For all $\nu \in (0,1)$, there exists $m > 3$, depending only on the data, such that*

$$\left|(x,t) \in Q\left(a_0 R^p, \frac{R}{2}\right) \ : \ \theta_\epsilon(x,t) < \mu^- + \frac{\omega}{2^m}\right| < \nu \left|Q\left(a_0 R^p, \frac{R}{2}\right)\right| \ .$$

Proof. Consider estimate (2.6) written for $(\theta_\epsilon - k)_-$, with $k = \mu^- + \frac{\omega}{4}$, and for a cutoff function $0 \le \xi \le 1$, defined in K_R, and verifying

$$\xi \equiv 1 \quad \text{in } K_{\frac{R}{2}} \;; \qquad \xi \equiv 0 \quad \text{on } |x| = R \;; \qquad |\nabla \xi| \le \frac{2}{R} \,.$$

Take $m > 3$ sufficiently large so that $0 < c = \frac{\omega}{2^m} < H_k^-$. The logarithmic function Ψ^- is well-defined and, since $H_k^- \le \frac{\omega}{4}$, the following inequalities hold

$$\Psi^- \le (m-2)\ln 2 \qquad \text{and} \qquad \left|\left(\Psi^-\right)'\right|^{2-p} \le \left(\frac{\omega}{2^m}\right)^{p-2} .$$

Then, from (2.6), we get for all $t \in (0, a_0 R^p)$, the estimate

$$\int_{K_R \times \{t\}} \left(\psi^-\right)^2 \xi^p \le C\,(m-2)\left|K_{\frac{R}{2}}\right| .$$

Next, integrate over the smaller set

$$\left\{x \in K_{\frac{R}{2}} : \theta_\epsilon(x,t) < \mu^- + \frac{\omega}{2^m}\right\} , \qquad \forall t \in (0, a_0 R^p)$$

where $\xi = 1$ and $\Psi^- \ge (m-3)\ln 2$, since $H_k^- \le \frac{\omega}{4}$. Consequently, for all $t \in (0, a_0 R^p)$,

$$\left|x \in K_{\frac{R}{2}} \;:\; \theta_\epsilon(x,t) < \mu^- + \frac{\omega}{2^m}\right| \le C\,\frac{m-2}{(m-3)^2}\left|K_{\frac{R}{2}}\right| .$$

The proof is complete if we choose m so large that $C\,\frac{m-2}{(m-3)^2} < \nu$. □

The next lemma provides a uniform lower bound for θ_ϵ within a smaller cylinder, through a specific choice of the value ν that appears in Lemma 3.2.

Lemma 3.3. *There exists $\nu_0 \in (0,1)$, depending only on the data, such that if*

$$\left|Q\left(a_0 R^p, \frac{R}{2}\right) \;:\; \theta_\epsilon(x,t) \le \mu^- + \frac{\omega}{2^m}\right| \le \nu_0 \left|Q\left(a_0 R^p, \frac{R}{2}\right)\right|$$

then

$$\theta_\epsilon(x,t) \ge \mu^- + \frac{\omega}{2^{m+1}} , \qquad \text{a.e. } (x,t) \in Q\left(a_0 R^p, \frac{R}{4}\right) .$$

Proof. Consider the decreasing sequences of real numbers

$$R_n = \frac{R}{4} + \frac{R}{2^{n+2}} \;; \qquad k_n = \mu^- + \frac{\omega}{2^{m+1}} + \frac{\omega}{2^{m+1+n}} , \qquad n = 0, 1, \ldots$$

and, in the energy estimates (2.5), take $\varphi = -(\theta_\epsilon - k_n)_- \, \xi_n^p$, where $0 \le \xi_n \le 1$ are smooth cutoff functions, defined in K_{R_n}, and verifying

$$\xi \equiv 1 \quad \text{in } K_{R_{n+1}} \;; \quad \xi \equiv 0 \quad \text{on } |x| = R_n \;; \quad |\nabla \xi_n| \le \frac{2^{n+3}}{R} .$$

Introduce the level

$$\bar{k}_n = \frac{k_n + k_{n+1}}{2} .$$

Since

$$\begin{aligned}\int_{K_{R_n}\times\{t\}} (\theta_\epsilon - k_n)_-^2\ \xi_n^p &= \int_{K_{R_n}\times\{t\}} (\theta_\epsilon - k_n)_-^p\ (\theta_\epsilon - k_n)_-^{2-p}\ \xi_n^p \\ &\geq (k_n - \bar{k}_n)^{2-p} \int_{K_{R_n}\times\{t\}} (\theta_\epsilon - \bar{k}_n)_-^p\ \xi_n^p \\ &= a_0\, 2^{-(n+3)p} \int_{K_{R_n}\times\{t\}} (\theta_\epsilon - \bar{k}_n)_-^p\ \xi_n^p\end{aligned}$$

and $(\theta_\epsilon - k_n)_-^p \leq \left(\frac{\omega}{2^m}\right)^p$, the referred energy estimates take the form

$$\sup_{0<t<a_0 R^p} \int_{K_{R_n}\times\{t\}} (\theta_\epsilon - \bar{k}_n)_-^p\ \xi_n^p + \frac{1}{a_0}\, 2^{-(n+3)p} \iint_{Q(a_0 R^p, R_n)} \left|\nabla(\theta_\epsilon - \bar{k}_n)_-\right|^p\ \xi_n^p$$

$$\leq C(p)\left(\frac{\omega}{2^m}\right)^p \frac{2^{2pn}}{R^p}\,\frac{1}{a_0} \iint_{Q(a_0 R^p, R_n)} \chi_{[(\theta_\epsilon - k_n)_- > 0]}\ .$$

Introducing the change of variable $z = \frac{t}{a_0}$, defining the new functions

$$\bar{\theta}_\epsilon(x,z) = \theta_\epsilon(x, a_0 z)\ ; \quad \bar{\xi}_n(x,z) = \xi_n(x, a_0 z)\ ,$$

and denoting $V^p = L^\infty(L^p) \cap L^p(W^{1,p})$, we arrive at

$$\left\|(\bar{\theta}_\epsilon - \bar{k}_n)_-\right\|_{V^p(Q(R^p, R_{n+1}))}^p \leq C(p)\, \frac{2^{2pn}}{R^p} \left(\frac{\omega}{2^m}\right)^p A_n\ ,$$

where

$$A_n := \int_0^{R^p} |A_n(z)|\ dz\ , \qquad A_n(z) := \left\{x \in K_{R_n} : (\bar{\theta}_\epsilon - k_n)_- > 0\right\}\ .$$

Since

$$\begin{aligned}\left(\frac{\omega}{2^m}\right)^p 2^{-(n+3)p} A_{n+1} &\leq \iint_{Q(R^p, R_{n+1})} (\bar{\theta}_\epsilon - \bar{k}_n)_-^p \\ &\leq C\, A_n^{1+\frac{p}{N+p}} \left\|(\bar{\theta}_\epsilon - \bar{k}_n)_-\right\|_{V^p(Q(R^p, R_{n+1}))}^p\ ,\end{aligned}$$

using Corollary 3.1 of [3, page 9], we conclude

$$A_{n+1} \leq C\, \frac{2^{3pn}}{R^p}\, A_n^{1+\frac{p}{N+p}}$$

and, consequently,

$$Y_{n+1} \leq C\, 2^{3pn}\, A_n^{1+\frac{p}{N+p}}\ , \qquad \text{for} \quad Y_n := \frac{A_n}{|Q(R^p, R_n)|}\ .$$

If $Y_0 \le C^{-\frac{N+p}{p}} 2^{-\frac{3(N+p)^2}{p}}$ then, by Lemma 4.1 of [3, page 12], $Y_n \to 0$ when $n \to \infty$ which completes the proof. Observe that, by the hypothesis,

$$Y_0 = \frac{\left|(x,z) \in Q(R^p, R) : \bar{\theta}_\epsilon(x,z) < \mu^- + \frac{\omega}{2^m}\right|}{|Q(R^p,R)|} \le \nu_0$$

so we just have to take

$$\nu_0 \equiv C^{-\frac{N+p}{p}} 2^{-\frac{3(N+p)^2}{p}} . \qquad \square$$

Now we can finally conclude the first iteration step in the proof of Proposition 3.1. Indeed, taking $\nu = \nu_0$ from Lemma 3.3, and determining the corresponding value m with the help of Lemma 3.2, we arrive at

$$\theta_\epsilon(x,t) \ge \mu^- + \frac{\omega}{2^{m+1}} , \quad \text{a.e.}(x,t) \in Q\left(a_0 R^p, \frac{R}{4}\right) ,$$

and then we conclude that

$$\operatorname*{ess\,osc}_{Q\left(a_0R^p, \frac{R}{4}\right)} \theta_\epsilon \le \left(1 - \frac{1}{2^{m+1}}\right) \omega = \sigma\,\omega .$$

Taking $C = 4$ in Proposition 3.1, we get $Q_1 \subset Q\left(a_0 R^p, \frac{R}{4}\right)$, and then

$$\operatorname*{ess\,osc}_{Q_1} \theta_\epsilon \le \operatorname*{ess\,osc}_{Q\left(a_0R^p, \frac{R}{4}\right)} \theta_\epsilon \le \sigma\,\omega = \omega_1 .$$

We can now repeat the whole process starting from Q_1.

Remark 3.4. Observe that we do not get a reduction on the t-direction since the cutoff functions ξ are independent of t.

Remark 3.5. The regularity result can be further extended; one can obtain continuity up to the lateral boundary Σ using a reasoning similar to the one presented in [2] and [8].

References

[1] DiBenedetto, E., *Continuity of weak solutions to certain singular parabolic equations*, Ann. Mat. Pura Appl. (4) **130** (1982), 131–176.

[2] DiBenedetto, E., *A boundary modulus of continuity for a class of singular parabolic equations*, J. Differential Equations **63** (1986), 418–447.

[3] DiBenedetto, E., Degenerate Parabolic Equations, Universitex, Springer-Verlag, New York, 1993.

[4] DiBenedetto, E., Urbano, J.M. and Vespri, V., *Current issues on singular and degenerate evolution equations*, Handbook of Differential Equations, Evolutionary Equations, vol.1, pp. 169–286, Elsevier/North-Holland, Amsterdam, 2004.

[5] Henriques, E. and Urbano, J.M., *On the doubly singular equation* $\gamma(u)_t = \Delta_p u$, Comm. Partial Differential Equations **30** (2005), 919–955.

[6] Henriques, E. and Urbano, J.M., *Intrinsic scaling for PDEs with an exponential nonlinearity*, Indiana Univ. Math. J., to appear.

[7] Urbano, J.M., *Continuous solutions for a degenerate free boundary problem*, Ann. Mat. Pura Appl. (4) **178** (2000), 195–224.

[8] Urbano, J.M., *A singular-degenerate parabolic problem: regularity up to the Dirichlet boundary* in: FBPs: Theory and Applications I, pp. 399–410, GAKUTO Int. Series Math. Sci. Appl. **13**, 2000.

Eurica Henriques
Departamento de Matemática
Universidade de Trás-os-Montes e Alto Douro
Quinta dos Prados, Apartado 1013
5000-911 Vila Real, Portugal
e-mail: `eurica@utad.pt`

José Miguel Urbano
Departamento de Matemática
Universidade de Coimbra
3001-454 Coimbra, Portugal
e-mail: `jmurb@mat.uc.pt`

International Series of Numerical Mathematics, Vol. 154, 241–250

Fast Reaction Limits and Liesegang Bands

D. Hilhorst, R. van der Hout, M. Mimura and I. Ohnishi

1. Introduction

The purpose of this study is to start understanding from a mathematical viewpoint experiments in which regularized structures with spatially distinct bands and rings of precipitated material were exhibited, with clearly visible scaling properties. This phenomenon has been originally observed by Liesegang [1] in 1896, after whom the name "Liesegang bands/rings" has been coined. Since then there have been a large number of contributions to the understanding of such precipitated pattern formation from experimental as well as theoretical viewpoints. However, as far as we know, there has not been any mathematical study of this problem apart from numerical simulations. In this note we introduce a one-dimensional reaction diffusion system which is a simplified model of the supersaturation model proposed by Keller and Rubinow [2] in 1981 and study the occurrence of precipitated bands in this system, by means of singular limit analysis.

2. The model

We study a model, for which the basic ideas have been proposed by Keller and Rubinov [2]. In a spatial domain Ω an immobile reactant $\mathcal{B}$ is present, with uniform concentration b_0. Starting at an initial time $t = 0$, the boundary $\partial\Omega$ is brought in contact with a different reactant $\mathcal{A}$, that penetrates into Ω through a diffusive process. Inside Ω, $\mathcal{A}$ and $\mathcal{B}$ react, to produce a third substance $\mathcal{C}$. This substance on the one hand diffuses through Ω, while on the other hand it crystallizes ("precipitates") to form an immobile product $\mathcal{D}$. This precipitation process starts as soon as the concentration c of $\mathcal{C}$ has surpassed a critical value $C_s > 0$; the precipitation rate is thought to depend linearly on c. In places where $\mathcal{D}$ has been formed, the precipitation process continues as long as c remains positive.

The purpose of our study is to show that such a reaction-diffusion system may give rise to precipitation regions where $\mathcal{D}$ is present. We restrict ourselves to a one-dimensional situation and assume that the space domain Ω is the semi-infinite slab $\{x > 0\}$. We also require that, for $t > 0$, $a(0,t) = a_0 > 0$ is constant and that $\mathcal{C}$ cannot leave Ω, that is, $\frac{\partial c}{\partial x}(0,t) = 0$. As for the precipitation reaction $\mathcal{C} \rightleftarrows \mathcal{D}$, we assume that the rate for the backward reaction $\mathcal{D} \to \mathcal{C}$ is negligible. This yields the problem

$$(\mathrm{P}_{\mathrm{k}}^*)\begin{cases} a_t = D_1 a_{xx} - kab & 0 < x < \infty,\ t > 0, \\ b_t = -kab & 0 < x < \infty,\ t > 0, \\ c_t = Dc_{xx} + kab - F(c,d) & 0 < x < \infty,\ t > 0, \\ d_t = F(c,d) & 0 < x < \infty,\ t > 0, \\ c_x(0,t) = 0 & t > 0, \\ a(0,t) = a_0 > 0 & t > 0, \\ a(x,0) = c(x,0) = d(x,0) = 0 & x > 0, \\ b(x,0) = b_0 > 0 & x > 0, \end{cases}$$

where D, D_1 and k are positive constants. We suppose that F is the discontinuous function

$$F(c,d) = \begin{cases} c \text{ if } c > C_s \text{ or } d > 0, \\ 0 \text{ if } c \le C_s \text{ and } d = 0, \end{cases}$$

where the positive constant C_s stands for the supersaturation concentration. Alternatively we can rewrite Problem $(\mathrm{P}_{\mathrm{k}}^*)$ in the form

$$(\mathrm{P}_{\mathrm{k}}^*)\begin{cases} a_t = D_1 a_{xx} - kab & 0 < x < \infty,\ t > 0, \\ b_t = -kab & 0 < x < \infty,\ t > 0, \\ c_t = Dc_{xx} + kab - c\tilde{H}\left((c - C_s)^+ + d\right) & 0 < x < \infty,\ t > 0, \\ d_t = c\tilde{H}\left((c - C_s)^+ + d\right) & 0 < x < \infty,\ t > 0, \\ a(0,t) = a_0 > 0 & t > 0, \\ c_x(0,t) = 0 & t > 0, \\ a(x,0) = c(x,0) = d(x,0) = 0 & x > 0, \\ b(x,0) = b_0 > 0 & x > 0, \end{cases}$$

where $\tilde{H}$ is the Heaviside function: $\tilde{H}(s) = 0$ if $s \le 0$, $\tilde{H}(s) = 1$ otherwise.

The equation for d in Problem (P_k^*) has been formulated to express the chemical assumption that $d_t = c$ if $c > C_s$ or if $d > 0$ and that $d_t = 0$ otherwise. The right-hand side of the d-equation is not Lipschitz in d and we cannot be sure that its solution is unique, even if $c - C_s \le 0$ everywhere. Chemical arguments imply that $d(x,t)$ can only be positive if its growth has been initiated by a positive value of $c(x,\tau) - C_s$ for some $\tau \le t$. This implies that we have to look for a solution such that $d(x,t) = 0$ if $c(x,\tau) \le C_s$ for all $\tau < t$. In view of these comments, we are able to reformulate again Problem (P_k^*) in a slightly simpler way, which in addition is more precise from a chemical point of view. Assuming, for the time being, that c is continuous, we introduce the function

$$w(x,t) = \int_0^t \left(c(x,s) - C_s)^+\right) ds. \tag{1}$$

We claim that, for the solution we are interested in, $w(x,0) > 0$ if and only if $(c(x,t) - C_s)^+ + d(x,t) > 0$. To see this, let us first assume that $w(\bar{x},\bar{t}) > 0$. Then, clearly, there are $t_1 < t_2 \leq \bar{t}$ such that $c(x,t) > C_s$ for all $t_1 < t < t_2$. The differential equation for d implies that $d(\bar{x},\bar{t}) > 0$, which in turn implies that $(c(\bar{x},\bar{t}) - C_s)^+ + d(\bar{x},\bar{t}) > 0$. Conversely, if $w(x,t) = 0$, it is clear that $c(x,\tau) \leq C_s$ for all $\tau \leq t$. In view of the preceding argument, we must now conclude that $d(x,t) = 0$ for the solution we are interested in. Thus, we may replace the argument $(c - C_s)^+ + d$ of the Heaviside function by the new argument w.

Note that, in (P_{k}^*), the first two equations can be solved separately. The behavior of a_k and b_k has been studied in [3,4] in the asymptotic limit $k \to \infty$. In particular, it has been shown in [3,4] that in this limit a behaves as the solution of a one-phase Stefan problem with melting boundary $\zeta(t) = \alpha\sqrt{t}$ for some positive constant α. It then remains to study the problem

$$(\mathrm{P^k}) \begin{cases} c_t = Dc_{xx} + kab - c\tilde{H}\left(\int_0^t (c(x,s) - C_s)^+)\, ds\right) & 0 < x < \infty,\ t > 0, \\ c_x(0,t) = 0 & t > 0, \\ c(x,0) = 0 & x > 0. \end{cases}$$

The organization of this note is as follows. In Section 2, we explain why the term ka_kb_k converges to a measure as k tends to infinity and give an explicit form of its limit.

In Section 3 we study the singular limit as $k \to \infty$ of the linear equation that one obtains when omitting the last term in the equation for c.

In Section 4 we indicate the main steps for the proof of the existence of a solution of Problem (P^k).

Further we present in Section 5 its limiting behavior as $k \to \infty$.

Finally we state our main results about Liesegang bands in Section 6. We refer to [5] and [6] for the complete proofs.

3. The singular limit of ka_kb_k as $k \to \infty$

Our main result is that

$$kab \to \frac{b_o\alpha}{2\sqrt{t}}\delta(x - \alpha\sqrt{t}) \qquad \text{as } k \to \infty, \tag{2}$$

where δ is the Dirac δ-function. It can be shown as follows. We first recall a result of [3,4]. For all $T > 0$, there holds

$$\int_0^T \int_{\mathbb{R}^+} a^k b^k \leqq \frac{C}{k},$$

where the constant C depends on T. One immediately deduces that there exists a Radon measure μ such that as $k \to \infty$

$$k a^k b^k \rightharpoonup \mu$$

along a subsequence in the sense of weak convergence of measures.

Multiplying the equation for b^k by a continuously differentiable test function with compact support in $(0, \infty) \times (0, T)$, and integrating by parts, one can show that the measure μ is concentrated on the free boundary ζ, or more precisely that

$$\mu = -b_0 \frac{\partial}{\partial t} \tilde{H}\left(x - \zeta(t)\right) = \zeta'(t) b_0 \delta(x - \zeta(t)),$$

$x > 0, t > 0$, which in turn implies (2).

4. The limiting behavior as $k \to \infty$ of the solution of a corresponding linear problem

The next step is to consider the linear problem

$$(\mathrm{Q^k}) \begin{cases} \psi_t = D\psi_{xx} + k a_k b_k & x > 0, t > 0, \\ \psi_x(0,t) = 0 & t > 0, \\ \psi(x,0) = 0 & x > 0. \end{cases}$$

We will prove in [5] that as $k \to \infty$ its solution ψ_k converges to the unique weak solution ψ of the problem

$$(\mathrm{Q}) \begin{cases} \psi_t = D\psi_{xx} + \frac{b_0 \alpha}{2\sqrt{t}} \delta(x - \alpha\sqrt{t}) & x > 0, t > 0, \\ \psi_x(0,t) = 0 & t > 0, \\ \psi(x,0) = 0 & x > 0. \end{cases}$$

We remark that Problem (Q) admits at most one solution and we prove its existence by constructing the solution explicitly. Indeed note that Problem (Q) is invariant by the rescaling $\{t = \mu^2 \tau, \quad x = \mu \xi\}$ so that a possible solution can be written in the form

$$\psi(t,x) = \Psi(\eta) \quad \text{with} \quad \eta = \frac{x}{\sqrt{t}},$$

where Ψ satisfies

$$D\Psi'' + \frac{\eta}{2}\Psi' + \frac{b_0 \alpha}{2} \delta(\eta - \alpha) = 0 \qquad \Psi(\infty) = 0, \ \Psi'(0) = 0.$$

We deduce that

$$\Psi(\eta) = \begin{cases} A(D, b_0, \alpha) \text{ when } \eta \le \alpha, \\ b_0 \alpha \dfrac{e^{\frac{\alpha^2}{4D}}}{2D} \displaystyle\int_\eta^\infty e^{-\frac{s^2}{4D}} ds \quad \text{when } \eta > \alpha, \end{cases}$$

where

$$A(D, b_0, \alpha) := \frac{b_0 \alpha e^{\frac{\alpha^2}{4D}}}{2D} \int_\alpha^\infty e^{-\frac{s^2}{4D}} ds.$$

5. Existence of a solution of Problem (P^k)

We extend the unknown concentration c^k to the whole domain $\mathbb{R} \times \mathbb{R}^+$ according to

$$c^k(-x, t) = c^k(x, t), \qquad x > 0, \ t > 0.$$

and a^k and b^k similarly. Problem ($\mathrm{P^k}$) becomes

$$(\mathrm{P^k}) \begin{cases} c_t = Dc_{xx} + kab - c\tilde{H}(\int_0^t (c - C_s)^+), & x \in \mathbb{R}, \ t > 0, \\ c(x, 0) = 0, & x \in \mathbb{R}. \end{cases}$$

In order to be able to work with partial differential equations with a bounded right hand side, we perform a change of the unknown function. We set

$$\psi^k(-x, t) = \psi^k(x, t), \qquad x > 0, \ t > 0,$$

and

$$\tilde{c}^k = c^k - \psi^k.$$

Problem ($\mathrm{P^k}$) can be rewritten as

$$(\mathrm{P^k}) \begin{cases} \tilde{c}_t = D\tilde{c}_{xx} - (\tilde{c} + \psi^k)\tilde{H}\left(\int_0^t (\tilde{c} + \psi^k - C_s)^+\right), & x \in \mathbb{R}, \ t > 0, \\ \tilde{c}(x, 0) = 0, & x \in \mathbb{R}. \end{cases}$$

To begin with we consider the boundary value problem

$$\begin{cases} \psi_t = D\psi_{xx} + kab, & x \in (-R, R), \ t > 0, \\ \psi(\pm R, t) = 0, & t > 0 \\ \psi(x, 0) = 0, & x \in (-R, R), \end{cases}$$

and denote its unique solution by ψ_R^k. We first prove the existence of a solution of the problem

$$(\mathrm{P_R^k}) \begin{cases} \tilde{c}_t = D\tilde{c}_{xx} - (\tilde{c} + \psi_R^k)\mathcal{H}_{\frac{1}{R}}\left(\int_0^t (\tilde{c} + \psi_R^k - C_s)^+\right), & x \in (-R, R), t > 0, \\ \tilde{c}(\pm R, t) = 0, & t > 0, \\ \tilde{c}(x, 0) = 0, & x \in (-R, R). \end{cases}$$

where $\mathcal{H}_\varepsilon$ is a smooth nondecreasing approximation of the Heaviside graph H such that $\mathcal{H}_\varepsilon(s) = H(s)$ for all $s > 0$ and $s < -\varepsilon$.

In order to prove the existence of a solution of Problem (P_R^k), we consider the map $T : \phi \to \gamma$, where γ is the unique solution of the auxiliary problem

$$(\mathrm{P_A})\begin{cases} \gamma_t = D\gamma_{xx} - (\gamma + \psi_R^k)\mathcal{H}_{\frac{1}{R}}\left(\int_0^t (\phi + \psi_R^k - C_s)^+\right), & x \in (-R,R), t > 0, \\ \gamma(\pm R, t) = 0, & t > 0, \\ \gamma(x,0) = 0, & x \in (-R,R). \end{cases}$$

We set

$$\mathcal{L}u = u_t - Du_{xx} + (u + \psi_R^k)\mathcal{H}_{\frac{1}{R}}\left(\int_0^t (\phi + \psi_R^k - C_s)^+\right),$$

and observe that

$$\mathcal{L}0 = \psi_R^k \mathcal{H}_{\frac{1}{R}}\left(\int_0^t (\phi + \psi_R^k - C_s)^+\right) \geqq 0,$$

and that

$$\mathcal{L}(-\psi_R^k) = (-\psi_R^k)_t - D(-\psi_R^k)_{xx} = -ka^k b^k \leqq 0.$$

We define

$$\mathcal{C} = \left\{u \in C([-R,R] \times [0,T]), -\psi_R^k \leqq u \leqq 0\right\},$$

and remark that T maps $\mathcal{C}$ into itself. Furthermore one can show that T is compact and continuous on $\mathcal{C}$ so that it follows from the Schauder fixed point theorem that the map T has a fixed point $\tilde{c}_R^k$ which is a classical solution of Problem $(\mathrm{P_A})$ with $\gamma = \phi = \tilde{c}_R^k$.

Finally letting R tend to ∞ we deduce the existence of a weak solution of Problem (P^k), where the Heaviside Function $\tilde{H}$ is replaced by the Heaviside graph H.

6. Singular limit as $k \to \infty$ of solutions of Problem (P^k)

In what follows we use the notation $Q_T = [0,\infty) \times [0,T)$. Next we consider the limit Problem (P) which we define by

$$(\mathrm{P})\begin{cases} c_t = Dc_{xx} + \frac{b_0\alpha}{2\sqrt{t}}\delta(x - \alpha\sqrt{t}) - c\,\mathcal{H}(x,t) & 0 < x < \infty,\ t > 0, \\ c_x(0,t) = 0 & t > 0, \\ c(x,0) = 0 & x > 0, \end{cases}$$

where

$$\mathcal{H}(x,t) \in H(\int_0^t (c - C_s)^+(x,\tau)d\tau), \tag{3}$$

and where H is the Heaviside graph

$$H(y) = \begin{cases} 0 & \text{when } y < 0, \\ [0,1] & \text{when } y = 0, \\ 1 & \text{when } y > 0. \end{cases}$$

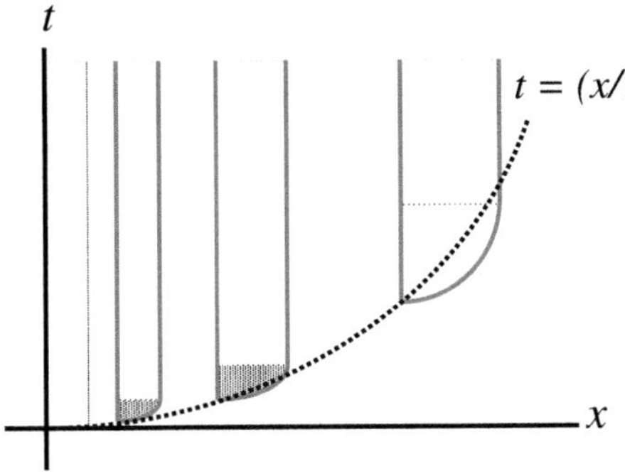

FIGURE 1. Time law for the Liesegang bands

Definition *A weak solution of Problem* (P) *is a function pair* $\{c, \mathcal{H}\}$ *with the properties*

(1) *for each* $T > 0$, $c - \psi \in C^{1+\gamma,\frac{1+\gamma}{2}}(\overline{Q_T}) \cap H^1(Q_T)$ *for all* $0 < \gamma < 1$*;*
(2) *for all* $\varphi \in C^1(\overline{Q_T})$ *such that* φ *vanishes for* $t = T$,

$$\int_{Q_T} (\psi - c)\varphi_t = \int_{Q_T} [D(\psi - c)_x \varphi_x - c\varphi\mathcal{H}]. \tag{4}$$

One can prove the following result.

Theorem 1 *There exists a subsequence* $\{c^{kn}\}$ *and a function* $c \in C^{1+\gamma,\frac{1+\gamma}{2}}([0,R] \times [0,T])$ *for all* $\gamma \in (0,1)$ *such that as* $k_n \to \infty$

$$c^{kn} - \psi^{kn} \to c - \psi \tag{5}$$

in $C^{1+\gamma,\frac{1+\gamma}{2}}([0,R] \times [0,T])$ *for all* $\gamma \in (0,1)$, R *and* T *positive. The pair* $(c, \mathcal{H})$*, where* $\mathcal{H}$ *is defined by* (3)*, is a weak solution of Problem (P) and the function* c *is such that* $0 \le c \le \psi$ *and* $c - \psi$ *is nonincreasing in time.*

In the next section, we discuss the existence and some properties of *precipitation* or *Liesegang* bands:

Definition *A Liesegang band is connected component in* (x,t)*-plane of the set* $\{w > 0\}$.

Note that a Liesegang band can exist only if

$$A(D, b_0, \alpha) > C_s, \tag{6}$$

which we assume from now on.

7. Properties of Liesegang bands

To begin with we consider the first Liesegang band. To do that, we define, for fixed $t > 0$,

$$\mathcal{P}(t) = \{x > 0 : \ w(y,t) > 0 \text{ for all } 0 < y < x\},$$

where w has been defined in (1).

$$S(t) = \sup\{x : x \in \mathcal{P}(t)\} \ \text{ if } \ \mathcal{P}(t) \neq \emptyset \ \text{ and } \ S(t) = 0 \ \text{ otherwise.}$$

We have that

Lemma 6 *$S(t) > 0$ for all $t > 0$. Moreover, $S(\cdot)$ is nondecreasing.*

Therefore there exists indeed a precipitation region. The continuity of c implies that

Lemma 7 *Suppose that t is such that $S(t+\epsilon) > S(t)$ for all $\epsilon > 0$. Then*

$$c(S(t),t) = C_s.$$

Next, we have

Lemma 8 *$S(\cdot)$ is continuous.*

Idea of the proof. Suppose, for contradiction, that $x_1 := S(t^{*-}) < S(t^{*+}) := x_2$. Then for every $x \in (x_1, x_2)$:

(i) $c(x,t) \leq C_s$ for $t \leq t^*$, and
(ii) There is a sequence $t_i \to t^{*+}$ such that $c(x,t_i) > C_s$.

Since the function $c - \psi$ is nonincreasing in time, this implies that $x > \alpha\sqrt{t^*}$; the strong maximum principle then yields a contradiction.

Our next step is to prove the following result. From now on we assume that the function $\mathcal{H}$ in the definition of a weak solution of Problem (P) coincides with the Heaviside function $\tilde{H}\left(\int_0^t (c - C_s)^+(x,\tau)d\tau\right)$.

Theorem 2 *Under the additional technical condition that $2C_s > A(D, b_0, \alpha)$, there is a time $T^* > 0$ such that $S(t) = S(T^*)$ for all $t \geq T^*$.*

Idea of the proof. Let $\beta > 0$ be defined by the relation $\psi(\beta) = C_s$. We have

$$\psi(x,t) \geq C_s \ \text{ whenever } \ x \leq \beta\sqrt{t}. \tag{7}$$

$$S(t) \leq \beta\sqrt{t}, \ \text{ and } \ \psi(x,t) \geq C_s \ \text{ whenever } \ x \leq S(t). \tag{8}$$

Let $0 < \gamma < 2C_s - A(D, b_0, \alpha)$ and suppose that $c(S(T^*), T^*) < \gamma$. Then we have:

$$C_s - c(S(T^*), T^*) \geq C_s - \gamma \geq A(D, b_0, \alpha) - C_s.$$

We claim that $S(t)$ cannot grow any more. Indeed suppose, for the purpose of contradiction, that $t^* > T^*$ and that $S(t^*) > S(T^*)$. Then there exists a $\bar{t} \in (T^*, t^*)$ such that

(i) $S(\bar{t}) = S(T^*)$ for every $T^* \leq t \leq \bar{t}$,
(ii) $c(S(T^*), \bar{t}) = C_s$.

This implies that

$$\begin{aligned} c(S(T^*),\bar{t}) - c(S(T^*),T^*) &> C_s - \gamma \\ &\geq A(D,b_0,\alpha) - C_s \\ &\geq \psi(S(T^*),\bar{t}) - \psi(S(T^*),T^*) \end{aligned}$$

(see (8)) and therefore that

$$c(S(T^*),\bar{t}) - \psi(S(T^*),\bar{t}) > c(S(T^*),T^*) - \psi(S(T^*),T^*).$$

This contradicts the fact that $c - \psi$ is nonincreasing in time.

Furthermore one can show that it indeed occurs that $c(S(T^*),T^*) < \gamma$.

Lemma 9 *For every* $0 < \gamma < C_s$, *there is a time* $t > 0$ *such that* $c(S(t),t) < \gamma$.

To conclude this note, we mention three additional results:

Theorem 3 *If* $2C_s > A(D,b_0,\alpha) > C_s$ *and if* $\frac{\alpha}{\sqrt{D}}$ *is sufficiently large, there are infinitely many distinct precipitation regions.*

Theorem 4 *If a new precipitation region germinates at* (x,t), *then* $x = \alpha\sqrt{t}$. This property is the so-called **time-law**.

Theorem 5 *In the* (x,t)*-plane, a Liesegang band can only extend to the right of the point of initiation.*

References

[1] R.E. Liesegang, *Chemische Fernwirkung,* Photo. Archiv. **800** (1896), 305–309.

[2] J.B. Keller, and S.I. Rubinow, *Recurrent precipitation and Liesegang rings,* J. Chem. Phys. **74** (1981), 5000–5007.

[3] D. Hilhorst, R. van der Hout, and L.A. Peletier, *The fast reaction limit for a reaction-diffusion system,* J. Math. Anal. Appl. **199** (1996), 349–373.

[4] D. Hilhorst, R. van der Hout, and L.A. Peletier, *Diffusion in the presence of fast reaction : the case of a general monotone reaction term,* J. Math. Sci. Univ. Tokyo **4**, (1997), 469–517.

[5] D. Hilhorst, R. van der Hout, M. Mimura and I. Ohnishi, *The singular limit of a problem for Liesegang bands,* in preparation.

[6] D. Hilhorst, R. van der Hout, M. Mimura and I. Ohnishi, *A mathematical study of Liesegang bands,* in preparation.

D. Hilhorst
Laboratoire de Mathématique
Université de Paris-Sud (Bât. 425)
F-91405 Orsay Cedex, France

R. van der Hout
Mathematical Institute
University of Leiden
POB 9512
NL-2300 RA Leiden, The Netherlands

M. Mimura
Department of Mathematics
School of Science and Technology
Meiji Institute for Mathematical Science and Technology
Meiji University
Higashimita, Tama-ku, Kawasaki, 214-8571 Japan

I. Ohnishi
Institute for Nonlinear Sciences and Applied Mathematics
Graduate School of Science
Hiroshima University
Higashi-Hiroshima, 739-8526 Japan

International Series of Numerical Mathematics, Vol. 154, 251–261

Numerical Modeling of Surfactant Effects in Interfacial Fluid Dynamics

Ashley J. James and John Lowengrub

Abstract. Surfactants are substances that preferentially accumulate at interfaces between two fluids, altering the local surface tension. An imposed flow can produce a non-uniform distribution of surfactant. In regions of high surfactant concentration the surface tension is low, so the interface offers less resistance to deformation and can become highly curved, allowing very small droplets or bubbles to pinch off. A numerical method to simulate interfacial surfactant mechanics within a volume of fluid method has been developed. To conserve surfactant, the surfactant mass and the interfacial surface area are tracked as the interface evolves, and then the surfactant concentration is reconstructed. The algorithm is coupled to an incompressible flow solver that uses a continuum method to incorporate both the normal and tangential components of the surface tension force into the momentum equation.

Keywords. surfactant, VOF, interfacial flow, surface tension.

1. Introduction

Surfactant plays a critical role in numerous important industrial and biomedical applications. For example, the formation of very small drops or bubbles by tip streaming relies on the presence of surfactant. The production of such tiny droplets is useful in drug delivery, industrial emulsification, liquid/liquid extraction, polymer blending and plastic production, and other applications.

Surfactants adhere to interfaces resulting in a lowered, non-uniform surface tension along the interface. This makes the capillary force non-uniform and introduces the Marangoni force. Interfacial surfactant is transported with the interface by convection, and may diffuse along the interface in the presence of a surfactant concentration gradient. Additionally, compression or stretching of the interface causes a corresponding increase or decrease in the concentration. The equation that governs these dynamics has been derived in various forms in [1, 2]. The motion of the surfactant and of the surrounding bulk fluids are coupled through the Marangoni force.

We simulate interface dynamics using the volume-of-fluid (VOF) method [3]. The main advantages of the method are that the interface shape is not constrained, changes in topology are handled automatically, and mass of each flow component is conserved exactly. The interface location is captured as it moves through the grid by tracking the local volume fraction. Flow discontinuities are smoothed and the surface tension force is distributed over a thin layer near the interface.

Continuum formulations of the interface governing equations have been implemented numerically primarily for clean drops, although Jan and Tryggvason [4] studied the effect of surfactants on rising bubbles using an immersed boundary/front tracking method and Ceniceros [5] used a hybrid level-set/front tracking method to study the effect of surfactants on capillary waves. Xu and Zhao [6] simulated surfactant transport on a deformable interface in conjunction with a level set method. They did not couple their method to a flow solver, but presented test cases in which a velocity field is prescribed. Renardy *et al.* [7] presented simulations of drops with surfactant using the VOF method. This work thus far has been limited to assuming a linear relation between the surfactant concentration and surface tension.

In the current paper we present a numerical method that incorporates surfactant dynamics in an axisymmetric, incompressible Navier-Stokes solver based on the VOF method for interface capturing. We focus on the case of insoluble surfactant. The surfactant mass is exactly conserved along the interface by our algorithm. An arbitrary equation of state relating the surfactant concentration to the surface tension may be used. Further details and verification are given in [8].

2. Governing equations

We assume that the flow is incompressible in both fluids, so the velocity, $\underline{u}$, is divergence free, $\nabla \cdot \underline{u} = 0$. The VOF method is used to track the interface between the two fluids, called fluid 1 and fluid 2. In this method a volume fraction, F, is defined in each grid cell as the fraction of the cell that contains fluid 1. The volume fraction evolution is governed by a convection equation that ensures the interface moves with the velocity of the fluid,

$$\frac{\partial F}{\partial t} + \underline{u} \cdot \nabla F = 0. \tag{2.1}$$

Surface tension is included in the momentum equation via the continuum surface force (CSF) method [9]. The momentum equation satisfies the stress balance boundary condition on the interface. The surface tension force is nonzero only near the interface. The VOF and CSF methods make it unnecessary to apply boundary conditions at the interface and one set of governing equations applies to the entire domain. Thus, the density and viscosity must be retained as variables in the momentum equation even though they are both constant in each fluid. Using inertial time and pressure scales, the momentum equation is

$$\rho\left(\frac{\partial \underline{u}}{\partial t}+\underline{u}\cdot\nabla\underline{u}\right)=-\nabla p+\frac{1}{Re}\nabla\cdot\left[\mu\left(\nabla\underline{u}+\nabla\underline{u}^{T}\right)\right]+\frac{1}{ReCa}\underline{F_S}, \qquad (2.2)$$

where ρ is the density, p is the pressure, Re is the Reynolds number, μ is the viscosity and Ca is the capillary number. The surface force $\underline{F_S}$ is

$$\underline{F_S}=\nabla\cdot\left[\sigma\left(\underline{\underline{I}}-\underline{nn}\right)\delta_\Sigma\right]=-\sigma\kappa\delta_\Sigma\underline{n}+\delta_\Sigma\nabla_S\sigma \qquad (2.3)$$

where σ is the surface tension, $\underline{n}$ is the unit vector normal to the interface, δ_Σ is the surface delta function, κ is the interface curvature and ∇_S is the surface gradient. The first term on the right-hand side of Eq. (2.3) is the capillary force and the second term is the Marangoni force. The normal vector and the surface delta function are determined from the gradient of the volume fraction,

$$\underline{n}=-\frac{\nabla F}{|\nabla F|}, \qquad \delta_\Sigma=|\nabla F|. \qquad (2.4)$$

The surfactant concentration evolution is governed by a convection-diffusion equation with a source term to account for interfacial stretching,

$$\frac{\partial \Gamma}{\partial t}+\underline{u}\cdot\nabla\Gamma=\frac{1}{Pe_S}\nabla_S^2\Gamma+\Gamma\underline{n}\cdot\nabla\underline{u}\cdot\underline{n}, \qquad (2.5)$$

where Γ is the interfacial surfactant concentration, Pe_S is the surface Peclet number, and ∇_S^2 is the surface Laplacian operator.

In our finite volume method, we do not solve Eq. (2.5) directly and instead relate the surfactant concentration in a finite volume to the ratio of the surfactant mass M and surface area A in that volume, i.e., $\Gamma = M/A$. The surfactant mass and surface area are tracked independently as described below. Siegel [10] has also proposed decomposing concentration into mass and area.

The equation governing A can be written in differential form as

$$\frac{DA}{Dt}=\frac{\partial A}{\partial t}+\underline{u}\cdot\nabla A=-A\left(\underline{n}\cdot\nabla\underline{u}\cdot\underline{n}\right). \qquad (2.6)$$

The left-hand side of equation (2.6) is the time rate of change of the area of a material element of the interface. The right-hand side represents changes in interfacial area due to stretching. The mass of surfactant on a material element of the interface can change if there is diffusion along the interface. The governing equation is

$$\frac{DM}{Dt}=\frac{\partial M}{\partial t}+\underline{u}\cdot\nabla M=\frac{A}{Pe_S}\nabla_S^2\Gamma. \qquad (2.7)$$

Equation (2.5) is regained by combining equations (2.6) and (2.7) with $\Gamma = M/A$.

Finally, an equation of state is given for the surface tension as a function of surfactant concentration. For example, the Langmuir equation of state is

$$\sigma=\frac{1+E\ln\left(1-x\Gamma\right)}{1+E\ln\left(1-x\right)}, \qquad (2.8)$$

where E is the surfactant elasticity and x is a measure of surfactant coverage. The scaling is constructed so that the equilibrium dimensionless concentration, $\Gamma = 1$, corresponds to the equilibrium dimensionless surface tension, $\sigma = 1$.

3. Computational method

3.1. Introduction

The axisymmetric governing equations are discretized using a finite-volume method, on a fixed, structured, uniform, staggered grid, in the $r - z$ plane. In the staggered grid arrangement the velocity components are defined on cell faces and all other variables are defined at cell centers. The continuity and momentum equations are discretized using second-order central differences, except for the surface stress, which is described in Section 3.6.

An explicit Euler time integration method is used, except that surfactant diffusion is discretized implicitly as described in Section 3.5. At each time step, first the velocity and pressure are updated, using a projection method, and then the volume fraction and the surfactant distribution are updated as described below.

The volume of fluid 1 in a grid cell at the beginning of a time step is the cell volume times the volume fraction. During a time step the volume flux of fluid 1 that moves between adjacent grid cells is computed. The volume fraction at the end of a time step is then the initial volume of fluid 1 minus the net volume flux out of the cell, divided by the cell volume. Thus, although equation (2.1) governs the evolution of the volume fraction, the method actually tracks fluid volumes.

Our approach to surfactant evolution is analogous to volume fraction evolution. Surfactant mass fluxes due to convection and diffusion are computed. The concentration in a grid cell is then the surfactant mass in the cell divided by the surface area of the interface in the cell. The surface area in a cell may vary, unlike the cell volume, so it is critical to accurately track the surface area.

Surface area evolution is described below, as well as volume fraction and surfactant mass evolution. However, the "reconstruction" of the interface and the surfactant concentration are described first.

3.2. Volume fraction and surfactant concentration reconstruction

To convect volumes of fluid while preventing smearing of the interface normal to itself it is first necessary to reconstruct the interface from the volume fraction field. This interface reconstruction locates where the volume of fluid 1 resides in the cell, rather than assuming both fluids are distributed uniformly. Convection of surfactant similarly suffers from excessive numerical diffusion if the surfactant is assumed to uniformly distributed along the interface, so its distribution is also reconstructed.

The volume fraction distribution in a cell is determined by approximating the interface in a cell as a straight line. The line segment approximation of the interface is defined independently in each cell, so the approximate interface need

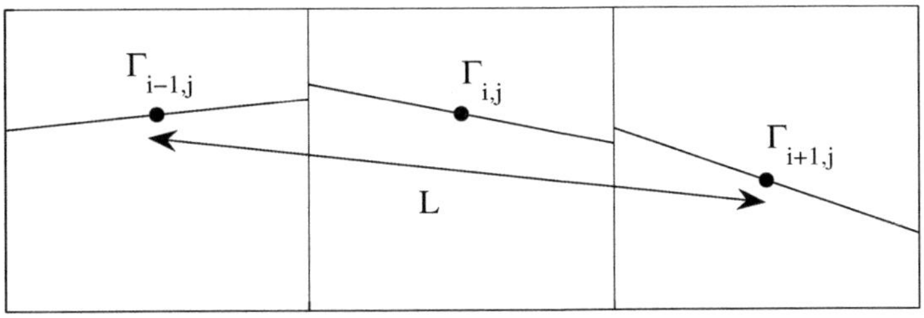

FIGURE 1. Computation of surface gradient of surfactant.

not be continuous from one cell to the next. First, the normal vector is computed as the volume fraction gradient using a finite difference method with a nine-point stencil. The normal vector defines the slope of the line. The intercept is calculated iteratively so that the volume of fluid 1 defined by the line equals that defined by the known cell volume fraction, $F_{i,j}$.

The surfactant concentration is reconstructed as a linear function of position, s, along the straight-line interface reconstruction, $\Gamma = (\nabla_S \Gamma)\, s + c$, where the surface gradient, $\nabla_S \Gamma$, is taken to be constant in each cell. As for the interface reconstruction, the function need not be continuous from one cell to the next. Since the concentration is only defined on the interface the surface gradient cannot be computed using a simple finite difference formula, as the normal vector is for the interface reconstruction. Instead the gradient is computed using only the two adjacent cells that contain an interface segment. This is illustrated in Figure 1 for the case in which the adjacent cells that contain an interface segment are cells $(i+1, j)$ and $(i-1, j)$. The procedure is analogous if other adjacent cells are used. The concentration gradient is the difference in concentration between the two cells divided by the distance, L, between their interface midpoints. For the case illustrated in Figure 1, for example

$$(\nabla_S \Gamma)_{i,j} = \frac{\Gamma_{i+1,j} - \Gamma_{i-1,j}}{L}. \tag{3.1}$$

The intercept, c, in the concentration reconstruction is then computed to ensure that the average concentration defined by the reconstruction equals the known average concentration in the cell, $\Gamma_{i,j}$.

3.3. Volume fraction evolution

The axisymmetric, conservative governing equation for the volume fraction is

$$\frac{\partial F}{\partial t} + \frac{1}{r}\frac{\partial}{\partial r}(ruF) + \frac{\partial}{\partial z}(vF) = F\left[\frac{1}{r}\frac{\partial}{\partial r}(ru) + \frac{\partial v}{\partial z}\right]. \tag{3.2}$$

where u is the radial velocity and v is the vertical velocity. The equation is split into radial and vertical directions using an intermediate volume fraction, $\widetilde{F}$,

$$\frac{\widetilde{F}_{i,j} - F^n_{i,j}}{\Delta t} + \frac{RF^n_{i+1/2,j} - RF^n_{i-1/2,j}}{2\pi r_i \Delta r \Delta z \Delta t} = \widetilde{F}_{i,j}\left(\frac{r_{i+1/2}u^{n+1}_{i+1/2,j} - r_{i-1/2}u^{n+1}_{i-1/2,j}}{r_i \Delta r}\right) \tag{3.3}$$

$$\frac{F^{n+1}_{i,j} - \widetilde{F}_{i,j}}{\Delta t} + \frac{\widetilde{ZF}_{i,j+1/2} - \widetilde{ZF}_{i,j-1/2}}{2\pi r_i \Delta r \Delta z \Delta t} = \widetilde{F}_{i,j}\left(\frac{v^{n+1}_{i,j+1/2} - v^{n+1}_{i,j-1/2}}{\Delta z}\right). \tag{3.4}$$

$RF_{i+1/2,j}$ is the volume flux of fluid 1 in the radial direction across the $(i+\frac{1}{2},j)$ face, and $ZF_{i,j+1/2}$ is the volume flux of fluid 1 in the vertical direction across the $(i,j+\frac{1}{2})$ face. The fluxes are calculated in one direction and used to update the volume fraction to the intermediate level. Then, using the intermediate volume fraction, the fluxes are calculated in the other direction and used to update the intermediate volume fraction to the next time level. The direction computed first is switched at each time step.

The volume flux is the amount of fluid 1 that passes through the face during the time step. This flux equals the amount of fluid 1 in the domain of dependence of the face, at the beginning of the time step. This is illustrated in Figure 2 for the case of flux across the $(i+\frac{1}{2},j)$ face with positive radial velocity, $u_{i+1/2,j}$. The domain of dependence is approximated by the region bounded by the face of interest, the two adjacent perpendicular grid lines, and a line parallel to the face of interest that is a distance of $U\Delta t$ away from the face, where U is the velocity normal to the face. The flux is the intersection of the domain of dependence and the portion of the cell volume that contains fluid 1, as defined by a straight-line reconstruction of the interface.

3.4. Surface area evolution

The interfacial area in a grid cell is governed by equation (2.6). The area is updated in three steps. Two account for convection, one step in each direction, and are taken in tandem with the volume fraction. An additional step to account for stretching is taken in between the two convective steps.

$$\widetilde{A}_{i,j} - A^n_{i,J} + RA^n_{i+1/2,j} - RA^n_{i-1/2,j} = 0 \tag{3.5}$$

$$\hat{A}_{i,j} = \widetilde{A}_{i,j} - \Delta t \widetilde{A}_{i,j}\left(\underline{n}\cdot\nabla\underline{u}\cdot\underline{n}\right)^{n+1}_{i,j} \tag{3.6}$$

$$A^{n+1}_{i,j} - \hat{A}_{i,J} + \hat{ZA}_{i,j+1/2} - \hat{ZA}_{i,j-1/2} = 0, \tag{3.7}$$

where $RA^n_{i+1/2,j}$ is the interfacial area flux in the radial direction across the $(i+\frac{1}{2},j)$ face, and $ZA^n_{i,j+1/2}$ is the interfacial area flux in the vertical direction across the $(i,j+\frac{1}{2})$ face. The stretching term in equation (3.6) is evaluated with 2nd order central differences. Convective fluxes of area are computed analogously to the volume fraction fluxes, and, as for the volume fraction, the direction computed

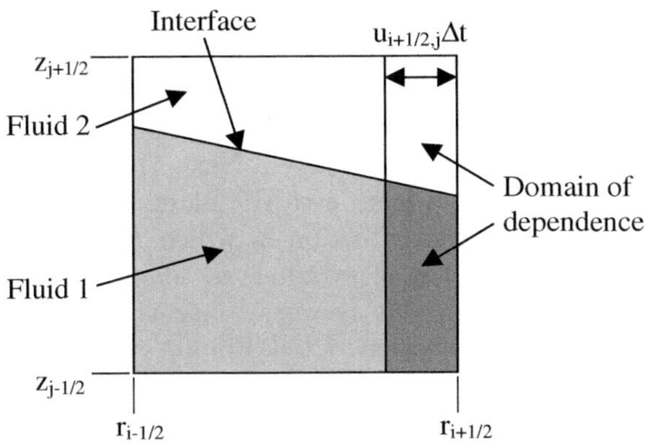

FIGURE 2. Calculation of volume flux. The volume flux, $RF_{i+1/2,j}$, is denoted by the darker gray area.

first is switched at each time step. The fluxes in one direction are used to update the area to an intermediate value in all cells, $\widetilde{A}$, at the same time the volume fraction is updated by convection in the same direction. The straight line interface reconstruction is then updated. Next, stretching is applied to update the area in all cells to $\hat{A}$. Finally, convective fluxes in the other direction complete the update of the area in all cells to the new time step, A^{n+1}. This is done in conjunction with the final update of the volume fraction by convection in the same direction.

The area flux is the area in the domain of dependence at the beginning of the time step. In Figure 2 this domain is the region $u_{i+1/2,j}\Delta t$ wide. The area of the straight line in the domain of dependence, A_{dod}, and the area of the straight line in the whole cell, A_{sl}, are computed from the interface geometry. Unlike in the volume fraction computation, A_{dod} is not an accurate representation of the area flux. However, the straight line does provide a good representation of which part of the cell the interface is in. Thus, we assume that the fraction of the actual area in the domain of dependence equals the fraction of the area of the straight line in the domain of dependence. Thus, the area flux is computed as the fraction of the area of the straight line in the domain of dependence, A_{dod}/A_{sl}, times the actual cell interfacial area, $A_{i,j}$.

3.5. Surfactant evolution

The evolution of the surfactant mass in a cell is governed by equation (2.7). At each time step, the mass equations are updated in three steps that correspond to convection and diffusion.

$$\widetilde{M}_{i,j} - M^n_{i,J} + RM^n_{i+1/2,j} - RM^n_{i-1/2,j} = 0 \tag{3.8}$$

$$\hat{M}_{i,j} - \widetilde{M}_{i,J} + \widetilde{ZM}_{i,j+1/2} - \widetilde{ZM}_{i,j-1/2} = 0 \tag{3.9}$$

$$M^{n+1}_{i,j} - \hat{M}_{i,j} = DR^{n+1}_{i+1/2,j} - DR^{n+1}_{i-1/2,j} + DZ^{n+1}_{i,j+1/2} - DZ^{n+1}_{i,j-1/2}. \tag{3.10}$$

First, the mass is updated in every cell to an intermediate level, $\widetilde{M}$, by convection in one direction, along the convection of volume fraction and interfacial area in the same direction. After this the interface approximation is reconstructed, the area is stretched, the average concentration is updated as $\Gamma = M/A$, and the concentration approximation is reconstructed. Next, the mass is updated in every cell by convection in the other direction to $\hat{M}$, along with convection of volume fraction and interfacial area. The direction in which F, A and M are convected first is switched at every time step to avoid skew. Then, once again, the interface approximation is reconstructed, the average concentration is updated, and the concentration approximation is reconstructed. Finally, the mass is updated in every cell to the next time level, M^{n+1}, by diffusion in both directions simultaneously.

The mass fluxed by convection through a cell face during a time step equals the mass in the domain of dependence at the beginning of the step, as for volume of fluid and interfacial area. Its computation is analogous to the area flux computation. A first approximation to the flux is the integral of the concentration over the straight line in the domain of dependence, M_{dod}. In computing M_{dod} it is crucial to use the linear reconstruction of the concentration, instead of simply its average value, to avoid excessive numerical diffusion. As for the area, this does not accurately represent the flux since M_{dod} is obtained using the straight line. However, M_{dod}/A_{dod} gives a consistent value for the average concentration on the portion of the interface that is convected. This is multiplied by the area flux to obtain a mass flux that is consistent with the area flux

$$\text{Mass Flux} = \left(\frac{M_{dod}}{A_{dod}}\right)\left(A_{i,j}\frac{A_{dod}}{A_{sl}}\right). \tag{3.11}$$

Next, the mass is updated to the new time step by diffusion in a single implicit step. Diffusion of surfactant across a cell face occurs only when there is an interface in both cells adjacent to the face. From Fick's Law, the radial flux across the face $(i+\frac{1}{2}, j)$, for example, is

$$DR^{n+1}_{i+1/2,j} = \frac{\Delta t}{Pe_S} 2\pi r_{i+1/2}\left(\frac{\Gamma^{n+1}_{i+1,j} - \Gamma^{n+1}_{i,j}}{L_{i+1/2,j}}\right), \tag{3.12}$$

where the surface gradient is approximated as the difference in the average concentration between the two adjacent cells, divided by L, the distance between the midpoints of the straight-line interface reconstructions in the cells. Note that equation (3.10) is implicit, since the concentration in the flux is evaluated at the new time step. In practice this is written as an equation for concentration by dividing by the area. Since the fluxes depend on $\Gamma^{n+1}_{i,j}$ this coupled system is solved iteratively for $\Gamma^{n+1}_{i,j}$. The surfactant mass is then updated as $M = \Gamma A$.

3.6. Surface tension force

Once the surfactant concentration distribution is known the average surface tension in each grid cell can be computed from the equation of state, (2.8). The surface stress, equation (2.3), can be written as

$$\underline{F_S} = \sigma\kappa\nabla F + \frac{\partial\sigma}{\partial s}|\nabla F|\underline{s}. \quad (3.13)$$

In the staggered grid arrangement the stress components are evaluated at cell faces. First, the curvature is computed in each grid cell center using standard methods. Next, the curvature is evaluated at each face as the average of the curvature in the two adjacent cells. The surface tension at cells faces is also computed as the average of the surface tension in the two adjacent cells, if both cells contain an interface segment. If only one of the adjacent cells contains an interface segment the surface tension in that cell is used as the surface tension at the face. If there is not an interface segment in either adjacent cell the surface tension at the face is set to zero and there is no surface force.

The surface gradient of the surface tension is non-zero only at faces for which both adjacent grid cells contain an interface segment. For such faces the gradient is computed exactly as the surface gradient of concentration is computed in evaluating surface diffusion. The magnitude of the volume fraction gradient at each face is computed using straightforward 2nd order finite difference approximations. Finally, the radial component of the surface stress, FR, becomes

$$FR_{i+1/2,j} = (\sigma\kappa)_{i+1/2,j}\left(\frac{F_{i+1,j} - F_{i,j}}{\Delta r}\right) + \left(\frac{\sigma_{i+1,j} - \sigma_{i,j}}{L_{i+1/2,j}}\right)|\nabla F|_{i+1/2,j}. \quad (3.14)$$

The vertical component is analogous.

4. Conclusions and future work

In this paper, we presented a volume of fluid method that accounts for an evolving surface distribution of insoluble surfactant and the associated Marangoni force in an axisymmetric geometry. The masses of the fluid components and of the surfactant are exactly conserved. An arbitrary equation of state relating the surfactant concentration to the surface tension may be used.

There are several directions we will pursue in the future. To resolve the wide-ranging length and time scales inherent in interfacial flows with surfactants, we will implement adaptive mesh refinement. This is necessary to resolve, for example, very small secondary drops that may pinchoff from the ends of a primary drop in the presence of surfactant (tip streaming). Additionally, coupling the VOF method to a level set method can improve the accuracy of the surface tension computation [11]. We have implemented a 2D coupled level-set/volume-of-fluid algorithm on an adaptive mesh for clean drops [12]. The method uses an arbitrary Lagrangian-Eulerian method to capture the interface evolution. An analogous method to track the surfactant is currently undergoing testing.

In additional future work, we will link the volume fraction and interface area advection routines to create a higher-order, self-consistent interface reconstruction. We will also simulate the transport of soluble surfactant in the fluid bulk and transfer of surfactant between the bulk and the interface. Finally, the simulations will be generalized to three dimensions.

Acknowledgment
This work utilized the resources of the Minnesota Supercomputing Institute.

References

[1] H.A. Stone, A simple derivation of the time-dependent convective-diffusion equation for surfactant transport along a deforming interface. *Phys. Fluids A.* **2** (1990) 111.

[2] H. Wong, D. Rumschitzki and C. Maldarelli, On the surfactant mass balance at a deforming fluid interface. *Phys. Fluids.* **8** (1996) 3203–3204.

[3] C.W. Hirt and B.D. Nichols, Volume of fluid (VOF) method for the dynamics of free boundaries. *J. Comput. Phys.* **39** (1981) 201–225.

[4] Y-J. Jan and G. Tryggvason, Computational studies of contaminated bubbles, Proceedings of a symposium on the dynamics of bubbles and vorticity near free surfaces. Ed. I. Sahin and G. Tryggvason, ASME, (1991) 46.

[5] H.D. Ceniceros, The effects of surfactants on the formation and evolution of capillary waves. *Phys. Fluids* **15** (2003) 245.

[6] J.-J. Xu and H.-K. Zhao, An Eulerian formulation for solving partial differential equations along a moving interface. *J. Sci. Comp.* **19** (2003) 573–594.

[7] Y.Y. Renardy, M. Renardy and V. Cristini, A new volume-of-fluid formulation for surfactants and simulations of drop deformation under shear at a low viscosity ratio. *Eur. J. Mech. B.* **21** (2002) 49–59.

[8] A.J. James and J. Lowengrub, A surfactant-conserving volume-of-fluid method for interfacial flows with insoluble surfactant. *J. Comput. Phys.* **201** (2004) 685–722.

[9] D.B. Kothe, R.C. Mjolsness and M.D. Torrey, RIPPLE: A computer program for incompressible flows with free surfaces. Los Alamos National Lab report LA-12007-MS. (1991).

[10] M. Siegel, private communication, 2000.

[11] M. Sussman and E.G. Puckett, A coupled level-set and volume-of-fluid method for computing 3D and axisymmetric incompressible two-phase flows. *J. Comp. Phys.* **162** (2000) 301–337.

[12] X. Yang, A.J. James, J. Lowengrub, X. Zheng and V. Cristini, An adaptive coupled level-set/volume-of-fluid interface capturing method for unstructured triangular grids. To appear in *J. Comp. Phys.*, 2006.

Ashley J. James
Department of Aerospace Engineering and Mechanics
University of Minnesota
e-mail: `ajames@aem.umn.edu`

John Lowengrub
Department of Mathematics
University of California, Irvine
e-mail: lowengrb@math.uci.edu

International Series of Numerical Mathematics, Vol. 154, 261–272

The Value of an American Basket Call with Dividends Increases with the Basket Volatility

Jörg Kampen

Abstract. We show the American Call on a basket (i.e., a weighted sum of assets) and with dividends increases with the basket volatility in a large class of multivariate continuous diffusion models. In case of a flat yield curve the same result holds for the American Put on a basket. The proof of our result is based on extensions of Hajek's mean stochastic comparison results to stochastic sums. We provide a simple proof of Hajek's result and show why the argument is much more involved in case of our extensions. We provide the main ideas of the proofs of our extensions based on heat kernel expansions.

Mathematics Subject Classification (2000). Primary 35K15; Secondary 60G40.

Keywords. Mean stochastic comparison, dependence of American call value on volatility.

1. American Call basket options with dividends

The American Call option on a basket of assets is the right to sell a certain basket of assets, i.e., a fixed weighted sum of underlyings, at strike price K during the time of the validity of the contract. Strike price, underlyings, and maturity time are written in the contract. The term "American" refers to the early exercise right of the option holder. This early exercise right leads to the feature of free boundaries in the typical continuous diffusion models of financial markets. Let us consider such a (for simplicity complete) market with n risky assets $S = (S_1, \ldots, S_n)$ which satisfy

$$dS_i = (r(S) - \delta_i(S))S_i dt + \sigma_i(S)S_i dW_i \tag{1.1}$$

in the risk-neutral measure, and where $S \to r(S)$, $S \to \sigma_i(S)$, and $S \to \delta_i(S)$ are bounded Lipschitz-continuous functions which model interest rates, volatilities, and dividends, respectively. Dividends are always nonnegative. Let $T > 0$ be the maturity time. We consider (1.1) on the time interval $[0, T]$ and assume that W is

This work was completed with the support of SFB 359 (DFG) and BMBF.

an n-dimensional Brownian motion which satisfies for all $0 \leq t \leq T$

$$\rho_{ij}(t) = \int_0^t [W_i, W_j](s)ds, \tag{1.2}$$

with constant ρ_{ij} modeling the correlations of the returns of the assets (we shall consider the extension to variable correlations below). In order to introduce the basket volatility we make the following observation on stochastic sums. If $F_1(t) = S_1(t) + \cdots + S_n(t)$ and assuming that $[F_1](t) > 0$, then

$$Z(t) = \int_0^t \frac{1}{\sqrt{[F_1](u)}} dF_1(u) \tag{1.3}$$

is a Brownian motion by Levy's theorem. We have

$$dF_1 = \sigma_B(S)dZ(t), \tag{1.4}$$

where we call

$$\sigma_B(S) = \sqrt{\sum_{ij} \rho_{ij}\sigma_i\sigma_j S_i S_j} \tag{1.5}$$

the basket volatility. We want to show that the value function of an American basket Call with dividends increases as the basket volatility increases. The value function is defined as the solution $(\delta t, S) \to V_C(\delta t, S)$ of the obstacle equation

$$\max\left\{\frac{\partial u}{\partial \delta t} - L_S u, f - u\right\} = 0, \tag{1.6}$$

in $(0, T] \times \mathbb{R}^n_+ \times \mathbb{R}^m_+$, and satisfies the initial condition (K is the strike price)

$$V_C(0, S) = f_C(S) = \left(\sum_i S_i - K\right)^+. \tag{1.7}$$

Here, δt denotes the time to maturity $T - t$ (hence the minus sign in the diffusion equation in (1.6)), and

$$L_S u = \frac{1}{2}\sigma\sigma^T : D_S^2 u + S_{\delta,r} \cdot \nabla_S u - ru, \tag{1.8}$$

where $S_{\delta,r} := ((r - \delta_1)S_1, \dots, (r - \delta_n)S_n)$ and $D_S^2 u = \left(S_i S_j \frac{\partial^2 u}{\partial S_i \partial S_j}\right)$. The volatility matrix $\sigma\sigma^T$, the volatilities σ_i, and covariances (ρ_{ij}) are related by

$$\sigma\sigma^T = \sigma_i \rho_{ij} \sigma_j. \tag{1.9}$$

Indicating the dependence of the basket volatility $\sigma_B(S) = \sqrt{\sum_{ij} \rho_{ij}\sigma_i\sigma_j S_i S_j}$ on $R = (\rho_{ij})$ by writing σ_B^R we observe that in our model of constant correlations we have

$$\sigma_B^R \leq \sigma_B^{R'} \text{ iff } R \leq R'. \tag{1.10}$$

As usual we say that $R \leq R'$ iff $R' - R$ is positive. Next recall that the exercise region is the contact set $E = \{(t, S) | V_C(t, S = f_C(S)\}$. Our main Theorem then is the following.

Theorem 1.1. *The American Call option basket value function V_C is monotone with respect to the basket volatility, i.e.,*

$$\sigma_B \uparrow \Rightarrow V_C^{\sigma_B} \uparrow \tag{1.11}$$

where

$$\sigma_B(S) = \sqrt{\sum_{ij} \rho_{ij}\sigma_i\sigma_j S_i S_j}.$$

This means that the exercise region shrinks with increasing basket volatility.

Remark 1.2. If dividends $\delta_i = 0$, then the American basket Call option value function V_C equals the European basket Call option value function.

The proof of Theorem 1 is easily reduced to the case of European basket Call option. The reason is that the value function of American Call options is the limit of the value functions of certain Bermudean Call options (as is well known). If $0 < T_1 < \cdots < T_n = T$ is the tenor structure of a Bermudean option and

$$D_k = \{T_1, \ldots, T_n\}, \tag{1.12}$$

then the value of a Bermudean Call option at time $t = 0$ (we consider the value at time $t = 0$ w.l.o.g.) is

$$V_{BC}(0, x) = \sup_{\tau \in D_k} E^x(e^{-r\tau} f_C(S(\tau))), \tag{1.13}$$

where the expectation is taken w.r.t. the risk neutral measure.

If $\Delta := \max_{k \in \{T_1,\ldots,T_n\}} T_{k+1} - T_k$, then $\Delta \to 0$ implies $V_{BC} \uparrow V_C$. Hence, the proof of Theorem 3 reduces to main stochastic comparison theorems of stochastic sums with convex nondecreasing data. If data are convex but not nondecreasing, then mean stochastic comparison results hold only with additional restrictions. One possible restriction is no drift at all, but this means that constant drift terms are also allowed (proof by coordinate transformation). Hence, in context of an American Put option, we can allow for a flat yield curve.

Theorem 1.3. *Let r be a constant function. The American basket Put option value function V_P is monotone with respect to the basket volatility, i.e.,*

$$\sigma_B \uparrow \Rightarrow V_P^{\sigma_B} \uparrow . \tag{1.14}$$

The elegance of stating theorems in terms of the basket volatility has to be paid off by some restriction of the model w.r.t. correlations. However, our results can be extended to financial market models where the assets $S = (S_1, \ldots, S_n)$ satisfy

$$\frac{dS_i}{S_i} = \mu_i(S)dt + \sigma(S)dW, \tag{1.15}$$

and where σ is an $n \times n$ matrix-valued bounded continuous function. Note that in the latter model correlations between the returns of the assets may depend on S. We state the following corollaries.

Corollary 1.4. *In the framework of the more general model class* (1.15)*, the American Call option basket value function* V_C *is monotone with respect to* $\sigma\sigma^T$*, i.e.,*

$$\sigma\sigma^T \uparrow \Rightarrow V_C^{\sigma_B} \uparrow . \tag{1.16}$$

This means that the exercise region shrinks with increasing volatility matrix $\sigma\sigma^T$.

Similarly,

Corollary 1.5. *Let the interest rates* r *be constant. In the framework of the more general model class* (1.15) *the American Put basket option value function* V_C *is monotone with respect to* $\sigma\sigma^T$*, i.e.,*

$$\sigma\sigma^T \uparrow \Rightarrow V_P^{\sigma_B} \uparrow . \tag{1.17}$$

This means that the exercise region shrinks with increasing basket volatility matrix $\sigma\sigma^T$.

This paper is organized as follows. In the next section 2 we state two mean stochastic comparison results. In Section 3 we recall some results on the WKB-expansion of the fundamental solution of parabolic equations. Finally, in Section 4 we provide a simplified proof of Hajek's result in the univariate case. In Section 5 we provide the main ideas of our extensions stated in Section 2.

2. Two mean stochastic comparison results

In this section we state two mean stochastic comparison results. The first one is for a general class of convex payoff and can be applied to option pricing in finance in case of a flat yield curve. The second theorem requires nondecreasing data, but allows for a quite general class of drift terms. It can be applied directly to continuous diffusion market models.

Theorem 2.1. *Let* $T > 0$, $f \in C(\mathbb{R})$ *be convex, and assume that* f *satisfies an exponential growth condition. Furthermore, let* X, Y *be semimartingales with* $x = X(0) = Y(0) \in \mathbb{R}^n$*, where*

$$X(t) = X(0) + \int_0^t \sigma(X(s))dW(s),$$

$$Y(t) = X(0) + \int_0^t \rho(Y(s))dW(s),$$

with $n \times n$*- matrix-valued bounded continuous functions* $x \to \sigma\sigma^T(x)$ *and* $y \to \rho\rho^T(y)$*. If* $\sigma\sigma^T \leq \rho\rho^T$*, then for all* $0 \leq t \leq T$

$$E^x\left(f\left(\sum_i X_i(t)\right)\right) \leq E^x\left(f\left(\sum_i Y_i(t)\right)\right).$$

Remark 2.2. Here, we say that $\sigma\sigma^T \leq \rho\rho^T$ if for all x $\sigma\sigma^T(x) \leq \rho\rho^T(x)$.

The second theorem includes drift terms but is more restrictive w.r.t. the data.

Theorem 2.3. *Let $T > 0$, $f \in C(\mathbb{R})$ nondecreasing, convex, and satisfying an exponential growth condition, X, Y semimartingales with $x = X(0) = Y(0) \in \mathbb{R}^n$, where*

$$X(t) = X(0) + \int_0^t \mu(X(s))ds + \int_0^t \sigma(X(s))dW(s),$$

$$Y(t) = X(0) + \int_0^t \nu(Y(s))ds + \int_0^t \rho(Y(s))dW(s),$$

with $n \times n$-matrix-valued bounded continuous functions $x \to \sigma\sigma^T(x)$ and $y \to \rho\rho^T(y)$. If $\mu \leq \nu$ are bounded continuous functions, and $\sigma\sigma^T \leq \rho\rho^T$, then for all $0 \leq t \leq T$

$$E^x\left(f\left(\sum_i X_i(t)\right)\right) \leq E^x\left(f\left(\sum_i Y_i(t)\right)\right).$$

3. Some results on the WKB-expansion of parabolic equations

We recall some results on the fundamental solution and its relation to continuous diffusion processes. The continuous diffusion

$$dX(\delta t) = \mu(X(\delta t))dt + \sigma(X(\delta t))dW(\delta t) \tag{3.1}$$

satisfies (under some standard assumptions)

$$P(X(\delta t) \in dy) = p(\delta t, x, y)dy, \tag{3.2}$$

where p is the fundamental solution of the parabolic equation

$$\frac{\partial u}{\partial \delta t} - \frac{1}{2}\sum_{ij}(\sigma\sigma^T)_{ij}(x)\frac{\partial^2 u}{\partial x_i \partial x_j} - \sum_i \mu_i(x)\frac{\partial u}{\partial x_i} = 0 \tag{3.3}$$

on the domain $\mathbb{R}^n \times (0, T)$ with the initial condition

$$p(0, x, y) = \delta_y(x), \tag{3.4}$$

where $\delta_y(x)$ denotes the delta distribution related to the Dirac delta distribution δ by $\delta_y(x) = \delta(x - y)$. The fundamental solution exists and is (strictly) positive on $\mathbb{R}^n \times (0, T)$ if the conditions

(A) The operator L is uniformly parabolic in $\mathbb{R}^n$, i.e., there exists $0 < \lambda < \Lambda < \infty$ such that for all $\xi \in \mathbb{R}^n \setminus \{0\}$

$$0 < \lambda|\xi|^2 \leq \sum_{i,j=1}^n (\sigma\sigma^T)_{ij}(x)\xi_i\xi_j \leq \Lambda|\xi|^2. \tag{3.5}$$

(B) The coefficients of L are bounded functions in $\mathbb{R}^n$ which are uniformly Hölder continuous of exponent α ($\alpha \in (0, 1)$).

hold.

Furthermore, if the additional condition

(C) the growth of all derivatives of the smooth coefficients functions

$$x \to (\sigma\sigma^T)_{ij}(x) \quad \text{and} \quad x \to b_i(x)$$

is at most of exponential order

holds, then the fundamental solution p has the representation

$$p(\delta t, x, y) = \frac{1}{\sqrt{4\pi\delta t}^n} \exp\left(-\frac{d^2(x,y)}{4\delta t} + \sum_{k\geq 0} c_k(x,y)\delta t^k\right), \tag{3.6}$$

where d is the Riemannian distance with respect to the line element

$$ds^2 = \sum_{ij} a^{ij} dx_i dx_j,$$

(a^{ij}) being the inverse of (a_{ij}) (cf. [7]), and c_k are global solutions of first-order equations converging pointwise to zero (cf. [3]).

4. A simple proof in case of convex value functions and the univariate case

Mean stochastic comparison results are easy to obtain if the larger value function is convex with respect to the spatial variables. We demonstrate this in the multivariate case, and apply the argument to the univariate case. If the diffusion coefficients are only assumed to depend on time (or, more generally, the larger value function is convex with respect to the spatial variables), then a multivariate extension of Hajek's result can be obtained as follows. Consider the process $X = (X_1, \ldots, X_n)$ which starts at some $x \in \mathbb{R}^n$ at time t and satisfies

$$dX(s) = \sigma(X(s))dZ(s), \tag{4.1}$$

where $\sigma(X(s))$ is a matrix-valued process (values in $\mathbb{R}(n \times n)$). Here, $Z(s)$ an n-dimensional Brownian motion. We assume that $x \to \Sigma(x) = \sigma\sigma^T(x)$ satisfies (A) and (B). Let $(x_1, \ldots, x_n) \to h(x_1, \ldots, x_n)$ be a convex function, which satisfies some exponential growth condition and assume that $\rho : [0, \infty) \to \mathbb{R}(n\times n)$ Lipschitz continuous deterministic function such that

$$\sigma\sigma^T(X(s)) \leq \rho(s)\rho^T(s) \text{ for all } s \in \mathbb{R}_+. \tag{4.2}$$

Then for all $T > t$ the value function

$$(t, x) \to u(t, x) = E^{(t,x)}(h(X(T))) \tag{4.3}$$

satisfies the final value problem

$$\frac{\partial u}{\partial t} + \frac{1}{2}\sum_{ij} \sigma(x)\sigma^T(x)_{ij} \frac{\partial^2 u}{\partial x_i \partial x_j} = 0, \tag{4.4}$$

with the final condition

$$u(T, x) = h(x). \tag{4.5}$$

Similarly, if $Y = (Y_1, \ldots, Y_n)$ with $Y(t) = x \in \mathbb{R}^n$ satisfies

$$dY(s) = \rho(s)dZ(s), \tag{4.6}$$

then the value function

$$(t, x) \to v(t, x) = E^{(t,x)}(h(X(T))) \tag{4.7}$$

satisfies the final value problem

$$\frac{\partial v}{\partial t} + \frac{1}{2}\sum_{ij} \rho\rho_{ij}^T(t)\frac{\partial^2 u}{\partial x_i \partial x_j} = 0, \tag{4.8}$$

with the final condition

$$v(T, x) = h(x). \tag{4.9}$$

Note that comparison holds for both problems (4.4), (4.8) by standard viscosity solution theory. We observe that the solution function v of the latter Cauchy problem is convex by the weak maximum principle. Now, plugging in v into the equation for u gives

$$\begin{aligned} &\tfrac{\partial v}{\partial t} + \tfrac{1}{2}\textstyle\sum_{ij} \sigma\sigma^T(x)_{ij}\tfrac{\partial^2 v}{\partial x_i \partial x_j} \\ = \quad &-\tfrac{1}{2}\textstyle\sum_{ij}\left(\rho(t)\rho^T(t)_{ij} - \sigma(x)\sigma^T(x)_{ij}\right)\tfrac{\partial^2 v}{\partial x_i \partial x_j}. \end{aligned} \tag{4.10}$$

Note that the right side of (4.10) equals

$$-\mathrm{Tr}\left(\left(\rho\rho^T - \sigma\sigma^T\right)D^2 v\right), \tag{4.11}$$

where $D^2 v$ denotes the Hessian of v. Since v is convex with respect to the spatial variables, and $\rho\rho^T \geq \sigma\sigma^T$, we have

$$\mathrm{Tr}\left(\left(\rho\rho^T - \sigma\sigma^T\right)D^2 v\right) \geq 0. \tag{4.12}$$

Hence, we have $v \geq u$ on $\mathbb{R}^n \times (0, T)$ by the comparison principle. Note that we consider final value problems here. The sign of the right side of (4.10) and changes if we consider initial value problems. We summarize

Proposition 4.1. *Let the processes X and Y with $X(0) = Y(0) = x$ satisfy (4.1) and (4.6) and $\sigma\sigma^T$ and $\rho\rho^T$ be Lipschitz continuous. Let f be a convex function which satisfies some exponential growth condition. Then for all $(t, x) \in (0, T) \times \mathbb{R}^n$ and $T \geq t$ we have*

$$E^{(t,x)}(f(X(T)) \leq E^{(t,x)}(f(Y(T))). \tag{4.13}$$

Next, in the univariate case we can prove that the value function is convex with respect to the spatial variable. Consider the case $n = 1$ and the (initial value) problem

$$\begin{cases} \frac{\partial V}{\partial \delta t} - \frac{1}{2}a(x)\frac{\partial^2 V}{\partial x^2} = 0 \text{ in } (0, T) \times \mathbb{R} \\ V(0, x) = h(x) \text{ for } x \in \mathbb{R}. \end{cases} \tag{4.14}$$

Then we get

$$\frac{\partial V}{\partial \delta t}(\delta t, x) = a(x)\frac{\partial^2 V}{\partial x^2} = \int_{\mathbb{R}} h(y)\frac{\partial p}{\partial \delta t}(\delta t, x, y)dy \tag{4.15}$$
$$= \int_{\mathbb{R}} h(y)\frac{\partial^2 (a(y)p(\delta t, x, y))}{\partial y^2}dy = \int_{\mathbb{R}} \left(\frac{d^2 h(y)}{dy^2}a(y)\right) p(\delta t, x, y)dy \geq 0$$

Since $a > 0$, it follows that

$$\frac{\partial^2 V}{\partial x^2}(\delta t, x) \geq 0. \tag{4.16}$$

An analogous argument for convexity in the spatial variable holds in case of $n = 1$ and the situation of Theorem 4 (where there is a drift term). Just use the forward Kolmogorov equation as in (4.15) but with drift term, and use that f is monotone non-decreasing. Hence, we have

Corollary 4.2. *Consider the case $n = 1$ in the situation of Theorem 3, where $f \in C(\mathbb{R})$ is convex assume that f satisfies an exponential growth condition. Then for all $(t, x) \in (0, T) \times \mathbb{R}^n$ and $T \geq t$ we have*

$$E^{(t,x)}(f(X(T)) \leq E^{(t,x)}(f(Y(T))). \tag{4.17}$$

The latter relation also holds in case $n = 1$ and the situation of Theorem 4.

5. Proof of Theorem 3 (main ideas)

In the situation of Theorem 3 convexity of the value function w.r.t. the spatial variables cannot be expected in general. Hence, we need a different argument in order to prove Theorem 3. We shall provide the basic ideas of this argument. If we abbreviate $(a_{ij}) = \sigma\sigma^T$, then the value function $V(t, x) = E^x(f(\sum_i X_i(t))$ with $V(0, x) = f(\sum_i^n x_i)$ satisfies the Cauchy problem

$$\begin{cases} \frac{\partial V}{\partial \delta t} - \frac{1}{2}\sum_{ij} a_{ij}(x)\frac{\partial^2 V}{\partial x_i \partial x_j} = 0 \text{ in } (0, T) \times \mathbb{R}^n \\ V(0, x) = f(\sum_i^n x_i) \text{ for } x \in \mathbb{R}^n. \end{cases} \tag{5.1}$$

Let $A = (a_{ij})$ and denote the solution of (5.1) by V_A. If the entries of A are replaced by the entries of some $A' = (a'_{ij})$ in (5.1), then we denote its solution by $V_{A'}$. We have to show

$$A \leq A' \Rightarrow V_A \leq V_{A'}. \tag{5.2}$$

We shall assume conditions (A), (B), and (C), and adding $\epsilon\|x\|^2$ to f for small ϵ if necessary, we assume that the second order partial derivatives of f are strictly positive. Having proved Theorem 3 in case of this additional restrictions, it is easy to see that it holds also under the conditions of Theorem 3. The first step now is to show that the time derivative of V_A is positive Using Kolmogorov's forward equation for the fundamental solution p of the diffusion equation in (5.1) we have

$$\begin{aligned}\frac{\partial V}{\partial \delta t}(\delta t, x) &= \int_{\mathbb{R}^n} h(y)\frac{\partial p}{\partial \delta t}(\delta t, x, y)dy \\ &= \int_{\mathbb{R}^n} h(y)\frac{1}{2}\sum_{ij}\frac{\partial^2(a_{ij}(y)p(\delta t, x, y))}{\partial y_i \partial y_j}dy \\ &= \int_{\mathbb{R}^n} \mathrm{Tr}\left(D^2h(y)A(y)\right)p(\delta t, x, y)dy.\end{aligned} \tag{5.3}$$

Since h is convex and $A = (a_{ij})$ is positive, we have

$$\mathrm{Tr}\left(D^2 f(\sum_i y_i)A(y)\right) > 0. \tag{5.4}$$

Note that this implies immediately that

$$\mathrm{Tr}\left(D^2 V(\delta t, x)A(x)\right) > 0, \tag{5.5}$$

because

$$\mathrm{Tr}\left(D^2 V(\delta t, x)A(x)\right) = \int_{\mathbb{R}^n} f''\left(\sum_{i=1}^n y_i\right)\mathrm{Tr}\left(A(y)\right)p(\delta t, x, y)dy, \tag{5.6}$$

and p exists and is strictly positive under the conditions (A) and (B). Next consider small δt, and the essential case where $n = 2$. We consider the standard WKB-expansion

$$p_A(\delta t, x, y) = \frac{1}{\sqrt{2\pi\delta t}^n\sqrt{\det A(y)}}\exp\left(-\frac{d^2(x,y)}{2\delta t} + c_0^0(x,y) + \sum_{k=1}^\infty c_k(x,y)\delta t^k\right), \tag{5.7}$$

where c_0^0 satisfies

$$-\frac{n}{2} + \frac{1}{2}Ld^2 + \frac{1}{2}\sum_i\left(\sum_j (a_{ij}(x) + a_{ji}(x))\frac{d^2_{x_j}}{2}\right)\frac{\partial c_0}{\partial x_i}(x,y) = 0, \tag{5.8}$$

but with the boundary condition

$$c_0^0(x,y) = 0 \text{ iff } x = y. \tag{5.9}$$

Now, define $\Delta x = (\Delta x_1, \Delta x_2) = (x - y)$. Then c_0^0 is of the form

$$c_0^{0y}(\Delta x) = c_{01}^{0y}\Delta x_1 + c_{02}^{0y}\Delta x_2 + O(|\Delta x|^2) = O(\Delta x), \tag{5.10}$$

where c_0^{0y}, c_{01}^{0y}, and c_{02}^{0y} are functions depending only in y. We have

$$\begin{aligned}\frac{\partial V_A}{\partial \delta t}(\delta t,x) &= \int_{\mathbb{R}^n} \operatorname{Tr}\left(D^2h(y)A(y)\right) p_A(\delta t,x,y)dy \\ &= \int_{\mathbb{R}^n} \frac{\operatorname{Tr}(D^2h(y)A(y))}{\det A(y)} \exp\left(\frac{-d_A^2(x,y)}{2\delta t} + \sum_{i\geq 0} c_i^y(x)\delta t^i\right) dy\end{aligned} \tag{5.11}$$

For simplicity we consider the essential case where $n = 2$. For $y \in \mathbb{R}^n$ and $\lambda_i^A(y), i = 1,2$ being the eigenvalues of $A(y) = (a_{ij}(y))$, we observe that the terms in (5.11) equal

$$\int_{\mathbb{R}^n} \frac{1}{\sqrt{2\pi\delta t}^n} \frac{\mu^A(y)(\lambda_1^A(y)+\lambda_2^A(y))}{\sqrt{\lambda_1^A(y)\lambda_2^A(y)}} \exp\left(\frac{-\Delta x_1^2}{2\lambda_1^A(y)\delta t} + \frac{-\Delta x_2^2}{2\lambda_2^A(y)\delta t} + O(\Delta x)\right) dy \tag{5.12}$$

for some function μ^A. We know that

$$\mu^A(y)(\lambda_1^A(y)+\lambda_2^A(y)) > 0, \tag{5.13}$$

hence μ^A is positive. Considering the derivative with respect to $\lambda_1^A(y)$ we get

$$\begin{aligned}&\int_{\mathbb{R}^n} \frac{\mu^A(y)}{\sqrt{2\pi\delta t}^n}\left(\left(\frac{1}{2\sqrt{\lambda_1^A(y)\lambda_2^A(y)}} - \frac{1}{2}\sqrt{\frac{\lambda_2^A(y)}{(\lambda_1^A(y))^3}}\right) + \left(\sqrt{\frac{\lambda_1^A(y)}{\lambda_2^A(y)}}\right.\right. \\ &\left.\left. + \sqrt{\frac{\lambda_2^A(y)}{\lambda_1^A(y)}}\right)\frac{1}{\lambda_1^A(y)^2}\frac{\Delta x_1^2}{\delta t}\right) \exp\left(\frac{-\Delta x_1^2}{2\lambda_1^A(y)\delta t} + \frac{-\Delta x_2^2}{2\lambda_2^A(y)\delta t} + O(\Delta x)\right) dy.\end{aligned} \tag{5.14}$$

The second bracket (in front of exp) in equation (5.14) becomes positive. If $\lambda_1^A(y) > \lambda_2^A(y)$ then

$$\frac{1}{2\sqrt{\lambda_1^A(y)\lambda_2^A(y)}} - \frac{1}{2}\sqrt{\frac{\lambda_2^A(y)}{(\lambda_1^A(y))^3}} > 0 \tag{5.15}$$

Furthermore, we know that $\mu^A(y)$ is positive. Now let w.l.o.g.

$$A = \begin{pmatrix} \lambda_1^A(y) & 0 \\ 0 & \lambda_2^A(y) \end{pmatrix}, A' = \begin{pmatrix} \lambda_1^{A'}(y) & 0 \\ 0 & \lambda_2^{A'}(y) \end{pmatrix}. \tag{5.16}$$

Now we can construct a homotopy $\overline{A} : [0,1] \times \mathbb{R}^n \to \mathbb{R}^{n\times n}$ from $A = \overline{A}(0)$ to $A' = \overline{A}(1)$. Considering transformed equations in new variables z with a large constant c

$$z_1 = cx_1, \ z_2 = x_2 \tag{5.17}$$

(if necessary) we can first assume that $\lambda_1^A > \lambda_2^A$ and $\lambda_1^{A'} > \lambda_2^{A'}$. Then

$$\int_{\mathbb{R}^n} \frac{\mu^{B(t)}(y)}{\sqrt{2\pi\delta t}^n}\left(\left(\frac{1}{2\sqrt{\lambda_1^{B(t)}\lambda_2^{B(t)}(y)}} - \frac{1}{2}\sqrt{\frac{\lambda_2^{B(t)}(y)}{(\lambda_1^{B(t)}(y))^3}}\right) + \left(\sqrt{\frac{\lambda_1^{B(t)}(y)}{\lambda_2^{B(t)}(y)}}\right.\right.$$

$$\left.\left. + \sqrt{\frac{\lambda_2^{B(t)}(y)}{\lambda_1^{B(t)}(y)}}\right)\frac{\Delta x_1^2}{\lambda_1^{B(t)}(y)^2\delta t}\right)\exp\left(\frac{-\Delta x_1^2}{2\lambda_1^{B(t)}(y)\delta t} + \frac{-\Delta x_2}{2\lambda_2^{B(t)}(y)\delta t} + O(\Delta x)\right) dy > 0 \tag{5.18}$$

for all $B(t), t \in [0,1]$, where $\lambda_i^{B(t)}(y)$ are the entries of

$$B(t) = \begin{pmatrix} (1-t)\lambda_1^A(y) + t\lambda_1^{A'}(y) & 0 \\ 0 & \lambda_2^A(y) \end{pmatrix}, A_M = \begin{pmatrix} \lambda_1^{A'}(y) & 0 \\ 0 & \lambda_2^A(y) \end{pmatrix}. \tag{5.19}$$

This shows that $B : [0,1] \to \times\mathbb{R}^n \to \mathbb{R}^{n\times n}$ is a homotopy from A to A_M such that for small δt and all x

$$s \to \frac{\partial V_{B(s)}}{\partial \delta t}(\delta t, x) \tag{5.20}$$

is monotone increasing. Hence,

$$V_A(\delta s, x) \le V_{A'}(\delta s, x) \text{ for all } \delta s \le \delta t \tag{5.21}$$

for small $\delta s \le \delta t_0 > 0$. Moreover, this $\delta t_0 > 0$ can be chosen uniformly in x. Since we assumed that the initial data are regular with bounded continuous second derivatives we know that

$$\frac{\partial V}{\partial \delta t} = \operatorname{Tr}\left(AD^2V\right) \tag{5.22}$$

is bounded. Hence, $\lambda_{\min} = \inf_y \lambda_{\min}^{A'}(y) - \lambda_{\min}^{A}(y) > 0$ implies that there exists $\delta t > 0$ such that

$$\inf_{x\in\mathbb{R}^n} \operatorname{Tr}\left(A'(y)D^2V(\delta t, x)\right) - \operatorname{Tr}\left(A(y)D^2V(\delta t, x)\right) > 0 \tag{5.23}$$

and this implies that

$$\frac{\partial V_A(\delta t, x)}{\partial \delta t} < \frac{\partial V_{A'}(\delta t, x)}{\partial \delta t} \tag{5.24}$$

for some $\delta t > 0$ which can be chosen independently of x. Hence,

$$0 < \delta t_0 := \arg \sup_{\delta s\in[0,T]} \left\{V_A(\delta t, x) \le V_{A'}(\delta t, x) | \text{for all } x \in \mathbb{R}^n\right\}. \tag{5.25}$$

Since $V_A, V_{A'}$ are continuous, we have $0 \le s \le \delta t_0$

$$V_A(s,.) \le V_{A'}(s,.). \tag{5.26}$$

The extension of the argument to global time involves an analysis of the parametric dependence of the higher order terms c_k of the WKB-expansion and is provided in [3].

6. Comments on literature

Mean stochastic comparison results can be used in order to identify optimal strategies of multivariate passport options (cf. [2], [3], [5]). Our Theorem 2 and Theorem 3 generalize Hajek's Theorem 3 in [1]. Hajek's Theorem (and that it can be useful in the financial context) was mentioned to me first by Peter Laurence.

References

[1] Hajek, B. *Mean stochastic comparison of diffusions,* Z. Wahrscheinlichkeitstheorie verw. Geb., vol. 68, pp. 315–329, 1985.

[2] Kampen, J., *On Mean Stochastic Comparison and Optimal Strategies of Multivariate Options on Traded Accounts* http://papers.ssrn.com (700541).

[3] Kampen, J., *The WKB-expansion of the fundamental solution and its applications* (book manuscript, submitted).

[4] Krylov, N.V., *Lectures on Elliptic and Parabolic Equations in Hölder Spaces,* Graduate Studies in Mathematics, Vol. 12, American Mathematical Society, 1996.

[5] Shreve, S.E., Vecer, J. *Options on a traded account: Vacation Calls, vacation Puts and passport options*, Finance & Stochastics, 2000.

[6] Stroock, D.W., Varadhan, S.N.R. *Multidimensional Diffusion Processes,* Springer 1979.

[7] Varadhan, S.R.S. *On the Behavior of the Fundamental Solution of the Heat Equation with variable coefficients* CPAM, 1967.

Jörg Kampen
Im Neuenheimer Feld 294
University of Heidelberg
D-69120 Heidelberg, Germany
e-mail: joerg.kampen@iwr.uni-heidelberg.de

International Series of Numerical Mathematics, Vol. 154, 273–282

Mathematical Modelling of Nutrient-limited Tissue Growth

J.R. King and S.J. Franks

Abstract. Free boundary problems associated with biological tissue growing under conditions of nutrient limitation are formulated. We consider two constitutive laws to describe the deformation of the tissue, Darcy flow and Stokes flow. Analysis by asymptotic methods, clarifying the model's stability properties, is then described for the two biologically-plausible limit cases: (1) the small-death-rate limit and (2) the thin-rim (fast-nutrient-consumption or large-tumour) limit. The former leads to some interesting variants on the Hele-Shaw squeeze film problem and the latter makes explicit a buckling instability associated with growth-induced compressive stresses.

Keywords. Tissue growth, singular perturbation methods, buckling instability.

1. Introduction

We summarise here an extension of the analysis of [4] corresponding to the incorporation of one of the most important effects neglected there, namely the slowing down of the growth (following the initial exponential growth phase) due to the limited availability of nutrient in the regions furthest from a nutrient source. We refer to [4] for background and further references. Here we follow [5] in describing nutrient transport by a reaction-convection-diffusion equation of the form

$$\frac{\partial c}{\partial t} + \nabla \cdot (c\mathbf{v}) = \nabla \cdot (D_c(n,m)\nabla c) - \lambda(c,n,m) \tag{1}$$

where $c(\mathbf{x},t)$ is the nutrient concentration, the nutrient diffusivity being D_c, $\mathbf{v}(\mathbf{x},t)$ is the macroscopic velocity field of the growing tissue and λ is the nutrient consumption rate. One of our prime motivations comes from tumour growth modelling and in this context $n(\mathbf{x},t)$ denotes the local volume fraction of live tumour cells,

The authors are grateful to EPSRC and AEA Technology for a CASE studentship held by the second author whilst much of this work was being undertaken. JRK also gratefully acknowledges funding from EPSRC grant no. GR/S31051/01.

$\rho(\mathbf{x}, t)$ that of the surrounding material (typically normal cells *in vivo* and fluid or gel, say, *in vitro*) and $m(\mathbf{x}, t)$ that of the necrotic material (i.e., the products of cell death); in the tissue-engineering context, n might denote the fraction of the void space occupied by tissue at a given point in a porous scaffold. We assume each of these phases to be incompressible and adopt the no-voids condition

$$n + \rho + m = 1, \tag{2}$$

which is why D_c and λ have not been made dependent upon ρ also. Neglecting cellular diffusion (i.e., in the 'sharp-interface' limit of [4]), we have in addition that

$$\begin{aligned} \frac{\partial n}{\partial t} + \nabla \cdot (n\mathbf{v}) &= k_m(c)n - k_d(c)n, \\ \frac{\partial \rho}{\partial t} + \nabla \cdot (\rho\mathbf{v}) &= 0, \\ \frac{\partial m}{\partial t} + \nabla \cdot (m\mathbf{v}) &= k_d(c)n, \end{aligned} \tag{3}$$

where k_m and k_d are the mitotic and death rates of the viable cells (k_m is an increasing function of c and k_d a decreasing one), and for brevity we neglect here any birth or death in the surrounding (ρ) material. For simplicity, we also neglect any volume change on cell death; moreover, in contrast to [6] we shall not include any mechanisms for the transport, consumption or further degradation of the necrotic material. From (2)–(3), we have

$$\nabla \cdot \mathbf{v} = k_m(c)n \tag{4}$$

and in one dimension, or with radial symmetry, (1)–(3) forms a closed system. In higher dimensions, however, we need to specify a constitutive law in addition; as in [4], we consider two cases, as follows.

(a) Darcy flow

Here

$$\mathbf{v} = -\frac{K}{\mu(n, m)} \nabla p, \tag{5}$$

introducing an extra unknown (the pressure field $p(\mathbf{x}, t)$), but closing the system; $\mu(n, m)$ is the viscosity of the growing medium and K is the permeability.

(b) Stokes flow

Here the stress tensor σ_{ij} satisfies for each i the momentum equations

$$\frac{\partial \sigma_{ij}}{\partial x_j} = 0 \tag{6}$$

and we adopt the constitutive law

$$\sigma_{ij} = -\left(p + \frac{2}{3}\mu(n, m)\frac{\partial v_k}{\partial x_k}\right)\delta_{ij} + \mu(n, m)\left(\frac{\partial v_i}{\partial x_j} + \frac{\partial v_j}{\partial x_i}\right), \tag{7}$$

where the pressure p is defined by

$$p = -\frac{1}{3}\sigma_{kk}. \tag{8}$$

We shall leave μ and D_c arbitrary, writing

$$\begin{array}{ccc} \mu(1,0) = \mu_n, & \mu(0,1) = \mu_m, & \mu(0,0) = \mu_\rho, \\ D_c(1,0) = D_c^{(n)}, & D_c(0,1) = D_c^{(m)}, & D_c(0,0) = D_c^{(\rho)}. \end{array} \tag{9}$$

For brevity, we neglect consumption of nutrient by the surrounding medium and by the necrotic material (noting that c will in any case typically be negligible in the necrotic core) and for definiteness take

$$\lambda(c,n,m) = k_c(c)n; \tag{10}$$

moreover, we note (for reasons explained in [5]) that (1) can in practice be approximated by its quasi-steady limit, i.e.,

$$\nabla \cdot (D_c(n,m)\nabla c) = k_c(c)n. \tag{11}$$

Having established this more general formulation, we shall henceforth restrict attention to the 'one-phase' case in which effects due to the surrounding (ρ) material are negligible, giving the simplified version of the model in Section 2 and providing asymptotic reductions in Sections 3 and 4.

2. The one-phase problem: formulation

2.1. Boundary conditions

We first focus on the case of *in vitro* growth, taking (as in [5]) the surroundings to be passive, so that we can set $\rho = 0$ in $\Omega(t)$ (and, since $m = 1 - n$ in $\Omega(t)$, we suppress the m dependence of μ and D_c) with

$$c = c^*, \qquad q_\nu = \mathbf{v} \cdot \boldsymbol{\nu} \qquad \text{on } \Gamma(t), \tag{12}$$

where $\Omega(t)$ is the tumour, having moving interface $\Gamma(t)$ with unit outward normal $\boldsymbol{\nu}$ and outward normal velocity q_ν, and c^* is a constant. We are thus assuming that the surroundings are well stirred (we note that c^* can readily be made a function of time in the analysis which follows) or possess a high nutrient diffusivity ($D_c^{(\rho)} \gg D_c^{(n)}$) or contain nutrient sources (such as vasculature) which maintain c at a constant level. In addition, we adopt stress free boundary conditions

$$\text{(a)} \qquad p = 0 \qquad \text{on } \Gamma(t), \tag{13}$$

$$\text{(b)} \qquad \sigma_{ij}\nu_j = 0 \qquad \text{on } \Gamma(t), \tag{14}$$

which correspond to taking the surrounding material to be inviscid.

2.2. Non-dimensionalisation

We define $k \equiv k_m(c^*)$ to be the 'nutrient-rich' mitotic rate and non-dimensionalise according to

$$\mathbf{x} = L\hat{\mathbf{x}}, \quad t = \hat{t}/k, \quad c = c^*\hat{c}, \quad \mathbf{v} = kL\hat{\mathbf{v}}, \quad q_\nu = kL\hat{q}_\nu;$$
$$(a)\ p = \mu_n k L^2 \hat{p}/K; \quad (b)\ p = \mu_n k \hat{p}, \quad \sigma_{ij} = \mu_n k \hat{\sigma}_{ij},$$

where L is a representative length scale. We then introduce

$$\hat{k}_m(\hat{c}) = k_m(c)/k, \quad \hat{k}_d(\hat{c}) = k_d(c)/k_d(c^*), \quad \hat{k}_c(\hat{c}) = k_c(c)/k_c(c^*),$$
$$\hat{\mu}(n) = \mu(n)/\mu_n, \quad \hat{D}_c(n) = D_c(n)/D_c^{(n)}$$

to give, dropping ^'s,

$$\begin{aligned}
\frac{\partial n}{\partial t} + \nabla\cdot(n\mathbf{v}) &= k_m(c)n - \sigma k_d(c)n, \\
\frac{\partial m}{\partial t} + \nabla\cdot(m\mathbf{v}) &= \sigma k_d(c)n, \\
n + m &= 1, \\
\nabla\cdot(D_c(n)\nabla c) &= \beta k_c(c)n \qquad \text{in } \Omega(t), \\
c = 1, \qquad q_\nu &= \mathbf{v}\cdot\boldsymbol{\nu}, \qquad \text{on } \Gamma(t),
\end{aligned} \tag{15}$$

where $k_m(1) = 1$, $k_c(1) = 1$ and $D_c(1) = 1$, and the dimensionless parameters σ and β are given in terms of dimensional quantities by $\sigma = k_d(c^*)/k_m(c^*)$ and $\beta = k_c(c^*)L^2/D_c^{(n)}c^*$. In addition

(a)
$$\mathbf{v} = -\frac{1}{\mu(n)}\nabla p \quad \text{in } \Omega(t), \qquad p = 0 \quad \text{on } \Gamma(t) \tag{16}$$

or

(b)
$$\frac{\partial \sigma_{ij}}{\partial x_j} = 0, \quad \sigma_{ij} = -\left(p + \frac{2}{3}\mu(n)\frac{\partial v_k}{\partial x_k}\right)\delta_{ij} + \mu(n)\left(\frac{\partial v_i}{\partial x_j} + \frac{\partial v_j}{\partial x_i}\right) \quad \text{in } \Omega(t),$$
$$\sigma_{ij}\nu_j = 0 \qquad \text{on } \Gamma(t), \tag{17}$$

where $\mu(1) = 1$ in each case. In view of (4), the latter constitutive law can also be expressed in the form

$$\sigma_{ij} = -\left(p + \frac{2}{3}k_m(c)\mu(n)n\right)\delta_{ij} + \mu(n)\left(\frac{\partial v_i}{\partial x_j} + \frac{\partial v_j}{\partial x_i}\right). \tag{18}$$

We note that

$$\frac{dn}{dt} = n(1 - \sigma - n)$$

holds at any material point on $\Gamma(t)$, so $n \to 1 - \sigma$ there as $t \to \infty$.

3. The small-death-rate limit, $\sigma \ll 1$

In this section we consider the first of two biologically plausible limit cases in which the formulations above can be significantly simplified, namely $\sigma \ll 1$, which corresponds to the apoptotic rate (the rate of natural cell death) being much smaller than the mitotic rate under nutrient rich conditions (care is of course required with this limit if k_d becomes too large for small c); this limit is discussed in the radially symmetric case in [5]. Since $\nabla \cdot \mathbf{v} = k_m(c)n$, it follows from (15) that

$$\frac{\partial n}{\partial t} + \mathbf{v} \cdot \nabla n = k_m(c)n(1-n) - \sigma k_d(c)n$$

so taking $n = 1 - \sigma$ at $t = 0$ in $\Omega(0)$, we have in the limit $\sigma \to 0$ the leading order solution

$$n = 1 \qquad \text{in } \Omega(t) \tag{19}$$

for $t = O(1)$ (as is obvious on physical grounds); to avoid a proliferation of suffices we shall omit the subscript $_0$ in denoting leading order solutions. Hence to leading order

$$\begin{aligned} \nabla \cdot \mathbf{v} = k_m(c), \quad \nabla^2 c = \beta k_c(c), \qquad &\text{in } \Omega(t), \\ c = 1, \quad q_\nu = \mathbf{v} \cdot \boldsymbol{\nu} \qquad &\text{on } \Gamma(t). \end{aligned} \tag{20}$$

with

$$(\text{a})\mathbf{v} = -\nabla p \quad \text{in } \Omega(t), \qquad\qquad p = 0 \quad \text{on } \Gamma(t) \tag{21}$$

or

$$\begin{aligned} (\text{b}) \quad & \frac{\partial \sigma_{ij}}{\partial x_j} = 0, \quad \sigma_{ij} = -\left(p + \frac{2}{3}\frac{\partial v_k}{\partial x_k}\right)\delta_{ij} + \left(\frac{\partial v_i}{\partial x_j} + \frac{\partial v_j}{\partial x_i}\right) \quad \text{in } \Omega(t), \\ & \sigma_{ij}\nu_j = 0 \quad \text{on } \Gamma(t). \end{aligned} \tag{22}$$

We now discuss the two constitutive laws separately.

(a) Darcy flow

Here we have the coupled moving boundary problem

$$\begin{aligned} \nabla^2 c = \beta k_c(c), \qquad & \nabla^2 p = -k_m(c) \quad \text{in } \Omega(t), \\ c = 1, \qquad & q_\nu = -\frac{\partial p}{\partial \nu} \quad \text{on } \Gamma(t). \end{aligned} \tag{23}$$

As explained in [5], $k_c(c) = k_m(c)$ seems in practice to be a reasonable assumption and if $\Gamma = \partial\Omega$, or for suitable boundary conditions on $\partial\Omega\backslash\Gamma$ if $\Gamma \neq \partial\Omega$, we then have

$$c = 1 - \beta p \tag{24}$$

and the leading-order problem reduces to

$$\begin{aligned} \nabla^2 c = \beta k_m(c) \qquad & \text{in } \Omega(t), \\ c = 1, \quad \beta q_\nu = \frac{\partial c}{\partial \nu} \qquad & \text{on } \Gamma(t). \end{aligned} \tag{25}$$

In the case $k_m(c) = c$, we have a Baiocchi transformation

$$w = \frac{1}{\beta}\int_0^t e^{-t'}\,(1 - c(\mathbf{x}, t'))\,dt' \quad \text{for } \mathbf{x} \in \Omega(0),$$
$$w = \frac{1}{\beta}\int_\omega^t e^{-t'}(1 - c(\mathbf{x}, t'))dt' \quad \text{for } \mathbf{x} \in \Omega(t)\backslash\Omega(0),$$

where we write $\Gamma(t)$ as $t = \omega(\mathbf{x})$, whereby w is given by

$$\nabla^2 w = \beta w - (1 - e^{-t}) \ \text{ for } \mathbf{x} \in \Omega(0), \quad \nabla^2 w = \beta w + e^{-t} \ \text{ for } \mathbf{x} \in \Omega(t)\backslash\Omega(0),$$
$$\text{at } t = \omega(\mathbf{x}) \qquad w = \frac{\partial w}{\partial \nu} = 0.$$

(b) Stokes flow

In this case we have

$$\nabla^2 c = \beta k_c(c), \quad \nabla \cdot \mathbf{v} = k_m(c), \quad \mathbf{0} = -\nabla\left(p - \frac{1}{3}k_m(c)\right) + \nabla^2\mathbf{v} \qquad \text{in } \Omega(t),$$
$$c = 1, \qquad q_\nu = \mathbf{v}\cdot\boldsymbol{\nu}, \qquad \sigma_{ij}\nu_j = 0 \qquad \text{on } \Gamma(t). \tag{26}$$

Various manipulations of (26) are possible, for example, if $k_c(c) = k_m(c)$; then writing

$$\mathbf{v} = \frac{1}{\beta}\nabla c + \mathbf{v}^\dagger, \qquad p = p^\dagger + \frac{4}{3}k_m(c)$$

leads to the usual Stokes flow equations

$$\nabla \cdot \mathbf{v}^\dagger = 0, \qquad \mathbf{0} = -\nabla p^\dagger + \nabla^2 \mathbf{v}^\dagger,$$

but the boundary conditions become non-standard.

4. The thin-rim (fast-nutrient-consumption or large-tumour) limit, $\beta \gg 1$

In the converse limit to the one we now discuss, i.e., as $\beta \to 0$, we have $c \sim 1$ and the nutrient-rich formulations of [4] thus apply. The limit as $\beta \to \infty$ also leads to significant simplifications, as we now describe.

(a) Darcy flow

Again denoting by $-\nu$ the inward normal distance from $\Gamma(t)$, the boundary-layer scalings

$$t = \sqrt{\beta}T, \quad \nu = \zeta/\sqrt{\beta}, \quad p = P/\beta, \quad \mathbf{v} = \mathbf{V}/\sqrt{\beta}, \quad q_\nu = Q_\nu/\sqrt{\beta}, \tag{27}$$

apply for Darcy flow, giving at leading order the one-dimensional 'travelling-wave' problem (in which Q_ν is of course independent of ζ)

$$\begin{aligned}
\frac{\partial}{\partial\zeta}\left((V_\nu - Q_\nu)n\right) &= k_m(c)n - \sigma k_d(c)n,\\
\frac{\partial}{\partial\zeta}\left((V_\nu - Q_\nu)m\right) &= \sigma k_d(c)n,\\
n + m &= 1,\\
\frac{\partial}{\partial\zeta}\left(D_c(n)\frac{\partial c}{\partial\zeta}\right) &= k_c(c)n,\\
V_\nu &= -\frac{1}{\mu(n)}\frac{\partial P}{\partial\zeta},\\
c = 1, \quad P = 0, \quad V_\nu &= Q_\nu \qquad \text{on } \zeta = 0,\\
n \to 0, \quad m \to 1, \quad \frac{\partial c}{\partial\zeta} &\to 0 \qquad \text{as } \zeta \to -\infty,
\end{aligned} \tag{28}$$

from which P decouples, with the solution then depending on V_ν and Q_ν only through $W_\nu \equiv V_\nu - Q_\nu$. The leading-order solutions W_ν, n, m and c depend only on ζ and not on T or on the tangential components of $\mathbf{x}$ (Q_ν depends on these, but is independent of ζ, being determined subsequently via the moving boundary problem formulated below). We note that the components of $\mathbf{V}$ in the directions tangential to Γ are of $O(1/\sqrt{\beta})$. Since

$$\frac{dW_\nu}{d\zeta} = k_m(c)n, \quad -\infty < \zeta < 0, \qquad W_\nu = 0 \quad \text{on } \zeta = 0 \tag{29}$$

it follows from (26) that

$$n = 1 - \sigma, \quad m = \sigma \qquad \text{on } \zeta = 0,$$

as is to be expected. As in the description of the large-time behaviour in [5], (26) determines a part of its solution the value of the constants U and c_∞, defined by

$$U = -\lim_{\zeta\to-\infty} W_\nu(\zeta), \qquad c_\infty = \lim_{\zeta\to-\infty} c(\zeta). \tag{30}$$

We note that the above formulation (apart of course from the initial value problem for P) holds irrespective of the choice of constitutive law (the leading-order problem being locally one-dimensional), so the quantities U and c_∞ defined by (30) are of more general relevance.

The quantities in (30) represent the information we need from the inner (boundary-layer) analysis. In the outer ($\nu = O(1)$) problem, i.e., in the necrotic core, we have $n \ll 1$ (indeed, n is exponentially small), $m \sim 1$ and by matching with (30) we obtain $c \sim c_\infty$ and, to leading order,

$$\begin{aligned}
\nabla\cdot\mathbf{V} = 0, \quad \mathbf{V} &= -\frac{1}{\mu_m}\nabla\bar{p} \qquad \text{in } \Omega(T),\\
\bar{p} = 0, \quad Q_\nu &= \mathbf{V}\cdot\boldsymbol{\nu} + U \quad \text{on } \Gamma(T),
\end{aligned} \tag{31}$$

where $p = \bar{p}/\sqrt{\beta}$ is the appropriate pressure scaling when $\mu_m = O(1)$. The second condition on $\Gamma(T)$ is a novel feature of this formulation, accounting for the $O(1)$

change in V_ν between the matching region and the true interface which results from cell division in the viable rim (i.e., in $\zeta = O(1)$). The solution to the problem (31) is of course the trivial one

$$\bar{p} = 0, \qquad \mathbf{V} = \mathbf{0}, \qquad Q_\nu = U, \tag{32}$$

the last of which implies that the outward normal velocity of the interface is constant; nevertheless, we record the full formulation (31) here in order to generalise it subsequently, in particular to the two-phase case in which the properties of the surrounding material cannot be ignored. We remark that there is also a distinguished limit $\mu_m = O(1/\sqrt{\beta})$ in which a Hele-Shaw type formulation results which contains a 'kinetic undercooling' regularisation.

(b) Stokes flow

Other than for p, which is of $O(1)$, the scalings (27) again apply in the boundary layer, and W_ν, n, m and c are determined, exactly as for Darcy flow, with the definitions (30) pertaining. The flow problem is rather more delicate, however, requiring the calculation of $O(1/\sqrt{\beta})$ terms in the momentum equations and requiring derivations of "viscous shell" equations, somewhat akin to those in [2] but with significant differences due to cell division. Denoting co-ordinates on $\Gamma(T)$ by τ and υ, using (27) we have (in an obvious notation)

$$\begin{aligned}
\sigma_{\tau\tau} &\sim -\left(p + \frac{2}{3}\mu(n)\frac{dW_\nu}{d\zeta}\right) + \frac{1}{\sqrt{\beta}}\sigma_{\tau\tau}^{(1)}, \\
\sigma_{\upsilon\upsilon} &\sim -\left(p + \frac{2}{3}\mu(n)\frac{dW_\nu}{d\zeta}\right) + \frac{1}{\sqrt{\beta}}\sigma_{\upsilon\upsilon}^{(1)}, \\
\sigma_{\nu\nu} &\sim -\left(p - \frac{4}{3}\mu(n)\frac{dW_\nu}{d\zeta}\right) + \frac{1}{\sqrt{\beta}}\sigma_{\nu\nu}^{(1)},
\end{aligned} \tag{33}$$

$$\sigma_{\tau\upsilon} \sim \frac{1}{\sqrt{\beta}}\sigma_{\tau\upsilon}^{(1)}, \quad \sigma_{\tau\nu} \sim \mu(n)\frac{\partial V_\tau}{\partial\zeta} + \frac{1}{\sqrt{\beta}}\sigma_{\tau\nu}^{(1)}, \quad \sigma_{\upsilon\nu} \sim \mu(n)\frac{\partial V_\upsilon}{\partial\zeta} + \frac{1}{\sqrt{\beta}}\sigma_{\upsilon\nu}^{(1)} \tag{34}$$

where p denotes the leading-order pressure; in conventional incompressible flow problems it follows from the continuity equation that V_ν is independent of ζ, but here we instead have (29) and some of the resulting balances accordingly differ. Nevertheless, we can read off the relevant asymptotic formulations of the momentum equations directly from [2], which we essentially follow in parametrising the free surface in such a way that lines of constant τ and υ are lines of curvature of $\Gamma(T)$; we write $\Gamma(T)$ as $\mathbf{x} = \mathbf{x}_\Gamma(\tau, \upsilon, T)$ and define

$$a_\tau = \left|\frac{\partial \mathbf{x}_\Gamma}{\partial \tau}\right|, \qquad a_\upsilon = \left|\frac{\partial \mathbf{x}_\Gamma}{\partial \upsilon}\right|.$$

At leading order the momentum equations then read

$$\frac{\partial}{\partial\zeta}\left(a_\tau a_\upsilon \sigma_{\tau\nu}^{(0)}\right) = \frac{\partial}{\partial\zeta}\left(a_\tau a_\upsilon \sigma_{\upsilon\nu}^{(0)}\right) = \frac{\partial}{\partial\zeta}\left(a_\tau a_\upsilon \sigma_{\nu\nu}^{(0)}\right) = 0,$$

where here and henceforth a_τ and a_υ denote leading-order quantities. Since $\sigma_{\tau\nu} = \sigma_{\upsilon\nu} = \sigma_{\nu\nu} = 0$ on $\Gamma(T)$, it follows that

$$\sigma_{\tau\nu}^{(0)} = \sigma_{\upsilon\nu}^{(0)} = \sigma_{\nu\nu}^{(0)} = 0 \tag{35}$$

and hence at $O(1)$

$$p = \frac{4}{3}\mu(n)\frac{dW_\nu}{d\zeta}, \qquad \sigma_{\tau\tau}^{(0)} = \sigma_{\upsilon\upsilon}^{(0)} = -2\mu(n)\frac{dW_\nu}{d\zeta}, \qquad \frac{\partial V_\tau}{\partial \zeta} = \frac{\partial V_\upsilon}{\partial \zeta} = 0. \tag{36}$$

At $O(1/\sqrt{\beta})$, we have (from [2], using (34) and (35))

$$\frac{\partial}{\partial \tau}\left(a_\upsilon \sigma_{\tau\tau}^{(0)}\right) + \frac{\partial}{\partial \zeta}\left(a_\tau a_\upsilon \sigma_{\tau\nu}^{(1)}\right) - \frac{\partial a_\upsilon}{\partial \tau}\sigma_{\upsilon\upsilon}^{(0)} = 0,$$

$$\frac{\partial}{\partial \upsilon}\left(a_\tau \sigma_{\upsilon\upsilon}^{(0)}\right) + \frac{\partial}{\partial \zeta}\left(a_\tau a_\upsilon \sigma_{\upsilon\nu}^{(1)}\right) - \frac{\partial a_\tau}{\partial \upsilon}\sigma_{\tau\tau}^{(0)} = 0,$$

$$\frac{\partial}{\partial \zeta}\left(a_\tau a_\upsilon \sigma_{\nu\nu}^{(1)}\right) - a_\tau a_\upsilon\left(\kappa_\tau \sigma_{\tau\tau}^{(0)} + \kappa_\upsilon \sigma_{\upsilon\upsilon}^{(0)}\right) = 0,$$

where κ_τ and κ_υ are the principal curvatures of $\Gamma(T)$ (our sign convention differing from that of [2]). Using (36) (observing that a_τ and a_υ are independent of ζ, while $\sigma_{\tau\tau}^{(0)}$ and $\sigma_{\upsilon\upsilon}^{(0)}$ are independent of τ and υ) we thus obtain

$$\frac{\partial}{\partial \zeta}\sigma_{\tau\nu}^{(1)} = \frac{\partial}{\partial \zeta}\sigma_{\upsilon\nu}^{(1)} = 0, \qquad \frac{\partial}{\partial \zeta}\sigma_{\nu\nu}^{(1)} = -4\kappa\mu(n)\frac{dW_\nu}{d\zeta},$$

where $\kappa = (\kappa_\tau + \kappa_\upsilon)/2$ is the leading-order mean curvature of $\Gamma(T)$. Hence

$$\sigma_{\tau\nu}^{(1)} = \sigma_{\upsilon\nu}^{(1)} = 0, \qquad \sigma_{\nu\nu}^{(1)} = 4\kappa\int_\zeta^0 k_m(c(\zeta'))\mu(n(\zeta'))n(\zeta')d\zeta', \tag{37}$$

We are now in a position to formulate the leading-order problem in the core, wherein n is again exponentially small and $c \sim c_\infty$. Writing

$$\sigma_{ij} = \bar{\sigma}_{ij}/\sqrt{\beta}, \qquad p = \bar{p}/\sqrt{\beta}$$

we thus obtain the standard Stokes flow problem

$$\nabla \cdot \mathbf{V} = 0, \quad \frac{\partial \bar{\sigma}_{ij}}{\partial x_j} = 0, \quad \bar{\sigma}_{ij} = -\bar{p}\delta_{ij} + \mu_m\left(\frac{\partial V_i}{\partial x_j} + \frac{\partial V_j}{\partial x_i}\right) \quad \text{in } \Omega(T); \tag{38}$$

this is subject to the moving boundary conditions

$$\bar{\sigma}_{ij}\nu_j = 2\gamma\kappa\nu_i, \qquad Q_\nu = \mathbf{V}\cdot\boldsymbol{\nu} + U \quad \text{on } \Gamma(T), \tag{39}$$

where we have matched with (37) and defined

$$\gamma = 2\int_{-\infty}^0 k_m(c(\zeta))\mu(n(\zeta))n(\zeta)d\zeta. \tag{40}$$

Cell division in the viable rim thus generates a surface-tension-like term, but with a negative coefficient of surface tension, $-\gamma$ (determined by (40)). This is unsurprising in that cell division generates compressive stresses in the rim, leading to the likelihood of "viscous-buckling" instabilities; the formulation (38)–(39) is of course ill-posed, and in practice buckling will presumably occur on a length scale comparable to the thickness of the viable rim (i.e., of $O(1/\sqrt{\beta})$), causing the thin-rim analysis above to cease to be valid. Were the U term absent from (39), the formulation would be equivalent to the time-reversal of the Stokes flow problem with positive surface tension, discussed in [1] for example, and for $\Gamma(T)$ initially analytic a solution can be expected to exist for some finite time, despite the problems ill-posedness.

5. Discussion

We have here summarised a relatively general moving-boundary formulation for the growth of soft tissue under constraints of nutrient limitation. Other asymptotic limits are noteworthy, including $\sigma \ll 1$ with $t = O(1/\sigma)$ which exhibits features of both of those that we have discussed. Worthwhile generalisations include the incorporation of multi-phase and elastic effects.

A particularly noteworthy result is the buckling instability mechanism which has been identified in the thin-rim limit. Similar effects can also be expected for other constitutive laws (including elasticity) and this growth-induced-stress instability mechanism can be added to the growth-driven one analysed in [3] and to the usual nutrient-limited fingering mechanism closely related to the familiar ones that can arise in Stefan and Hele-Shaw problems.

References

[1] L.J. Cummings, S.D. Howison, J.R. King, *Two-dimensional Stokes and Hele-Shaw flows with free surfaces.* Euro. J. Appl. Math. **10** (1999), 635–680.

[2] P.D. Howell, *Extensional thin layer flows.* D. Phil Thesis, University of Oxford, 1994.

[3] S.J. Franks, J.R. King, *Interactions between a uniformly proliferating tumour and its surroundings: uniform material properties.* Math. Med. Biol. **20** (2003), 47–89.

[4] J.R. King, S.J. Franks, *Mathematical analysis of some multi-dimensional tissue-growth models.* Euro. J. Appl. Math. **15** (2004), 273–295.

[5] J.P. Ward, J.R. King, *Mathematical modelling of avascular-tumour growth.* IMA J. Math. Appl. Med. Biol. **14** (1997), 39–69.

[6] J.P. Ward, J.R. King, *Mathematical modelling of avascular-tumour growth II: Modelling growth saturation.* IMA J. Math. Appl. Med. Biol. **16** (1999), 171–211.

J.R. King
Centre for MathematicalMedicine, School of Mathematical Sciences
University of Nottingham, Nottingham NG7 2RD, UK
e-mail: `John.King@nottingham.ac.uk`

S.J. Franks
Health and Safety Laboratory
Harpur Hill, Buxton SK17 9JN, UK
e-mail: `Susan.Franks@hsl.gov.uk`

International Series of Numerical Mathematics, Vol. 154, 283–290

Asymptotic Hysteresis Patterns in a Phase Separation Problem

Pavel Krejčí

Abstract. A non-smooth temperature-driven phase separation model with conserved energy and a large set of equilibria is shown to develop spontaneously two different time scales as time tends to infinity. The temperature evolution becomes slower and slower, while the microevolution on the unknown phase interface keeps its own independent characteristic speed. In the large time limit, the temperature becomes uniform in space, there exists a partition of the physical body into at most three constant limit phases, and the phase separation process has a hysteresis-like character.

Mathematics Subject Classification (2000). Primary 80A22; Secondary 35K50; 35B40; 73N40.

Keywords. Phase-field system, asymptotic phase separation, energy, entropy, hysteresis.

Introduction

In this contribution, we present a model illustrating how hysteresis may arise in systems with multiple equilibria and multiple time scales. The *rate-independence*, as one of the prominent features of hysteresis, results from the competition between a slow observer, who tries to move the system away from an equilibrium, and the underlying physical mechanism, which is so fast that it attains a new equilibrium position "almost instantaneously". If more than one equilibrium exists under a given external loading, such a process may become irreversible and manifests a true hysteresis behavior on the observer's time scale.

A simple example involving a first order singularly perturbed ODE with a small parameter at the time derivative, a non-monotone right-hand side, and a time dependent forcing, has been investigated in detail in [2]. It was shown there that if the coefficient in front of the time derivative tends to zero (that is, the internal time becomes infinitely fast), then the limit input-output mapping defines a hysteresis operator in the space of regulated functions.

The philosophy of the present paper is different. We do not prescribe different time scales in our problem. Instead, we show that in a phase-field system, the time scale doubling may occur as a result of the interplay between the heat diffusion (which causes a tendency to a slow temperature stabilization), and the non-convexity of the free energy with respect to the order parameter, which makes phase mixtures unstable and thus produces sharp phase interfaces in the limit with fast motions on the microscale.

We give here merely a commented survey of mathematical results obtained in [4, 5]. Note only that the proofs are not straightforward. They require a thorough analysis of the local fast dynamics of the order parameter, classified according to all possible (a priori unknown) limit temperature scenarios, and combined with the global energy conservation and entropy growth principles. An interested reader may find the detailed argument as well as a more complete reference list in [5].

2. Mathematical model

Let us consider the following phase-field system for the state variables θ (the *absolute temperature*) and χ (an *order parameter* characterizing the physical phase).

$$\partial_t(c_V\theta + \Lambda(\chi)) - \kappa\Delta\theta = 0\,, \tag{2.1}$$

$$\mu\partial_t\chi + \lambda(\chi) + \partial I_{[0,1]}(\chi) \ni \frac{L}{\theta_c}\,\theta\,, \tag{2.2}$$

$$\Lambda(\chi) \;=\; L\chi + \alpha\chi(1-\chi)\,, \quad \lambda(\chi) = \Lambda'(\chi) \;=\; L + \alpha - 2\alpha\chi \tag{2.3}$$

for $(x,t) \in \Omega \times (0,\infty)$, where $\Omega \subset \mathbb{R}^N$ with $N \geq 1$, representing the physical body, is an open bounded domain with Lipschitzian boundary, $I_{[0,1]}$ is the indicator function of the interval $[0,1]$ (that is, $I_{[0,1]}(\chi) = 0$ if $\chi \in [0,1]$, $I_{[0,1]}(\chi) = +\infty$ if $\chi \notin [0,1]$), and $\partial I_{[0,1]}$ is its (maximal monotone) subdifferential. The specific heat c_V, heat conductivity κ, latent heat L, phase relaxation coefficient μ, mean phase transition temperature θ_c, and undercooling/overheating parameter $\alpha < L$ are assumed to be positive constants. The exact shape (2.3) of Λ has only been chosen for simplicity. The argument remains valid if Λ is any *strictly concave increasing function* in $C^2([0,1])$.

The thermodynamic derivation of the system (2.1)–(2.3) is based on the choice of the *free energy density* F in the form

$$\begin{aligned} F(\theta,\chi) &= c_V\theta(1-\log\theta) + \Lambda(\chi) + I_{[0,1]}(\chi) - \frac{L\theta}{\theta_c}\,\chi \\ &= c_V\theta(1-\log\theta) + L\chi\left(1-\frac{\theta}{\theta_c}\right) + \alpha\chi(1-\chi) + I_{[0,1]}(\chi)\,, \end{aligned} \tag{2.4}$$

hence it is of double obstacle type with respect to χ as in [6, Sect. VII.3], with two local minima at $\chi = 0$ and $\chi = 1$ in the temperature range

$$1 - \frac{\alpha}{L} \;<\; \frac{\theta}{\theta_c} \;<\; 1 + \frac{\alpha}{L}\,. \tag{2.5}$$

Beyond this interval, only one local minimum persists, namely $\chi = 1$ for high temperatures, and $\chi = 0$ for low temperatures. The values of χ outside $[0,1]$ are

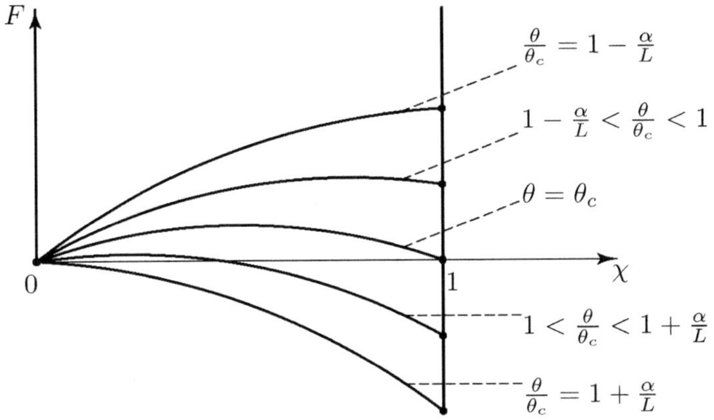

FIGURE 1. The "phase component" of the free energy at different temperatures.

not accessible due to the term $I_{[0,1]}(\chi)$ in the free energy. Figure 1 shows the shape of $F(\theta, \chi)$ for different values of θ, not accounting for the purely caloric component $c_V\theta(1 - \log\theta)$, which only produces vertical shifts in the diagram.

The order parameter χ can thus be interpreted as a characterization of the phase: the body Ω is in high temperature phase at point x and time t if $\chi(x, t) = 1$, and in low temperature phase if $\chi(x, t) = 0$, while the intermediate values of χ correspond to a mixture of both. Intuitively, the mixtures can be expected to be unstable because of the concave character of the free energy in the open interval $(0, 1)$.

Similarly to the general scheme in [1], we associate with the free energy density F given by (2.4) the densities of *internal energy* U and *entropy* S in the form

$$U(\theta, \chi) = c_V\theta + \Lambda(\chi) + I_{[0,1]}(\chi)\,, \tag{2.6}$$

$$S(\theta, \chi) = c_V \log\theta + \frac{L}{\theta_c}\,\chi\,. \tag{2.7}$$

Using the identity

$$\partial_t I_{[0,1]}(\chi(t)) \;=\; \xi(t)\,\partial_t\chi(t) \;=\; 0 \quad \text{a.e.}, \tag{2.8}$$

which holds for every absolutely continuous function χ and every measurable selection $\xi(t) \in \partial I_{[0,1]}(\chi(t))$, we may interpret Eq. (2.1) as the energy balance

$$\partial_t U + \operatorname{div}\mathbf{q} \;=\; 0 \tag{2.9}$$

with Fourier heat flux $\mathbf{q} = -\kappa\nabla\theta$. Eq. (2.2) describes, similarly as in [6, Sect. V.1], the phase relaxation dynamics

$$\mu\partial_t\chi \;\in\; -\delta_\chi F(\theta, \chi)\,, \tag{2.10}$$

where δ_χ denotes here, by an abuse of notation, alternatively the derivative with respect to χ of the components of $F(\theta, \chi)$, which are differentiable, and the subdifferential for the components which are convex in χ.

Using (2.8), we easily check that every solution of (2.1)–(2.14) with the properties stated in Theorem 3.1 below ($\theta > 0$, in particular) satisfies the entropy balance equation

$$\partial_t S + \operatorname{div} \frac{\mathbf{q}}{\theta} = \frac{\mu}{\theta}(\partial_t \chi)^2 + \frac{\kappa |\nabla \theta|^2}{\theta^2}, \tag{2.11}$$

where the entropy production term on the right-hand side is non-negative in agreement with the Second principle of thermodynamics.

We couple the system (2.1)–(2.3) with the homogeneous Neumann boundary condition

$$\frac{\partial \theta}{\partial n} = 0 \qquad \text{on } \partial\Omega \times (0, \infty) \tag{2.12}$$

and initial conditions

$$\theta(x, 0) = \theta^0(x), \quad \chi(x, 0) = \chi^0(x), \tag{2.13}$$

with given functions

$$\left.\begin{array}{ll} \theta^0 \in W^{1,2}(\Omega) \cap L^\infty(\Omega), & \inf\operatorname{ess}\{\theta^0(x)\,;\, x \in \Omega\} > 0, \\ \chi^0 \in L^\infty(\Omega), & \chi^0(x) \in [0,1] \text{ a.e.} \end{array}\right\} \tag{2.14}$$

The main results of [4, 5] are Theorems 3.1, 3.2 below on the convergence of (θ, χ) towards an equilibrium $(\theta_\infty, \chi_\infty)$ as $t \to \infty$. The limit temperature θ_∞ is constant in space. If it is contained in the interval given by (2.5), the range of χ_∞ consists of three points at most: the two pure phases $\chi = 0$ and $\chi = 1$, and possibly one unstable intermediate phase. If it is lower or higher, then only one phase (the cold one or the hot one, respectively) persists.

3. Main results

The exact values of the physical constants in (2.1)–(2.3) are not relevant for the qualitative behavior of the solution. We therefore assume that

$$c_V = \kappa = L = \mu = \theta_c = 1, \quad 0 < \alpha < 1, \tag{3.1}$$

$$|\Omega| = 1, \quad \text{where } |\cdot| \text{ denotes the Lebesgue measure in } \mathbb{R}^N. \tag{3.2}$$

In other words, system (2.1)–(2.3) now reads

$$\partial_t(\theta + \Lambda(\chi)) - \Delta\theta = 0, \tag{3.3}$$

$$\partial_t \chi + \lambda(\chi) + \partial I_{[0,1]}(\chi) \ni \theta, \tag{3.4}$$

$$\Lambda(\chi) = \chi + \alpha\chi(1 - \chi), \quad \lambda(\chi) = 1 + \alpha - 2\alpha\chi. \tag{3.5}$$

This is a special case of the system

$$\partial_t(\theta + F_1[w]) - \Delta\theta = 0, \tag{3.6}$$

$$\mu(\theta)\, \partial_t w + f_1[w] + \theta f_2[w] = 0 \tag{3.7}$$

with hysteresis operators f_1, f_2, F_1, which was investigated in [4]. Indeed, (3.3)–(3.5) can be transformed into (3.6)–(3.7) by introducing an auxiliary function

$$w(x,t) \;=\; \int_0^t (\theta(x,\tau) - \lambda(\chi(x,\tau)))\,d\tau\,. \tag{3.8}$$

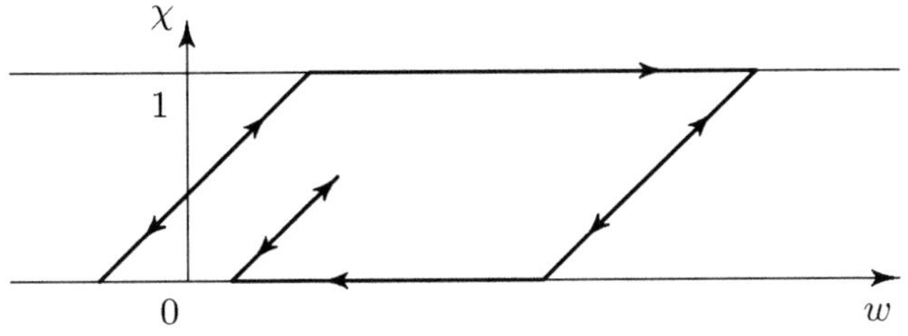

FIGURE 2. A diagram of the stop $\chi = \mathfrak{s}[\chi^0, w]$ with $\chi^0 = 0$.

Then the inclusion

$$\partial_t\chi + \partial I_{[0,1]}(\chi) \;\ni\; \partial_t w\,, \quad \chi(0) = \chi^0 \tag{3.9}$$

defines the so-called *stop operator* $\chi = \mathfrak{s}[\chi^0, w]$ (see Figure 2), and we obtain (3.6)–(3.7) with $F_1[w] = \Lambda(\mathfrak{s}[\chi^0, w])$, $f_1[w] = \lambda(\mathfrak{s}[\chi^0, w])$, $f_2[w] = -1$, $\mu(\theta) = 1$, see [3] for details.

The main result in [4] was Theorem 2.1, which reads (with respect to the present notation) as follows.

Theorem 3.1. *Let $\Omega \subset \mathbb{R}^N$ be an open bounded domain with Lipschitzian boundary, and let θ^0, χ^0 satisfying* (2.14) *be given. Then system* (3.3)–(3.5), (2.12)–(2.13) *admits a unique global solution $(\theta, \chi) \in [L^\infty(\Omega\times(0,\infty))]^2$ such that $\partial_t\chi \in L^\infty(\Omega\times(0,\infty))$, $\partial_t\theta, \Delta\theta \in L^2(\Omega\times(0,\infty))$, $\theta(x,t) > 0, \chi(x,t) \in [0,1]$ a. e. in $\Omega\times(0,\infty)$, and the function*

$$V(t) \;:=\; \int_\Omega \left(|\nabla\theta|^2 + |\partial_t\chi|^2\right)(x,t)\,dx \tag{3.10}$$

has the property

$$\int_0^\infty V(t)\,dt < \infty\,, \qquad \operatorname*{Var}_{[0,\infty)} V^2 < \infty\,, \qquad \lim_{t\to\infty}\,\sup_{s>t}\operatorname{ess} V(s) = 0\,. \tag{3.11}$$

Note that $V(t)$ may be discontinuous. This makes the proof of the convergence of $V(t)$ towards zero technically complicated, and special dissipation properties of hysteresis operators have to be taken into account.

The total energy $\mathcal{E}(t)$ and entropy $\mathcal{S}(t)$ are given by the respective formulas

$$\mathcal{E}(t) = \int_\Omega (\theta + \Lambda(\chi))\,(x,t)\,dx \tag{3.12}$$

$$\mathcal{S}(t) = \int_\Omega (\log\theta + \chi)\,(x,t)\,dx\,. \tag{3.13}$$

Integrating (3.3) and (2.11) over Ω and using the boundary condition (2.12) we obtain

$$\mathcal{E}(t) = \mathcal{E}(0) \quad =: \ \mathcal{E}_0 \tag{3.14}$$

$$\dot{\mathcal{S}}(t) = \int_\Omega \left(\frac{(\partial_t \chi)^2}{\theta} + \frac{|\nabla\theta|^2}{\theta^2} \right)(x,t)\, dx \quad \text{a. e.} \tag{3.15}$$

We further have $\log\theta \le \theta$, $\chi \le \Lambda(\chi)$ for all admissible arguments, hence $\mathcal{S}(t)$ is a bounded non-decreasing function, and there exists $\mathcal{S}_\infty \le \mathcal{E}_0$ such that

$$\mathcal{S}(t) \nearrow \mathcal{S}_\infty \quad \text{as} \ \ t \to \infty\,. \tag{3.16}$$

Since θ is uniformly bounded from below and from above by positive constants, we also see that the quantity $V(t)$ in Theorem 3.1 is actually nothing but the entropy production. In other words, Theorem 3.1 says that it tends to zero sufficiently fast as t tends to infinity.

The above balance principles for $\mathcal{E}(t)$ and $\mathcal{S}(t)$ play a crucial role in the proof of the following main result of [5].

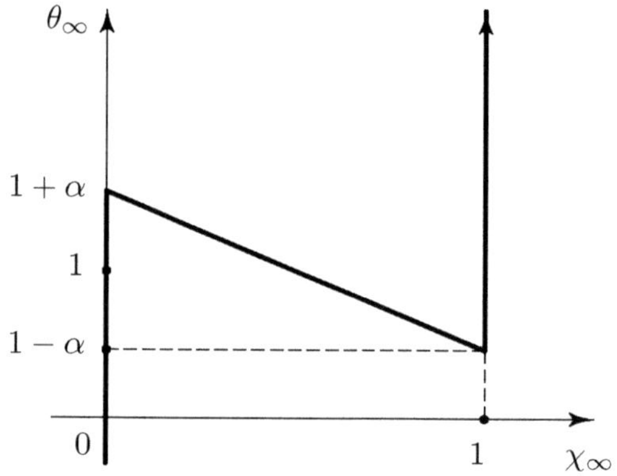

FIGURE 3. A diagram of the equilibrium set (3.19).

Theorem 3.2. *In addition to the hypotheses of Theorem* 3.1, *assume that* $\Omega \subset \mathbb{R}^N$ *is of class* C^2 *if* $N \ge 4$. *Then there exist a constant* $\theta_\infty > 0$ *and a function* $\chi_\infty \in L^\infty(\Omega)$ *such that the solution to* (3.3)–(3.5), (2.12)–(2.13) *has the properties*

$$\lim_{t\to\infty} \sup_{x\in\Omega} |\theta(x,t) - \theta_\infty| \ = \ 0\,, \tag{3.17}$$

$$\lim_{t\to\infty} \chi(x,t) \ = \ \chi_\infty(x) \qquad a.\ e.\,, \tag{3.18}$$

$$\theta_\infty \ \in \ \lambda(\chi_\infty(x)) + \partial I_{[0,1]}(\chi_\infty(x)) \qquad a.\ e. \tag{3.19}$$

The convergence (3.17) does not indeed follow directly from Theorem 3.1 and some deeper results from the theory of analytical semigroups have to be used. What we can conclude however from Theorem 3.1 is that the temperature variations become infinitely slow for large times. This is in contrast with the order parameter equation (3.4) which has unit characteristic speed independently of how slow the right-hand side is. This produces the difference in the time scales for θ and for χ. If we equip the observer with the slower and slower "temperature clock", the phase transition dynamics described by Eq. (3.4) will eventually look like "infinitely fast hysteresis jumps" in the $\chi - \theta$ plane between the two stable vertical branches as on Fig. 4.

Condition (3.19) means that $\chi_\infty(x) = 1$ a. e. if $\theta_\infty > 1 + \alpha$, and $\chi_\infty(x) = 0$ a. e. if $\theta_\infty < 1 - \alpha$. For $1 - \alpha \le \theta_\infty \le 1 + \alpha$ (see Figure 3), the domain Ω is decomposed into $\Omega = A_\infty \cup B_\infty \cup C_\infty$, with $\chi_\infty(x) = 0$ for $x \in A_\infty$, $\chi_\infty(x) = 1$ for $x \in C_\infty$, and $\chi_\infty(x) = (1+\alpha-\theta_\infty)/(2\alpha)$ for $x \in B_\infty$. In fact, the intermediate value of χ between 0 and 1 is unstable with respect to small perturbations and is actually unlikely to persist for $t \to \infty$ except for some particular cases, like for instance:

- $\theta^0(x) = \bar{\theta} = $ const., $\bar{\theta} \in \lambda(\chi^0(x)) + \partial I_{[0,1]}(\chi^0(x))$ a. e. Then the solution remains constant in time $\theta(x,t) = \theta^0(x)$, $\chi(x,t) = \chi^0(x)$ independently of the distribution of $\chi^0(x)$ (time-independent solutions).
- $\theta^0(x) = \bar{\theta} = $ const., $\chi^0(x) = \bar{\chi} = $ const. If $\mathcal{E}_0$ is such that

$$1 + \alpha \ < \ \mathcal{E}_0 \ < \ 2 - \alpha\,, \tag{3.20}$$

then the function $\Gamma(\chi) = \mathcal{E}_0 - \lambda(\chi) - \Lambda(\chi)$ has only one null point χ_∞ in $(0,1)$, $\Gamma(0) > 0$, $\Gamma(1) < 0$, hence the solution $\chi^*(t)$ of the differential equation

$$\dot{\chi}^*(t) = \Gamma(\chi^*(t))\,, \quad \chi^*(0) = \bar{\chi}\,, \tag{3.21}$$

stays in $(0,1)$ for all $t > 0$, $\lim_{t\to\infty} \chi^*(t) = \chi_\infty$, and $\chi(x,t) = \chi^*(t)$, $\theta(x,t) = \mathcal{E}_0 - \Lambda(\chi^*(t))$ is a solution to (3.3)–(3.5), (2.12)–(2.13) which entirely lies in the unstable region (space-independent solutions).

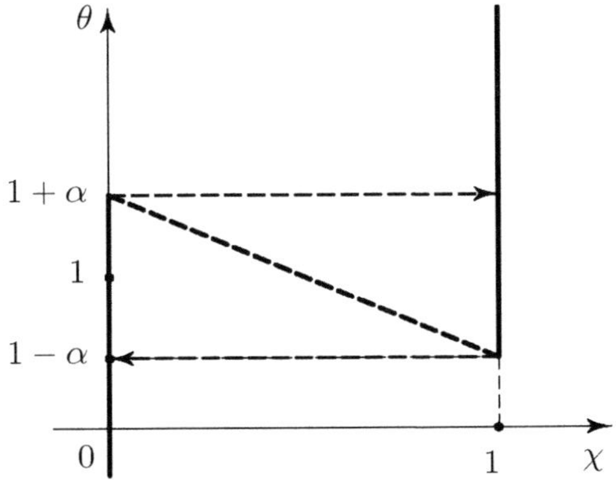

FIGURE 4. A typical limit hysteresis behavior.

The two above examples seem to be quite isolated, and we make the following conjecture.

Conjecture. For a generic set of initial data (for example a set of second Baire's category like in [2, Remark 5.3]), the Lebesgue measure of the set B_∞ is zero.

Acknowledgment

The author wishes to acknowledge helpful and stimulating discussions with Pierluigi Colli, Philippe Laurençot, Hana Petzeltová, Elisabetta Rocca, Jürgen Sprekels, and Songmu Zheng.

References

[1] M. Brokate and J. Sprekels, *Hysteresis and Phase Transitions.* Appl. Math. Sci., Vol. 121, Springer-Verlag, New York, 1996.

[2] P. Krejčí, *Hysteresis in singularly perturbed problems.* In: Singular Perturbations and Hysteresis, M. Mortell, R. O'Malley, A. Pokrovskii, V. Sobolev, eds., SIAM, Philadelphia, 2005, 73–100.

[3] P. Krejčí and J. Sprekels, *A hysteresis approach to phase-field models.* Nonlin. Anal. **39** (2000), 569–586.

[4] P. Krejčí, J. Sprekels, and S. Zheng, *Asymptotic behaviour of a phase-field system with hysteresis.* J. Differential Eq. **175** (2001), 88–107.

[5] P. Krejčí and S. Zheng, *Pointwise asymptotic convergence of solutions for a phase separation model*, Discrete Contin. Dyn. Syst. **16** (2006), 1–18.

[6] A. Visintin: *Models of Phase Transitions.* Birkhäuser, Boston, 1996.

Pavel Krejčí
Weierstrass Institute for Applied Analysis and Stochastics
Mohrenstr. 39
D-10117 Berlin, Germany

and

Mathematical Institute
Czech Academy of Sciences
Žitná 25
CZ-11567 Praha 1, Czech Republic
e-mail: `krejci@wias-berlin.de`, `krejci@math.cas.cz`

International Series of Numerical Mathematics, Vol. 154, 291–305

Obstacle Problems for Monotone Operators with Measure Data

Chiara Leone

Abstract. The aim of this work is to study obstacle problems associated to monotone operators when the forcing term is a bounded Radon measure. Existence, uniqueness, stability results, and properties of the solutions are investigated.

1. Introduction

Given a bounded open set Ω of $\mathbb{R}^N$, $N \geq 2$, and a strictly monotone elliptic operator A in divergence form, we study obstacle problems for the operator A in Ω with homogeneous Dirichlet boundary conditions on $\partial\Omega$, when the forcing term μ is a measure on Ω and the obstacle ψ is an arbitrary function on Ω.

Obstacle problems when the forcing term F belongs to the dual of the energy space, determined by the growth assumptions on the operator, have been studied as part of the theory of Variational Inequalities (for which we refer to well-known books such as [20] and [35]). In this frame the problem consists in finding a function $u \in W_0^{1,p}(\Omega)$ (the energy space) which is above the obstacle ψ, such that

$$\begin{cases} \langle A(u), v-u\rangle \geq \langle F, v-u\rangle, \\ \forall v \in W_0^{1,p}(\Omega),\ v \geq \psi. \end{cases} \tag{1.1}$$

For such problems a wide abstract theory has been developed, and we know that, if there exists at least a function $z \in W_0^{1,p}(\Omega)$ above the obstacle, then there exists one and only one solution.

Among all classical results, we recall that the solution of (1.1) is also characterized as the smallest function $u \in W_0^{1,p}(\Omega)$ such that

$$\begin{cases} A(u) - F \geq 0 & \text{in } D'(\Omega), \\ u \geq \psi & \text{in } \Omega, \end{cases} \tag{1.2}$$

or, equivalently, u is the smallest function $u \in W_0^{1,p}(\Omega)$, greater than or equal to ψ, such that

$$\begin{cases} A(u) - F = \lambda & \text{in } \Omega, \\ u = 0 & \text{on } \partial\Omega, \end{cases} \tag{1.3}$$

for some nonnegative measure λ of $W^{-1,p'}(\Omega)$.

Trying to extend this theory to problems where the forcing term is a measure various difficulties arise.

We recall that, already in the case of equations, the term $\langle \mu, u \rangle$ has not always a meaning when μ is a measure and $u \in W_0^{1,q}(\Omega)$, $q \leq N$. On the other hand, simple examples show that the solution of equations with measure data cannot be expected to belong to the energy space $W_0^{1,p}(\Omega)$. When $N \geq 2$, indeed, the solution of the Laplace equation in a ball with the Dirac mass at the center as datum, does not belong to $H_0^1(\Omega)$, but only to $W_0^{1,q}(\Omega)$, with $q < \frac{N}{N-1}$. Hence the classical formulation (1.1) of the variational inequality fails.

Also the use of the characterization (1.2) to define the obstacle problem with measure data is not possible because another problem arises: an example by J. Serrin (see [32] and, for more details, [31]) shows that, when A is a particular linear elliptic operator with discontinuous coefficients, the homogeneous equation

$$\begin{cases} Au = 0 & \text{in } D'(\Omega), \\ u = 0 & \text{on } \partial\Omega, \end{cases}$$

has a nontrivial solution v which does not belong to $H_0^1(\Omega)$. The function $u \equiv 0$ is obviously the unique solution in $H_0^1(\Omega)$.
So (1.2) in general does not determine the solution of the obstacle problem: indeed, with such A and v, if we choose $\psi \equiv -\infty$, and if u were the minimal supersolution, then we would have $u \leq u + tv$ a.e. in Ω, for any $t \in \mathbb{R}$, which implies $v \equiv 0$, i.e., a contradiction.

To overcome these difficulties, when the forcing term is a measure, we introduce a formulation for obstacle problems, based on a suitable notion of solution to equations with measure data, which ensures existence and uniqueness results.

We briefly recall that, in the linear case, i.e., $A(u) = -\mathrm{div}(A(x)\nabla u)$, where A is a uniformly elliptic matrix with $L^\infty(\Omega)$ coefficients, the problem of finding a solution of

$$\begin{cases} A(u) = \mu & \text{in } \Omega \\ u = 0 & \text{on } \partial\Omega, \end{cases} \tag{1.4}$$

when μ is a bounded Radon measure on Ω was investigated by G. Stampacchia, who introduced and studied in [33] a notion of solution using duality and regularity arguments. This allowed him to prove both existence and uniqueness results.

Stampacchia's framework, which heavily relies on a duality argument, cannot be extended to the general nonlinear case, except in the case $p = 2$, where Stampacchia's ideas continue to work if the operator is strongly monotone and Lipschitz continuous. In this setting, indeed, we can use the notion of solution, namely the reachable solution, considered by F. Murat in [28] to solve uniquely the Dirichlet problem (1.4), when μ is a bounded Radon measure.

Finally, in the general nonlinear monotone case, when μ is a bounded Radon measure vanishing on all sets of p-capacity zero (the capacity defined starting from $W_0^{1,p}(\Omega)$: see Section 2), we may use the notion of entropy solution introduced in [2] and [6], which ensures us that (1.4) has a unique entropy solution.

Using these types of solutions we give a definition for unilateral problems with measure data quite similar to the characterization given by (1.3) in the classical setting.

We say that a function u solves the Obstacle Problem when the forcing term μ is a bounded Radon measure and the obstacle ψ is an arbitrary function on Ω, if u is the smallest function with the following properties: $u \geq \psi$ in Ω and u is the unique solution (in an appropriate sense), of a problem of the form

$$\begin{cases} A(u) = \mu + \lambda & \text{in } \Omega, \\ u = 0 & \text{on } \partial\Omega, \end{cases} \tag{1.5}$$

for some bounded Radon measure $\lambda \geq 0$. More precisely, if A is linear, we assume that u is the solution of (1.5) in the sense of Stampacchia; if $p = 2$ and A is strongly monotone and Lipschitz continuous, we assume that u is the reachable solution of (1.5); finally, when A is a general nonlinear strictly monotone operator, and μ, λ are bounded Radon measures vanishing on sets of p-capacity zero, then u is the entropy solution.

The measure λ which corresponds to the solution u of the unilateral problem relative to A, μ, and ψ, is called the obstacle reaction associated with u.

The outline of the paper is as follows.

After giving the definitions and some preliminary results we consider the questions of existence and uniqueness of solutions to Obstacle Problems with measure data. We study also some properties of the obstacle reactions associated with the solutions of the Obstacle Problems, obtaining the so called Lewy-Stampacchia inequality, and we investigate the interaction between obstacles and data, and in particular the complementarity conditions. Finally we will give some stability results, dealing with the convergence properties of the solutions to the Obstacle Problem, under simultaneous perturbation of the operator, the forcing term, and the obstacle.

The results we propose here are contained in some already published papers (see [22], [21], and [23], as well as [13] and [12]). We decided to give the results with no proof, referring to those papers for the proofs.

For a different approach to Obstacle Problems with measure data see also [5], [3], [4], [29], [30], and more recently [8] and [9].

2. Assumptions and preliminary results

Given p and p' two real numbers, with $p > 1$ and $\frac{1}{p} + \frac{1}{p'} = 1$, let $a : \Omega \times \mathbb{R}^N \to \mathbb{R}^N$ be a Carathéodory function such that for almost every x in Ω and for every ξ, η in $\mathbb{R}^N$ $(\xi \neq \eta)$:

$$|a(x,\xi)| \leq C_0[k_0(x) + |\xi|^{p-1}], \tag{2.1}$$

$$a(x,\xi)\xi \geq C_1|\xi|^p - k_1(x), \tag{2.2}$$

$$(a(x,\xi) - a(x,\eta))(\xi - \eta) > 0, \tag{2.3}$$

$$a(x,0) = 0, \tag{2.4}$$

where C_0 and C_1 are two positive real constants, k_0 is a nonnegative function in $L^{p'}(\Omega)$ and k_1 is a nonnegative function in $L^1(\Omega)$. Thanks to hypotheses (2.1)–(2.4) the operator $A : u \mapsto -\mathrm{div}(a(x, \nabla u))$ maps $W_0^{1,p}(\Omega)$ into its dual $W^{-1,p'}(\Omega)$ and for every $F \in W^{-1,p'}(\Omega)$ there exists a unique function $u \in W_0^{1,p}(\Omega)$ such that

$$\begin{cases} A(u) = F & \text{in } \Omega \\ u = 0 & \text{on } \partial\Omega, \end{cases} \tag{2.5}$$

in the weak sense (see, e.g., [25]).

Let $M_b(\Omega)$ the space of Radon measures μ on Ω whose total variation $|\mu|$ is bounded on Ω. As usual we identify $M_b(\Omega)$ with the dual of the Banach space $C_0(\Omega)$ of continuous functions that are zero on the boundary, so that the duality is $\langle \mu, u \rangle = \int_\Omega u\, d\mu$, for every u in $C_0(\Omega)$ and the norm is $\|\mu\|_{M_b(\Omega)} = |\mu|(\Omega)$.

In order to include in our analysis also the case of thin obstacles, it is convenient to introduce the notion of p-capacity. Given a compact set $K \subseteq \Omega$, its p-capacity with respect to Ω is given by

$$C_p(K) = \inf \left\{ \int_\Omega |\nabla z|^2\, dx : z \in C_0^\infty(\Omega), z \geq \chi_K \right\},$$

where χ_K is the characteristic function of K. This definition can be extended to any open subset A of Ω in the following way:

$$C_p(A) = \sup \{ C_p(K) : K \text{ compact}, K \subseteq A \}.$$

Finally, it is possible to define the p-capacity of any set $B \subseteq \Omega$ as:

$$C_p(B) = \inf \{ C_p(A) : A \text{ open}, B \subseteq A \}.$$

A property holds C_p-quasi everywhere (abbreviated as C_p-q.e.) when it holds up to sets of p-capacity zero.

A function $v : \Omega \to \overline{\mathbb{R}}$ is C_p-quasi continuous (resp. C_p-quasi upper semicontinuous) if, for every $\varepsilon > 0$ there exists a set E such that $C_p(E) < \varepsilon$ and $v_{|\Omega \setminus E}$ is continuous (resp. upper semicontinuous) in $\Omega \setminus E$. We recall also that if u and v are C_p-quasi continuous and $u \leq v$ a.e. in Ω then also $u \leq v$ C_p-q.e. in Ω. A function $u \in W_0^{1,p}(\Omega)$ always has a C_p-quasi continuous representative, which is uniquely defined (and finite) up to a set of p-capacity zero. In the sequel we shall always identify u with its C_p-quasi continuous representative, so that the pointwise values of u are defined C_p-quasi everywhere.

In the sequel $M^p_{b,0}(\Omega)$ will be the special subspace of $M_b(\Omega)$ of all measures which vanish on all sets of p-capacity zero. Moreover, we denote the positive cones of $M_b(\Omega)$ and $M^p_{b,0}(\Omega)$ by $M^+_b(\Omega)$ and $M^{p,+}_{b,0}(\Omega)$, respectively.

It is well known that, if μ belongs to $W^{-1,p'}(\Omega)\cap M_b(\Omega)$, then μ is in $M^p_{b,0}(\Omega)$, every u in $W^{1,p}_0(\Omega)\cap L^\infty(\Omega)$ is summable with respect to μ and

$$\langle \mu, u\rangle = \int_\Omega u\, d\mu,$$

where $\langle\cdot,\cdot\rangle$ denotes the duality pairing between $W^{-1,p'}(\Omega)$ and $W^{1,p}_0(\Omega)$, while in the right hand side u denotes the C_p-quasi continuous representative and, consequently, the pointwise values of u are defined μ-almost everywhere.
A self-contained presentation of all these notions can be found, for instance, in [19].

Let us recall here the following result, that is the analogous of the Lebesgue decomposition theorem and can be proved in the same way (see Lemma 2.1 in [18]).

Proposition 2.1. *For every measure $\mu \in M_b(\Omega)$ there exists a unique pair of measures (μ_a, μ_s), with $\mu_a \in M^p_{b,0}(\Omega)$ and μ_s concentrated on a set of p-capacity zero, and $\mu = \mu_a + \mu_s$. If μ is nonnegative, so are μ_a and μ_s.*

Let us fix a function $\psi : \Omega \to \overline{\mathbb{R}}$, and the corresponding convex set

$$K_\psi(\Omega) := \{z \;\; C_p - \text{quasi continuous in } \Omega : \;\; z \geq \psi \; C_p - \text{q.e. in } \Omega\}.$$

For every $j > 0$ we define the truncation function $T_j : \mathbb{R} \mapsto \mathbb{R}$ by

$$T_j(t) = \begin{cases} t & \text{if } |t| \leq j \\ j\,\mathrm{sign}(t) & \text{if } |t| > j. \end{cases}$$

Let us consider the space $T^{1,p}_0(\Omega)$ of all functions $u : \Omega \mapsto \overline{\mathbb{R}}$ which are almost everywhere finite and such that $T_j(u) \in W^{1,p}_0(\Omega)$ for every $j > 0$. It is easy to see that every function $u \in T^{1,p}_0(\Omega)$ has a C_p-quasi continuous representative with values in $\overline{\mathbb{R}}$, that will always be identified with the function u. Moreover, for every $u \in T^{1,p}_0(\Omega)$ there exists a measurable function $\Phi : \Omega \mapsto \mathbb{R}^N$ such that $\nabla T_j(u) = \Phi\chi_{\{|u|\leq j\}}$ a.e. in Ω (see Lemma 2.1 in [2]). This function Φ, which is unique up to almost everywhere equivalence, will be denoted by ∇u. Note that ∇u coincides with the distributional gradient of u whenever $u \in T^{1,p}_0(\Omega)\cap L^1_{\mathrm{loc}}(\Omega)$ and $\nabla u \in L^1_{\mathrm{loc}}(\Omega, \mathbb{R}^N)$.

In order to study the elliptic problem

$$\begin{cases} A(u) = \mu & \text{in } \Omega, \\ u = 0 & \text{on } \partial\Omega \end{cases} \tag{2.6}$$

when μ is a bounded Radon measure, we cannot use the variational formulation , since, in general, the term $\langle \mu, v\rangle$ has not always meaning when μ is a measure and $v \in W^{1,r}_0(\Omega)$, with $r \leq N$. On the other hand the solution cannot be expected to

belong to the energy space $W_0^{1,p}(\Omega)$, as simple examples show. Thus, it is necessary to change the functional setting in order to prove existence result.
In the linear case, i.e., if $p = 2$ and $a(x, \nabla u) = A(x)\nabla u$, where A is a $N \times N$ matrix such that

$$|A(x)\xi| \le C_0|\xi| \quad \text{and } A(x)\xi\xi \ge C_1|\xi|^2, \quad \forall \xi \in \mathbb{R}^N, \text{ for a.e. } x \in \Omega, \tag{2.7}$$

with C_0 and C_1 two positive real constants, problem (2.6) was studied by G. Stampacchia in [33].

Definition 2.2. Assume that Ω satisfies the following regularity condition: there exists a constant $\gamma > 0$ such that

$$\text{meas}(B_r(x) \setminus \Omega) \ge \gamma\, \text{meas}(B_r(x)),$$

for every $x \in \partial\Omega$ and for every $r > 0$, where $B_r(x)$ denotes the open ball with centre x and radius r. A function $u_\mu \in L^1(\Omega)$ is a solution in the sense of Stampacchia (also called solution by duality) of the equation (2.6) if

$$\int_\Omega u_\mu g\, dx = \int_\Omega u_g^*\, d\mu, \qquad \forall g \in L^\infty(\Omega), \tag{2.8}$$

where u_g^* is the solution of

$$\begin{cases} A^* u_g^* = g & \text{in } \Omega \\ u_g^* \in H_0^1(\Omega), \end{cases}$$

and A^* is the adjoint of A.

Existence and uniqueness of u_μ are proved in [33]. It is easy to prove that, for every $j > 0$, $T_j(u_\mu) \in H_0^1(\Omega)$ and

$$\int_\Omega |\nabla T_j(u_\mu)|^2\, dx \le j|\mu|(\Omega). \tag{2.9}$$

In the case $p = 2$, if A is strongly monotone and Lipschitz continuous, Stampacchia's ideas continue to work. In this special case, indeed, F. Murat (in [28]) proved the existence and uniqueness of a solution (called reachable solution) of (2.6).
Assuming that Ω is a regular set (in the sense of Definition 2.2), we consider $a : \Omega \times \mathbb{R}^N \to \mathbb{R}^N$ a Carathéodory function such that for every ξ, η in $\mathbb{R}^N$ ($\xi \ne \eta$), and for almost every $x \in \Omega$,

$$|a(x,\xi) - a(x,\eta)| \le C_0|\xi - \eta|, \tag{2.10}$$

$$(a(x,\xi) - a(x,\eta))(\xi - \eta) \ge C_1|\xi - \eta|^2, \tag{2.11}$$

$$a(x, 0) = 0, \tag{2.12}$$

where C_0 and C_1 are two positive real constants.

Definition 2.3. We say that a function $u \in T_0^{1,p}(\Omega)$ is a reachable solution of the problem (2.6) if there exist two sequences μ_n and u_n such that

(i) $\mu_n \in M_b(\Omega) \cap W^{-1,p'}(\Omega)$ and μ_n converges to μ in the $*$-weak topology of $M_b(\Omega)$;
(ii) $u_n \in W_0^{1,p}(\Omega)$ and u_n solves the Dirichlet problem (2.5) relative to the datum μ_n;
(iii) u_n converges to u a.e. in Ω.

Remark 2.4. Note that this definition can be given also for a general monotone operator A (see [15], where the existence of a reachable solution has been proved in this generality).

Actually, in the general nonlinear case, when $\mu \in M^p_{b,0}(\Omega)$, other types of solutions to (2.6) have been proposed. The notion of entropy solution, of SOLA, and of renormalized solution were introduced respectively in [2], [11], and [26]. These three frameworks, which are actually equivalent, are successful since they allow to prove existence and uniqueness results.

Definition 2.5. Let $\mu \in M^p_{b,0}(\Omega)$; we say that a function $u \in T_0^{1,p}(\Omega)$ is an entropy solution of (2.6) if

$$\int_\Omega a(x, \nabla u)\nabla T_k(u - \varphi)\, dx = \int_\Omega T_k(u - \varphi)\, d\mu, \tag{2.13}$$

for every $\varphi \in W_0^{1,p}(\Omega) \cap L^\infty(\Omega)$ and for every $k > 0$.

Remark 2.6. Let us point out that these three types of solutions coincide when each make sense. This means, for example, that if A is linear the Stampacchia's solution is also the reachable solution, as well as, if $p = 2$ and A is strongly monotone and Lipschitz continuous, and μ is a measure in $M^p_{b,0}(\Omega)$, then the reachable solution is also the entropy solution.

Thanks to these notions of solutions we will arrive to a suitable definition of obstacle problems with measure data (see [13], [22], and [21]).
Roughly speaking, we choose the minimum element among those functions v above the obstacle, such that $A(v)-\mu$ is not only nonnegative in the sense of distributions but it is actually a nonnegative bounded Radon measure, and the equation is solved in the sense of Stampacchia or in the sense of the reachable solutions, if A is linear or A is strongly monotone and Lipschitz continuous, respectively, and in the sense of entropy in the general nonlinear case when μ is a bounded Radon measure vanishing on all sets of p-capacity zero.

Actually, to define the solution for obstacle problems with measure data, we will distinguish two cases. The first one regards the case where μ is a general bounded Radon measure. This case can be treated only considering $p = 2$ and the operator A Lipschitz continuous and strongly monotone. The second case deals with bounded Radon measures μ vanishing on all sets of p-capacity zero, so that we are able to handle with a general nonlinear monotone operator A. In this case, we are not able to drop the assumption that μ is absolutely continuous with respect to the p-capacity, since in our approach we need a suitable notion of solution to equations with measure data which ensures existence and uniqueness results. As a matter of fact, for a general monotone operator A the question of existence of a solution to (2.6) has been faced in [16] where the authors extend the notion of renormalized solution (see [26]) to the case of a general measure $\mu \in M_b(\Omega)$. In that paper they proved the existence of such a solution and introduced other equivalent definitions, which show that all the renormalized solutions are constructed by approximating, in an appropriate way, the measure μ (with respect to the $*$-weak convergence of measures), so that they are reachable solutions. On the other hand the question of uniqueness of a reachable solution, in this case, is still an open problem.

Definition 2.7. Let $p = 2$ and let a satisfy (2.10)–(2.12); we say that u is the solution of the Obstacle Problem with datum $\mu \in M_b(\Omega)$ and the obstacle ψ (denoted by $OP(A, \mu, \psi)$) if

1. there exists a measure $\lambda \in M_b^+(\Omega)$ such that u is the reachable solution of (2.6) relative to $\mu + \lambda$, and $u \in K_\psi(\Omega)$.
2. for any $\nu \in M_b^+(\Omega)$ such that the reachable solution v of (2.6) relative to $\mu + \nu$ is in $K_\psi(\Omega)$, we have $u \leq v$ a.e. in Ω.

Remark 2.8. This definition includes, of course, also the case where A is linear. In this case, as we already observed in Remark 2.6, the notion of reachable solution and of Stampacchia's solution are equivalent.

Definition 2.9. Let a satisfy (2.1)–(2.4); we say that u is the solution of the Obstacle Problem with datum $\mu \in M_{b,0}^p(\Omega)$ and the obstacle ψ (denoted by $OP_0(A, \mu, \psi)$) if

1. there exists a measure $\lambda \in M_{b,0}^{p,+}(\Omega)$ such that u is the entropy solution of (2.6) relative to $\mu + \lambda$, and $u \in K_\psi(\Omega)$.
2. for any $\nu \in M_{b,0}^{p,+}(\Omega)$ such that the entropy solution v of (2.6) relative to $\mu + \nu$ is in $K_\psi(\Omega)$, we have $u \leq v$ a.e. in Ω.

By definition, it is clear that, if such a solution (of $OP(A,\mu,\psi)$ or $OP_0(A,\mu,\psi)$) exists, it is unique.

3. Main results

First of all we want to treat the question of existence (the uniqueness being implicit in the definition itself) of solutions to $OP(A, \mu, \psi)$ and $OP_0(A, \mu, \psi)$.

We have the following theorems (see [22] and [21] for the proofs).

Theorem 3.1. *Let $p = 2$ and let a satisfy* (2.10)–(2.12). *If $\psi : \Omega \to \overline{\mathbb{R}}$ is such that*

$$\psi \leq u_\rho \ C_2 - \text{q.e. in } \Omega, \tag{3.1}$$

where $\rho \in M_b(\Omega)$ *and* u_ρ *is the reachable solution of*

$$\begin{cases} A(u) = \rho & \text{in } \Omega, \\ u = 0 & \text{on } \partial\Omega; \end{cases}$$

then, for every $\mu \in M_b(\Omega)$, *there exists a unique solution of* $OP(A, \mu, \psi)$ *in the sense of Definition* 2.7. *Moreover the corresponding obstacle reaction* λ *satisfies*

$$\|\lambda\|_{M_b(\Omega)} \leq \|(\mu - \rho)^-\|_{M_b(\Omega)}.$$

Theorem 3.2. *Let* a *satisfy* (2.1)–(2.4) *and let* $\psi : \Omega \to \overline{\mathbb{R}}$ *be such that*

$$\psi \leq u_\rho \; C_p - \text{q.e. in } \Omega, \tag{3.2}$$

where $\rho \in W^{-1,p'}(\Omega) \cap M_b(\Omega)$ *and* u_ρ *is the variational solution of*

$$\begin{cases} A(u) = \rho & \text{in } \Omega, \\ u = 0 & \text{on } \partial\Omega; \end{cases}$$

then, for every $\mu \in M^p_{b,0}(\Omega)$, *there exists a unique solution of* $OP_0(A, \mu, \psi)$ *in the sense of Definition* 2.9. *Moreover the corresponding obstacle reaction* λ *satisfies*

$$\|\lambda\|_{M_b(\Omega)} \leq \|(\mu - \rho)^-\|_{M_b(\Omega)}.$$

The proofs of existence are based on an approximation technique. The obstacle reactions associated with the solutions for regular data are shown to satisfy an estimate on the masses, which allows to pass to the limit and obtain the solution in the general case.

In our setting we are able to prove (see [22] and [21]) the Lewy-Stampacchia inequality: first proved in [24] it has been extended by various authors to different cases. It has become a powerful tool for proving existence and regularity results.

Theorem 3.3. *Let* $p = 2$ *and let* a *satisfy* (2.10)–(2.12). *Let* $\mu \in M_b(\Omega)$ *and* u *be the solution of* $OP(A, \mu, u_\rho)$ (u_ρ *defined in* (3.1)). *If we denote by* λ *the obstacle reaction associated with* u, *it holds*

$$\lambda \leq (\mu - \rho)^-.$$

Theorem 3.4. *Let* a *satisfy* (2.1)–(2.4), $\mu \in M^p_{b,0}(\Omega)$ *and* u *be the solution of* $OP_0(A, \mu, u_\rho)$ (u_ρ *defined in* (3.2)). *If we denote by* λ *the obstacle reaction associated with* u, *it holds*

$$\lambda \leq (\mu - \rho)^-.$$

Furthermore, as in the classical framework, we study the interaction between obstacles and data, and in particular the complementarity conditions. More precisely, the following theorem (see [22] for the proof) shows that the solution u of $OP_0(A, \mu, \psi)$, when $\mu \in M^p_{b,0}(\Omega)$, is the only entropy solution of (2.6) relative to $\mu + \lambda$ such that $u = \psi$ λ-almost everywhere in Ω, and $u \geq \psi$ C_p-q.e. in Ω. We also find a more technical characterization of the solution of the Obstacle Problem, which turns out to be similar to (1.1).

Theorem 3.5. *Let a satisfy (2.1)–(2.4), $\mu \in M^p_{b,0}(\Omega)$ and ψ satisfy (3.2); then the following statements are equivalent:*

(1) *u is the solution of $OP_0(A,\mu,\psi)$ and λ is the associated obstacle reaction;*

(2) *$u \geq \psi$ C_p-q.e. in Ω, $\lambda \in M^{p,+}_{b,0}(\Omega)$, u is the entropy solution of (2.6) relative to $\mu+\lambda$, and*

$$\begin{cases} \int_\Omega T_k(u-\varphi)\,d\lambda \leq \int_\Omega T_k(v-\varphi)\,d\lambda, \\ \forall \varphi \in W^{1,p}_0(\Omega)\cap L^\infty(\Omega),\ \forall v \in \mathcal{T}^{1,p}_0(\Omega),\ v \geq \psi\ C_p\text{-q.e. in } \Omega; \end{cases} \tag{3.3}$$

(3) *$u \geq \psi$ C_p-q.e. in Ω, $\lambda \in M^{p,+}_{b,0}(\Omega)$, u is the entropy solution of (2.6) relative to $\mu+\lambda$, and*

$$u = \psi\ \lambda\text{-a.e. in } \Omega.$$

Remark 3.6. Observe that if ψ is C_p-q.e. upper bounded, we can consider in (3.3) $\varphi \in W^{1,p}_0(\Omega)\cap L^\infty(\Omega)$, $\varphi \geq \psi$ C_p-q.e. in Ω and $v=\varphi$, so that, taking into account that u is the entropy solution of (2.6) relative to $\mu+\lambda$, for every $k>0$, u satisfies

$$\int_\Omega a(x,\nabla u)\nabla T_k(u-\varphi)\,dx \leq \int_\Omega T_k(u-\varphi)\,d\mu, \tag{3.4}$$

which turns out to be quite similar to the usual variational formulation (1.1). The previous formula was already obtained in [4] when the datum μ is a function in $L^1(\Omega)$.

This fact is non longer true when we pass to consider general data in $M_b(\Omega)$ (when $p=2$ and A is strongly monotone and Lipschitz continuous). The following example, which is a variant of an example studied by L. Orsina and A. Prignet, shows that the solution of the obstacle problem with right-hand side measure does not touch the obstacle, though it is not the solution of the equation.

Example. Let $N \geq 2$, Ω be the ball $B_1(0)$, $A=-\Delta$, $\mu=-\delta_0$ (the Dirac mass at the origin), and $\psi=-h$ a negative constant. Let u be the solution of $OP(-\Delta,-\delta_0,-h)$, then $-\Delta u = -\delta_0+\lambda$ (and the equation is satisfied in the sense of Stampacchia or, equivalently, in the setting of reachable solutions). We want to show that $\lambda=\delta_0$.

Taking $\nu=\delta_0$ in condition 2 of Definition 2.7, we obtain $u \leq 0$ C_2-q.e. in Ω. As $u \geq -h$, we have $u=T_h(u)$ and hence $u \in H^1_0(\Omega)$ by (2.9). This implies that the measure $-\delta_0+\lambda$ belongs to $M_b(\Omega)\cap H^{-1}(\Omega)$, which is contained in $M^2_{b,0}(\Omega)$. In other words $\lambda=\delta_0+\lambda_0$, with $\lambda_0 \in M^2_{b,0}(\Omega)$. Since λ is nonnegative and $\delta_0 \perp \lambda_0$, the measure λ_0 is nonnegative. Then $u=u_{\lambda_0}\geq 0$, and finally $u=0$. Thus the solution can be far above the obstacle, but the obstacle reaction is nonzero, and is exactly δ_0.

The previous example can be explained by the following theorem that shows that, when the obstacle is controlled from above and from below in an appropriate way, it is possible to "isolate" the effect of the singular negative part of the data. Namely, the reaction λ will be written as $\lambda=\lambda_0+\mu^-_s$, where λ_0 belongs to $M^{2,+}_{b,0}(\Omega)$. Moreover the "regular part" λ_0 is concentrated on the coincidence set

$\{x \in \Omega : u(x) = \psi(x)\}$ whenever ψ is C_2-quasi upper semicontinuous, and a complementarity condition holds.

Theorem 3.7. *Let $p = 2$, and let a satisfy (2.10)–(2.12). Let $\mu \in M_b(\Omega)$ and let $\psi : \Omega \to \overline{\mathbb{R}}$ be C_2-quasi upper semicontinuous satisfying*

$$u_{-\rho-\tau} \leq \psi \leq u_\rho,$$

where $\rho \in M^2_{b,0}(\Omega)$ and $\tau \in M_b(\Omega)$, with $\tau \perp \mu_s^-$ (here $u_{-\rho-\tau}$ and u_ρ are the reachable solutions of (2.6) relative to $-\rho-\tau$ and ρ, respectively). Let u and u_0 be the solutions of $OP(A,\mu,\psi)$ and $OP(A,\mu_a+\mu_s^+,\psi)$, respectively, and λ and λ_0 be the corresponding obstacle reactions. Then $u = u_0$ a.e. in Ω, $\lambda_0 \in M^{2,+}_{b,0}(\Omega)$, $\lambda = \lambda_0 + \mu_s^-$, and $u = \psi$ λ_0-a.e. in Ω.

In [12] this theorem was proved in the linear case, investigating the behavior of the potential of two mutually singular measures near their singular points. Actually in [21] we extend this result to our (nonlinear) context giving alternative proofs.

Finally we deal with the behavior of the Obstacle Problem in the sense of Definition 2.9 under perturbation of the operator, of the forcing term, and of the obstacle.

The study of the properties of the solutions to the Obstacle Problem when the operators vary is based on a notion of convergence for strictly monotone operators, called G-convergence. Actually, to deal with this type of convergence we have to restrict our class of functions a satisfying (2.1)–(2.4).

In particular, given two constants $c_0, c_1 > 0$ and two constants α and β, with $0 \leq \alpha \leq 1 \wedge (p-1)$ and $p \vee 2 \leq \beta < +\infty$, we consider the family $L(c_0, c_1, \alpha, \beta)$ of Carathéodory functions $a(x,\xi) : \Omega \times \mathbb{R}^N \to \mathbb{R}^N$ such that:

$$|a(x,\xi) - a(x,\eta)| \leq c_0(1+|\xi|+|\eta|)^{p-1-\alpha}|\xi-\eta|^\alpha, \tag{3.5}$$

$$(a(x,\xi) - a(x,\eta))(\xi-\eta) \geq c_1(1+|\xi|+|\eta|)^{p-\beta}|\xi-\eta|^\beta, \tag{3.6}$$

$$a(x,0) = 0, \tag{3.7}$$

for almost every $x \in \Omega$, for every $\xi, \eta \in \mathbb{R}^N$.

Definition 3.8. We say that a sequence of functions $a_h(x,\xi) \in L(c_0,c_1,\alpha,\beta)$ G-converges to a function $a(x,\xi)$ satisfying the same hypotheses (possibly with different constants $\tilde{c}_0, \tilde{c}_1, \tilde{\alpha}, \tilde{\beta}$) if for any $F \in W^{-1,p'}(\Omega)$, the solution u_h of

$$\begin{cases} A_h(u_h) = F & \text{in } \Omega \\ u = 0 & \text{on } \partial\Omega \end{cases}$$

satisfies

$$u_h \rightharpoonup u \text{ weakly in } W^{1,p}_0(\Omega)$$

and

$$a_h(x, \nabla u_h) \rightharpoonup a(x, \nabla u) \text{ weakly in } L^{p'}(\Omega)^N,$$

where u is the unique solution of (2.5).

The following theorem justifies the definition of G-convergence.

Theorem 3.9. *Any sequence $a_h(x,\xi)$ of functions belonging to $L(c_0, c_1, \alpha, \beta)$ admits a subsequence which G-converges to a function $a(x,\xi) \in L(\tilde{c}_0, \tilde{c}_1, \frac{\alpha}{\beta-\alpha}, \beta)$, where $\tilde{c}_0, \tilde{c}_1$ depend only on $N, p, \alpha, \beta, c_0, c_1$*

This compactness theorem was obtained by L. Tartar (see [34] and Theorem 1.1 of [17]) in the case of nonlinear monotone operators defined from $H_0^1(\Omega)$ into $H^{-1}(\Omega)$, when $p = 2$ and the functions $a_h \in L(c_0, c_1, 1, 2)$, and then extended in the version of Theorem 3.9 in [10] (see Theorem 4.1).

Concerning the perturbation of the obstacles we consider a notion of convergence for sequences of convex sets introduced by U. Mosco in [27].

Definition 3.10. Let K_h be a sequence of subsets of a Banach space X. The strong lower limit

$$s - \liminf_{h\to+\infty} K_h$$

of the sequence K_h is the set of all $v \in X$ such that there exists a sequence $v_h \in K_h$, for h large, converging to v strongly in X.

The weak upper limit

$$w - \limsup_{h\to+\infty} K_h$$

of the sequence K_h is the set of all $v \in X$ such that there exists a sequence v_k converging to v weakly in X and a sequence of integers h_k converging to $+\infty$, such that $v_k \in K_{h_k}$.

The sequence K_h converges to the set K in the sense of Mosco, shortly $K_h \overset{M}{\to} K$, if

$$s - \liminf_{h\to+\infty} K_h = w - \limsup_{h\to+\infty} K_h = K.$$

Mosco proved that this type of convergence is the right one for the stability of variational inequalities with respect to obstacles. This is the main result of his theory.

Theorem 3.11. *Let $K_{\psi_h} := K_{\psi_h}(\Omega) \cap W_0^{1,p}(\Omega)$ and $K_\psi := K_\psi(\Omega) \cap W_0^{1,p}(\Omega)$ be nonempty. Then*

$$K_{\psi_h} \overset{M}{\to} K_\psi$$

if and only if, for any $F \in W^{-1,p'}(\Omega)$,

$$u_h \to u \text{ strongly in } W_0^{1,p}(\Omega),$$

where u_h and u are the solutions of $VI(A, F, \psi_h)$ (the variational inequality relative to A, F, and ψ_h) and $VI(A, F, \psi)$ (the variational inequality relative to A, F, and ψ), respectively.

Several stability results can be proved as corollaries of this theorem by Mosco. In particular, the strong convergence

$$\psi_h \to \psi \text{ strongly in } W^{1,p}(\Omega)$$

easily implies the convergence of K_{ψ_h} to K_ψ in the sense of Mosco, but the weak convergence

$$\psi_h \rightharpoonup \psi \text{ weakly in } W^{1,r}(\Omega), \ r > p,$$

also implies the same result (see [7], [1]).

Now, we consider a sequence a_h of functions in $L(c_0, c_1, \alpha, \beta)$, a sequence of measures $\rho_h \in M_b(\Omega) \cap W^{-1,p'}(\Omega)$, and the variational solution u_{ρ_h} of

$$\begin{cases} A_h(u_{\rho_h}) = \rho_h & \text{in } \Omega \\ u_{\rho_h} \in W_0^{1,p}(\Omega). \end{cases}$$

We assume that

$$\sup_h \|\rho_h\|_{M_b(\Omega)} < +\infty \tag{3.8}$$

and that the function ψ_h satisfies:

$$\psi_h \leq u_{\rho_h} \ C_p\text{-q.e. in } \Omega. \tag{3.9}$$

Moreover we suppose that

$$\psi \leq 0 \ C_p\text{-q.e. in } \Omega. \tag{3.10}$$

Theorem 3.12. *Let a_h be a sequence in $L(c_0, c_1, \alpha, \beta)$, which G-converges to a function a, and let A_h and A be the operators associated to a_h and a, respectively. Let us assume* (3.8), (3.9), *and* (3.10), *with K_{ψ_h} converging to K_ψ in the sense of Mosco. Finally, we consider μ_h, $\mu \in M_{b,0}^p(\Omega)$ such that*

$$\mu_h(B) \to \mu(B), \quad \textit{for every Borel set } B \subseteq \Omega.$$

Then the solutions u_h and u of the obstacle problems $OP_0(A_h, \mu_h, \psi_h)$ and $OP_0(A, \mu, \psi)$, respectively, satisfy

$$T_j(u_h) \rightharpoonup T_j(u) \textit{ weakly in } W_0^{1,p}(\Omega), \textit{ for every } j > 0,$$

$$a_h(x, \nabla u_h) \rightharpoonup a(x, \nabla u) \textit{ weakly in } L^q(\Omega)^N, \textit{ for every } q < \tfrac{N}{N-1},$$

$$\int_\Omega a_h(x, \nabla u_h) \nabla T_j(u_h)\, dx \to \int_\Omega a(x, \nabla u) \nabla T_j(u)\, dx, \textit{ for every } j > 0.$$

Remark 3.13. By formal modifications we can prove Theorem 3.12 (see [23]) replacing (3.10) with (3.2) and

$$\psi \leq M \ C_p\text{–q.e. in } \Omega,$$

where M is a positive constant.

Theorem 3.12 is the analogous of Theorem 3.1 of [14] proved in the classical setting of variational inequalities.

References

[1] H. Attouch, C. Picard, Problèmes variationnels et théorie du potential non linéaire, *Ann. Faculté Sci. Toulouse* **1** (1979), 89–136.

[2] L. Benilan, L. Boccardo, T. Galloüet, R. Gariepy, M. Pierre, J.L. Vazquez, An L^1-theory of existence and uniqueness of nonlinear elliptic equations, *Ann. Scuola Norm. Sup. Pisa* **22** (1995), 240–273.

[3] L. Boccardo, G.R. Cirmi, Nonsmooth unilateral problems. *Nonsmooth optimization: methods and applications (Erice, 1991)*, F. Giannessi ed., 1–10, Gordon and Breach, Amsterdam 1992.

[4] L. Boccardo, G.R. Cirmi, Existence and uniqueness of solution of unilateral problems with L^1-data, *J. Convex Anal.* **6** (1999), 195–206

[5] L. Boccardo, T. Galloüet, Problèmes unilatéraux avec données dans L^1, *C. R. Acad. Sci. Paris Sér. I Math.* **311** (1990), 617–619.

[6] L. Boccardo, T. Galloüet, L. Orsina, Existence and uniqueness of entropy solutions for nonlinear elliptic equations with measure data, *Ann. Inst. H. Poincaré* **13** (1996), 539–551.

[7] L. Boccardo, F. Murat, Nouveaux résultats de convergence dans des problèmes unilatéraux, *Nonlinear partial differential equations and their applications* (Collège de France séminaire), Pitman Res. Notes Math. **60** (1962), 64–85.

[8] H. Brezis, S. Serfaty, A variational formulation for the two-sided obstacle problem with measure data, *Commun. Contemp. Math.* **4** (2002), 357–374.

[9] H. Brezis, A. Ponce, Reduced measures for obstacle problems, *Adv. Differential Equations*, **10** (2005), 1201–1234.

[10] V. Chiadò Piat, A. Defranceschi, Asymptotic behaviour of quasi linear problems with Neumann boundary conditions on perforated domains, *Appl. Anal.* **36** (1990), 65–87.

[11] A. Dall'Aglio, Approximated solutions of equations with L^1 data, *Ann. Mat. Pura Appl.* **170** (1996), 207–240.

[12] P. Dall'Aglio, G. Dal Maso, Some properties of the solutions of obstacle problems with measure data, *Ricerche Mat.* **48** (1999), 99–116.

[13] P. Dall'Aglio, C. Leone, Obstacle problems with measure data and linear operators, *Potential Anal.* **17** (2002), 45–64.

[14] G. Dal Maso, A. Defranceschi, Convergence of unilateral problems for monotone operators, *J. Analyse Math.* **53** (1989), 269–289.

[15] G. Dal Maso, A. Malusa, Some properties of reachable solutions of nonlinear elliptic equations with measure data, *Ann. Scuola Norm. Sup. Pisa Cl. Sci.* **25** (1997), 375–396.

[16] G. Dal Maso, F. Murat, L. Orsina, A. Prignet, Renormalized solutions of elliptic equations with general measure data, *Ann. Scuola Norm. Sup. Pisa Cl. Sci.* **28** (1999), 741–808.

[17] T. Del Vecchio, On the homogenization of a class of pseudomonotone operators in divergence form, *Boll. Un. Mat. Ital.* **7** (1991), 369–388.

[18] M. Fukushima, K. Sato, S. Taniguchi, On the closable part of pre-Dirichlet forms and the fine support of the underlying measures, *Osaka J. Math.* **28** (1991), 517–535.

[19] J. Heinonen, T. Kilpeläinen, O. Martio, *Nonlinear potential theory of degenerate elliptic equations.* Clarendon Press, Oxford, 1993.

[20] D. Kinderlehrer, G. Stampacchia, *An introduction to Variational Inequalities and their applications.* Academic, New York, 1980.

[21] C. Leone, On a class of nonlinear obstacle problems with measure data, *Comm. Partial Differ. Equations* **25** (2000), 2259–2286.

[22] C. Leone, Existence and uniqueness of solutions for nonlinear obstacle problems with measure data, *Nonlinear Anal.* **43** (2001), 199–215.

[23] C. Leone, Stability results for obstacle problems with measure data, *Ann. Inst. H. Poincaré*, **22** (2005), 679–704.

[24] H. Lewy, G. Stampacchia, On the smoothness of superharmonics which solve a minimum problem, *J. Analyse Math.* **23** (1970), 227–236.

[25] J.L. Lions, *Quelques méthodes de résolution des problèmes aux limites non linéaires.* Dunod, Gauthier-Villars, Paris, 1969.

[26] P.L. Lions, F. Murat, Solutions renormalisées d'équations elliptiques non linéaires, manuscript.

[27] U. Mosco, Convergence of convex sets and of solutions of variational inequalities, *Adv. in Math.* **3** (1969), 510–585.

[28] F. Murat, Équations elliptiques non linéaires avec second membre L^1 ou mesure, *Actes du 26ème Congrés National d'Analyse Numérique*, Les Karellis, France (1994), A12-A24.

[29] P. Oppezzi, A.M. Rossi, Existence of solutions for unilateral problems with multivalued operators, *J. Convex Anal.* **2** (1995), 241–261.

[30] P. Oppezzi, A.M. Rossi, Esistenza di soluzioni per problemi unilaterali con dato misura o L^1, *Ricerche Mat.* **45**, 491–513.

[31] A. Prignet, Remarks on the existence and uniqueness of solutions of elliptic problems with right-hand side measures, *Rend. Mat.* **15** (1995), 321–337.

[32] J. Serrin, Pathological solutions of elliptic differential equations, *Ann. Scuola Norm. Sup. Pisa* **18** (1964), 385–387.

[33] G. Stampacchia, Le problème de Dirichlet pour les équations elliptiques du second ordre à coefficients discontinus, *Ann. Inst. Fourier Grenoble* **15** (1965), 189–258.

[34] L. Tartar, Cours Peccot au Collège de France, Paris 1977.

[35] G.M. Troianiello, *Elliptic differential equations and obstacle problems.* Plenum Press, New York, 1987.

Chiara Leone
Dipartimento di Matematica "R. Caccioppoli"
Università degli Studi di Napoli "Federico II"
Via Cintia
I-80126 Napoli, Italy
e-mail: `chileone@unina.it`

International Series of Numerical Mathematics, Vol. 154, 307–316

Piecewise Constant Level Set Method for Interface Problems

Hongwei Li and Xue-Cheng Tai

Abstract. We apply the Piecewise Constant Level Set Method (PCLSM) to interface problems, especially for elliptic inverse and multiphase motion problems. PCLSM allows using one level set function to represent multiple phases, and the interfaces are represented implicitly by the discontinuity of a piecewise constant level set function. The inverse problem is solved using a variational penalization method with total variation regularization of the coefficient, while the multiphase motion problem is solved by an Additive Operator-Splitting scheme.

Mathematics Subject Classification (2000). Primary 99Z99; Secondary 00A00.

Keywords. Level set, piecewise constant, inverse problem, multiphase motion, TV regularization.

1. Introduction

Traditional level set idea was first proposed by Osher and Sethian[8]. Since then, the level set method plays a great role in dealing with interface problems. By level set method, one does not evolve the interfaces, instead one just evolves the level set function. The main advantage of level set approach is that the interfaces are implicitly represented by a level set function, and so complicated topological changes can be dealt naturally and easily. In [4, 3, 2], some variants to the traditional level set method were proposed. In this work, we are trying to show the applications of the PCLSM of [4, 3, 2] for some interface problems. For traditional level set methods, one needs to reinitialize the level set function to be a signed distance function during the iterations, and cautions must be taken with respect to the discretization of Heaviside and Dirac functions. The piecewise constant level set method does not need to care about these issues [4].

2. Piecewise Constant Level Set Method Formulations

The essential idea of the PCLSM of [4] is to use a piecewise constant level set function to identify the interfaces separating the subdomains. Assume that we need to partition the domain Ω into subdomains Ω_i, $i = 1, 2, \ldots, n$ and the number of subdomains is a priori known. In order to identify the subdomains, we try to identify a piecewise constant level set function ϕ such that

$$\phi = i, \quad \text{in } \Omega_i, \quad i = 1, 2, \ldots, n. \tag{2.1}$$

Thus, for any given partition $\{\Omega_i\}_{i=1}^n$ of the domain Ω, it corresponds to a unique PCLS function ϕ which takes the values $1, 2, \ldots, n$. Associated with such a level set function ϕ, the characteristic functions of the subdomains are given as

$$\psi_i = \frac{1}{\alpha_i} \prod_{j=1, j\neq i}^{n} (\phi - j), \quad \alpha_i = \prod_{k=1, k\neq i}^{n} (i - k). \tag{2.2}$$

If ϕ is given as in (2.1), then we have $\psi_i(x) = 1$ for $x \in \Omega_i$, and $\psi_i(x) = 0$ elsewhere. We can use the characteristic functions to extract geometrical information for the subdomains and the interfaces between the subdomains. For example,

$$\text{Length}(\partial\Omega_i) = \int_\Omega |\nabla\psi_i| dx, \quad \text{Area}(\Omega_i) = \int_\Omega \psi_i dx. \tag{2.3}$$

Define

$$K(\phi) = (\phi - 1)(\phi - 2)\cdots(\phi - n) = \prod_{i=1}^{n} (\phi - i). \tag{2.4}$$

At every point in Ω, the level set function ϕ should satisfy

$$K(\phi) = 0. \tag{2.5}$$

This level set idea has been used for image segmentation in [4, 9, 10] and inverse problems involving shape identification in [11]. Fast algorithms have been also developed for this method for image segmentation in [9, 10].

3. PCLSM for elliptic inverse problem

We try to use PCLSM for an inverse problem. Consider the partial differential equation:

$$-\nabla \cdot (q(x)\nabla u) = f, \ x \in \Omega \subset R^2, \quad u(x) = 0, \ x \in \partial\Omega. \tag{3.1}$$

Suppose we have some observations of the solution u, and we want to recover the coefficient $q(x)$ by using the observations. In [1], the standard level set method has been applied to elliptic inverse problems.

Due to the ill-posedness of the problem, output-least-square method is often used for recovering $q(x)$. Assume that $u_d \in L^2(\Omega)$ is an observation for u, and let K be the set of admissible coefficients

$$K = \{q(x) \mid q(x) \in L^\infty(\Omega) \cap TV(\Omega), \quad 0 < \underline{q}(x) \leqslant q(x) \leqslant \bar{q}(x) < \infty\}. \tag{3.2}$$

with $\underline{q}(x)$ and $\bar{q}(x)$ known a priori, and TV denotes the total variation norm. We solve the following minimization problem for the output-least-square method to find the parameter $q(x)$.

$$\min_{q\in K} F(q), \quad F = \int_\Omega \frac{1}{2}|u(q) - u_d|^2 dx + \beta R(q), \tag{3.3}$$

above $R(q) = \int_\Omega |\nabla q| dx$ is the total variation norm of q, β is the regularization parameter and $u(q)$ is the solution of (3.1) with a given q. We assume that $q(x)$ is piecewise constant and represent $q(x)$ by piecewise constant level set function

$$q(x) = \sum_{i=1}^{n} c_i \psi_i(x). \tag{3.4}$$

Incorporating it into (3.3), and letting $G(c_i, q) = F(q(c_i, \phi))$, we then need to solve

$$\min_{\substack{c_i,\phi \\ K(\phi)=0}} G(c_i, \phi), \quad G(c_i, \phi) = \int_\Omega \frac{1}{2}|u(q(c_i, \phi) - u_d|^2 + \beta R(q). \tag{3.5}$$

To deal with the constraint, we use the common penalization method

$$\min_{c_i,\phi} L, \quad L = G + W = \int_\Omega \frac{1}{2}|u - u_d|^2 + \beta R(q) + \frac{1}{2\mu}\int_\Omega K^2(\phi) dx. \tag{3.6}$$

The following algorithm is used to solve (3.6).

Algorithm 1. Choose initial values for ϕ^0 and $\bar{c}^0$. For $k = 1, 2, \ldots,$ do

1. Find $\bar{c}^{k+1} = \{c_i^{k+1}\}_{i=1}^n$, such that

$$\bar{c}^{k+1} = \arg\min_{\bar{c}} L(\bar{c}^k, \phi^k). \tag{3.7}$$

2. Find ϕ^{k+1} such that

$$\phi^{k+1} = \arg\min_{\phi} L(\bar{c}^{k+1}, \phi^k). \tag{3.8}$$

3. Check convergence, if converged, stop; else goto 1.

Above $\arg\min L(\cdot)$ denotes the minimizer of $L(\cdot)$. In order to find a minimizer of $L(\cdot)$ with respect to c_i, $i = 1, 2, \ldots, n$, we use a gradient based method with line search. Usually, we update c_i after each 5–10 outer iterations. The most difficult part of the above algorithm for our model problem is the second step – minimizing ϕ, so we concentrate on the minimizing of ϕ. At the minimizer, we should have

$$\frac{\partial L}{\partial \phi} = \frac{\partial G}{\partial \phi} + W'(\phi) = 0. \tag{3.9}$$

To solve (3.9), we can instead solve the following evolution differential equation to steady state

$$\phi_t + \frac{\partial L}{\partial \phi} = 0. \tag{3.10}$$

According to the operator splitting scheme [5, 6], we can solve (3.10) in the following way:

For $l = 1, 2, \ldots$, until convergence, do

$$\frac{\phi^{l+1/2} - \phi^l}{\tau} + \frac{\partial G}{\partial \phi}(\bar{c}^l, \phi^{l+1/2}) = 0, \tag{3.11}$$

$$\frac{\phi^{l+1} - \phi^{l+1/2}}{\tau} + W'(\phi^{l+1}) = 0. \tag{3.12}$$

where τ is a pseudo time-step. Notice that (3.12) can be written as

$$\phi - \phi^{l+1/2} + \alpha_2 K(\phi) K'(\phi) = 0. \tag{3.13}$$

where α_2 is a parameter that should be chosen properly. Note that (3.13) is a polynomial of ϕ, and we will use Newton method to solve it. The two minimization problems in Algorithm 1 are never solved exactly. A fixed number of gradient iterations is used for solving (3.7). A fixed number of iteration of (3.11)–(3.12) is used for solving (3.8).

3.1. Numerical experiments for elliptic inverse problem

We take the examples in [1] to testify the efficiency of our **Algorithm 1**. Let $\Omega = (0,1) \times (0,1)$, $f = 20\pi^2 \sin(\pi x) \sin(\pi y)$. Let $u^\star$ be the exact finite element solution for the exact q and σ be the noise level. We get the observed solution $u_d = u^\star + \sigma ||u^\star||_{L^2} / ||R_d||_{L^2} R_d$. Where R_d is a finite element function with nodal values being uniform random numbers between $[-1, 1]$ with zero mean.

The domain Ω consists of a rectangular mesh with uniform mesh size $h = 1/64$ for both x and y directions. In all the figures, the dotted lines in the background show the true level set curves and the dashed lines are the computed level set curves.

In this example, the exact coefficient $q(x)$ is given in Figure 1, i.e., $q(x) = 2$ inside the two closed curves and $q(x) = 1$ outside the curves. See Fig. 2 for the numerical results. We see that only 300 iterations are needed to recover $q(x)$ accurately.

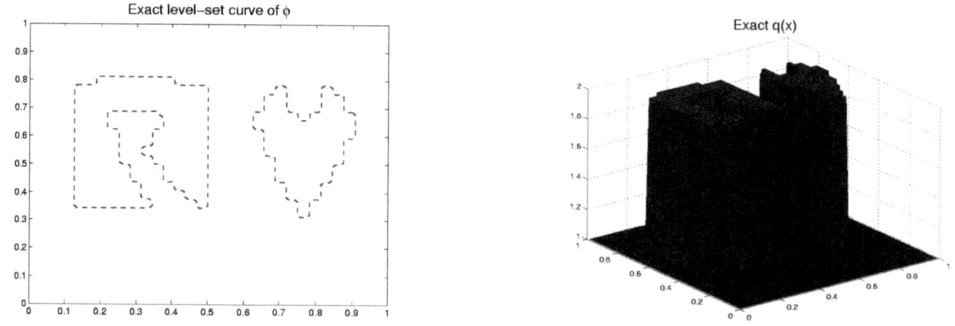

(a) exact level set curve (b) exact q(x)

FIGURE 1. The exact $q(x)$ and the location of the discontinuity

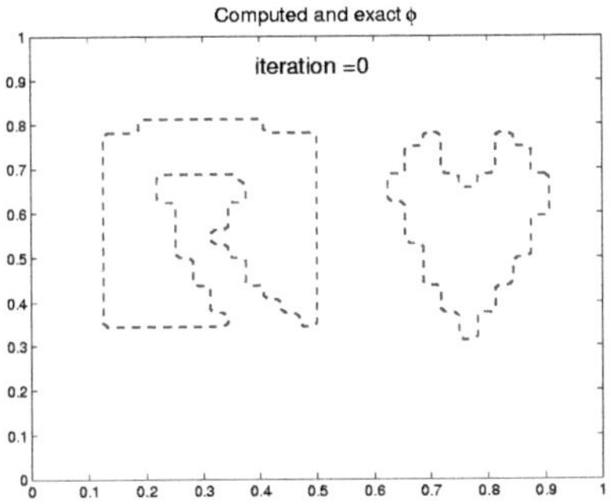

(a) Initial

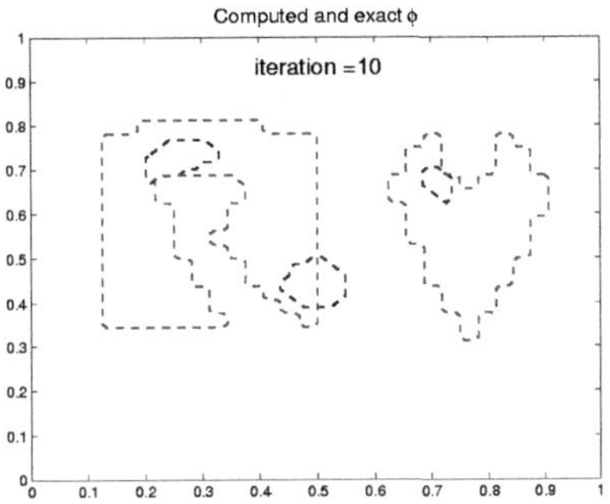

(b) 10 iterations

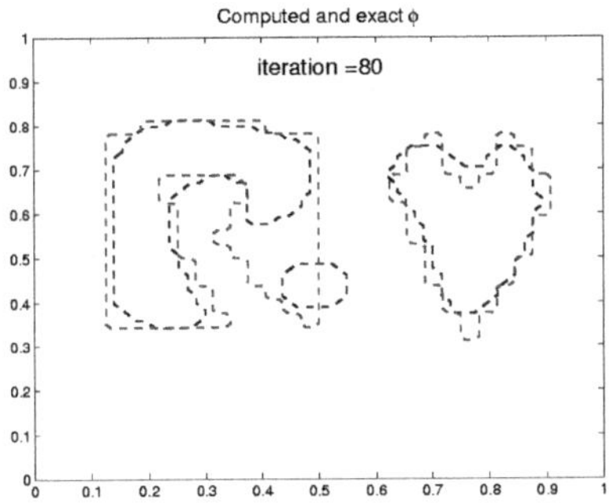

(c) 80 iterations

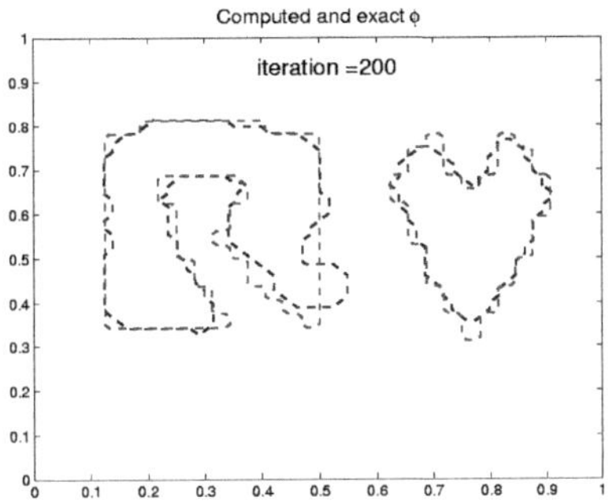

(d) 200 iterations

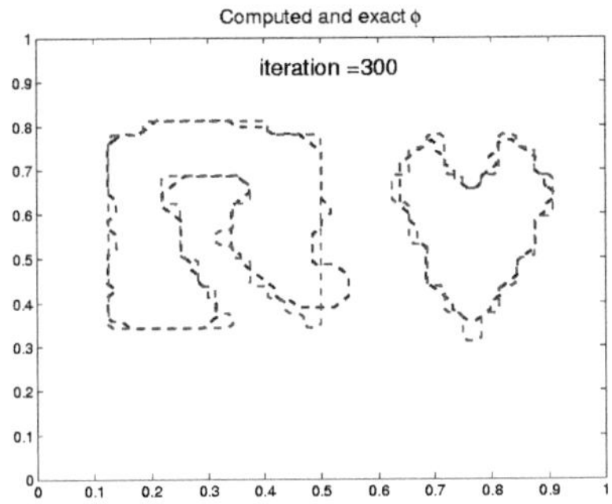

(e) 300 iterations

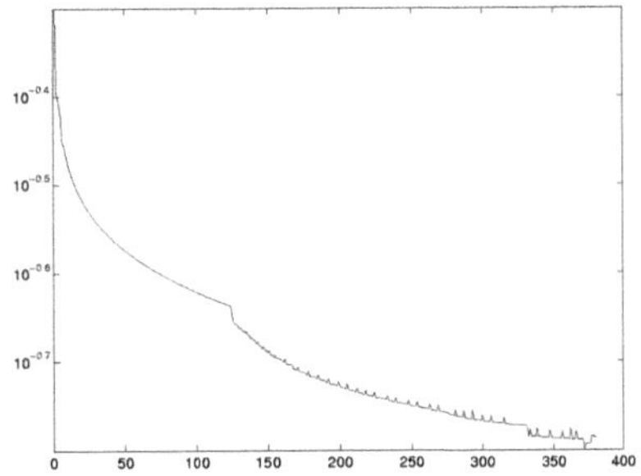

(f) The error $||q - q^k||_{L^2(\Omega)}$ v.s. iteration number

FIGURE 2. The computed solution at different iterations , with $\sigma = 1\%$. Initial $q_i = [1.2,\ 1.8]$, Initial level set function $\phi = 1.5$

4. PCLSM for multiphase motion problem

Usually, the multiphase motion problem involves curves meeting at a point with prescribed angles. Each interface Γ_{ij}, separates regions Ω_i and Ω_j and moves with a normal velocity

$$v_{ij} = f_{ij}\kappa_{ij} + (e_i - e_j). \tag{4.1}$$

where κ_{ij} is the local curvature, f_{ij} is the constant surface tension of Γ_{ij}, and e_i corresponds to the bulk energy. This model problem can be obtained by associating an energy functional E to the motion, which involves the length of each interface and the area of each subregion, i.e.,

$$\begin{aligned} E &= E_1 + E_2 \\ E_1 &= \sum_{1\leqslant i<j\leqslant n} f_{ij}\text{Length}(\Gamma_{ij}) \\ E_2 &= \sum_{1\leqslant i\leqslant n} e_i\text{Area}(\Omega_i). \end{aligned} \tag{4.2}$$

By minimizing this energy functional, the internal interfaces are driven to equilibrium. Our method is especially inspired by [7] and [13].

In the following, the PCLSM will be used to solve the motion by mean curvature problem. For simplicity, let us consider problem (4.2) with

$$e_i = 0, \quad f_{ij} = 1. \tag{4.3}$$

We want to emphasize that there is no problem to apply PCLSM for general setting for (4.2). Under condition (4.3), the problem (4.2) reduces to the model problem:

$$\min_{\Gamma_{ij}} \sum_{1\leqslant i<j\leqslant n} \text{Length}(\Gamma_{ij}). \tag{4.4}$$

There are different ways to find the curves that minimize the above energy functional. Under the condition that Γ_{ij} is the interface between Ω_i and Ω_j and $\{\Omega_i\}_{i=1}^n$ are represented by (2.1), we see that

$$\sum_{i=1}^{n}\int_\Omega |\nabla\psi_i|dx = 2\sum_{1\leqslant i<j\leqslant n} \text{Length}(\Gamma_{ij}).$$

Thus, If we use our PCLSM for (4.4), then we need to find a function ϕ that solves the following constrained minimization problem:

$$\min F, \quad F = \sum_{i=1}^{n}\int_\Omega |\nabla\psi_i|dx, \quad \text{subject to } K(\phi) = 0 \text{ and } \phi|_{\partial\Omega} = g. \tag{4.5}$$

Usually, the Neumann boundary condition is supposed. However, in this paper, we would like to try Dirichlet boundary conditions, which should produce a constrained motion. By using the same penalization technique and gradient method, we found that the equation we need to solve is

$$\phi_t + \frac{\partial F}{\partial\phi} + W'(\phi) = 0. \tag{4.6}$$

Applying the operator-splitting scheme again, we need to solve the following two equations alternatively

$$\phi_t + \frac{\partial F}{\partial \phi}(\phi) = 0, \tag{4.7}$$

$$\phi_t + W'(\phi) = 0. \tag{4.8}$$

The first equation is trying to minimize the energy functional and the second equation is trying to enforce that the minimizer is taking the values $1, 2, \ldots, n$.

we have tried to solve the first equation is the Additive version of Operator Splitting (AOS) scheme of [5, 6, 12]. Note that

$$\nabla \psi_i = \psi_i'(\phi)\nabla\phi. \tag{4.9}$$

and

$$\frac{\partial F}{\partial \phi} = -\sum_{i=1}^{n} \nabla \cdot \left(\frac{\nabla \psi_i}{|\nabla \psi_i|}\right)\psi_i' = -\sum_{i=1}^{n} \nabla \cdot \left(\operatorname{sign}(\psi_i')\frac{\nabla \phi}{|\nabla \phi|}\right)\psi_i'. \tag{4.10}$$

For two-dimensional problems, we have

$$\frac{\partial F}{\partial \phi} = -\sum_{i=1}^{n} \psi_i' \left(\operatorname{sign}(\psi_i')\frac{\phi_x}{|\nabla \phi|}\right)_x - \sum_{i=1}^{n} \psi_i' \left(\operatorname{sign}(\psi_i')\frac{\phi_y}{|\nabla \phi|}\right)_y. \tag{4.11}$$

If we apply the AOS [5, 6] and do some standard linearization, we need to solve

$$\frac{\tilde{\phi}^{k+1/4} - \phi^k}{\tau} - \sum_{i=1}^{n} \psi_i'(\phi^k)\left(\operatorname{sign}(\psi_i'(\phi^k))\frac{\tilde{\phi}_x^{k+1/4}}{|\nabla \phi^k|}\right)_x = 0, \tag{4.12}$$

$$\frac{\tilde{\phi}^{k+1/2} - \phi^k}{\tau} - \sum_{i=1}^{n} \psi_i'(\phi^k)\left(\operatorname{sign}(\psi_i'(\phi^k))\frac{\tilde{\phi}_y^{k+1/2}}{|\nabla \phi^k|}\right)_y = 0. \tag{4.13}$$

Then, set

$$\phi^{k+1/2} = \frac{1}{2}(\tilde{\phi}^{k+1/4} + \tilde{\phi}^{k+1/2}). \tag{4.14}$$

When the value of $\phi^{k+1/2}$ is obtained, we solve (3.12) to get ϕ^{k+1}. The two equations (4.12)–(4.13) can be solved efficiently on lines parallel to the x and y-axes.

4.1. Numerical experiments for multiphase motion problem

In all the experiments, we take $\Omega = (0,1) \times (0,1)$ and use Dirichlet boundary conditions. The domain Ω is divided into square elements with uniform mesh size $h = hx = hy = 1/64$.

In this example, we test our algorithm on the well-known triple-junction problem which involves three phases. The boundary and initial values are: $\phi^0|_\Omega = 1.0$, $g(0,[0,1/2]) = g([0,1],0) = 1$, $g(0,[1/2,1]) = g([0,1],1) = 3$, $g(1,[0,1]) = 2$.

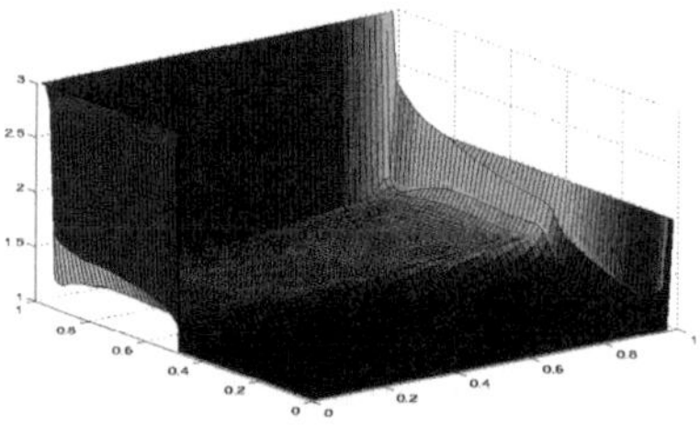

(a) 10 iterations

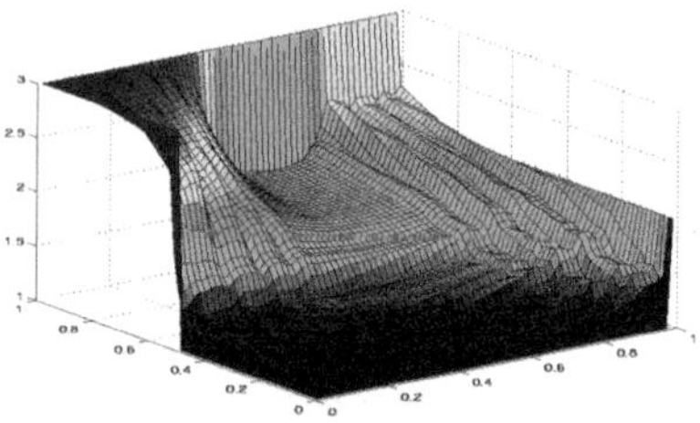

(b) 100 iterations

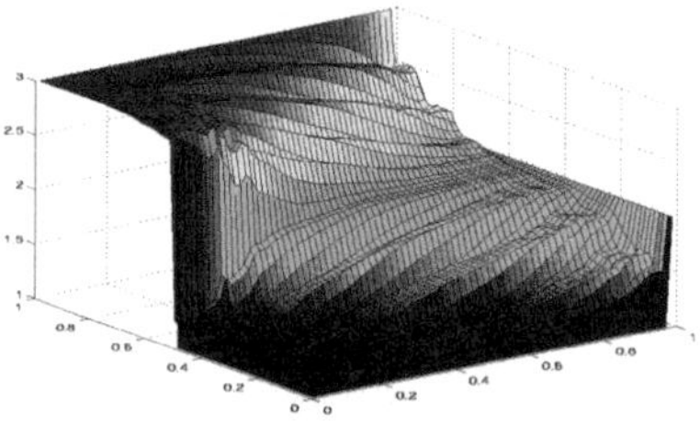

(c) 200 iterations

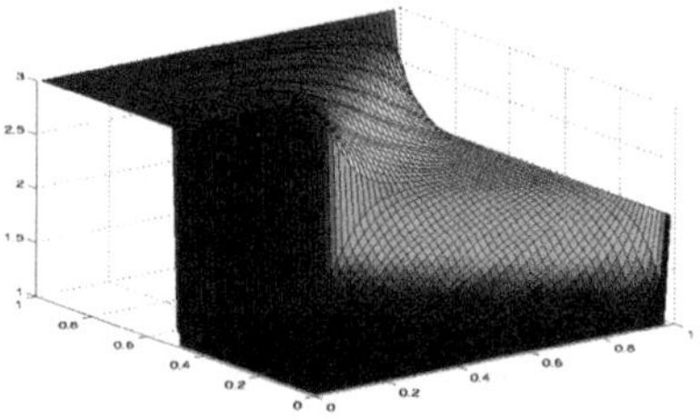

(d) 300 iterations

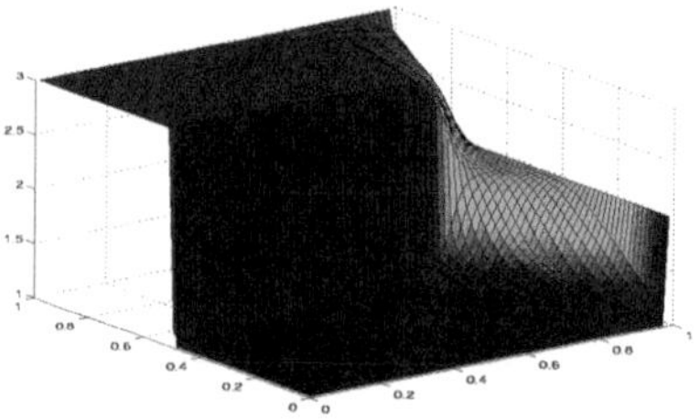

(e) 1000 iterations

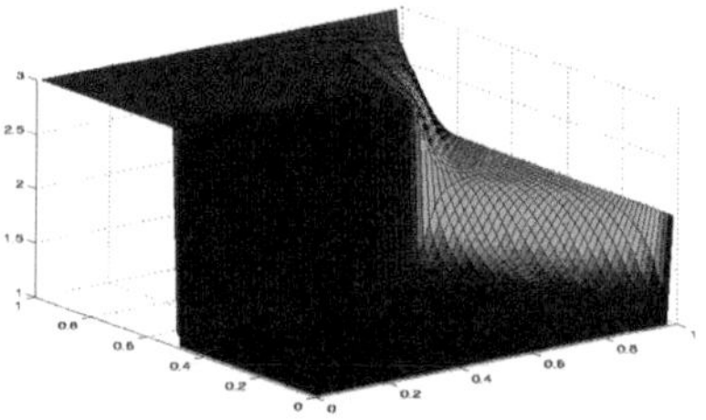

(f) 2000 iterations

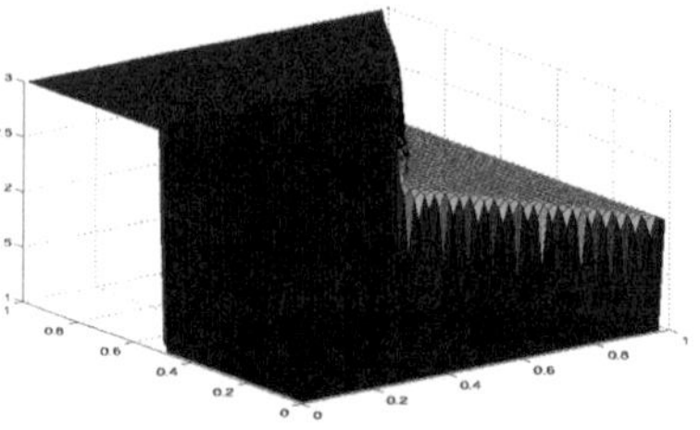

(g) 2100 iterations

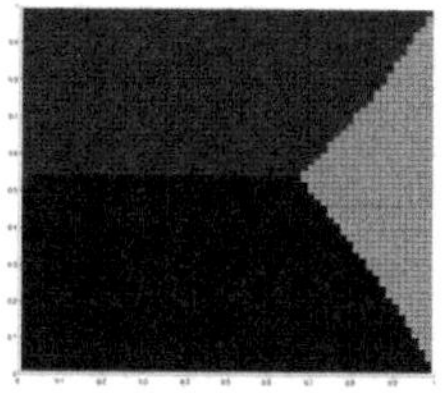

(h) The computed solution

FIGURE 3. Example 2: Computed solution for 3 phases.

For this test problem, the real triple junction point should be at $(1-1/2\sqrt{3}, 1/2)$ which is approximately $(0.7118, 0.5)$. The three interface curves should be straight lines and the three angles around the triple junction point should satisfy the classical angle condition, i.e., $(\frac{2\pi}{3}, \frac{2\pi}{3}, \frac{2\pi}{3})$. We see that algorithm needs only about 2100 iteration steps to approximate the real solution accurately.

5. Conclusion

The purpose of this work is to show that the PCLSM of [4] can be used for interface problems coming from mean curvature motion and elliptic inverse problem. The experiments given here plus the tests done in [4, 3, 2] reveals the potential to use the PCLSM for a large class of interface problems. For more details about the PCLSM, we refer to [4, 3, 2].

References

[1] T.F. Chan and X.-C. Tai. Level set and total variation regularization for elliptic inverse problems with discontinuous coefficients. *Journal of Computational Physics*, 193:40–66, 2003.

[2] M. Lysaker J. Lie and X.-C. Tai. Piecewise constant level set methods and image segmentation. In Ron Kimmel, Nir Sochen, and Joachim Weickert, editors, *Scale Space and PDE Methods in Computer Vision: 5th International Conference, Scale-Space 2005*, volume 3459, pages 573–584. Springer-Verlag, Heidelberg, April 2005.

[3] J. Lie, M. Lysaker, and X.-C. Tai. A binary level set model and some applications to image processing. *IEEE Trans. Image Process.*, to appear. Also as UCLA, Applied Math., CAM04-31, 2004.

[4] J. Lie, M. Lysaker, and X.-C. Tai. A variant of the levelset method and applications to image segmentation. *Math. Comp.*, to appear. Also as UCLA, Applied Math. CAM-03-50, 2003.

[5] T. Lu, P. Neittaanmäki, and X.-C. Tai. A parallel splitting up method and its application to Navier-Stokes equations. *Appl. Math. Lett.*, 4(2):25–29, 1991.

[6] T. Lu, P. Neittaanmäki, and X.-C. Tai. A parallel splitting-up method for partial differential equations and its applications to Navier-Stokes equations. *RAIRO Modél. Math. Anal. Numér.*, 26(6):673–708, 1992.

[7] B. Merriman, J. Bence, and S. Osher. Motion of multiple junctions: A level set approach. *J. Comput. Phys.*, 112(2):334, 1994.

[8] S. Osher and J.A. Sethian. Fronts propagating with curvature dependent speed: Algorithms based on Hamilton-Jacobi formulations. *J. Comput. Phys.*, 79:12–49, 1988.

[9] X.-C. Tai and C. Yao. Fast piecewise constant level set method with Newton updating. Technical report, UCLA, Applied Math. CAM-report, 2005.

[10] Xue-Cheng Tai, Oddvar Christiansen, Ping Lin, and Inge Skjaelaaen. A remark on the mbo scheme and some piecewise constant level set methods. Cam-report-05-24, UCLA, Applied Mathematics, 2005.

[11] Xue-Cheng Tai and Hongwei Li. Dynamic constraint and regularization for applying piecewise constant level set method to elliptic inverse problems. Technical report, UCLA, Applied Mathematics, 2005.

[12] J. Weickert, B. H. Romeny, and M. A. Viergever. Efficient and reliable schemes for nonlinear diffusion filtering. *IEEE Trans. Image Process.*, 7:398 409, 1998.

[13] H. Zhao, T. Chan, B. Merriman, and S. Osher. A variational level set approach to multiphase motion. *J. Comput. Phys.*, 127:179, 1996.

Hongwei Li
Center for Integrated Petroleum Research
University of Bergen, Norway
e-mail: hongwei.li@cipr.uib.no

Xue-Cheng Tai
Department of Mathematics
University of Bergen, Norway
e-mail: tai@mi.uib.no

International Series of Numerical Mathematics, Vol. 154, 317–326

Dynamics of a Moving Reaction Interface in a Concrete Wall

Adrian Muntean and Michael Böhm

Abstract. We formulate a 1D partly dissipative moving-boundary reaction-diffusion system that describes the penetration of a reaction front into a concrete wall. We state the well-posedness of the model and the existence of non-trivial upper and lower bounds for the concentrations, speed of the interface, and shut-down time of the process. A numerical example illustrates the typical behavior of concentrations and interface penetration in a real-world application.

Mathematics Subject Classification (2000). Primary 35R35; Secondary 35K57, 74F25, 35D05.

Keywords. Moving-boundary problem, reaction-diffusion system, corrosion, porous media, concrete carbonation.

1. Statement of the problem and a mathematical model

Carbon dioxide, which is present under normal atmospheric conditions and also emitted as industrial output, attacks reinforced concrete structures by destroying their protection against corrosion. The process is called *carbonation*. The core reaction can be described as

$$CO_2(g \to aq) + Ca(OH)_2(aq) \overset{H_2O}{\longrightarrow} CaCO_3(aq) + H_2O. \tag{1.1}$$

Its impact on concrete microstructure is significant and possible repairs are often expensive. Therefore there is need of models capable to predict the depth of CO_2 penetration in concrete structures accurately. We consider the carbonation penetration in a wall whose chemistry, humidity level, and microstructure are known [6]. Experiments show that the zone of reaction is narrowly confined to the interface between the unreacted solid and the product layer, i.e., the region where calcium carbonate precipitates to the solid matrix. See [5, 6, 8] for more details on this motivating practical problem.

This work was completed with the partial support of DFG-SPP1122 *Prediction of the Course of Physicochemical Damage Processes Involving Mineral Materials*.

We idealize the reaction front by a surface $\Gamma(t)$. Let the positive x-axis be directed normally to $\Gamma(t)$ and into the uncarbonated part. The basic geometry is sketched in Figure 1 (a). At $t = 0$, we assume that the origin located at $x = 0$ is behind the reaction interface $\Gamma(t)$. Assuming that the reactants, which depend only on the real variables x and t, are available to reaction, we expect that the reaction interface moves as $x = s(t)$ for $t \in S_T :=]0, T[$ such that $s(0) = s_0$, where $T \in]0, +\infty]$, $s_0 \in]0, L[$, and $L \in]0, +\infty[$ are given, see Figure 1 (b). We denote the mass concentration of the reactants and products as follows: $\bar{u}_1 := [CO_2(aq)]$, $\bar{u}_2 := [CO_2(g)]$, $\bar{u}_4 := [CaCO_3(aq)]$, and $\bar{u}_5 := [H_2O]$ are the chemical species present in the region $\Omega_1(t) := [0, s(t)[$; $\bar{u}_3 := [Ca(OH)_2(aq)]$ and $\bar{u}_6 := [H_2O]$ are species present in $\Omega_2(t) :=]s(t), L]$. For ease of notation, we use the

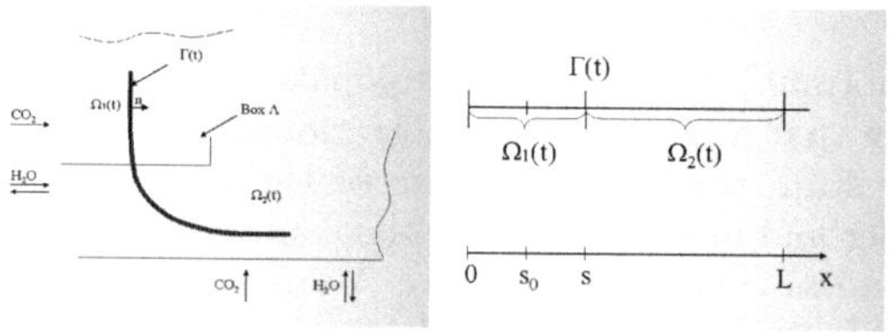

(a) (b)

FIGURE 1. (a) Basic geometry for the P_Γ model. The box A is the region which our model refers to. (b) Schematic 1D geometry. The reactants of (1.1) are spatially segregated at any time t.

set of indices $\mathcal{I} := \mathcal{I}_1 \cup \{4\} \cup \mathcal{I}_2$, where $\mathcal{I}_1 := \{1, 2, 5\}$ points out the active concentrations in $\Omega_1(t)$, and $\mathcal{I}_2 := \{3, 6\}$ refers to the active concentrations living in $\Omega_2(t)$. Specifically, we take into account that $CaCO_3(aq)$ is not transported in $\Omega := \Omega_1(t) \cup \Gamma(t) \cup \Omega_2(t)$, therefore the only partly dissipative character of the model. Then, we are led to discuss the *moving-boundary* problem of determining the concentrations $\bar{u}_i(x, t), i \in \mathcal{I}$ and the interface position $s(t)$ which satisfy for all $t \in S_T$ the equations

$$\left\{\begin{array}{rcl} (\phi\phi_w \bar{u}_i)_{,t} + (-D_i \nu_{i2} \phi\phi_w \bar{u}_{i,x})_x & = & +f_{i,\mathrm{Henry}}, x \in \Omega_1(t), i \in \{1, 2\}, \\ (\phi\phi_w \bar{u}_3)_{,t} + (-D_3 \phi\phi_w \bar{u}_{3,x})_x & = & +f_{\mathrm{Diss}}, x \in \Omega_2(t), \\ (\phi\phi_w \bar{u}_4)_{,t} & = & +f_{\mathrm{Prec}} + f_{\mathrm{Reac}\,\Gamma}, x = s(t) \in \Gamma(t), \\ (\phi \bar{u}_5)_{,t} + (-D_5 \phi \bar{u}_{5,x})_x & = & +f_{\mathrm{Reac}\,\Gamma}, x \in \Omega_1(t), \\ (\phi \bar{u}_6)_{,t} + (-D_6 \phi \bar{u}_{6,x})_x & = & 0, x \in \Omega_2(t). \end{array}\right. \tag{1.2}$$

The initial and boundary conditions are $\phi\phi_w \nu_{i2} \bar{u}_i(x, 0) = \hat{u}_{i0}$, $i \in \mathcal{I}, x \in \Omega(0)$, $\phi\phi_w \nu_{i2} \bar{u}_i(0, t) = \lambda_i$, $i \in \mathcal{I}_1$, $\bar{u}_{i,x}(L, t) = 0$, $i \in \mathcal{I}_2$, $x \in \Omega_2(t)$, where $t \in S_T$. Specific to our problem, we impose the following interface conditions

$$\left\{\begin{array}{rcl} [j_1 \cdot n]_{\Gamma(t)} & = & -\tilde{\eta}_\Gamma(s(t), t) + s'(t)[\phi\phi_w \bar{u}_1]_{\Gamma(t)}, \\ [j_i \cdot n]_{\Gamma(t)} & = & s'(t)[\phi\phi_w \nu_{i2} \bar{u}_i]_{\Gamma(t)}, i \in \{2, 5, 6\}, \\ [j_3 \cdot n]_{\Gamma(t)} & = & -\tilde{\eta}_\Gamma(s(t), t) + s'(t)[\phi\phi_w \bar{u}_3]_{\Gamma(t)}, \end{array}\right. \tag{1.3}$$

$$s'(t) = \alpha \frac{\tilde{\eta}_\Gamma(s(t),t)}{\phi\phi_w \bar{u}_3(s(t),t)}, s(0) = s_0, \tag{1.4}$$

where $\nu_{12} = \nu_{32} := 1$, $\nu_{22} := \frac{\phi_a}{\phi_w}$, $\nu_{52} = \nu_{62} := \frac{1}{\phi_w}$, $\nu_{i\ell} := 1$ $(i \in \mathcal{I}, \ell \in \mathcal{I} - \{2\})$, $j_i := -D_i \nu_{i\ell} \phi\phi_w \bar{u}_i$ $(i, \ell \in \mathcal{I}_1 \cup \mathcal{I}_2)$ are the corresponding diffusive fluxes, and $\alpha > 0$. Here D_i, L, and s_0 are strictly positive constants, λ_i are prescribed in agreement with the environmental conditions to which Ω – a part of the wall (cf. Figure 1 (b)) – is exposed [6]. The initial conditions $\hat{u}_{i0} > 0$ are determined by the chemistry of the cement. We assume $\hat{u}_{i0} = \lambda_i$. The hardened mixture of aggregate, cement, and water (i.e., the concrete) imposes ranges for the porosity $\phi > 0$ and also for the water and air fractions, $\phi_w > 0$ and $\phi_a > 0$, see [6]. Since the active concentrations are small, the constant porosity assumption is valid. The productions terms $f_{i,\text{Henry}}$, f_{Diss}, f_{Prec}, and $f_{\text{Reac}\,\Gamma}$ are sources or sinks by Henry-like interfacial transfer mechanisms, dissolution, precipitation, and carbonation reactions. Typical examples are (cf. [4, 8]):

$$\begin{cases} f_{i,\text{Henry}} := (-1)^i P_i(\phi\phi_w \bar{u}_1 - Q_i \phi\phi_a \bar{u}_2)(P_i > 0, Q_i > 0), i \in \{1,2\}, \\ f_{\text{Diss}} := -S_{3,\text{diss}}(\phi\phi_w \bar{u}_3 - u_{3,eq}), S_{3,\text{diss}} > 0, f_{\text{Prec}} := 0, f_{\text{Reac}\,\Gamma} := \tilde{\eta}_\Gamma. \end{cases} \tag{1.5}$$

Let $\bar{u}$ denote the vector of concentrations $(\bar{u}_1, \ldots, \bar{u}_6)^t$ and M_Λ be the set of parameters $\Lambda := (\Lambda_1, \ldots, \Lambda_m)^t$ that are needed to describe the reaction rate. For our purposes, it suffices to assume that M_Λ is a non-empty compact subset of $\mathbb{R}_+^m$. We introduce the function

$$\bar{\eta}_\Gamma : \mathbb{R}^6 \times M_\Lambda \to \mathbb{R}_+ \text{ by } \bar{\eta}_\Gamma(\bar{u}(x,t), \Lambda) := k\phi\phi_w \bar{u}_1^{\bar{p}}(x,t))\bar{u}_3^{\bar{q}}(x,t), x = s(t). \tag{1.6}$$

In (1.6) $m := 3$ and $\Lambda := \{\bar{p}, \bar{q}, k\phi\phi_w\} \in \mathbb{R}_+^3$. We define the rate of reaction (1.1) $\tilde{\eta}_\Gamma(s(t),t)$, which arises in (1.3)–(1.5), by

$$\tilde{\eta}_\Gamma(s(t),t) := \bar{\eta}_\Gamma(\bar{u}(s(t),t), \Lambda). \tag{1.7}$$

(1.6) is a *special* choice of $\bar{\eta}_\Gamma$ and represents the classical power-law ansatz. Other choices of $\bar{\eta}_\Gamma$ may be employed, too. See [4, 8], e.g.

Note that some equations are valid in $\Omega_1(t)$, while others act in $\Omega_2(t)$ or at $\Gamma(t)$! All of the three space domains are varying in time and they are *a priori* unknown. The interface conditions require some explanation. The term $\tilde{\eta}_\Gamma(s(t),t) \approx \alpha s'(t)$ denotes the number of moles per volume that are transported by diffusion to the reaction interface. For convenience, we assume here $\tilde{\eta}_\Gamma(s(t),t) = s'(t)$. The expression $\pm\phi\phi_w \bar{u}(s(t),t)s'(t)$ accounts for the mass flux induced by the motion of the boundary $\Gamma(t)$ in order to preserve conservation of mass. The conditions (1.3) express jumps in the gradients of concentrations across $\Gamma(t)$. They are typical interface relations for a surface reaction mechanism, i.e., the classical Rankine-Hugoniot jump relations cf. [1], Section 1.2.E, e.g. The non-local law (1.4) governs the dynamics of the reaction interface. The latter interface condition is derived via first principles in the 1D case and for simple 2D geometries in [4, 8]. This kind of relationship is missing in the approach by Brieger and Wittmann [5]. It is however needed to complete the model formulation and allows the determination of the interface location once the reactants at $\Gamma(t)$ are known. The setting is applicable when the reaction rate is very rapid and the diffusion of the gaseous CO_2 is sufficiently slow, or in other terms, when the characteristic time of the carbonation reaction is much smaller than the characteristic time of diffusion of the fastest species. This difference in the characteristic times causes the concentrations of the active chemical species and their gradient to have a jump at $\Gamma(t)$. The magnitude of the jump typically depends on the concentration itself. The system (1.2)–(1.7) forms the *sharp-interface carbonation model* P_Γ, or shortly (P_Γ). The model P_Γ consists of a coupled semi-linear system of parabolic equations that has a moving *a priori* unknown internal boundary $\Gamma(t)$, where the reaction (1.1) is assumed to take place. The coupling between the equations and the nonlinearities comes from the influence of the chemical reaction on the transport part and also from the dependence of the moving regions $\Omega_1(t)$ and $\Omega_2(t)$ on $s(t)$. Other nonlinearities might be introduced by different assumptions on the production terms.

We present results on the well-posedness of the sharp-interface model P_Γ and prove some useful estimates. To do this, we firstly fix the moving boundary by means of Landau-like transformations. Then we define the weak solutions to the transformed model and state its well-posedness locally in time. Since upper and lower bounds on the weak solution and on the shut-down time of the process are available, the maximum estimates become uniformly in time and the weak solution can be extended up to a global one, see the results and remarks in Sections 2 and 4. In Section 3, a simulation example shows the typical behavior of active concentrations and interface penetration into a real concrete wall. The model shows qualitatively good results when the numerical solution is compared with measured penetration depth profiles.

2. Main results

For each $i \in \mathcal{I}_1 \cup \mathcal{I}_2$, we denote $H_i := L^2(a,b)$, where we set $[a,b] := [0,1]$ for $i \in \mathcal{I}_1$ and $[a,b] := [1,2]$ for $i \in \mathcal{I}_2$. Moreover, $\mathbb{H} := \prod_{i \in \mathcal{I}_1 \cup \mathcal{I}_2} H_i$, $V_i = \{u \in H^1(a,b) : u_i(a) = 0\}, i \in \mathcal{I}_1$, $V_i := H^1(a,b), i \in \mathcal{I}_2$, and $\mathbb{V} = \prod_{i \in \mathcal{I}_1 \cup \mathcal{I}_2} V_i$, see [8]. In addition, $|\cdot| := ||\cdot||_{L^2(a,b)}$ and $||\cdot|| := ||\cdot||_{H^1(a,b)}$. If $(X_i : i \in \mathcal{I})$ is a sequence of given sets X_i, then $X^{|\mathcal{I}_1 \cup \mathcal{I}_2|}$ denotes the product $\prod_{i \in \mathcal{I}_1 \cup \mathcal{I}_2} X_i := X_1 \times X_2 \times X_3 \times X_5 \times X_6$. Note that sometimes $u(1)$ and $u_{,y}(1)$ replace $u(1,t)$ and $u_{,y}(1,t)$, respectively.

We re-formulate the model P_Γ in terms of macroscopic quantities by performing the transformation of all concentrations into volume-based concentrations via $\hat{u}_i := \phi\phi_w \bar{u}_i, i \in \{1,3,4\}, \hat{u}_2 := \phi\phi_a \bar{u}_2, \hat{u}_i := \phi \bar{u}_i, i \in \{5,6\}$. We map (P_Γ) onto a domain with fixed boundaries. To this effect, we employ the Landau transformations $(x,t) \in [0,s(t)] \times \bar{S}_T \longmapsto (y,\tau) \in [a,b] \times \bar{S}_T$, $y = \frac{x}{s(t)}$ and $\tau = t$, for $i \in \mathcal{I}_1$, $(x,t) \in [s(t),L] \times \bar{S}_T \longmapsto (y,\tau) \in [a,b] \times \bar{S}_T$, $y = a + \frac{x-s(t)}{L-s(t)}$ and $\tau = t$, for $i \in \mathcal{I}_2$. We re-label τ by t and introduce the new concentrations, which act in the auxiliary y-t plane by $u_i(y,t) := \hat{u}_i(x,t) - \lambda_i(t)$ for all $y \in [a,b]$ and $t \in S_T$. Thus, the model equations are reduced to

$$(u_i+\lambda_i)_{,t}-\frac{1}{s^2(t)}(D_iu_{i,y})_{,y}=f_i(u+\lambda)+y\frac{s'(t)}{s(t)}u_{i,y}, i\in\mathcal{I}_1, \tag{2.1}$$

$$(u_i+\lambda_i)_{,t}-\frac{1}{(L-s(t))^2}(D_iu_{i,y})_{,y}=f_i(u+\lambda)+(2-y)\frac{s'(t)}{L-s(t)}u_{i,y}, i\in\mathcal{I}_2,$$

where u is the vector of concentrations $(u_1,u_2,u_3,u_5,u_6)^t$ and $\lambda:=(\lambda_1,\ \lambda_2,\ \lambda_3,\ \lambda_5,\ \lambda_6)^t$ represents the boundary data. The transformed initial, boundary, and interface conditions are

$$u_i(y,0)=0, i\in\mathcal{I}_1\cup\mathcal{I}_2, u_i(a,t)=0, i\in\mathcal{I}_1, u_{i,y}(b,t)=0, i\in\mathcal{I}_2, \tag{2.2}$$

$$\frac{-D_1}{s(t)}u_{1,y}(1)=\eta_\Gamma(1,t)+s'(t)(u_1(1)+\lambda_1), \frac{-D_2}{s(t)}u_{2,y}(1)=s'(t)(u_2(1)+\lambda_2), \tag{2.3}$$

$$\frac{-D_3}{L-s(t)}u_{3,y}(1)=\eta_\Gamma(1,t)-s'(t)(u_3(1)+\lambda_3), \tag{2.4}$$

$$\frac{D_5}{s(t)}u_{5,y}(1)-\frac{D_6}{L-s(t)}u_{6,y}(1)=s'(t)(u_5(1)+\lambda_5-u_6(1)-\lambda_6), \tag{2.5}$$

where $\eta_\Gamma(1,t)$ denotes the reaction rate that acts in the y-t plane. This is defined by

$$\eta_\Gamma(1,t):=\bar{\eta}_\Gamma(\bar{u}(ys(t),t)+\lambda(t),\Lambda), y\in[0,1], t\in S_T \tag{2.6}$$

for a given $\Lambda\in M_\Lambda$. Finally, two ode's

$$s'(t)=\eta_\Gamma(1,t) \text{ and } \hat{u}_4'(t)=f_4(\hat{u}(s(t),t)) \text{ a.e. } t\in S_T, \tag{2.7}$$

complete the model formulation, where $\hat{u}:=(\hat{u}_1,\hat{u}_2,\hat{u}_3,\hat{u}_4,\hat{u}_5,\hat{u}_6)^t$. We also assume

$$s(0)=s_0>0, \hat{u}_4(s(0),0)=\hat{u}_{40}\ge 0. \tag{2.8}$$

The transformed model equations are collected in (2.1)–(2.8). Let $\varphi:=(\varphi_1,\varphi_2,\varphi_3,\varphi_5,\varphi_6)^t\in\mathbb{V}$ be an arbitrary test function, and take $t\in S_T$. To write the weak formulation of (2.1)–(2.8) in a compact form, we introduce the notation:

$$\begin{cases} a(s,u,\varphi) & :=\frac{1}{s^2}\sum_{i\in\mathcal{I}_1}(D_iu_{i,y},\varphi_{i,y})+\frac{1}{(L-s)^2}\sum_{i\in\mathcal{I}_2}(D_iu_{i,y},\varphi_{i,y}),\\ b_f(u,\varphi) & :=\sum_{i\in\mathcal{I}_1\cup\mathcal{I}_2}(f_i(u),\varphi_i),\\ e(s,s',u,\varphi) & :=\frac{1}{s}\sum_{i\in\mathcal{I}_1}g_i(s,s',u(1))\varphi_i(1)+\frac{1}{L-s}\sum_{i\in\mathcal{I}_2}g_i(s,s',u(1))\varphi_i(1),\\ h(s,s',u_{,y},\varphi) & :=\frac{s'}{s}\sum_{i\in\mathcal{I}_1}(yu_{i,y},\varphi_i)+\frac{s'}{L-s}\sum_{i\in\mathcal{I}_2}((2-y)u_{i,y},\varphi_i),\end{cases} \tag{2.9}$$

for any $u\in\mathbb{V}$ and $\lambda\in W^{1,2}(S_T)^{|\mathcal{I}_1\cup\mathcal{I}_2|}$. Furthermore, $v_4(t):=\hat{u}_4(s(t),t)$ for $t\in S_T$. The term $a(\cdot)$ incorporates the diffusive part of the model, $b_f(\cdot)$ comprises volume productions, $e(\cdot)$ sums up reaction terms acting on $\Gamma(t)$, and $h(\cdot)$ is a non-local term due to fixing of the domain. For our application (see (1.5) and (1.6)), the interface terms $g_i(i\in\mathcal{I}_1\cup\mathcal{I}_2)$ are given by

$$\begin{cases} g_1(s,s',u):=\eta_\Gamma(1,t)+s'(t)u_1(1), & g_2(s,s',u):=s'(t)u_2(1),\\ g_3(s,s',u):=-\eta_\Gamma(1,t)+s'(t)u_3(1), & g_5(s,s',u):=g_6(s,s',u)=0,\end{cases} \tag{2.10}$$

whereas the volume terms f_i $(i \in \mathcal{I})$ are defined as

$$\left\{\begin{array}{llllll} f_1(u) & := & P_1(Q_1u_2 - u_1), & f_4(\hat{u}) & := & +\tilde{\eta}_\Gamma(s(t),t), \\ f_2(u) & := & -P_2(Q_2u_2 - u_1), & f_5(u+\lambda) & := & +\eta_\Gamma(1,t), \\ f_3(u) & := & S_{3,\mathrm{diss}}(u_{3,eq} - u_3), & f_6(u) & := & 0. \end{array}\right. \tag{2.11}$$

Set $M_{\eta_\Gamma} := \sup_{u(y,t)\in\mathcal{K}}\{\eta_\Gamma(1,t) : y \in [a,b], t \in S_T\}$ $\left(\mathcal{K} := \prod_{i\in\mathcal{I}}[0,k_i]\right)$, where

$$\left\{\begin{array}{lll} k_i & := & \max\{u_{i0}(y) + \lambda_i(t), \lambda_i(t) : y \in [a,b], t \in \bar{S}_T\}, i = 1,2,3, \\ k_4 & := & \max\{\hat{u}_{40}(x) + M_{\eta_\Gamma}T : x \in [0,s(t)], t \in \bar{S}_T\}, \\ k_j & := & \max\{u_{i0}(y) + \lambda_i(t) + M_{\eta_\Gamma}T : y \in [a,b], t \in \bar{S}_T\}, j = 5,6. \end{array}\right. \tag{2.12}$$

Definition 2.1. (Local Weak Solution) We call the triple (u, v_4, s) a *local weak solution* to (2.1)–(2.8) if there is a $S_\delta :=]0,\delta[$ with $\delta \in]0,T]$ such that

$$v_4 \in W^{1,4}(S_\delta), \quad s \in W^{1,4}(S_\delta), \tag{2.13}$$

$$u \in W_2^1(S_\delta;\mathbb{V},\mathbb{H}) \cap [\bar{S}_\delta \mapsto L^\infty(a,b)]^{|\mathcal{I}_1\cup\mathcal{I}_2|} \cap L^\infty(S_\delta;C^{0,\frac{1}{2}-}([a,b]^{|\mathcal{I}_1\cup\mathcal{I}_2|})), \tag{2.14}$$

$$\left\{\begin{array}{l} (u'(t),\varphi) + a(s,u,\varphi) + e(s,s',u,\varphi) = b_f(u(t)+\lambda(t),\varphi) \\ +h(s,s',u_{,y},\varphi) - (\lambda'(t),\varphi) \quad \text{for all } \varphi \in \mathbb{V}, \text{ a.e. } t \in S_\delta, \\ \quad s'(t) = \eta_\Gamma(1,t), \quad v_4'(t) = f_4(\hat{u}(s(t),t)) \text{ a.e. } t \in S_\delta, \\ \qquad u(0) = u_0 \in \mathbb{H}, s(0) = s_0, v_4(0) = \hat{u}_{40}. \end{array}\right. \tag{2.15}$$

There is some freedom in choosing the exact structure of the reaction rate η_Γ. The only assumptions that are needed are the following:

(A) There exists a positive constant $C_\eta = C_\eta(\Lambda, u, \lambda, T)$ such that

$$\bar{\eta}_\Gamma(\bar{u}(s(t),t),\Lambda) \le C_\eta \bar{u}(s(t),t) \text{ for all } t \in S_T.$$

(B) The reaction rate η_Γ (defined cf. (2.6)) is locally Lipschitz with respect to all variables. More precisely, let $(u^{(i)}, v_4^{(i)}, s^{(i)})$ be two solutions corresponding to the sets of data $\mathcal{D}_i := (u_0^{(i)}, \lambda^{(i)}, \ldots, \Lambda^{(i)})^t$, where $i \in \{1,2\}$. Set $\Delta u := u^{(2)} - u^{(1)}$, $\Delta v_4 := v_4^{(2)} - v_4^{(1)}$, $\Delta\lambda := \lambda_2^{(2)} - \lambda^{(1)}$, $\Delta u_0 := u_0^{(2)} - u_0^{(1)}$, and $\Delta\Lambda := \Lambda^{(2)} - \Lambda^{(1)}$. The Lipschitz condition on $\Delta\eta_\Gamma := \Delta\bar{\eta}_\Gamma = \bar{\eta}_\Gamma(\bar{u}^{(2)},\Lambda^{(2)}) - \bar{\eta}_\Gamma(\bar{u}^{(1)},\Lambda^{(1)})$ reads: There exists a positive constant $c_L = c_L(\mathcal{D}_1,\mathcal{D}_2)$ such that the inequality $|\Delta\eta_\Gamma| \le c_L(|\Delta u| + |\Delta\Lambda|)$ holds locally pointwise. For a particular choice of $\bar{\eta}_\Gamma$, Λ, and hence c_L, see (1.6).

(C1) $k_3 \ge \max\{|u_{3,eq}(t)| : t \in \bar{S}_T\}$;

(C2) $P_1Q_1k_2 \le P_1k_1$; $P_2k_1 \le P_2Q_2k_2$;

(C3) $Q_2 > Q_1$.

Theorem 2.2 (Local Existence and Uniqueness, [8]). *Assume the hypotheses* (A)–(C2) *and let the following conditions* (2.16)–(2.20) *be satisfied:*

$$u_{3,eq} \in L^2(S_T), \lambda \in W^{1,2}(S_T)^{|\mathcal{I}_1 \cup \mathcal{I}_2|}, \lambda(t) \geq 0 \text{ a.e. } t \in \bar{S}_T, \tag{2.16}$$

$$u_0 \in L^\infty(a,b)^{|\mathcal{I}_1 \cup \mathcal{I}_2|}, u_0(y) + \lambda(0) \geq 0 \text{ a.e. } y \in [a,b], \tag{2.17}$$

$$\hat{u}_{40} \in L^\infty(0, s_0), \hat{u}_4(x,0) \geq 0 \text{ a.e. } x \in [0, s_0], \tag{2.18}$$

$$\min\{\min_{\bar{S}_T}\{u_{3,eq}(t)\}, S_{3,diss}, P_1, Q_1, P_2, Q_2\} > 0, \tag{2.19}$$

$$0 < s_0 \leq s(t) \leq L_0 < L \text{ for all } t \in \bar{S}_T. \tag{2.20}$$

Then the following assertions hold:

(a) *There exists a* $\delta \in]0, T[$ *such that the problem* (2.1)–(2.8) *admits a unique local solution on* S_δ *in the sense of Definition* 2.1*;*

(b) $0 \leq u_i(y,t) + \lambda_i(t) \leq k_i$ *a.e.* $y \in [a,b]$ $(i \in \mathcal{I}_1 \cup \mathcal{I}_2)$ *for all* $t \in S_\delta$*. Moreover,* $0 \leq \hat{u}_4(x,t) \leq k_4$ *a.e.* $x \in [0, s(t)]$ *for all* $t \in S_\delta$*;*

(c) $v_4, s \in W^{1,\infty}(S_\delta)$.

Remark 2.3. Let $(u^{(i)}, v_4^{(i)}, s^{(i)})(i \in \{1,2\})$ be two local weak solutions on S_δ. Correspondingly, let $(u_0^{(i)}, \lambda^{(i)}, \Lambda^{(i)})$ be the initial, boundary and reaction data. Then the function $\mathbb{H} \times W^{1,2}(S_\delta)^{|\mathcal{I}_1 \cup \mathcal{I}_2|} \times M_\Lambda \to W_2^1(S_\delta, \mathbb{V}, \mathbb{H}) \times W^{1,4}(S_\delta)^2$ that maps $(u_0, \lambda, \Lambda)^t$ into $(u, v_4, s)^t$ is Lipschitz in the following sense: There exists a constant $c = c(\delta, s_0, \hat{u}_{40}, L, k_i) > 0$ such that

$$\begin{aligned} &||\Delta u||^2_{W_2^1(S_\delta,\mathbb{V},\mathbb{H}) \cap L^\infty(S_\delta,\mathbb{H})} + ||\Delta v_4||^2_{W^{1,4}(S_\delta) \cap L^\infty(S_\delta)} + ||\Delta s||^2_{W^{1,4}(S_\delta) \cap L^\infty(S_\delta)} \\ &\quad \leq c \left(||\Delta u_0||^2_{\mathbb{H} \cap L^\infty([a,b]^{|\mathcal{I}_1 \cup \mathcal{I}_2|})} + ||\Delta \lambda||^2_{W^{1,2}(S_\delta) \cap L^\infty(S_\delta)} + \max_{M_\Lambda} |\Delta \Lambda|^2 \right). \end{aligned} \tag{2.21}$$

Furthermore, we were able to show that small changes in the concentration fields induce small displacements of the position of the reaction front, for details see [8]. A straightforward consequence of this aspect is that the main output of the model (the penetration curve *versus* time, see Figure 2 (c), e.g.) is stable with respect to small perturbations in the reaction rate structure.

Proposition 2.4 (Strict Lower Bounds, [8]). *Assume that the hypotheses of Theorem* 2.2 *are satisfied. In addition, if the restriction* (C3) *holds and the initial and boundary data are strictly positive, then there exists a range of parameters such that the positivity estimates stated in Theorem* 2.2(b) *are strict for all times.*

Theorem 2.2 and Remark 2.3 report on the well-posedness of (2.1)–(2.8) with respect to the time interval S_δ. In the sequel, we extend this local well-posed solution up to a global solution. Firstly, we assume that the hypotheses of Proposition 2.4 hold. In this case, for an *arbitrary* $L_0 \in]s_0, L[$ there is a $T_{\text{fin}} < +\infty$ such that $s(T_{\text{fin}}) = L_0$. Thus, T_{fin} is the time when $\Gamma(t)$ has penetrated all of $]s_0, L_0[$. We refer to it as the *final carbonation time* or *shut-down time of the (carbonation) process.* Physically reasonable restrictions on the life span of the weak solution (hence, on T_{fin}) are given in Proposition 2.6 (iii). See also [10, 12, 13] for some closely related scenarios. The next results are direct consequences of Theorem 2.2 and Proposition 2.4.

Proposition 2.5 (Strict Monotonicity of the Reaction Interface). *If the hypotheses of Proposition 2.4 are satisfied, then the position $s \in W^{1,\infty}(S_\delta)$ of the interface $\Gamma(t)$ is strictly monotonic increasing on S_δ.*

Proposition 2.6 (Basic Estimates). *Let (u, v_4, s) be the unique local solution to (2.1)–(2.8) that fulfills the hypotheses of Proposition 2.4. Then the following estimates hold:*

(i) $\eta_{\min} < s'(t) < \eta_{\max}$ *for all* $t \in S_\delta$;
(ii) $s_0 \leq s(t) \leq s_0 + \eta_{\max} t$ *for all* $t \in S_\delta$;
(iii) $\frac{L_0 - s_0}{\eta_{\max}} < T_{\text{fin}} < \frac{L_0 - s_0}{\eta_{\min}}$,

where $\eta_{\min}$ and $\eta_{\max}$ denote uniform lower and upper bounds on $\tilde{\eta}_\Gamma$.

Proof. By Theorem 2.2 (b) and Proposition 2.4, (i) and (ii) are straightforward. The equation for s' in (2.7) leads to $T_{\text{fin}} - t_0 = \int_{s(t_0)}^{s(T_{\text{fin}})} \frac{1}{\tilde{\eta}_\Gamma(s)} ds$ ($t_0 \in [0, T_{\text{fin}}[$), see [2]. We apply the mean-value theorem and estimate the reaction rate $\tilde{\eta}_\Gamma$ from below by using the non-trivial uniform lower bounds on the reactants (i.e., on u_1 and u_3), and afterwards from above, by means of the corresponding maximum estimates. In this way, it yields (iii). □

Theorem 2.7 (Global Solvability). *Assume that the hypotheses of Proposition 2.4 are satisfied. Then the time interval $S_{T_{\text{fin}}} :=]0, T_{\text{fin}}[$ of (global) solvability of (2.1)–(2.8) is finite and is characterized by*

$$T_{\text{fin}} = s^{-1}(L_0). \tag{2.22}$$

Proof. The finiteness of the length of $S_{T_{\text{fin}}}$ is a consequence of Proposition 2.6 (iii). The uniform maximum estimates of concentrations together with the nonnegativity of concentrations imply that the region $\prod_{i\in\mathcal{I}}[0, k_i] \times [s_0, s_0 + T_{\text{fin}} M_{\eta_\Gamma}]$, which confines the graph of (u, v_4, s), remains invariant along the *physically relevant* interval $]0, T_{\text{fin}}[$ of existence of the weak solution. The strictly positive constant M_{η_Γ} is defined by (2.12), while the value of T_{fin} obeys the *a priori* estimate pointed out in Proposition 2.6 (iii). Note that the invariant region is independent of u, s, x, or t and that T_{fin} can be *a posteriori* calculated via (2.22). By the strict monotonicity of s (cf. Proposition 2.5) and $W^{1,\infty}(S_{T_{\text{fin}}}) \hookrightarrow C(\bar{S}_{T_{\text{fin}}})$, we obtain (2.22). □

3. Numerical example

We consider an 18 years old concrete wall made of the cement PZ35F, whose chemistry and outdoor exposure conditions are described in [6], Table 3.1. The weak formulation (2.15) allows us to approximate the underlying moving-boundary problem by using the finite element method. The examples shown in Figures 2 and 3 are obtained with a uniform 1D Galerkin scheme. For more numerical simulations of carbonation scenarios and details on the numerical scheme, see [4, 7, 8, 11].

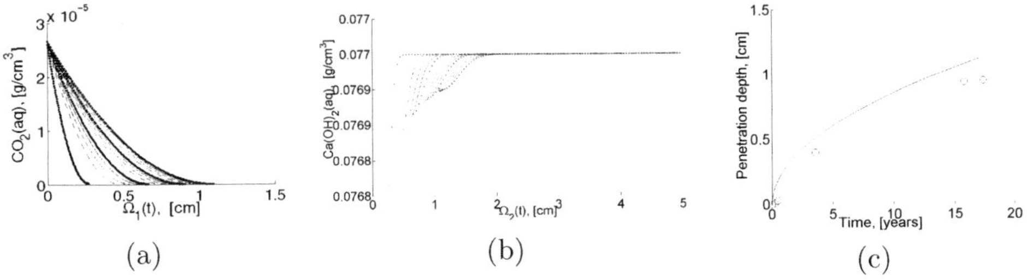

FIGURE 2. (a)+(b) CO_2(aq) and $Ca(OH)_2$(aq) profiles vs. space. Each curve refers to time $t = i$ years, $i \in \{1, \ldots, 18\}$. (c): Interface location vs. the experimental points " ∘ " (see [6]) after $T_{\text{fin}} = 18$ years of exposure.

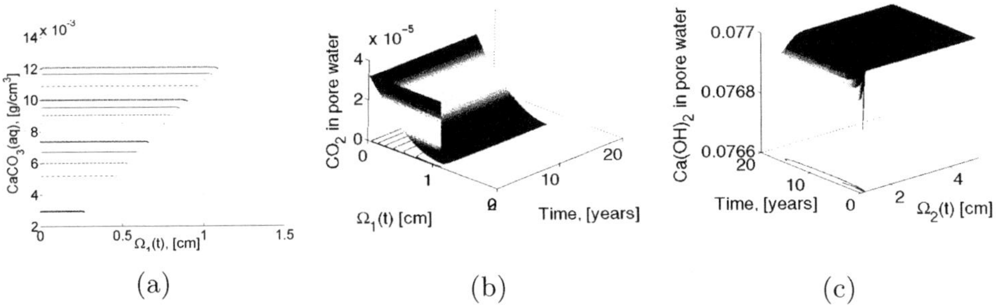

FIGURE 3. (a) $CaCO_3$(aq) profiles vs. space. Each curve refers to time $t = i$ years, $i \in \{1, \ldots, 18\}$. (b)+(c) Concentration of $CO_2(aq)$ and $Ca(OH)_2(aq)$ vs. time and space.

Observe that steep concentration gradients arise near $\Gamma(t)$ (Figure 3 (c), e.g.) and the calculated interface location is in the experimental range (Figure 2 (c)).

4. Remarks

Proofs for the statements in Section 2 are given in [8]. The energy approach and Banach fixed-point principle (see also [3]) were used to obtain the local well-posedness result. The main effort was concentrated on proving the positivity and the maximum estimates of the concentrations. The structure of the problem (2.1)–(2.8) allowed us to obtain non-trivial uniform lower estimates for all concentrations (Proposition 2.4). It is straightforward to deduce a strict lower bound for the speed of the interface (see Proposition 2.6 (i)) which shows, e.g., that the reaction interface $\Gamma(t)$ does not stop before it reaches the end of the geometry. The positivity of s' also ensures that a backward movement of $\Gamma(t)$ is impossible. This fact contradicts Brieger and Wittmann [5], who suggested that $\Gamma(t)$ may follow a forward-backward movement. A non-trivial *a priori* upper bound of the final time of carbonation $T_{\rm fin}$ is available cf. Proposition 2.6 (iii). This margins the time needed to completely carbonate the domain $[s_0, L_0]$. Another merit of the bound on $T_{\rm fin}$ is that it supports the extending of the local solution (defined on S_δ) up to a global solution (defined on $S_{T_{\rm fin}}$), see Theorem 2.7. It is noteworthy that for sufficiently large time t and width of the sample L_0, the interface position has a sublinear growth in time (cf. Proposition 2.6 (ii)). This is only a rough estimate. Since the reaction-rate structure (1.7) affects the scaling of the penetration depth and of the width of the reaction front, improved asymptotic bounds are expected.

References

[1] V. Alexiades, A.D. Solomon, *Mathematical Modeling of Melting and Freezing Processes.* Hemisphere Publishing Group, 1993.

[2] P. Binding, *The differential equation $x' = f \circ x$.* J. Diff. Eqs. **31**(1979), 183–199.

[3] M. Böhm, G. Rosen, *Global weak solutions and uniqueness for a moving boundary problem for a coupled system of quasilinear diffusion-reaction equations arising as a model of chemical corrosion of concrete surfaces.* Humboldt University, 1997.

[4] M. Böhm, J. Kropp, A. Muntean, *On a prediction model for concrete carbonation based on moving interfaces – Interface concentrated reactions.* Berichte aus der Technomathematik 03-03, University of Bremen, Germany, 2003.

[5] L.M. Brieger, F.H. Wittmann, *Numerical simulation of concrete carbonation.* Werkstoffwissenschaften und Bausanierung, Int. Koll. Esslingen, Germany, 1986, 635–640.

[6] D. Bunte, *Zum Karbonatisierungsbedingten Verlust der Dauerhaftigkeit von Außenbauteilen aus Stahlbeton.* Dissertation, TU Braunschweig, Germany, 1994.

[7] A. Muntean, M. Böhm, *On a prediction model for the service life of concrete structures based on moving interfaces.* In Proceedings of ICLODC, (edited by Stangenberg, F., et al.), 209-218, Ruhr-Universität Bochum, Germany, 2004.

[8] A. Muntean, *A Moving-Boundary Problem: Modeling, Analysis and Simulation of Concrete Carbonation*, PhD thesis, University of Bremen, Cuvillier Verlag, Göttingen, Germany.

[9] P.J. Ortoleva, *Geochemical Self-Organization.* Oxford University Press, 1994.

[10] A. Pawell, K.D. Krannich, *Dissolution effects in porous media.* SIAM J. Appl. Math. **1** (1996), 89–118.

[11] A. Schmidt, A. Muntean, M. Böhm, *Numerical experiments with self-adaptive finite element simulations in 2D for the concrete carbonation.* Berichte aus der Technomathematik 05-01, University of Bremen, Germany, 2005.

[12] E. Vannini, *A reaction-diffusion problem with a reaction front.* Math. Meth. Appl. Sci. **21** (1998), 417–432.

[13] W. Xie, *The Stefan problem with a kinetic condition at the free boundary.* SIAM J. Math. Anal. **21** (1990), 362–373.

Adrian Muntean and Michael Böhm
Centre for Industrial Mathematics, Department of Mathematics (FB3)
University of Bremen, Postfach 330440
D-28334 Bremen, Germany
e-mail: {muntean,mbohm}@math.uni-bremen.de

International Series of Numerical Mathematics, Vol. 154, 327–337

Adaptive Finite Elements with High Aspect Ratio for Dendritic Growth of a Binary Alloy Including Fluid Flow Induced by Shrinkage

Jacek Narski and Marco Picasso

Abstract. An adaptive phase field model for the solidification of binary alloys in two space dimensions is presented. The fluid flow in the liquid due to different liquid/solid densities is taken into account. The unknowns are the phase field, the alloy concentration and the velocity/pressure in the liquid.

Continuous, piecewise linear finite elements are used for the space discretization, a semi-implicit scheme is used for time discretization. An adaptive method allows the number of degrees of freedom to be reduced, the mesh triangles having high aspect ratio whenever needed.

Numerical results are presented for dendritic growth of four dendrites.

1. Introduction

In recent years, considerable progress has been made in numerical simulation of solidification processes at microscopic scale [1]. Although sharp interface [2, 3] and level-set models [4, 5] have proved to be efficient, the phase field method emerged as a method of choice in order to simulate dendritic growth in binary alloys [6, 7, 8, 9, 10, 11]. In phase field models, the location of the solid and liquid phases in the computational domain is described by introducing an order parameter, the phase field, which varies smoothly from one in the solid to zero in the liquid through a slightly diffused interface. The main difficulty when solving numerically phase field models is due to the very rapid change of the phase field (and also of the concentration field in alloys) across the diffuse interface, whose thickness has to be taken very small (between 1 and 10 nm) to correctly capture the physics of the phase transformation. A high spatial resolution is therefore needed to describe the smooth transition. In order to reduce the computational time and the number of grid points adaptive isotropic finite elements [12, 13] have been used. Further reduction of the number of nodes has been achieved using adaptive finite elements with high aspect ratio [14, 15].

Jacek Narski is supported by the Swiss National Science Foundation.

The influence of inter-dendritic liquid flow has already been taken into account in dendritic simulations [16, 17, 18, 19, 20]. Also, the inter-dendritic liquid flow induced by shrinkage – that is to say the flow induced by the fact that solid and liquid densities are different – has been considered [21, 22, 23, 24, 25]. The goal of this paper is to take into account the inter-dendritic liquid flow due to shrinkage, using adaptive finite elements with high aspect ratio as in [14, 15]. Numerical results show that the method is capable of predicting micro-porosity.

The outline of the paper is the following. In the following section, we present the model and briefly discuss the numerical method. Numerical results are presented in section 3, where the pressure drop due to shrinkage is observed during solidification of four dendrites.

2. The model

The equations governing the solidification process are derived using a volume averaging technique in a similar way as in [16, 17]. The key idea is to develop two sets of equations (for the solid and liquid phases) and transform them into one set using averaging over small volume and introducing average quantities.

In the following we present and discuss the averaged mass, momentum and species conservation equation for binary alloy undergoing a solid/liquid phase transition. As in [21, 22, 23, 24, 25], we take into account the fact that the solid and liquid densities are different.

2.1. Mass conservation

The solidification of a binary alloy in a bounded domain Ω of $\mathbb{R}^2$ between time 0 and t_{end} is considered. Let $\phi : \Omega \times (0, t_{\text{end}}) \longrightarrow \mathbb{R}$ be the phase field describing presence of solid ($\phi = 1$) or liquid ($\phi = 0$). The phase field ϕ varies smoothly but rapidly from zero to one in a thin region of width δ, the so-called solid/liquid diffused interface. Let ρ_s and ρ_l be the constant solid and liquid densities (for most alloys $\rho_l < \rho_s$). Then the average density $\rho : \Omega \times (0, t_{\text{end}}) \longrightarrow \mathbb{R}$ is defined by

$$\rho = \rho_s \phi + \rho_l (1 - \phi). \tag{1}$$

Let $\boldsymbol{v}_s$, $\boldsymbol{v}_l$ be the solid and liquid velocities, respectively. In this model, the solid velocity $\boldsymbol{v}_s$ is a known constant (in most cases $\boldsymbol{v}_s = \mathbf{0}$), whereas $\boldsymbol{v}_l$ is unknown. Then, the average velocity $\boldsymbol{v} : \Omega \times (0, t_{\text{end}}) \longrightarrow \mathbb{R}^2$ is defined by

$$\rho \boldsymbol{v} = \rho_s \phi \boldsymbol{v}_s + \rho_l (1 - \phi) \boldsymbol{v}_l. \tag{2}$$

Averaging the mass conservation equation in the solid and liquid regions yields

$$\frac{\partial}{\partial t}\Big(\rho_s \phi + \rho_l (1 - \phi)\Big) + \text{div}\, \Big(\rho_s \phi \boldsymbol{v}_s + \rho_l (1 - \phi) \boldsymbol{v}_l\Big) = 0,$$

which can be rewritten, using (1) and (2):

$$\frac{\partial \rho}{\partial t} + \text{div}\, (\rho \boldsymbol{v}) = 0. \tag{3}$$

It should be noted that, in the sharp interface limit (that is to say when the width of the solid-liquid interface δ tends to zero), then the phase field ϕ is the characteristic function of the solid so that the density ρ becomes a step function and (3) has to be understood in the sense of distributions. Then, the following relation holds on the solid/liquid interface:

$$[\rho]V + [\rho \boldsymbol{v} \cdot \boldsymbol{n}] = 0 \tag{4}$$

where $[\cdot]$ denotes the jump of the inside quantity across the interface, V is the normal velocity of the solid/liquid interface and the vector $\boldsymbol{n}$ denotes the normal to the interface. For instance, when solid is not moving ($\boldsymbol{v}_s = 0$), this condition reduces to

$$\boldsymbol{v}_l \cdot \boldsymbol{n} = \frac{\rho_s - \rho_l}{\rho_l} V, \tag{5}$$

see Figure 1.

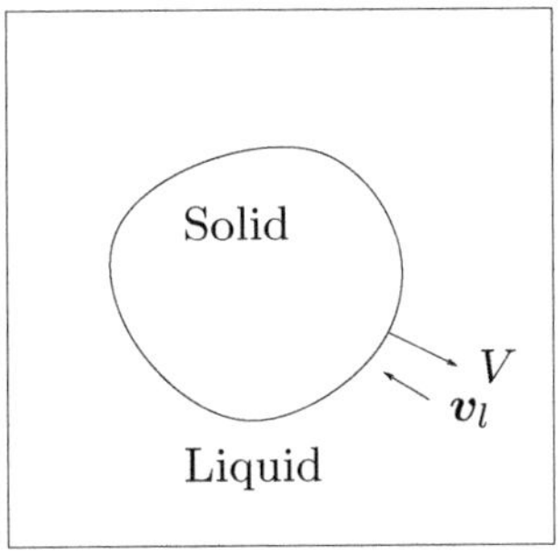

FIGURE 1. Solidification of a solid seed with density ρ_s larger than the liquid density ρ_l: as the solid/liquid interface moves with normal velocity V toward the boundary of the calculation domain Ω, liquid flows with velocity $\boldsymbol{v}_l$ toward the solid.

At this point it should be noted that certain geometrical configurations are incompatible with the mass conservation equation (3). This is the case when a liquid region is surrounded by a solid region, see Figure 2 where examples of compatible and not compatible configurations are shown.

2.2. Momentum conservation

Averaging the momentum conservation equation in the solid and liquid regions yields

$$\frac{\partial}{\partial t}\Big(\rho_s \phi \boldsymbol{v}_s + \rho_l (1-\phi)\boldsymbol{v}_l\Big) + \mathrm{div}\,\Big(\rho_s \phi \boldsymbol{v}_s \otimes \boldsymbol{v}_s + \rho_l (1-\phi)\boldsymbol{v}_l \otimes \boldsymbol{v}_l\Big)$$
$$- \mathrm{div}\,\Big(\phi \sigma_s + (1-\phi)\sigma_l\Big) = 0,$$

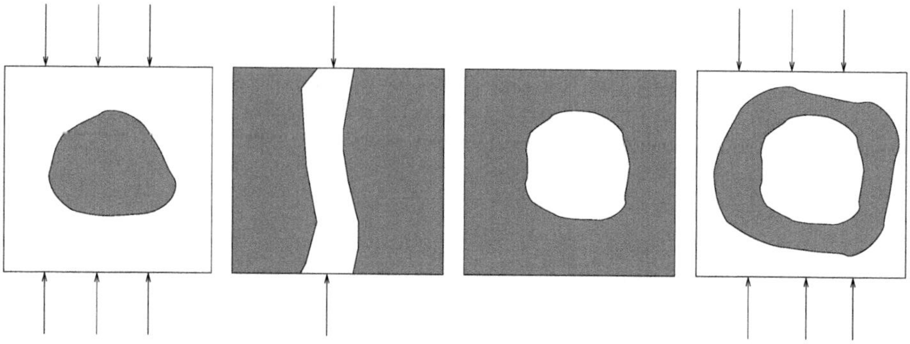

FIGURE 2. Examples of solidification configurations that are compatible and not compatible with equation (3). The first two configurations are compatible with (3) since some external liquid can enter the computational domain in order to feed all liquid regions. The last two configurations are not compatible with (3) since external liquid cannot feed the liquid region located at the center of the computational domain.

where σ_s, σ_l are the solid and liquid stress tensors, respectively. Our goal is to obtain a momentum equation for the unknown average velocity $\boldsymbol{v}$. First, the definitions (1) (2) together with (3) are used to eliminate the liquid velocity $\boldsymbol{v}_l$ in the momentum equation:

$$\rho\frac{\partial \boldsymbol{v}}{\partial t} + (\rho\boldsymbol{v}\cdot\nabla)\boldsymbol{v} + \operatorname{div}\left(\rho\frac{\phi\rho_s}{(1-\phi)\rho_l}(\boldsymbol{v}-\boldsymbol{v}_s)\otimes(\boldsymbol{v}-\boldsymbol{v}_s)\right) - \operatorname{div}\left(\phi\sigma_s + (1-\phi)\sigma_l\right) = 0.$$

Moreover, the solid mechanical deformation is neglected, thus $\sigma_s = 0$. Also, given a penalty parameter $\varepsilon \ll 1$, the penalty term

$$\frac{1}{\varepsilon}\phi^2(\boldsymbol{v}-\boldsymbol{v}_s)$$

is added to the momentum equation in order to force the average velocity field $\boldsymbol{v}$ to equal the solid velocity $\boldsymbol{v}_s$ in the solid region ($\phi = 1$). Finally, the liquid stress tensor $(1-\phi)\sigma_l$ is replaced by $-pI + 2\mu_l\epsilon(\boldsymbol{v})$ where p is the average pressure, μ_l the liquid viscosity and $\epsilon(\boldsymbol{v}) = 1/2(\nabla\boldsymbol{v} + \nabla\boldsymbol{v}^T)$ the rate of deformation tensor of the average velocity $\boldsymbol{v}$. Therefore, the momentum equation can be rewritten in the whole computational domain Ω as:

$$\rho\frac{\partial \boldsymbol{v}}{\partial t} + (\rho\boldsymbol{v}\cdot\nabla)\boldsymbol{v} + \operatorname{div}\left(\rho\frac{\phi\rho_s}{(1-\phi)\rho_l}(\boldsymbol{v}-\boldsymbol{v}_s)\otimes(\boldsymbol{v}-\boldsymbol{v}_s)\right) - 2\operatorname{div}\left(\mu_l\epsilon(\boldsymbol{v})\right) + \nabla p + \frac{1}{\varepsilon}\phi^2(\boldsymbol{v}-\boldsymbol{v}_s) = 0. \quad (6)$$

It should be stressed that in the liquid far from the solid/liquid interface ($\phi = 0$), the mass and momentum equations (3) (6) reduce to the incompressible Navier-Stokes equations whereas in the solid region ($\phi = 1$), due to the penalty term, the velocity equals the solid velocity $\boldsymbol{v}_s$ as ε becomes small.

2.3. Species conservation

We proceed as in [20], eq. (33) and (34). Let C be the average massic concentration of the alloy defined by

$$\rho C = \rho_s \phi C_s + \rho_l (1-\phi) C_l,$$

where C_s, C_l are the solid and liquid concentrations, respectively. Introducing the constant partition coefficient k

$$k = \frac{C_s}{C_l}$$

yields

$$C_l = \frac{\rho C}{k\rho_s\phi + \rho_l(1-\phi)} \quad \text{and} \quad C_s = \frac{k\rho C}{k\rho_s\phi + \rho_l(1-\phi)}. \tag{7}$$

Averaging the species conservation gives

$$\frac{\partial}{\partial t}\Big(\rho_s\phi C_s + \rho_l(1-\phi)C_l\Big) + \nabla\cdot\Big(\rho_s\phi C_s\boldsymbol{v}_s + \rho_l(1-\phi)C_l\boldsymbol{v}_l\Big) \\ - \nabla\cdot\Big(D_s\rho_s\phi\nabla C_s + D_l\rho_l(1-\phi)\nabla C_l\Big) = 0,$$

where D_s, D_l are the constant solid and liquid diffusion coefficients. Using (7) and (2), the quantities C_s, C_l and $\boldsymbol{v}_l$ can be eliminated to obtain

$$\frac{\partial(\rho C)}{\partial t} + \operatorname{div}\left(\frac{\rho C}{k\rho_s\phi + \rho_l(1-\phi)}\Big(\rho\boldsymbol{v} + (k-1)\rho_s\phi\boldsymbol{v}_s\Big)\right) \\ - \operatorname{div}\left(D(\phi)\nabla(\rho C) + D(\phi)\frac{\rho C(\rho_l - k\rho_s)}{k\rho_s\phi + \rho_l(1-\phi)}\nabla\phi\right) = 0,$$

where we have set

$$D(\phi) = \frac{k\rho_s\phi D_s + \rho_l(1-\phi)D_l}{k\rho_s\phi + \rho_l(1-\phi)}.$$

Introducing the volumic concentration $c = \rho C$, the above equation writes

$$\frac{\partial c}{\partial t} + \operatorname{div}\left(\frac{c}{k\rho_s\phi + \rho_l(1-\phi)}\Big(\rho\boldsymbol{v} + (k-1)\rho_s\phi\boldsymbol{v}_s\Big)\right) \\ - \operatorname{div}\Big(D(\phi)\nabla c + \tilde{D}(c,\phi)\nabla\phi\Big) = 0, \tag{8}$$

where we have set

$$\tilde{D}(c,\phi) = D(\phi)\frac{c(\rho_l - k\rho_s)}{k\rho_s\phi + \rho_l(1-\phi)}.$$

2.4. Phase-field

As in [17], we consider a standard phase-field equation for the solid phase moving with constant velocity $\boldsymbol{v}_s$:

$$\frac{1}{\mu_k}\left(\frac{\partial\phi}{\partial t}+\boldsymbol{v}_s\cdot\nabla\phi\right)=\Gamma\left(\operatorname{div}\big(A(\nabla\phi)\nabla\phi\big)-\frac{\phi(1-\phi)(1-2\phi)}{\delta^2}\right)$$
$$+\left(T_m+m_l\frac{c}{k\rho_s\phi+\rho_l(1-\phi)}-T\right)\frac{\phi(1-\phi)}{\delta}. \quad (9)$$

Here μ_k denotes the kinetic mobility, Γ is the Gibbs-Thomson coefficient. The term $\operatorname{div}(A(\nabla\phi)\nabla\phi)$ is the functional derivative (that is to say the Fréchet derivative) of the surface energy

$$\frac{1}{2}\int_\Omega a\big(\theta(\nabla\phi(\boldsymbol{x}))\big)^2|\nabla\phi(\boldsymbol{x})|^2d\boldsymbol{x},$$

where a is the real-valued function defined by

$$a(\theta)=1+\bar{a}\cos(4\theta),$$

with $\bar{a}$ the anisotropy parameter and where $\theta(\xi)$ denotes the angle between a vector $\xi\in\mathbb{R}^2\setminus\{0\}$ and the first component of the orthonormal Cartesian basis (e_1,e_2), that is

$$\cos\theta(\xi)=\frac{\xi\cdot e_1}{\|\xi\|}.$$

Therefore the matrix $A(\cdot)$ in (9) is defined for $\xi\in\mathbb{R}^2\setminus\{0\}$ by

$$A(\xi)=\begin{pmatrix} a^2(\theta(\xi)) & -a(\theta(\xi))a'(\theta(\xi)) \\ a(\theta(\xi))a'(\theta(\xi)) & a^2(\theta(\xi)) \end{pmatrix},$$

The term $\phi(1-\phi)(1-2\phi)$ in (9) is the derivative of the double well which forces the phase field to values close to zero or one. Finally, the last term in (9) is a source term accounting for the energy due to the solid-liquid phase transformation, where T_m is the melting temperature of the pure substance, m_l is the slope of the liquid line in the equilibrium phase diagram. The temperature field T is a known linear function of space and time

$$T(\boldsymbol{x},t)=T_0+t\dot{T}+G\boldsymbol{d}\cdot\boldsymbol{x} \quad (10)$$

where $\dot{T}$ and G are given constants and $\boldsymbol{d}$ is a selected solidification direction in space.

It should be noted that the velocity field $\boldsymbol{v}$ is not present in equation (9) so that, in the sharp interface limit, the classical Gibbs-Thomson relation between the interface velocity V, the curvature and the concentration c is recovered. Therefore, the effect of the fluid motion on Gibbs-Thomson relation is neglected. We refer to [24, 25] for more general models including such effects.

2.5. Summary of the model

The goal of the present model is to find the phase field $\phi : \Omega \times (0, t_{\text{end}}) \to \mathbb{R}$, the volumic concentration $c : \Omega \times (0, t_{\text{end}}) \to \mathbb{R}$, the velocity $\boldsymbol{v} : \Omega \times (0, t_{\text{end}}) \to \mathbb{R}^2$ and the pressure $p : \Omega \times (0, t_{\text{end}}) \to \mathbb{R}$, satisfying equations (9), (8), (6) and (3). Natural boundary conditions apply on the boundary of the calculation domain Ω for ϕ, c and $\boldsymbol{v}$. Moreover, initial conditions at time $t = 0$ must be prescribed for ϕ, c and $\boldsymbol{v}$.

Existence and uniqueness of solutions for this model in the absence of liquid flow and for sufficiently small $\bar{a}$ (small anisotropy) are proved in [26]. A posteriori error estimation and adaptive finite elements are presented in [14]. Existence and convergence of solutions in presence of liquid flow is an open problem.

2.6. Numerical method

Equations (9), (8), (6) and (3) are discretized in time using an order one semi-implicit scheme. Space discretization is based on continuous, piecewise linear finite elements on triangular adapted meshes. In order to reduce the number of degrees of freedom, the triangles may have large aspect ratio whenever needed. The refinement/coarsening criterion is based on a posteriori error estimates and the adaptive algorithm has already been presented for elliptic problems [27], parabolic problems [28], Stokes problem [29], dendritic growth [14] and coalescence [15].

3. Numerical experiments

The liquid flow due to shrinkage is now computed around four dendrites. We place $2 \times 2 = 4$ dendritic seeds in a square Ω of size $0.001\ m$ and let the system evolve observing the liquid flow and the pressure drop during the process. The distance between the seeds is $0.0002\ m$. The temperature is given by (10) with $T_0 = 993.8\ K$, $\dot{T} = 10\ K/s$ and $G = 0$. The values of the physical parameters are given in Table 1.

T_m	k	D_s	D_l	Γ
$1000\ K$	0.5	$5\ 10^{-10}\ m^2/s$	$5\ 10^{-9}\ m^2/s$	$5\ 10^{-7}\ Km$

$\bar{a}$	μ_k	m_l	ρ_s	ρ_l	μ_l
0.04	$0.0015\ m/(Ks)$	$-260\ K$	$1000\ kg/m^3$	$950\ kg/m^3$	$0.014\ kg/(ms)$

TABLE 1. Values of the physical parameters.

The solidification shrinkage causes the liquid to flow toward the center of the square. When the dendrites are sufficiently big, one can observe negative pressure appearing in the almost closed central region of the computational domain Ω. The velocity field is changing significantly during the experiment, the maximum observed velocity being $0.0001\ m/s$. In the beginning, where distance between dendrites is big, the maximum velocity is small, namely $0.00001\ m/s$. However, when neighboring dendrite tips are sufficiently close to each other, the velocity changes rapidly in order to supply the central region of the domain with new material. When dendrites grow further, the spacing between dendrite tips of neighboring dendrites gets smaller and that limits the amount of liquid that can pass toward the center, which causes the pressure drop. Furthermore, the concentration in the center gets much higher than in the rest of the domain. Figures 3 and 5 show the evolution of the system in the case of 4 dendrites.

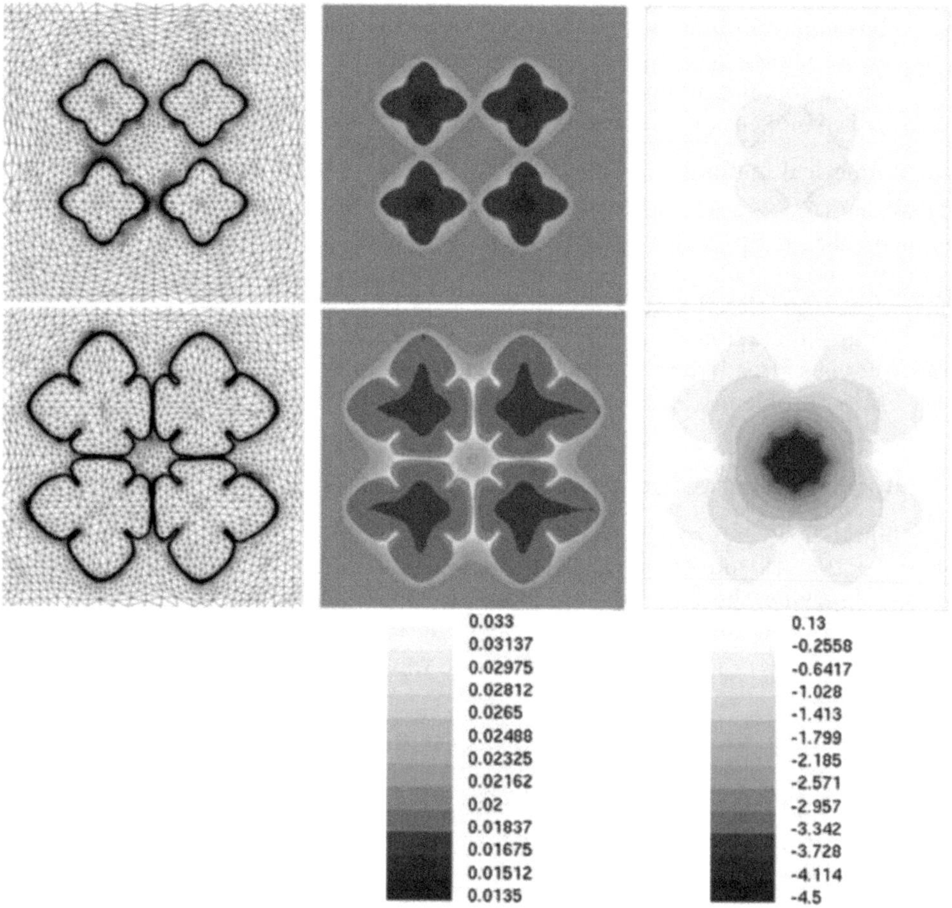

FIGURE 3. Dendritic growth of 4 dendrites at times $t = 0.15$ and 0.225s. Left: adapted meshes (39442 vertexes at final time). Middle: concentration. Right: pressure.

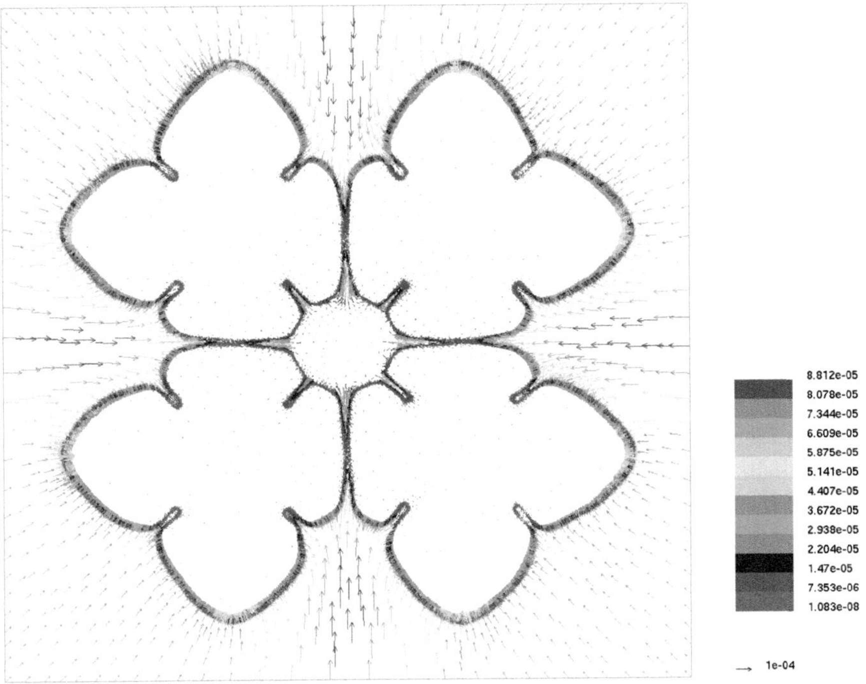

FIGURE 4. Dendritic growth of 4 dendrites at time $t = 0.225\ s$. Velocity field.

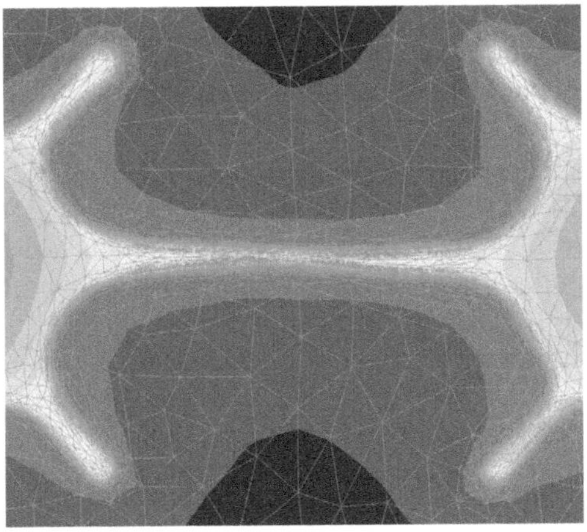

FIGURE 5. Dendritic growth of 4 dendrites at time $t = 0.225\ s$. Zoom of the concentration and adapted mesh. Same scale as in Figure 3.

References

[1] W.J. Boettinger, S.R. Coriell, A.L. Greer, A. Karma, W. Kurz, M. Rappaz, and R. Trivedi. Solidification microstructures: Recent developments, future directions. *Acta Materialia*, 48(1):43–70, 2000.

[2] D. Juric and G. Tryggvason. A front-tracking method for dendritic solidification. *J. Comput. Phys.*, 123(1):127–148, 1996.

[3] A. Jacot and M. Rappaz. A pseudo-front tracking technique for the modelling of solidification microstructures in multi-component alloys. *Acta Mater.*, 50(8):1909–1926, 2002.

[4] M. Fried. A level set based finite element algorithm for the simulation of dendritic growth. *Comput. Vis. Sci.*, 7(2):97–110, 2004.

[5] F. Gibou, R. Fedkiw, R. Caflisch, and S. Osher. A level set approach for the numerical simulation of dendritic growth. *J. Sci. Comput.*, 19(1-3):183–199, 2003.

[6] J.B. Collins and H. Levine. Diffuse interface model of diffusion-limited crystal growth. *Phys. Rev. B*, 31(9):6119–6122, 1985.

[7] G. Caginalp and W. Xie. Phase-field and sharp-interface alloy models. *Phys. Rev. E (3)*, 48(3):1897–1909, 1993.

[8] R. Kobayashi. Modeling and numerical simulations of dendritic crystal growth. *Physica D*, 63:410–423, 1993.

[9] A. Karma. Phase-field model of eutectic growth. *Phys. Rev. E*, 49:2245–2250, 1994.

[10] J.A. Warren and W.J. Boettinger. Prediction of dendritic growth and microsegregation patterns in a binary alloy using the phase-field model. *Acta Metall. Mater.*, 43(2):689–703, 1995.

[11] L.Q. Chen. Phase-field models for microstructure evolutions. *Annual Rev. Mater. Res.*, 32:163–194, 2002.

[12] A. Schmidt. Computation of three dimensional dendrites with finite elements. *J. Comput. Phys.*, 125(2):293–312, 1996.

[13] N. Provatas, N. Goldenfeld, and J. Dantzig. Adaptive mesh refinement computation of solidification microstructures using dynamic data structures. *J. Comput. Phys.*, 148(1):265–290, 1999.

[14] E. Burman and M. Picasso. Anisotropic, adaptive finite elements for the computation of a solutal dendrite. *Interfaces Free Bound.*, 5(2):103–127, 2003.

[15] E. Burman, A. Jacot, and M. Picasso. Adaptive finite elements with high aspect ratio for the computation of coalescence using a phase-field model. *J. Comput. Phys.*, 195(1):153–174, 2004.

[16] J. Ni and C. Beckermann. A volume-averaged 2-phase model for transport phenomena during solidification. *Metall. Trans. B*, 22(3):349–361, 1991.

[17] C. Beckermann, H.J. Diepers, I. Steinbach, A. Karma, and X. Tong. Modeling melt convection in phase-field simulations of solidification. *J. Comput. Phys.*, 154(2):468–496, 1999.

[18] B. Nestler, A.A. Wheeler, L. Ratke, and C. Stöcker. Phase-field model for solidification of a monotectic alloy with convection. *Phys. D*, 141(1-2):133–154, 2000.

[19] J.H. Jeong, N. Goldenfeld, and J. Dantzig. Phase field model for three-dimensional dendritic growth with fluid flow. *Phys. Rev. E*, 64(4):041602, 2001.

[20] W.J. Boettinger, J.A. Warren, C. Beckermann, and A. Karma. Phase-field simulation of solidification. *Annu. Rev. Mater. Res.*, 32:163–194, 2002.

[21] D.M. Anderson, G.B. McFadden, and A.A. Wheeler. A phase-field model of solidification with convection. *Phys. D*, 135(1-2):175–194, 2000.

[22] D.M. Anderson, G.B. McFadden, and A.A. Wheeler. A phase-field model with convection: sharp-interface asymptotics. *Phys. D*, 151(2-4):305–331, 2001.

[23] J.C. Heinrich and D.R. Poirier. Convection modeling in directional solidification. *C. R. Mecanique*, 332:429–445, 2004.

[24] M. Griebel, W. Merz, and T. Neunhoeffer. Mathematical modeling and numerical simulation of freezing processes of a supercooled melt under consideration of density changes. *Comp. Vis. Sci.*, 1(4):201–219, 1999.

[25] Y. Sun and C. Beckermann. Diffuse interface modeling of two-phase flows based on averaging: mass and momentum equations. *Phys. D*, 198(3-4):281–308, 2004.

[26] E. Burman and J. Rappaz. Existence of solutions to an anisotropic phase-field model. *Math. Methods Appl. Sci.*, 26(13):1137–1160, 2003.

[27] M. Picasso. Numerical study of the effectivity index for an anisotropic error indicator based on Zienkiewicz-Zhu error estimator. *Comm. Numer. Methods Engrg.*, 19(1):13–23, 2003.

[28] M. Picasso. An anisotropic error indicator based on Zienkiewicz-Zhu error estimator: application to elliptic and parabolic problems. *SIAM J. Sci. Comput.*, 24(4):1328–1355 (electronic), 2003.

[29] M. Picasso. An adaptive algorithm for the stokes problem using continuous, piecewise linear stabilized finite elements and meshes with high aspect ratio. *Appl. Numer. Math.*, 54(3-4):470–490, 2005.

Jacek Narski and Marco Picasso
Institut d'Analyse et Calcul Scientifique
Ecole Polytechnique Fédérale de Lausanne
CH-1015 Lausanne, Switzerland
e-mail: {jacek.narski,marco.picasso}@epfl.ch

International Series of Numerical Mathematics, Vol. 154, 339–349

A Free Boundary Problem for Nonlocal Damage Propagation in Diatomites

C. Nitsch

Abstract. A new model for fluid flow in diatomites [3] motivates the study of a new degenerate parabolic system. We provide numerical as well as analytical evidence that there exists a free boundary, which represents the interface between the pristine rock and the damaged one.

Keywords. Nonlocal damage mechanics, oil engineering, nonlinear parabolic system, degenerate parabolic system.

1. Introduction

We present a work devoted to the study of a new model of fluid flow in porous media arising in petroleum engineering. The formulation and analysis of the model is motivated by the difficulties to extract oil from oil bearing diatomaceous formations. Important examples of such oil fields can be found in California. The problem is particularly challenging because of the peculiarities of diatomaceous rocks: high porosity, low permeability and fragility. Because of the high porosity the diatomite oil reservoirs are often very rich, but, in view of the low permeability, the wells placed in this kind of oil reservoirs have a very low production rate unless the hydraulic fracture technique is applied. This technique consists of the injection of high pressure fluid from the wells inside the reservoir in order to induce fractures which increase the rock permeability. However, because of the rock fragility, a long term fracturing process causes subsidence phenomena, with very serious consequences for the safety of the wells themselves. Therefore the damaging process on the one hand is necessary to increase the production, on the other hand it has to be monitored and predicted to avoid the collapse of the wells.

A first serious attempt to give a realistic mathematical description which takes into account both the damage accumulation process and the fluid flux, has been recently made by Barenblatt, Patzek, Prostokishin & Silin in [3]. In the next section we review their model. It contains several nonlinearities which cannot be chosen properly without further study. In our work we restrict ourselves to power type nonlinearities, which will be used to identify the dependance of the solutions on the exponents of the powers.

2. Physical background

Hydraulic fracturing was developed in the 1950s and gave to producers the possibilities to extract more oil out of newly discovered and existing fields. Powerful pumps at the surface inject a fluid (the so-called "fracture fluid") into the reservoir rocks. The pressure exerted by the fluid exceeds the compressive stress of the rock, opening fractures which constitute paths of increased permeability.

In diatomaceous oil bearing formations like Lost Hills, the hydraulic fracturing is performed by injecting water. Due to the peculiarities of the diatomite the microstructure of this rock has to change to get any appreciable fluid flow. Actually during field operations the stress in the rock leads to the collapse of wall pores resulting in a network of microcracks which increases the permeability of the diatomite. Eventually the microcrack net connects with the macrofracture. This microcrack accumulation motivated the development of a new model based on the continuum damage mechanics approach. According to this branch of mechanics the damage is usually denoted by ω and is a scalar field which takes values from 0 to 1. The pristine rock in a point is represented by $\omega = 0$ and the completely damaged rock by $\omega = 1$. The intermediate values $(0 < \omega < 1)$ represents partially damage states.

We derive a model which solves simultaneously the macroscopic fluid flow and the microstructural changes of diatomite. We start from the filtration equation of the fluid in the diatomite stratum assuming that the word "fluid" refers to both species (oil and water that we do not distinguish). The diatomite is considered as a weakly compressible elastic porous medium and the fluid as a weakly compressible fluid, so its density ρ is a linear function of the pressure. The diatomaceous stratum is supposed for simplicity homogeneous, with constant height and depth, and bounded from above and below by impermeable rocks. Moreover we assume that inside the reservoir, during geologic times, the pressure p of the fluid and the mean geostatic stress[1] σ, assumed respectively the constant values p_i and σ_i. Since the deposit is deep if we perturb the fluid pressure p from its equilibrium initial value p_i, the sum $p + \sigma$ remains constant and is equal to $p_i + \sigma_i$ during the whole process. Under such hypotheses, following [3] (see also [2]), we obtain from the continuity equation

$$\partial_t(m\rho) + \nabla(\rho\mathbf{u}) = 0 \qquad (\mathbf{u} \text{ is the filtration velocity})$$

[1] The mean geostatic stress is one third of the first invariant of the stress tensor i.e.: $1/3(\sigma_x + \sigma_y + \sigma_z)$,

and from Darcy's law

$$\mathbf{u} = -\frac{k}{\mu}\nabla p \qquad (\mu \text{ is the fluid viscosity})$$

the equation for the pressure:

$$\partial_t p = \nabla(\mathcal{K}\nabla p). \tag{2.1}$$

Here the "piezo diffusivity" coefficient $\mathcal{K}$ is defined as

$$\mathcal{K} = \frac{k}{\mu m c},$$

where c is a constant taking into account the compressibility coefficient of the fluid and the compressibility coefficient of the rock porosity. Equation (2.1) is defined in two spatial dimensions, and all the quantities involved (pressure, porosity, permeability, compressibility etc.) have to be interpreted as averaged on the height of the diatomaceous stratum.

The key idea in [3] was to consider the permeability as a function of the rock damage alone, $k = k(\omega)$.

The basic equation now becomes

$$\partial_t p = \nabla(\mathcal{K}(\omega)\nabla p). \tag{2.2}$$

Here $\mathcal{K}(\omega)$ is an increasing function of ω, and by hypothesis, $\mathcal{K}(0) = 0$. In order to perform a numerical and analytical investigation, we restrict ourselves to power type functional relation $\mathcal{K}(\omega) = K\omega^\alpha$, $\alpha > 0$.

To complete the problem formulation it is now necessary to add an equation for the damage accumulation. Initially the rock is considered pristine ($\omega = 0$), with the possible exception of small regions around the wells which appear during the drilling. When the water is pumped into the wells and starts to filtrate in the diatomaceous rock, the pressure in the pores eventually increases above a certain critical value I and microcracks start to appear. The exact value of I is unknown and has to be determined by infield experiments. We claim that it has to be not less then p_i, since during geological time no damage has been accumulated.

The basic assumption is that locally the damage accumulation rate $\partial_t\omega$ is proportional to a certain power of $(p - I)_+$, and also proportional to the fraction of undamaged rock $1-\omega$. From classical continuum damage mechanics, the process of damage accumulation is governed locally by a kinetic equation of the type:

$$\frac{d\omega}{dt} = A(1-\omega)(p-I)_+^\gamma,$$

where A is a constant. In addition to this bulk mechanism, we consider a nonlocal damage diffusion process. We expect in fact, that fluid wedging takes place in the microcracks. Again we focus on qualitative evaluation of the equations and we choose a very simple nonlocal damage evolution equation of the form:

$$\partial_t\omega = \left[\nabla(D(\omega,p)\nabla\omega) + A(1-\omega)(p-I)_+^\gamma\right]_+ \tag{2.3}$$

where the positive part on the right-hand side avoids a nonphysical damage healing. In particular we will use the following expression for the damage diffusivity coefficient:

$$D(\omega, p) = \Lambda^2 \omega^\mu (p - I)_+^\beta,$$

where Λ is constant.

The physical model, up to a space-time scaling, leads to the following mathematical formulation

$$\begin{cases} \omega_t = \operatorname{div}(\omega^\mu (p-I)_+^\beta \nabla\omega) + A(1-\omega)(p-I)_+^\gamma & \text{in } \mathbb{R}^2 \times \mathbb{R}^+ \\ p_t = \operatorname{div}(\omega^\alpha \nabla p) & \text{in } \mathbb{R}^2 \times \mathbb{R}^+, \end{cases} \tag{2.4}$$

with the following initial conditions

$$\omega(\mathbf{x}, 0) = \omega_0(\mathbf{x}) \qquad p(\mathbf{x}, 0) = p_0(\mathbf{x}) \qquad \text{in } \mathbb{R}^2. \tag{2.5}$$

We shall refer to (2.4) and (2.5) as the *2D formulation* of the diatomite problem.

We assume that $p = 0$ correspond to the rest pressure in the undamaged zone, hence also $I \geq 0$. The constants $\alpha, \beta, \gamma, \mu$ and a satisfy

$$\alpha, \beta, \gamma, A > 0 \quad \mu \geq 0.$$

The case we are interested in is when the initial data ω_0 and p_0 are compactly supported, and have the same support. Outside such domain the rock is pristine and the pressure is the rest pressure.

We also consider the one-dimensional version

$$\begin{cases} \omega_t = (\omega^\mu (p-I)_+^\beta \omega_x)_x + A(1-\omega)(p-I)_+^\gamma & \text{in } \mathbb{R} \times \mathbb{R}^+ \\ p_t = (\omega^\alpha p_x)_x & \text{in } \mathbb{R} \times \mathbb{R}^+ \end{cases} \tag{2.6}$$

with initial data

$$\omega(x, 0) = \omega_0(x) \qquad p(x, 0) = p_0(x) \qquad \text{in } \mathbb{R}. \tag{2.7}$$

We will refer to (2.6) and (2.7) as the *1D formulation* of the diatomite problem.

Both formulations present some mathematical difficulties. The nonnegativity of $\partial_t \omega$ makes the equation fully nonlinear while the diffusivity coefficients ω^α and $\omega^\mu (p - I)_+$ enforce degeneracy.

In [3] it was conjectured that this degeneracy leads to a free boundary problem. Both ω and p remain compactly supported and have the same support for all $t > 0$, and there exists an a priori unknown front which separates the damaged and undamaged rocks. In the following sections we shall see that both numerical and analytical evidence confirm such a conjecture.

3. Numerical simulations

The first numerical experiment presented was performed using the 2D formulation. We have chosen for this example $\alpha = 2$, $\gamma = 5$, $\beta = A = 1$, $\mu = 0$. The domain is a square of vertices $(\pm 1, \pm 1)$. The initial damage and pressure are zero except inside a circle of radius 0.1 and centered in $(0, 0)$ (see Figure 1 when $t = 0$) where pressure and damage are prescribed to be respectively 8 and 1. No flux boundary conditions for both damage and pressure are prescribed on the sides of the square.

These conditions actually models a well injecting fluid in the middle of a square oil field. The results are shown in Figure 1. The simulation runs from $t = 0$ to $t = 0.6$. The idea is to show the sharp front which separates damaged and undamaged regions with the strong indication that it coincides with a pressure front.

In order to catch the behavior of p and ω across such an interface, we turn our attention to the 1D formulation. This time the system was studied on the half line $x \geq 0$ again with initial damage pressure set to zero, and the following boundary conditions: $\partial_x \omega(0,t) = 0$ and $p(0,t) = 1$.

We solved the problem numerically for several values of the parameters involved and selected some specific values in order to exhibit different types of behavior.

The results clearly show that the behavior of the solutions across the boundary is strongly dependent on the constants involved. In Figure 2(a) we set $\alpha = 2$, $\beta = 0.5$, $A = 1$, $\gamma = \mu = 1$ and $I = 0.7$. The simulation suggests that the damage is discontinuous across the free boundary. The pressure seems to decrease linearly on the left side of the boundary toward a value which is reasonably close to I. In Figure 2(b) we set $\alpha = 0.5$, $\beta = 2$, $A = 1$, $\gamma = \mu = 1$ and $I = 0.5$. In this case ω seems to go smoothly to zero crossing the boundary. In Figure 2(c) we set $\alpha = 2$, $\beta = 2$, $A = 1$, $\gamma = \mu = 3$ and $I = 0.5$. It seems that p jumps across the free boundary to a value greater than I, while ω is again continuous.

These results show that when $I > 0$, p is discontinuous across the interfaces. What is more interesting is that even $(p-I)_+$ and ω may be sometimes discontinuous. This opens a nontrivial problem: when ω is discontinuous and $I > 0$ the term $\omega^\alpha p_x$ is ambiguous being the product of a measure and a discontinuous function. An investigation of the travelling waves solution of system 2.6 in [5] confirmed that the behavior of ω and p strongly depends on the parameters involved and in particular there exists travelling waves for which ω or $(p-I)_+$ have jumps across the interface. Nevertheless for those travelling waves the product of ω and $(p-I)_+$ happens to be always zero on the interface. Hence at least one of the two functions has to be continuous. The last condition is in perfect agreement with the results of the next section.

4. Analytical results

We studied existence of solutions of the following system

$$\begin{cases} \omega_t = (\omega^\mu (p-I)_+^\beta \omega_x)_x + A(1-\omega)(p-I)_+^\gamma & \text{in } \mathbb{R} \times \mathbb{R}^+ \\ p_t = (\omega^\alpha p_x)_x & \text{in } \mathbb{R} \times \mathbb{R}^+ \end{cases} \tag{4.1}$$

with initial data

$$\omega(x,0) = \omega_0(x) \qquad p(x,0) = p_0(x) \qquad \text{in } \mathbb{R}. \tag{4.2}$$

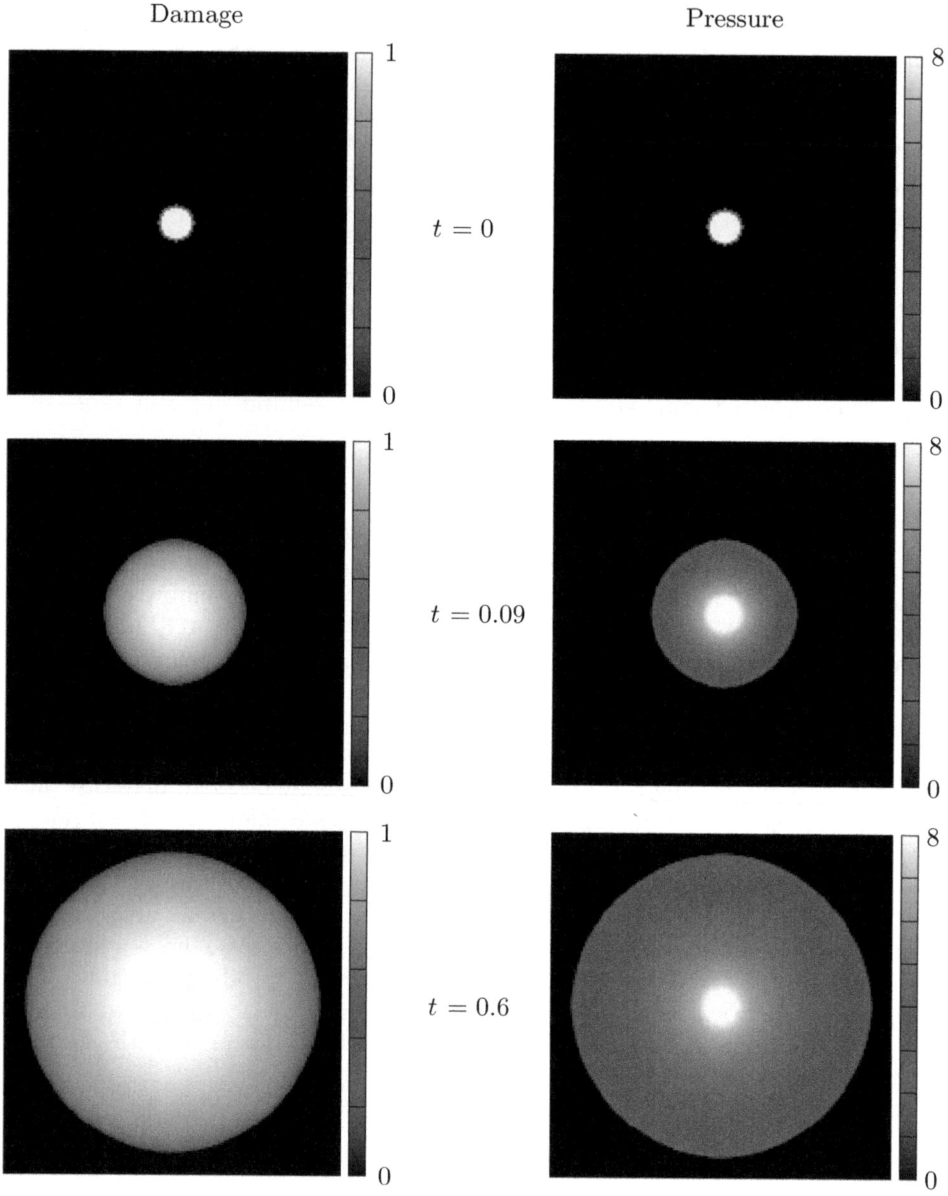

FIGURE 1. The damage and pressure fields at three different instants $t = 0$, 0.09, and 0.6 which demonstrate the damage and pressure extension.

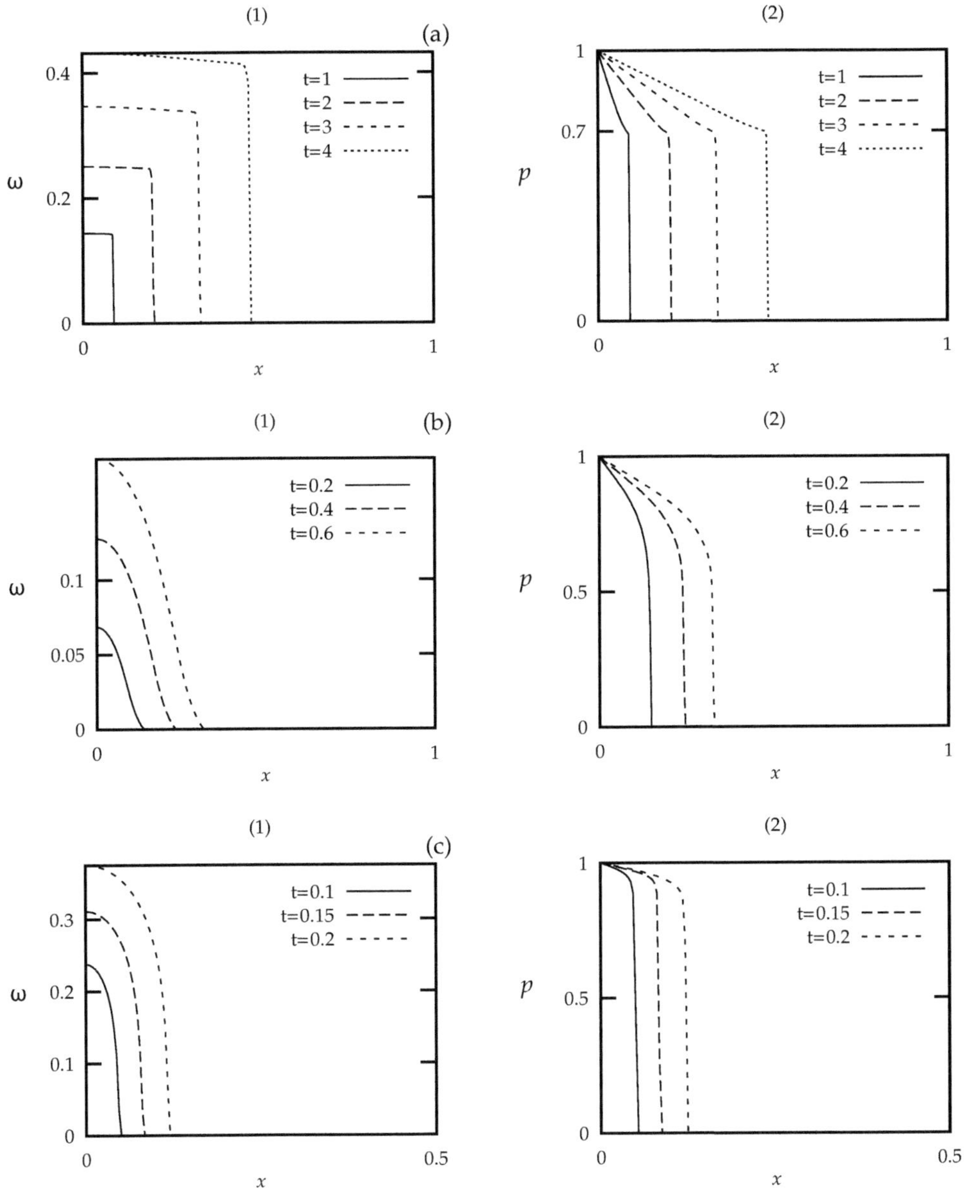

FIGURE 2. Numerical experiment (1) damage, (2) pressure
(a) $\alpha = 2$, $\beta = 0.5$, $\gamma = \mu = 1$; $I = 0.8$
(b) $\alpha = 0.5$, $\beta = 2$, $\gamma = \mu = 1$; $I = 0.6$
(c) $\alpha = 2$, $\beta = 2$, $\gamma = 1$, $\mu = 3$; $I = 0.6$

In this section we do not take into account the physical constraint of no damage healing $\partial_t\omega \geq 0$. The analytical study of this simplified problem is already quite complicated, and experience with other system containing this type of constraint [8] suggest that from an analytical point of view the constraint $\partial_t\omega \geq 0$ is mostly a technical complication.

Throughout the section we shall assume that ω_0 and p_0 are BV functions with compact support. Namely they satisfy the hypotheses

$$\begin{cases} p_0, \omega_0 \in BV(\mathbb{R}), \\ p_0 \geq 0, \quad 0 \leq \omega_0 \leq 1 \quad \text{in } \mathbb{R}, \end{cases} \tag{4.3}$$

and

$$\begin{cases} \omega_0(x) > 0 \quad \text{and} \quad p_0(x) > 0 \qquad \text{if } a < x < b, \\ \omega_0(x) = p_0(x) = 0 \qquad \text{if } x < a \text{ or } x > b. \end{cases} \tag{4.4}$$

We also make a mere technical assumption (it allow us to use, at a certain point, lap number theorem):

$$\begin{cases} \text{there exists } \bar\rho > 0 \text{ such that for any } \rho \in [0, \bar\rho) \\ \text{the sets } \{x : \omega_0(x) > \rho\} \text{ and } \{x : p_0(x) > I \pm \rho\} \\ \text{are intervals (if they are non empty).} \end{cases} \tag{4.5}$$

Here, if $g : D \subseteq \mathbb{R}^n \to [0, \infty)$ with $n \geq 1$, we denote by $\{g > 0\}$ the interior of the set $\left\{x \in D : \liminf_{\rho\to 0} \fint_{B_\rho(x)\cap D} g > 0\right\}$, where we have used the notation

$$\fint_{B_\rho(x)\cap D} f := \frac{1}{|B_\rho(x) \cap D|} \int_{B_\rho(x)\cap D} f,$$

$B_\rho(x)$ being the ball of radius ρ centered in x.

We begin specifying what we mean by solution.

Definition 4.1. A pair $(\omega, p) \in X_T^2 := [L^\infty((0,T); BV(\mathbb{R})) \cap L^\infty([0,T] \times \mathbb{R}) \cap H^1((0,T); H^{-1}(\mathbb{R}))]^2$ for all $T > 0$ is a solution of system (4.1)–(4.2) if

(i) $\omega_x \in L^2_{loc}(\{p > I\})$, $p_x \in L^2_{loc}(\{\omega > 0\})$,
(ii) $\omega^\alpha p_x^2 \in L^1(\{\omega > 0\})$ and $\omega^\mu (p - I)^\beta \omega_x^2 \in L^1(\{p > I\})$,
(iii) for all $T > 0$ and $\phi, \psi \in H^1(\mathbb{R} \times [0,T])$ with compact support (i.e., for some $M > 0$, $\phi = \psi = 0$ for a.e. $x > |M|$) and for all $t \in [0,T]$

$$\int_{\mathbb{R}} \phi(t)p(t) - \int_{\mathbb{R}} \phi(0)p_0 = \int_0^t \int_{\mathbb{R}} \phi_t p - \int\int_{\{\omega>0\}\cap Q_t} \phi_x \omega^\alpha p_x$$

and

$$\int_{\mathbb{R}} \psi(t)\omega(t) - \int_{\mathbb{R}} \psi(0)\omega_0$$
$$= \int_0^t \int_{\mathbb{R}} \psi_t \omega - \iint_{\{p>I\}\cap Q_t} \psi_x \omega^\mu (p-I)^\beta \omega_x + A \int_0^t \int_{\mathbb{R}} \psi (p-I)_+^\gamma (1-\omega).$$

Here $Q_T := \mathbb{R} \times (0,T]$ for all $T > 0$.

Since X_T^2 is embedded in $[C([0,T]; L^2_{loc}(\mathbb{R}))]^2$ (see Proposition 4.3 and its proof in [4] for more details) the integral equalities in (iii) are well defined for all $t \in [0,T]$. We observe the difference between the usual definition of weak solution and the one given above. Two integrals in Definition 4.1 (iii) are restricted to the sets $\{\omega > 0\}$ and $\{p > I\}$. We notice that without such a restriction these integrals might even be ill defined. Indeed we anticipated that across the free boundary we might deal with cases in which both p and ω have jumps.

It is possible to prove the following result.

Theorem 4.2. *Let hypotheses* (4.3), (4.4) *and* (4.5) *be satisfied. Then Problem* (4.1)–(4.2) *has a solution* (ω, p) *which satisfies the following properties:*

(i) *There exist* $a, b \in C([0,\infty))$ *such that* $a(0) = a$, $b(0) = b$, $a(t)$ *is nonincreasing,* $b(t)$ *is nondecreasing, and*
$$\omega(x,t) > 0 \text{ and } p(x,t) > 0 \qquad \text{if } a(t) < x < b(t),$$
$$\omega(x,t) = p(x,t) = 0 \qquad \text{if } x < a(t) \text{ or } x > b(t);$$
(ii) *for a.e.* $t > 0$ $\omega(p-I)_+ \to 0$ *as* $x \searrow a(t)$ *and* $x \nearrow b(t)$;
(iii) *if* $p_0 \geq I$ *in* (a,b) *then* $p(x,t) \geq I$ *if* $a(t) < x < b(t)$.

We give a brief overview of the proof of the Theorem 4.2 and we refer the reader interested to the details to [4].

Our result relies on several steps.

A candidate solution is first provided by an approximation argument. We replace, for $\varepsilon > 0$, the degenerate system (4.1)–(4.2) with

$$\begin{cases} p_t = ((\omega^\alpha + \varepsilon)p_x)_x & \text{in } \mathbb{R}\times\mathbb{R}^+ \\ \omega_t = ((\omega^\mu (p-I)_+^\beta + \varepsilon)\omega_x)_x + A(1-\omega)(p-I)_+^\gamma & \text{in } \mathbb{R}\times\mathbb{R}^+ \\ p(x,0) = p_0(x), \quad \omega(x,0) = \omega_0(x) & \text{for } x \in \mathbb{R}. \end{cases} \tag{4.6}$$

A weak solution of (4.6) is defined in a standard way (requiring that $0 \leq \omega_0 \leq 1$, $p_0 \geq 0$ and $\omega_x, p_x \in L^2_{loc}(\mathbb{R}\times[0,\infty))$). Existence of solutions is provided by the following proposition

Proposition 4.3. *Let* ω_0 *and* p_0 *satisfy*

$$\begin{cases} 0 \leq \omega_0 \leq 1, \ p_0 \geq 0 \quad \text{in } \mathbb{R}, \\ \omega_0, \ p_0 \in BV(\mathbb{R}) \cap L^q(\mathbb{R}) \quad (1 \leq q \leq \infty). \end{cases} \tag{4.7}$$

Then, for all $\varepsilon > 0$*, Problem* (4.6) *has a weak solution* $(\omega_\varepsilon, p_\varepsilon)$ *satisfying*

(i) $0 \le \omega_\varepsilon \le 1$ *and* $0 \le p_\varepsilon \le C$ *in* $\mathbb{R} \times [0, \infty)$*;*

(ii) *for all* $T > 0$ *there exists* $C_T > 0$ *such that*

$$\int_0^T \int_{\mathbb{R}} (\omega_\varepsilon^\mu (p_\varepsilon - I)^\beta + \varepsilon)\omega_{\varepsilon x}^2 dxdt + \int_0^T \int_{\mathbb{R}} (\omega_\varepsilon^\alpha + \varepsilon) p_{\varepsilon x}^2 dxdt \le C_T; \tag{4.8}$$

(iii) ω_ε *and* p_ε *are uniformly bounded in*

$$L^\infty((0,T); L^q(\mathbb{R})) \cap H^1((0,T); H^{-1}(\mathbb{R})) \cap L^\infty((0,T); BV(\mathbb{R}))$$

for all $1 \le q \le \infty$ *and* $T > 0$*;*

(iv) *for all* $T > 0$

$$(\omega_\varepsilon (p_\varepsilon - I)_+)^\nu$$

is uniformly bounded in $L^2((0,T); H^1(\mathbb{R}))$ *if* $\nu \ge \frac{1}{2} \min\{\alpha + 1, \beta + 1, \mu + 2\}$.

A candidate solution of problem (4.1)–(4.2) is then supplied by compactness arguments. It is straightforward to prove that there exists a couple

$$\omega, p \in C([0,T]; L^q(\mathbb{R})) \cap H^1((0,T); H^{-1}(\mathbb{R})) \cap L^\infty((0,T); BV(\mathbb{R})),$$

with $1 \le q < \infty$, $0 \le \omega \le 1$ and $0 \le p \le C$, such that

$$(\omega_{\varepsilon_n}, p_{\varepsilon_n}) \to (\omega, p) \quad \text{in } [C([0,T]; L^2_{loc}(\mathbb{R}))]^2, \tag{4.9}$$

for some sequence $\varepsilon_n \to 0$.

The main difficulty is now to prove that (ω, p) is a solution of (4.1)–(4.2). In particular the proof of (i), (ii) and (iii) in Theorem 4.2 is quite delicate. The existence and continuity of the two functions $a(t)$ and $b(t)$ rely on an iteration technique based on integral estimates and a generalized version of Stampacchia's Lemma, which has been introduced and developed in [6, 7] in the context of the thin film equation and which was applied here for the first time to a system of equations.

Our results suggest several open problems. We noticed that very little is known about the behavior of ω and p across the free boundary. However from (iv) in Proposition 4.3 we can claim that $\omega(x,t)(p(x,t) - I)_+ \to 0$ as $x \searrow a(t)$ and $x \nearrow b(t)$ for a.e. $t > 0$. Therefore at least the continuity of ω or $(p - I)_+$ is guaranteed, in perfect agreement with numerical experiments and travelling wave analysis.

The existence of solutions for the one dimensional formulation without neglecting the positive part at the right-hand side of the equation of ω is another open problem. Also the uniqueness of the solution, as well as the existence of a solution in the two dimensional formulation. The latter is particularly difficult, since it is not even known if the BV-estimate ((iii) in Proposition 4.3) for the solutions of the approximating system 4.6 continues to hold in 2D.

Acknowledgment

The author express sincere gratitude to G.I. Barenblatt, M. Bertsch and R. Dal Passo for continuous support and many stimulating discussions. The author was supported in part by the Office of Science, Office of Advanced Scientific Computing Research, Mathematical, Information, and Computational Sciences Division, Applied Mathematical Sciences Subprogram, of the U.S. Department of Energy, under Contract No. DE-AC03-76SF00098. He thanks the Lawrence Berkeley National Laboratory, where he carried out most of his research, for the great hospitality.

References

[1] G.I. Barenblatt, M. Bertsch, and C. Nitsch, *Nonlocal damage accumulation and fluid flow in diatomites.* Comm. App. Math. and Comp. Sci. **1** (2005), no. 1, 143–168.

[2] G.I. Barenblatt, V.M. Entov, and Ryzhik V.M., *Theory of fluid flows through natural rocks*, Kluwer Academic Publishers, 1990.

[3] G.I. Barenblatt, T.W. Patzek, V.M. Prostokishin, and D.B. Silin, *Oil deposit in diatomites: a new challenge for subterranean mechanics*, Society of Petroleum Engineers Thirteenth Symposium on Oil Recovery, 2002.

[4] M. Bertsch R. Dal Passo, and C. Nitsch, *A system of degenerate parabolic nonlinear PDE's: a new free boundary problem*, Interfaces Free Bound. **7** (2005), 255–276.

[5] M. Bertsch and C. Nitsch, *Traveling wave solutions of a nonlinear degenerate parabolic system from petroleum engineering*, Preprint (2006).

[6] R. Dal Passo, L. Giacomelli, and G. Grün, *A waiting time phenomena for thin film equations*, Ann. Scuola Norm. Sup. Pisa, XXX (2001), 437–463.

[7] ______, *Waiting time phenomena for degenerate parabolic equations – A unifying approach*, Geometric Analysis and Nonlinear Partial Differential Equations (S. Hildebrant and H. Karcher, eds.), Springer-Verlag, 2003, pp. 637–648.

[8] C. Nitsch, *A nonlinear parabolic system arising in damage mechanics under chemical aggression*, Nonlinear Anal. **61** (2005), no. 5, 695–713.

C. Nitsch
Dipartimento di Matematica e Applicazioni "R. Caccioppoli"
Università degli Studi di Napoli "Federico II"
Via Cintia, Monte S. Angelo
I-80126 Napoli, Italy
e-mail: carlo.nitsch@dma.unina.it

International Series of Numerical Mathematics, Vol. 154, 351–360

Concentrating Solutions for a Two-dimensional Elliptic Problem with Large Exponent in Nonlinearity

Angela Pistoia

Abstract. We study the existence of positive and sign-changing solutions to the boundary value problem $-\Delta u = |u|^{p-1}u$ in a bounded smooth domain Ω in $\mathbb{R}^2$, with homogeneous Dirichlet boundary condition, when p is a large exponent. We find topological conditions on Ω which ensure the existence of a positive solution concentrating at exactly m points as $p \to \infty$. In particular, for a non-simply connected domain such a solution exists for any given $m \geq 1$. Moreover, for p large enough, we prove the existence of two pairs of solutions which change sign exactly once and whose nodal lines intersect the boundary of Ω.

Mathematics Subject Classification (2000). Primary 35J60; Secondary 35B33.

Keywords. Large exponent, concentrating solutions, nodal solutions, Green's function, finite-dimensional reduction.

1. Introduction

In this paper we collect some results, obtained in collaboration with Pierpaolo Esposito and Monica Musso in [12] and [13], about the problem

$$-\Delta u = |u|^{p-1}u \text{ in } \Omega, \ u = 0 \text{ on } \partial\Omega. \tag{1.1}$$

where Ω is a smooth bounded domain in $\mathbb{R}^2$ and $p > 1$.

It is well known that for any $p > 1$ problem (1.1) has a least energy solution u_p which is obtained by minimizing the Rayleigh quotient:

$$I_p(u) = \frac{\int_\Omega |\nabla u|^2}{\left(\int_\Omega |u|^{p+1}\right)^{\frac{2}{p+1}}}, \ u \in \mathrm{H}^1_0(\Omega) \setminus \{0\}.$$

The author is supported by M.U.R.S.T., project "Metodi variazionali e topologici nello studio di fenomeni non lineari".

In [17, 18] the authors show that such a least energy solution has L^∞-norm bounded and bounded away from zero uniformly in p, for p large. Furthermore, up to subsequence, the renormalized energy density $p|\nabla u_p|^2$ concentrates as a Dirac delta around a critical point ξ_0 of the Robin function $H(x,x)$, where H is the regular part of the Green function of the Laplacian in Ω with homogeneous Dirichlet boundary condition. Namely, the Green function $G(x,y)$ is the solution of the problem

$$-\Delta_x G(x,y) = \delta_y(x) \ x \in \Omega, \quad G(x,y) = 0 \ x \in \partial\Omega$$

and $H(x,y)$ is the regular part defined as

$$H(x,y) = G(x,y) - \frac{1}{2\pi} \log \frac{1}{|x-y|}.$$

In [14] the authors show that the concentration point of the least energy solution is the maximum point of the Robin function.

In [1, 10] the authors give a further description of the asymptotic behavior of u_p, as $p \to \infty$, by identifying a limit profile problem of Liouville-type:

$$-\Delta u = e^u \ \text{ in } \mathbb{R}^2, \ \int_{\mathbb{R}^2} e^u < +\infty \tag{1.2}$$

and showing that $\|u_p\|_\infty \to \sqrt{e}$ as $p \to +\infty$. The only solutions to this problem are given by

$$U_{\delta,\xi}(y) = \log \frac{8\delta^2}{(\delta^2 + |y-\xi|^2)^2} \quad y \in \mathbb{R}^2,$$

with $\delta > 0$ and $\xi \in \mathbb{R}^2$ (see [8]).

As far as it concerns the existence of sign-changing solutions to (1.1), in [5, 7] it was proved the existence of a least energy nodal solution $\bar{u}_p$ to (1.1) which has exactly two nodal domains. More precisely, if $J_p : \mathrm{H}^1_0(\Omega) \to \mathbb{R}$ is defined by $J_p(u) = \frac{1}{2}\int_\Omega |\nabla u|^2 dx - \frac{1}{p+1}\int_\Omega |u|^{p+1} dx$ and $N_p = \{u \in \mathrm{H}^1_0(\Omega) \mid u^+ \neq 0,\ u^- \neq 0,\ J_p'(u)(u^+) = J_p'(u)(u^-) = 0\}$, then it holds $J_p(\bar{u}_p) = \min_{N_p} J_p$.

In [12, 13] the authors build positive and sign-changing solutions for problem (1.1) that, up to a suitable normalization, look like a sum or a difference of concentrated solutions for the limit profile problem (1.2) centered at several points $\xi_1, \ldots, \xi_m$, as $p \to \infty$.

Our main results are the following.

The first result concerns the existence of positive solutions in a not simply connected domain.

Theorem 1.1. *If Ω is not simply connected, then for any k there exists $p_k > 0$ such that for any $p \geq p_k$ problem (1.1) has at least a family of positive solutions concentrating at k different points of Ω.*

The second result concerns the existence of positive solutions in a simply connected domain.

Theorem 1.2. *If Ω is a h-dumbbell with thin handles, then there exists $p_h > 0$ such that for any $p \geq p_h$ problem (1.1) in Ω has at least $2^h - 1$ different families of positive solutions concentrating at different points in Ω, i.e., h families of solutions concentrating at 1 point , $\frac{h(h-1)}{2}$ families of solutions concentrating at 2 points, ..., 2 families of solutions concentrating at $h-1$ points, 1 family of solutions concentrating at h points.*

The third result concerns the existence and the profile of sign-changing solutions in a general domain.

Theorem 1.3. *There exists $p_0 > 0$ such that for any $p \geq p_0$ problem (1.1) has at least two pairs of sign-changing solution $\pm u_p^i$, $i = 1, 2$, with one positive peak and one negative peak. Moreover $\Omega \setminus \{x \in \Omega \,:\, u_p^i(x) = 0\}$ has exactly two connected components and $\overline{\{x \in \Omega \,:\, u_p^i(x) = 0\}} \cap \partial\Omega \neq \emptyset$.*

We would like to compare problem (1.1) with a widely studied problem which has some analogies with it.

In higher dimension the problem equivalent to problem (1.1) is the slightly sub-critical problem

$$-\Delta u = |u|^{\frac{4}{N-2}-\epsilon} u \text{ in } \Omega, \;\; u = 0 \text{ on } \partial\Omega, \tag{1.3}$$

where Ω is a smooth bounded domain in $\mathbb{R}^N$, $N \geq 3$ and ϵ is a positive parameter. Indeed, in dimension $N \geq 3$, the embedding of $H_0^1(\Omega)$ in $L^{p+1}(\Omega)$ is compact for every $p < \frac{N+2}{N-2}$ and the minimum of the Rayleigh quotient corresponding to problem (1.3) is achieved by a positive function u_ϵ, called least energy solution, which, after a multiplication by a suitable positive constant, is a solution to (1.3). It is well known that, as ϵ goes to 0, the least energy solution u_ϵ concentrates around a point, which is a critical point of the Robin function of the corresponding Green function (see [4, 14, 15, 19]). Also the converse is true: around any *stable* critical point of the Robin function one can build a family of positive solutions for (1.3) concentrating precisely there (see [19, 20]). In [4] and [16] the authors showed that also for problem (1.3) there exist positive solutions with concentration in multiple points and, as in the problem that we are considering in the present paper, the points of concentration are given by critical points of a certain function defined in terms of both the Green function and Robin function. However, as far as it concerns positive solutions, the analogies between problems (1.1) and (1.3) break down here. Indeed, while for (1.1) one can find positive solutions with an arbitrarily large number of condensation points in any given not simply connected domain Ω as proved in Theorem 1.1, in [4] the authors proved that positive solutions to (1.3) can have at most a finite number of peaks which depends on Ω.

On the other hand, as far as it concerns sign-changing solutions, results similar to those in Theorem 1.3 have been obtained in [6] for problem (1.3). In [6] the authors prove that, if ϵ is small enough, problem (1.3) has at least N pairs of sign changing solutions $\pm u^i_\epsilon$, $i = 1, \dots, N$, such that, as ϵ goes to zero, u^i_ϵ blows up positively at a point ξ^i_1 and blows up negatively at a point ξ^i_2, with $\xi^i_1, \xi^i_2 \in \Omega$, $\xi^i_1 \neq \xi^i_2$. Moreover, the set $\Omega \setminus \overline{\{x \in \Omega \ : \ u^i_\epsilon(x) = 0\}}$ has exactly two connected components and the nodal set $\overline{\{x \in \Omega \ : \ u^i_\epsilon(x) = 0\}}$ intersects the boundary of Ω, provided $H(\xi^i_1, \xi^i_1) = H(\xi^i_2, \xi^i_2)$. Here H denotes the Robin function of the Green function of the Dirichlet Laplacian.

2. Finite-dimensional reduction

The proof of all our results relies on a Lyapunov-Schmidt procedure (as in the pioneering papers [2, 3]), which is briefly sketched in the following. The detailed proofs can be found in [12, 13].

Let $PU_{\delta,\xi}$ denote the projection of $U_{\delta,\xi}$ onto $\mathrm{H}^1_0(\Omega)$, namely $\Delta PU_{\delta,\xi} = \Delta U_{\delta,\xi}$ in Ω, $PU_{\delta,\xi} = 0$ on $\partial\Omega$. Secondly, let $U(y) := U_{1,0}(y)$. Let V be the radial solution to

$$\Delta V + e^U V = e^U v \quad \text{in } \mathbb{R}^2, \ v := \frac{1}{2} U^2$$

and let W be the radial solution to

$$\Delta W + e^U W = e^U w \quad \text{in } \mathbb{R}^2,$$
$$w := VU - \frac{1}{2}(V)^2 - \frac{1}{3}U^3 - \frac{1}{8}U^4 + \frac{1}{2}VU^2.$$

For any $\delta > 0$ and $\xi \in \mathbb{R}^2$, we define

$$V_{\delta,\xi}(x) := V\left(\frac{x-\xi}{\delta}\right), \ W_{\delta,\xi}(x) := W\left(\frac{x-\xi}{\delta}\right), \ x \in \Omega.$$

Then ${}_{\delta,\xi}$ solves

$$\Delta V_{\delta,\xi} + e^{U_{\delta,\xi}} V_{\delta,\xi} = e^{U_{\delta,\xi}} v_{\delta,\xi} \quad \text{in } \mathbb{R}^2, \ v_{\delta,\xi}(x) := v^\eta\left(\frac{x-\xi}{\delta}\right)$$

and $W_{\delta,\xi}$ solves

$$\Delta W_{\delta,\xi} + e^{U_{\delta,\xi}} W_{\delta,\xi} = e^{U_{\delta,\xi}} w_{\delta,\xi} \quad \text{in } \mathbb{R}^2, \ w_{\delta,\xi}(x) := w^\eta\left(\frac{x-\xi}{\delta}\right).$$

Let $PV_{\delta,\xi}$ and $PW_{\delta,\xi}$ denote the projection onto $\mathrm{H}^1_0(\Omega)$ of $V_{\delta,\xi}$ and $W_{\delta,\xi}$, respectively.

Let $\xi_1, \dots, \xi_k$ be k different points in Ω. Define now

$$U_\xi(x) := \sum_{i=1}^k \frac{a_i}{\gamma \mu_i^{\frac{2}{p-1}}} \left(PU_{\delta_i,\xi_i}(x) + \frac{1}{p} PV^{\xi_i}_{\delta_i,\xi_i}(x) + \frac{1}{p^2} PW^{\xi_i}_{\delta_i,\xi_i}(x) \right) \tag{2.1}$$

where $a_i \in \{-1, +1\}$,

$$\gamma := p^{\frac{p}{p-1}} e^{-\frac{p}{2(p-1)}}, \tag{2.2}$$

so that $\lim_{p\to+\infty} \frac{\gamma}{p} = \frac{1}{\sqrt{e}}$. Moreover, concentration parameters $\delta_1, \dots, \delta_k$ satisfy

$$\delta_i = \mu_i e^{-p/4} \tag{2.3}$$

with $\mu_i := \mu_i(p,\xi)$, $i = 1, \dots, k$, given by

$$\log(8\mu_i^4) = \left(8\pi H(\xi_i,\xi_i) + 8\pi \sum_{j\neq i} a_i a_j \frac{\mu_i^{\frac{2}{p-1}}}{\mu_j^{\frac{2}{p-1}}} G(\xi_i,\xi_j)\right)\left(1 - \frac{\nu}{4p} - \frac{\omega}{4p^2}\right) + \frac{\log \delta_i}{p}\left(\nu + \frac{\omega}{p}\right). \tag{2.4}$$

for some positive constants ν and ω.

A direct computation shows that, for p large, μ_i thus satisfies

$$\mu_i \sim e^{-\frac{3}{4}} e^{2\pi H(\xi_i,\xi_i) + 2\pi \sum_{j\neq i} a_i a_j G(\xi_j,\xi_i)}.$$

We point out that with this choice of parameters μ_i the profile of U_ξ in a neighborhood of the concentration point ξ_i is

$$U_\xi(\delta_i y + \xi_i) \sim a_i \sqrt{e}\left[1 + \frac{1}{p}U(y) + \frac{1}{p^2}V(y) + \frac{1}{p^3}W(y)\right].$$

We will look for a solution to (1.1) of the form $u = U_\xi + \phi$, where ϕ is a higher order term in the expansion of u and $\xi \in O_\epsilon$, where, for some $\epsilon > 0$,

$$O_\epsilon := \left\{\xi \in \Omega^k \mid (\text{dist } \xi_i, \partial\Omega) \geq 2\epsilon,\ |\xi_i - \xi_j| \geq 2\epsilon,\ i,j = 1,\dots,k,\ i \neq j\right\}.$$

It is useful to rewrite problem (1.1) in terms of ϕ, namely

$$\begin{cases} L(\phi) = -\left[R + N(\phi)\right] & \text{in } \Omega, \\ u = 0 \quad \text{on } \partial\Omega. \end{cases} \tag{2.5}$$

where

$$L(\phi) := L(p,\xi,\phi) = \Delta\phi + g_p'(U_\xi)\,\phi, \tag{2.6}$$

$$R := R(p,\xi) = \Delta U_\xi + g_p(U_\xi), \tag{2.7}$$

$$N(\phi) := N(p,\xi,\phi) = g_p(U_\xi + \phi) - g_p(U_\xi) - g_p'(U_\xi)\,\phi. \tag{2.8}$$

Here $g_p(s) := |s|^{p-1}s$.

A first step to solve (2.5), or equivalently (1.1), consists in studying the invertibility properties of the linear operator L. In order to do so we introduce the weighted L^∞ norm defined by

$$\|h\|_* := \sup_{x\in\Omega}\left|\left(\sum_{i=1}^k \frac{\delta_i}{(\delta_i^2 + |x-\xi_i|^2)^{\frac{3}{2}}}\right)^{-1} h(x)\right| \quad \text{for any } h \in L^\infty(\Omega).$$

With respect to this norm, the error term $R(p,\xi)$ given in (2.7) can be estimated in the following way.

Lemma 2.1. *Let $\epsilon > 0$ be fixed. There exists $c > 0$ and $p_0 > 0$ such that for any $\xi \in O_\epsilon$ and $p \geq p_0$ we have $\|\Delta U_\xi + g_p(U_\xi)\|_* \leq \frac{c}{p^4}$.*

Next, we will solve the following projected linear problem: given $h \in C^{0,\alpha}(\Omega)$, find a function ϕ and constants $c_{i,j}$, for $i = 1, \dots, k$, $j = 1, 2$, such that

$$\begin{cases} L(\phi) = h + \sum\limits_{i,j} c_{i,j} e^{U_{\delta_i,\xi_i}} Z_{i,j}, & \text{in } \Omega, \\ \phi = 0 \quad \text{on } \partial\Omega, \\ \int\limits_{\Omega} e^{U_{\delta_i,\xi_i}} Z_{i,j} \phi = 0 & \text{if } i = 1, \dots, k,\ j = 1, 2, \end{cases} \tag{2.9}$$

Here we set, for $i = 1, \dots, k$ and $j = 1, 2$,

$$Z_{i,j}(x) := z_j \left(\frac{x - \xi_i}{\delta_i} \right), \quad \text{with} \quad z_j(y) := \frac{\partial U}{\partial y_j}(y) = \frac{4 y_j}{1 + |y|^2}.$$

This linear problem is uniquely solvable, for p sufficiently large, in the set of functions with bounded $\|\cdot\|_*$-norm.

Lemma 2.2. *Let $\epsilon > 0$ be fixed. There exist $c > 0$ and $p_0 > 0$ such that for any $p > p_0$ and $\xi \in \mathcal{O}_\epsilon$ there is a unique solution ϕ to problem (2.9) which satisfies $\|\phi\|_\infty \le cp\|h\|_*$.*

Let us now introduce the following nonlinear auxiliary problem

$$\begin{cases} \Delta(U_\xi + \phi) + g_p(U_\xi + \phi) = \sum\limits_{i,j=1}^{2} c_{i,j} e^{U_{\delta_i,\xi_i}} Z_{i,j} & \text{in } \Omega, \\ \phi = 0 \quad \text{on } \partial\Omega, \\ \int\limits_{\Omega} e^{U_{\delta_i,\xi_i}} Z_{i,j} \phi = 0 & \text{if } i = 1, \dots, k,\ j = 1, 2, \end{cases} \tag{2.10}$$

for some coefficients $c_{i,j}$. The following result holds.

Proposition 2.3. *Let $\epsilon > 0$ be fixed. There exist $c > 0$ and $p_0 > 0$ such that for any $p > p_0$ and $\xi \in \mathcal{O}_\epsilon$ problem (2.10) has a unique solution $\phi_p(\xi)$ which satisfies $\|\phi_p(\xi)\|_\infty \le \frac{c}{p^3}$. Furthermore, the function $\xi \to \phi_p(\xi)$ is a C^1 function in $L^\infty(\Omega)$ and in $H_0^{1,2}(\Omega)$.*

After problem (2.10) has been solved, we find a solution to problem (2.5) (hence to the original problem (1.1)) if we find a point ξ such that coefficients $c_{ij}(\xi)$ in (2.10) satisfy $c_{ij}(\xi) = 0$ for $i = 1, \dots, k$, $j = 1, 2$.

Let us introduce the energy functional $J_p : H_0^1(\Omega) \to \mathbb{R}$ given by

$$J_p(u) := \frac{1}{2} \int\limits_{\Omega} |\nabla u|^2 dx - \frac{1}{p+1} \int\limits_{\Omega} |u|^{p+1} dx, \tag{2.11}$$

whose critical points are solutions to (1.1). We also introduce the finite-dimensional restriction $\widetilde{J}_p : \Omega^k \to \mathbb{R}$ given by

$$\widetilde{J}_p(\xi) := J_p\big(U_\xi + \phi_p(\xi)\big).$$

The following result holds.

Lemma 2.4. *If $\xi \in \Omega^k$ is a critical point of $\widetilde{J}_p$, then $U_\xi + \phi_p(\xi)$ is a critical point of J_p, namely a solution to problem (1.1).*

Next we need to write the expansion of $\widetilde{J}_p$ as p goes to $+\infty$,

Lemma 2.5. *It holds*

$$\widetilde{J}_p(\xi) = k\frac{4\pi e}{p} + k\frac{8\pi e(\log 8 - 1)}{p^2} - \frac{32\pi^2 e}{p^2}\Phi(\xi) + R_p(\xi),$$

where $R_p = O(p^{-3})$ C^1*-uniformly with respect to* ξ *in compact sets of* $M_k := \Omega^k \setminus \Delta$, *where* $\Delta = \{\xi \in \Omega^k \ : \ \xi_i = \xi_j \text{ for some } i,\ j\}$ *and the function* $\Phi_k : M_k \to \mathbb{R}$ *is defined by*

$$\Phi_k(\xi_1, \dots, \xi_k) = \sum_{i=1}^{k} H(\xi_i, \xi_i) + \sum_{\substack{i,j=1 \\ i \neq j}}^{k} a_i a_j G(\xi_i, \xi_j). \tag{2.12}$$

Finally by Lemma 2.5 and Lemma 2.4 we easily deduce the following result.

Theorem 2.6. *Assume* $\xi^* = (\xi_1^*, \dots, \xi_k^*) \in M$ *is a critical point of* Φ_k *stable with respect to* C^1*-perturbations. Then there exists* $p_0 > 0$ *such that for any* $p \geq p_0$ *there exists a solution to problem* (1.1) *which concentrates at points* $\xi_1^*, \dots, \xi_k^*$.

3. Proof of main results

By Theorem 2.6, it is clear that looking for positive k-peaks solutions to (1.1) means looking for C^1-stable critical points of the function Φ_k defined in (2.12) with $a_1 = \dots = a_k = +1$, namely

$$\Phi_k(\xi_1, \dots, \xi_k) = \sum_{i=1}^{k} H(\xi_i, \xi_i) + \sum_{\substack{i,j=1 \\ i \neq j}}^{k} G(\xi_i, \xi_j).$$

Proof of Theorem 1.1. We argue as in [9]. Let γ be a curve in Ω around the hole. Let $c^* := \inf_{H \in H} \sup_{z \in \gamma^k} \Phi_k\big(H(1, z)\big)$, where $H := \big\{H : [0,1] \times \gamma^k \to M_k$ continuous, $H(0, \cdot) = \mathrm{id}\big\}$. Then c^* is a critical value of Φ_k and there exists $\xi^* \in M_k$ which is a C^1-stable critical point of Φ_k. Finally, the claim follows by Theorem 2.6. □

Proof of Theorem 1.2. We argue as in [11]. Let h be an integer. By h-dumbbell domain with thin handles we mean the following: let $\Omega_0 = \Omega_1 \cup \dots \cup \Omega_h$, with $\Omega_1, \dots, \Omega_h$ smooth bounded domains in $\mathbb{R}^2$ such that $\bar{\Omega}_i \cap \bar{\Omega}_j = \emptyset$ if $i \neq j$. Assume that $\Omega_i \subset \{(x_1, x_2) \in \mathbb{R}^2 \ : \ a_i \leq x_1 \leq b_i\}$, and $\Omega_i \cap \{x_2 = 0\} \neq \emptyset$, for some $b_i < a_{i+1}$ and $i = 1, \dots, h$. Let $C_\varepsilon = \{(x_1, x_2) \in \mathbb{R}^2 \ : \ |x_2| \leq \varepsilon,\ x_1 \in (a_1, b_h)\}$, for some $\varepsilon > 0$. We say that Ω_ε is a h-dumbbell with thin handles if Ω_ε is a smooth simply connected domain such that $\Omega_0 \subset \Omega_\varepsilon \subset \Omega_0 \cup C_\varepsilon$, for some $\varepsilon > 0$.

Set, for $1 \leq k \leq h$,

$$\Phi_k^\varepsilon(\xi) = \sum_{i=1}^{k} H_{\Omega_\varepsilon}(\xi_i, \xi_i) + \sum_{\substack{i,j=1 \\ i \neq j}}^{k} G_{\Omega_\varepsilon}(\xi_i, \xi_j),\ \ \xi \in \Omega_\varepsilon^k.$$

It is easy to check that Φ_k^0 has $\frac{h!}{(k-h)!k!}$ local maximum points in Ω_0^k, i.e., the absolute maximum point in the connected component $\Omega_{i_1} \times \cdots \times \Omega_{i_k}$, for any $\{i_1, \ldots, i_k\} \subset \{1, \ldots, h\}$ with $i_l \neq i_m$, $l \neq m$. Then if $\varepsilon \sim 0$, Φ_k^ε has $\frac{h!}{(k-h)!k!}$ local maximum points in Ω_ε^k, which are C^1-stable critical points of Φ_k^ε. Finally, the claim follows by Theorem 2.6. □

By Theorem 2.6, it is clear that looking for sign-changing solutions to (1.1) with one positive peak and one negative peak means looking for C^1-stable critical points of Φ_2 defined in (2.12) with $a_1 = +1$ and $a_2 = -1$, namely

$$\Phi_2(\xi_1, \xi_2) = H(\xi_1, \xi_1) + H(\xi_2, \xi_2) - 2G(\xi_1, \xi_2).$$

Proof of Theorem 1.3. First of all, we note that $\Phi_2(\xi_1, \xi_2) \to -\infty$ as (ξ_1, ξ_2) approaches ∂M_2 and $\Phi_2(\xi_1, \xi_2) = \Phi_2(\xi_2, \xi_1)$. Therefore Φ_2 has at least cat $\widetilde{M}_2$ critical points which are C^1-stable. Here $\widetilde{M}_2$ denote the quotient manifold M_2 with respect to the equivalence $(\xi_1, \xi_2) \sim (\xi_2, \xi_1)$. It is easy to see that cat $\widetilde{M}_2 \geq 2$.

By Theorem 2.6 it follows that, provided p is large enough, there exist pairs $u_p^{(i)}$ and $-u_p^{(i)}$ of solutions to (1.1) for $i = 1, \ldots, \mathrm{cat}\, \widetilde{M}_2$. For sake of simplicity, set $u_p = u_p^{(i)}$ for some i. We know the profile of u_p, more precisely

$$\begin{aligned} u_p(x) = & \frac{1}{\gamma_p \mu_{1p}^{\frac{2}{p-1}}} \left(PU_{\delta_{1p}, \xi_{1p}}(x) + \frac{1}{p} PV_{\delta_{1p}, \xi_{1p}} \right) \\ & - \frac{1}{\gamma_p \mu_{2p}^{\frac{2}{p-1}}} \left(PU_{\delta_{2p}, \xi_{2p}}(x) + \frac{1}{p} PV_{\delta_{2p}, \xi_{2p}} \right) + \hat{\phi}_p(x) \end{aligned}$$

where γ_p, δ_{ip} and μ_{ip} are given in (2.2), (2.3) and (2.4), respectively. Moreover $\xi_{ip} \to \xi_i^*$, $\xi_1^*, \xi_2^* \in \Omega$, $\xi_1^* \neq \xi_2^*$. Finally $\|\hat{\phi}_p\|_{\mathrm{L}^\infty(\Omega)} \leq \frac{C}{p^3}$.

Let us prove that $\{x \in \Omega \,:\, u_p(x) \neq 0\}$ has exactly two connected components. We remark that there exists $r > 0$, independent of p, such that $B(\xi_{1p}, r) \subset \Omega_p^+ := \{x \in \Omega \,:\, u_p(x) > 0\}$, $B(\xi_{2p}, r) \subset \Omega_p^- := \{x \in \Omega \,:\, u_p(x) < 0\}$ and $\|u_p\|_{\mathrm{L}^\infty(\Omega \setminus \{B(\xi_{1p}, r) \cup B(\xi_{2p}, r)\})} \leq \frac{c}{p}$. Assume there exists a third connected component $\omega_p \subset \Omega \setminus \{B(\xi_{1p}, r) \cup B(\xi_{2p}, r)\}$. Then u_p solves

$$-\Delta u_p = |u_p|^{p-1} u_p \ \text{ in } \omega_p, \ u_p = 0 \ \text{ on } \partial\omega_p.$$

It follows that

$$\|u_p\|^2_{\mathrm{H}_0^1(\omega_p)} \leq \||u_p|^{p-1}\|_{\mathrm{L}^{\frac{p}{p-2}}(\omega_p)} \|u_p\|^2_{\mathrm{L}^p(\omega_p)} \leq S_p^2 \|u_p\|^{p-1}_{\mathrm{L}^{\frac{p(p-1)}{p-2}}(\omega_p)} \|u_p\|^2_{\mathrm{H}_0^1(\omega_p)}.$$

Finally a contradiction easily arises, since $S_p^2 \sim 8\pi e p$ and $p\|u_p\|^{p-1}_{\mathrm{L}^{\frac{p(p-1)}{p-2}}(\omega_p)} = o(1)$.

Let us prove that $\overline{\{x \in \Omega \,:\, u_p(x) = 0\}} \cap \partial\Omega \neq \emptyset$. We can prove that

$$p u_p(x) \to 8\pi\sqrt{e}\,[G(x, \xi_1^*) - G(x, \xi_2^*)] \ \text{ in } C^1_{loc}(\overline{\Omega} \setminus \{\xi_1^*, \xi_2^*\}).$$

Then it follows that

$$p\frac{\partial u_p}{\partial \nu}(x) \to 8\pi\sqrt{e}\frac{\partial}{\partial \nu}\left[G(x,\xi_1^*) - G(x,\xi_2^*)\right] \text{ uniformly on } \partial\Omega.$$

On the other hand, the fact that $\int_{\partial\Omega} \frac{\partial}{\partial \nu}\left[G(x,\xi_1^*) - G(x,\xi_2^*)\right] = 0$, implies that $\frac{\partial}{\partial \nu}\left[G(\cdot,\xi_1^*) - G(\cdot,\xi_2^*)\right]$ changes sign on $\partial\Omega$. Therefore $\frac{\partial u_p}{\partial \nu}$ changes sign on $\partial\Omega$ and finally u_p changes sign in a neighborhood of $\partial\Omega$. □

References

[1] Adimurthi, M. Grossi, *Asymptotic estimates for a two-dimensional problem with polynomial nonlinearity.* Proc. Amer. Math. Soc. **132** (2004), no. 4, 1013–1019.

[2] A. Bahri, *Critical points at infinity in some variational problems,* Pitman Research Notes in Mathematics Series **182** (1989), Longman.

[3] A. Bahri, J.M. Coron, *On a nonlinear elliptic equation involving the critical Sobolev exponent: the effect of the topology of the domain.* Comm. Pure Appl. Math. **41** (1988), no. 3, 253–294.

[4] A. Bahri, Y.Y. Li, O. Rey, *On a variational problem with lack of compactness: the topological effect of the critical points at infinity.* Calc. Var. Partial Differential Equations **3** (1995), no. 1, 67–93.

[5] T. Bartsch, *Critical point theory on partially ordered Hilbert spaces.* J. Funct. Anal. **186** (2001), 117–152.

[6] T. Bartsch, A.M. Micheletti, A. Pistoia, *On the existence and the profile of nodal solutions of elliptic equations involving critical growth.* Calc. Var. Partial Differential Equations **26** (2006), no. 3, 265–282.

[7] T. Bartsch, T. Weth, *A note on additional properties of sign-changing solutions to superlinear elliptic equations.* Top. Meth. Nonlin. Anal. **22** (2003), 1–14.

[8] W. Chen, C. Li, *Classification of solutions of some nonlinear elliptic equations.* Duke Math. J. **63** (1991), 615–623.

[9] M. del Pino, M. Kowalczyk, M. Musso, *Singular limits in Liouville-type equations.* Calc. Var. Partial Differential Equations **24** (2005), no. 1, 47–81.

[10] K. El Mehdi, M. Grossi, *Asymptotic estimates and qualitative properties of an elliptic problem in dimension two.* Adv. Nonlinear Stud. **4** (2004), no. 1, 15–36.

[11] P. Esposito, M. Grossi, A. Pistoia, *On the existence of blowing-up solutions for a mean field equation.* Ann. IHP Analyse Non Linéaire **22**, no. 2, 227–257.

[12] P. Esposito, M. Musso, A. Pistoia, *Concentrating solutions for a planar elliptic problem involving nonlinearities with large exponent.* Journal of Differential Equations (to appear).

[13] P. Esposito, M. Musso, A. Pistoia, *On the existence and profile of nodal solutions for a two-dimensional elliptic problem with large exponent in nonlinearity* Journal of London Mathematical Society (to appear).

[14] M. Flucher, J. Wei, *Semilinear Dirichlet problem with nearly critical exponent, asymptotic location of hot spots.* Manuscripta Math. **94** (1997), no. 3, 337–346.

[15] Z.C. Han, *Asymptotic approach to singular solutions for nonlinear elliptic equations involving critical Sobolev exponent.* Ann. IHP Analyse Non Linéaire **8** (1991), 159–174.

[16] M. Musso, A. Pistoia, *Multispike solutions for a nonlinear elliptic problem involving the critical Sobolev exponent.* Indiana Univ. Math. J. **51** (2002), no. 3, 541–579.

[17] X. Ren, J. Wei, *On a two-dimensional elliptic problem with large exponent in nonlinearity.* Trans. Amer. Math. Soc. **343** (1994), no. 2, 749–763.

[18] X. Ren, J. Wei, *Single point condensation and least energy solutions.* Proc. Amer. Math. Soc. **124** (1996), 111–120.

[19] O. Rey, *The role of the Green function in a nonlinear elliptic equation involving the critical Sobolev exponent.* J. Funct. Anal. **89** (1990), 1–52.

[20] O. Rey, *A multiplicity result for a variational problem with lack of compactness.* Nonlinear Anal. T.M.A. **13** (1989), 1241–1249.

Angela Pistoia
Dipartimento Me. Mo. Mat.
via A. Scarpa 16
I-00161 Roma, Italy
e-mail: `pistoia@dmmm.uniroma1.it`

International Series of Numerical Mathematics, Vol. 154, 361–368

Existence of Weak Solutions for the Mullins-Sekerka Flow

Matthias Röger

Abstract. We prove the long-time existence of solutions for the Mullins-Sekerka flow. We use a time discrete approximation which was introduced by Luckhaus and Sturzenhecker [Calc. Var. PDE 3 (1995)] and pass in a new weak formulation to the limit.

Mathematics Subject Classification (2000). Primary 49Q20; Secondary 80A22.

Keywords. Free boundaries, Mullins-Sekerka flow, varifolds.

1. Introduction

The Mullins-Sekerka flow as a variant of the classical Stefan problem describes phase transitions in materials with small specific heat constant. A geometric condition on the phase interface, known as Gibbs-Thomson law, accounts for surface tension effects and allows to model superheating and undercooling effects.

We consider a time interval $(0,T)$, an open bounded region $\Omega \subset \mathbb{R}^3$ representing the body of the material and we set $\Omega_T := (0,T) \times \Omega$. The state variables are the relative temperature

$$u : \Omega_T \to \mathbb{R},$$

($u = 0$ denoting the melting point) and a phase function

$$\mathcal{X} : \Omega_T \to \{0,1\},$$

dividing Ω at a time $t \in (0,T)$ into a *liquid phase* $\{\mathcal{X}(t,.) = 1\}$ and a *solid phase* $\{\mathcal{X}(t,.) = 0\}$ separated by the *phase interface*, that is their common boundary.

In a non-dimensional form the energy balance reads

$$\partial_t \mathcal{X} - \Delta u = f, \tag{1.1}$$

The research of the author was supported by DFG Sonderforschungsbereich 611 and by the European Community's Human Potential Programme under contract HPRN-CT-2002-00274, FRONTS-SINGULARITIES..

where f is a given heat source. The Mullins-Sekerka flow is obtained coupling (1.1) with the Gibbs-Thomson law

$$H(t,.) = u(t,.) \quad \text{on the phase interface,} \tag{1.2}$$

where $H(t,.)$ denotes the scalar mean curvature of the phase interface, which we take positive for convex liquid phases. An initial condition for $\mathcal{X}$ and a boundary condition for u on $\partial\Omega$ are added.

Existence of classical solutions for the Mullins-Sekerka problem locally in time was proved by Chen, Hong and Li [CHY96] and by Escher and Simonett [ES97]. Chen [Che96] obtains solutions globally in time as limit of a certain Cahn-Hilliard model, but there the Gibbs-Thomson law is satisfied in a rather weak and complex formulation. Luckhaus and Sturzenhecker obtained in [LS95] solutions for a time-discrete approximation of the Mullins-Sekerka flow and give a comparatively simple formulation of the Gibbs-Thomson law, but they could prove the convergence to solutions of (1.1), (1.2) only under an additional condition on the approximations.

Our goal is to give a general existence result for the Mullins-Sekerka problem in space dimension $n = 3$ using a rather natural generalization of the Gibbs-Thomson law.

In this contribution we introduce our result and sketch the proof. For more details and a complete proof we refer to [Rög05].

2. Weak formulations of the Gibbs-Thomson law

A suitable weak formulation of the problem is crucial to obtain long-time existence. We consider for a moment a time-independent situation. Weak formulations of the Gibbs-Thomson law use the fact that the mean curvature gives the first variation of the area functional:

Let $h_s(\Gamma), -\varepsilon < s < \varepsilon$ be a smooth variation of a hypersurface $\Gamma \subset \mathbb{R}^n$ with the variation field $\xi = \frac{d}{ds}\Big|_{s=0} h_s$. Then

$$\frac{d}{ds}\Big|_{s=0} \int_{h_s(\Gamma)} 1 \, d\mathcal{H}^{n-1} = \int_\Gamma \operatorname{div}_{T_x\Gamma} \xi(x) \, d\mathcal{H}^{n-1}(x) = \int_\Gamma -\vec{H} \cdot \xi \, d\mathcal{H}^{n-1} \tag{2.1}$$

holds. One strategy to generalize the mean curvature vector is to formulate the first variation of the area functional, given by the first equality in (2.1). The advantage is that only 'first derivatives' (the tangential planes) of the surface enter this expression.

For a phase and a temperature functions

$$\mathcal{X} \in \mathrm{BV}(\Omega; \{0,1\}), \quad u \in \mathring{\mathrm{H}}^{1,2}(\Omega).$$

in [LS95] a *BV-formulation of the Gibbs-Thomson law* was introduced, that is

$$\int_0^T \int_\Omega \left(\nabla \cdot \xi - \frac{\nabla \mathcal{X}}{|\nabla \mathcal{X}|} \cdot D\xi \frac{\nabla \mathcal{X}}{|\nabla \mathcal{X}|} \right) (t,.) |\nabla \mathcal{X}|(t,.) \, dt = \int_{\Omega_T} \nabla \cdot (u\xi) \mathcal{X}$$

for all $\xi \in C_c^\infty(\Omega_T;\mathbb{R}^3)$. On the left-hand side we recognize the first variation of the area functional, given by the integral over the tangential divergence of the vector-field ξ. Formally, using the Gauss-Green Theorem on the right-hand side and the identity (2.1) we obtain $H(t) = u(t)$.

Let us now investigate how this formulation behaves if we pass in approximate Gibbs-Thomson equations to the limit. We consider a sequence $h \to 0$ and $\mathcal{X}^h \in \mathrm{BV}(\Omega;\{0,1\}))$, $u^h \in \mathring{\mathrm{H}}^{1,2}(\Omega)$ such that

$$H^h \quad = \quad u^h \qquad \text{for } h > 0$$

holds in the BV-formulation. Moreover we assume that

$$\begin{aligned} \mathcal{X}^h &\to \mathcal{X} \quad \text{in } L^1(\Omega), \\ u^h &\rightharpoonup u \quad \text{in } H^{1,2}(\Omega), \end{aligned}$$

and that the total surface area is preserved,

$$\int_\Omega |\nabla \mathcal{X}^h| \quad \to \quad \int_\Omega |\nabla \mathcal{X}|. \tag{2.2}$$

Then an argument going back to Reshetnyak (see [AFP00] Theorem 2.39) shows that

$$H \quad = \quad u$$

holds in the BV-formulation.
Thus, under the above assumptions we can pass to the limit within the BV-formulation. The condition (2.2) that no surface area is lost is crucial and it was this condition which had to be assumed for the time-discrete approximations in [LS95] to obtain solutions for the Mullins-Sekerka flow.

Difficulties which arise if (2.2) does not hold are captured in the following example. Assume two solid parts of approximations $\mathcal{X}^h$ which merge to one when letting $h \to 0$.

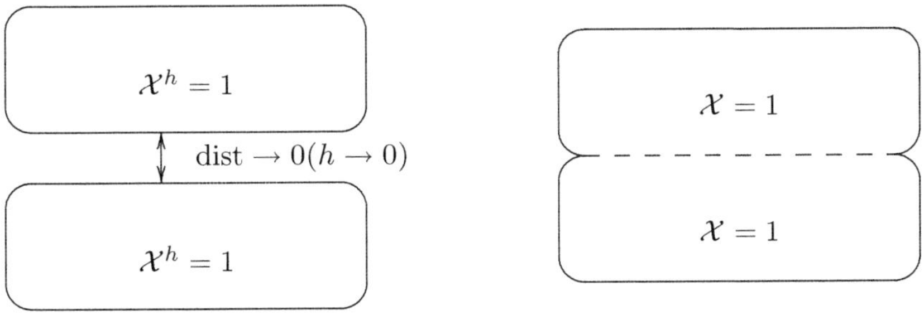

A part of the boundary, indicated by the dashed line, has ceased to separate two different phases. We call this part the *hidden boundary*, whereas the *phase interface* represents the physically relevant part of the boundary. Cusp singularities occur due to the cancellation of phase interfaces. As shown in [Sch97] the BV-formulation of the Gibbs-Thomson law breaks down.

We follow an idea of Schätzle [Sch01] to consider the surface measure of the phase interfaces and we use the concept of integral varifolds introduced by Almgren [Alm65]. We briefly recall the basic definitions and refer for details to the book of Simon [Sim83].

Remark 2.1. A rectifiable varifold μ is a Radon measure on Ω such that for μ-almost all $x \in \Omega$ a $(n-1)$-dimensional tangential plane $T_x\mu$ (in the sense of geometric measure theory) exists, and an *integral varifold*, if in addition the multiplicity of μ is μ-almost everywhere integer-valued. For a rectifiable varifold μ we define the *first variation* by

$$\delta\mu(\xi) \quad := \quad \int_\Omega \operatorname{div}_{T_x\mu} \xi(x)\, d\mu(x) \quad \text{for } \xi \in C^1_c(\Omega;\mathbb{R}^n).$$

We say that μ is of *locally bounded first variation* with *mean curvature vector* $\vec{H}_\mu$, if $\vec{H}_\mu \in \mathrm{L}^1_{loc}(\mu)$ and

$$\delta\mu(\xi) \quad = \quad \int_\Omega -\vec{H}_\mu \cdot \xi\, d\mu \quad \text{for all } \xi \in C^1_c(\Omega;\mathbb{R}^n).$$

To prove the existence of solutions we will use a result of Schätzle [Sch01] on the convergence in approximate Gibbs-Thomson equations. However, in our case the control about hidden boundaries is quite delicate and we have to focus on the physically relevant part of the boundary.
For this reason we use a generalized formulation of the Gibbs-Thomson law which is based on the following definition of mean curvature for a general phase interface.

Definition 2.2. Let $E \subset \Omega$ with $\mathcal{X}_E \in \mathrm{BV}(\Omega)$ and let there exist an integral $(n-1)$-varifolds μ on Ω, such that

$$\partial^* E \subset \operatorname{spt}(\mu),$$

μ has locally bounded first variation with mean curvature vector $\vec{H}_\mu$,

$$\vec{H}_\mu \in \mathrm{L}^s_{\mathrm{loc}}(\mu),\ s > n-1,\ s \geq 2.$$

Then we call

$$\vec{H} \quad := \quad \vec{H}_\mu|_{\partial^* E}$$

the generalized mean curvature vector of $\partial^* E$.

The above definition was proposed and justified in [Rög04]. The essential boundary $\partial^* E$ represents the phase interface, whereas the set $\operatorname{spt}(\mu) \setminus \partial^* E$ can be seen as a hidden boundary. Under the above assumptions the varifolds mean curvature restricted to the phase interface is a property of the phase interface itself and independent of the location of hidden boundaries.

3. Main result

Our solutions of the Mullins-Sekerka problem satisfy the Gibbs-Thomson in the sense, that for almost all times a generalized mean curvature for the phase interface exists and is given pointwise almost everywhere by equation (1.2).

Theorem 3.1. *Let $n = 3$ and data*

$$\begin{aligned} \mathcal{X}_0 &\in \mathrm{BV}(\Omega; \{0, 1\}), \\ f &\in \mathrm{L}^2(\Omega) \end{aligned}$$

be given. Then there exists

$$\begin{aligned} \mathcal{X} &\in \mathrm{L}^\infty(0, T; \mathrm{BV}(\Omega; \{0, 1\})), \\ u &\in \mathrm{L}^2(0, T; \mathring{\mathrm{H}}^{1,2}(\Omega)), \end{aligned}$$

such that

$$\int_{\Omega_T} \mathcal{X}\partial_t\varphi + \int_\Omega \mathcal{X}_0\varphi(0, .) - \int_{\Omega_T} \nabla u \cdot \nabla\varphi = -\int_{\Omega_T} f\varphi \tag{3.1}$$

for all $\varphi \in C_c^\infty([0, T) \times \Omega)$. For almost all $t \in (0, T)$ a generalized mean curvature $H(t)$ of $\partial^*\{\mathcal{X}(t, .) = 1\}$ exists and satisfies

$$H(t, .) = u(t, .) \tag{3.2}$$

$\mathcal{H}^2$-almost everywhere on $\partial^*\{\mathcal{X}(t, .) = 1\}$.

Our formulation of the Gibbs-Thomson law generalizes the BV-formulation. Nevertheless we remark that our solution concept does not include a weak formulation of a boundary condition on the angle between the phase interface and the fixed boundary $\partial\Omega$. In [Rög05] more general boundary conditions for u are considered. Time-dependent heat sources $f \in L^2(\Omega_T)$ can also be included.

4. Time-discretization

In [LS95] the gradient flow structure of the Mullins-Sekerka problem was used to construct for a given time step $h > 0$ iteratively step functions in time which solve a time-discrete version of (1.1), (1.2).

Proposition 4.1 ([LS95]). *There exists*

$$u^h : (0, T) \to \mathring{\mathrm{H}}^{1,2}(\Omega), \quad \mathcal{X}^h : (0, T) \to \mathrm{BV}(\Omega; \{0, 1\}),$$

such that

$$\mathcal{X}_t^h = \mathcal{X}_0 \quad \text{for} \quad 0 \le t < h$$

and such that the approximate energy-balance

$$\frac{1}{h}(\mathcal{X}_t^h - \mathcal{X}_{t-h}^h) = \Delta u_t^h + f, \tag{4.1}$$

and the approximate Gibbs-Thomson law

$$H_t^h \quad = \quad u_t^h \tag{4.2}$$

hold, where H_t^h *denotes the mean curvature of* $\partial^*\{\mathcal{X}_t^h = 1\}$ *and* (4.2) *is satisfied in the BV-formulation. Moreover an energy estimate and an estimate for time-differences are satisfied,*

$$\int_\Omega |\nabla \mathcal{X}_t^h| + \frac{1}{4}\int_0^t \int_\Omega |\nabla u^h|^2 \quad \leq \quad \int_\Omega |\nabla \mathcal{X}_0| + C(T)\|f\|_{L^2(\Omega)}, \tag{4.3}$$

for all $h < t < T$ *and*

$$\int_\tau^T \int_\Omega |\mathcal{X}_t^h - \mathcal{X}_{t-\tau}^h| \, d\mathcal{L}^3 \, dt \leq C\tau^{\frac{1}{4}} \tag{4.4}$$

for all $0 < \tau < T$.

As a corollary we obtain the following compactness result.

Corollary 4.2. *There exists a subsequence* $h \to 0$ *and functions*

$$\mathcal{X} \in \mathrm{L}^\infty(0, T; \mathrm{BV}(\Omega)), \quad u \in \mathrm{L}^2(0, T; \mathring{\mathrm{H}}^{1,2}(\Omega)),$$

such that

$$u^h \rightharpoonup u \quad \text{weakly in } \mathrm{L}^2(0, T; \mathring{\mathrm{H}}^{1,2}(\Omega)), \tag{4.5}$$

$$\mathcal{X}^h \to \mathcal{X} \quad \text{in } \mathrm{L}^1(\Omega_T), \tag{4.6}$$

and such that for almost all $t \in (0, T)$

$$\mathcal{X}^h(t) \to \mathcal{X}(t) \quad \text{in } \mathrm{L}^1(\Omega), \tag{4.7}$$

$$\sup_{h>0} \|\mathcal{X}^h(t,.)\|_{\mathrm{BV}(\Omega)} < \infty, \tag{4.8}$$

$$\liminf_{h\to 0} \|u^h(t,.)\|_{\mathring{\mathrm{H}}^{1,2}(\Omega)} < \infty. \tag{4.9}$$

5. Convergence to solutions

We sketch here the proof of Theorem 3.1. Let $u, \mathcal{X}$ and a subsequence $h \to 0$ be given as above. Letting $h \to 0$ in equations (4.1) and (4.2) we will show that $u, \mathcal{X}$ are weak solution of the Mullins-Sekerka problem. Whereas the energy balance is derived in a standard way the main effort is the passage to a limit in the approximate Gibbs-Thomson law (4.2). One difficulty is that we have to argue pointwise in time. Due to the lack of strong $\mathrm{L}^1(\Omega_T)$-compactness of the approximate temperatures, we have to consider any limit point of $(u^h(t,.))_{h>0}$ in the weak-$\mathring{\mathrm{H}}^{1,2}(\Omega)$ topology and have to identify their traces on $\partial^*\{\mathcal{X}(t,.) = 1\}$ with the trace of the weak limit u in (4.5).

We associate to the surface measure of the phase interfaces an integral 2-varifolds with density one by defining

$$\mu_t^h(\eta) \quad := \quad \int_\Omega \eta \, |\nabla \mathcal{X}_t^h| \qquad \text{for } \eta \in C_c^0(\Omega).$$

For the first variation of μ_t^h we obtain, recalling (4.2)

$$\delta\mu_t^h(\xi) = \int_\Omega \left(\nabla \cdot \xi - \frac{\nabla \mathcal{X}_t^h}{|\nabla \mathcal{X}_t^h|} \cdot D\xi \frac{\nabla \mathcal{X}_t^h}{|\nabla \mathcal{X}_t^h|}\right) |\nabla \mathcal{X}_t^h| = \int_\Omega \mathcal{X}_t^h \, \nabla \cdot (u_t^h \xi)$$

for all $\xi \in C_c^1(\Omega, \mathbb{R}^3)$.

The first step in the proof of Theorem 3.1 is to show that the phase interfaces $\partial^*\{\mathcal{X}(t,.) = 1\}$ have a generalized mean curvature. Since this mean curvature is determined by the phase interface itself, it is given by the strong convergence in (4.6), (4.7). Even more, also any limit of the first variations $(\delta\mu_t^h)_{h>0}$ is determined by (4.6).

Lemma 5.1. *For almost all $t \in (0,T)$ the phase interface $\partial^*\{\mathcal{X}(t,.) = 1\}$ has a generalized mean curvature $\vec{H}(t)$ in the sense of Definition 2.2 with*

$$\vec{H}(t) \quad \in \quad \mathrm{L}^4_{\mathrm{loc}}(|\nabla \mathcal{X}(t,.)|). \tag{5.1}$$

For any subsequence $(h_i)_{i\in\mathbb{N}}$ with $\sup_{i\in\mathbb{N}} \|u^{h_i}(t,.)\|_{\mathring{\mathrm{H}}^{1,2}(\Omega)} < \infty$ we obtain

$$\delta\mu_t^{h_i}(\xi) \;\rightarrow\; \int_\Omega -\vec{H}(t) \cdot \xi \, |\nabla \mathcal{X}(t,.)| \tag{5.2}$$

for all $\xi \in C_c^1(\Omega, \mathbb{R}^3)$.

Proof. See [Rög05] □

To prove that the mean curvature of the phase interface is given as trace of the weak limit u in (4.5) the difficulty is that the pointwise convergence (5.2) holds only for time-dependent subsequences. By an argument similar to a refined dominated convergence theorem in [PS93] we conclude the following convergence.

Lemma 5.2. *There is a subsequence $h \to 0$, such that*

$$\int_0^T \delta\mu_t^h(\xi(t,.))\, dt \quad \rightarrow \quad \int_0^T \Big(\int_\Omega -\vec{H}(t) \cdot \xi(t,.) \, |\nabla \mathcal{X}(t,.)|\Big)\, dt \tag{5.3}$$

for all $\xi \in L^2(0,T; C_c^1(\Omega))$.

Proof. See [Rög05]. □

The Gibbs-Thomson law now follows. We obtain from (4.5), (4.6) and Lemma 5.2 that

$$\begin{aligned}\int_{\Omega_T} \mathcal{X} \nabla \cdot (u\xi) &= \lim_{h\to 0} \int_{\Omega_T} \mathcal{X}^h \nabla \cdot (u^h \xi) = \lim_{h\to 0} \int_0^T \delta\mu_t^h(\xi(t,.))\, dt \\ &= \int_0^T \Big(\int_\Omega -\vec{H}(t) \cdot \xi(t,.) \, |\nabla \mathcal{X}(t,.)|\Big)\, dt\end{aligned}$$

and deduce that for almost all $t \in (0,T)$ and all $\xi \in C_c^1(\Omega;\mathbb{R}^3)$

$$\int_\Omega \mathcal{X}(t,.)\nabla\cdot\Big(u(t,.)\xi\Big) \;=\; \int_\Omega -\vec{H}(t)\cdot\xi\,|\nabla\mathcal{X}(t,.)|$$

holds, which implies by the Gauss-Green Theorem (3.2).

References

[Alm65] F.J. Almgren. The theory of varifolds. *Princeton notes*, 1965.

[AFP00] L. Ambrosio, N. Fusco, and D. Pallara. *Functions of Bounded Variation and Free Discontinuity Problems.* Oxford Science Publications, 2000.

[Che96] X. Chen. Global asymptotic limit of solutions of the Cahn-Hilliard equation. *J. Differential Geom.*, 44(2):262–311, 1996.

[CHY96] X. Chen, J. Hong, and F. Yi. Existence, uniqueness, and regularity of classical solutions of the Mullins-Sekerka problem. *Comm. Partial Differential Equations*, 21(11-12):1705–1727, 1996.

[ES97] J. Escher and G. Simonett. Classical solutions for Hele-Shaw models with surface tension. *Adv. Differential Equations*, 2(4):619–642, 1997.

[LS95] S. Luckhaus and T. Sturzenhecker. Implicit time discretization for the mean curvature flow equation. *Calc. Var. Partial Differ. Equ.*, 3, No.2:253–271, 1995.

[PS93] P.I. Plotnikov and V.N. Starovoitov. The Stefan problem with surface tension as a limit of a phase field model. *Diff. Eq.*, 29, No. 3:395–404, 1993.

[Rög04] M. Röger. Solutions for the Stefan problem with Gibbs-Thomson law by a local minimisation. *Interfaces Free Bound.*, 6(1):105–133, 2004.

[Rög05] M. Röger. Weak solutions for the Mullins-Sekerka flow. *Siam J. on Math. Analysis*, to appear, 2005.

[Sch97] R. Schätzle. A counterexample for an approximation of the Gibbs–Thomson law. *Advances in Math. Sciences and Appl.*, 7, No. 1:25–36, 1997.

[Sch01] R. Schätzle. Hypersurfaces with mean curvature given by an ambient Sobolev function. *J. Differential Geometry*, 58:371–420, 2001.

[Sim83] L. Simon. Lectures on Geometric Measure Theory. Proceedings of the Centre for Mathematical Analysis Australian National University Vol. 3, 1983.

Matthias Röger
Dept. of Mathematics and Computing Science
Eindhoven University of Technology
Den Dolech 2
NL-5600 MB Eindhoven, The Netherlands
e-mail: `mroeger@win.tue.nl`

International Series of Numerical Mathematics, Vol. 154, 369–380

Existence and Approximation Results for General Rate-independent Problems via a Variable Time-step Discretization Scheme

Riccarda Rossi

Abstract. In this note, we prove an existence and approximation result for a class of *state-dependent* rate-independent problems (which have already been investigated in [6]), by passing to the limit in a time-discretization scheme with suitably constructed variable-time steps.

Mathematics Subject Classification (2000). Primary 47H15; Secondary 73E99.

Keywords. Rate-independent models, hysteresis, existence, variable time-step approximation.

1. Introduction

In this paper, we present an existence and approximation result for the Cauchy problem for the doubly nonlinear evolution equation:

$$\partial\Psi(z(t), \dot{z}(t)) + \partial\mathcal{E}(t, z(t)) \ni 0 \quad \text{in } Z, \quad t \in (0, T). \tag{1.1}$$

Here, Z is a (separable) reflexive Banach space, and the symbol $\langle \cdot, \cdot \rangle$ denotes the duality pairing between Z' and Z. Throughout the paper, we will always assume the two proper functionals $\mathcal{E} : [0, T] \times Z \to \mathbb{R}$ and $\Psi : Z \times Z \to [0, +\infty)$ to fulfil

$$\begin{aligned} \mathcal{E}(t, \cdot) : Z \to \mathbb{R} \quad &\text{is convex and l.s.c.} \quad \text{for a.e. } t \in (0, T), \\ \mathcal{E}(\cdot, z) : [0, T] \to \mathbb{R} \quad &\text{is differentiable} \quad \forall\, z \in Z, \end{aligned} \tag{1.2}$$

$$\Psi(z, \cdot) : Z \to [0, +\infty) \text{ is convex and 1-positively homogeneous } \forall z \in Z, \tag{1.3}$$

and the symbol ∂ in (1.1) denotes the subdifferential of both functionals w.r.t. their second variables, i.e., for $t \in [0, T]$ and $z, v \in Z$ we have

$$\begin{aligned} \xi \in \partial\mathcal{E}(t, z) \;&\Leftrightarrow\; \mathcal{E}(t, \hat{z}) - \mathcal{E}(t, z) \geq \langle \xi, \hat{z} - z \rangle \quad \forall \hat{z} \in Z, \\ \omega \in \partial\Psi(z, v) \;&\Leftrightarrow\; \Psi(z, \hat{v}) - \Psi(z, v) \geq \langle \omega, \hat{v} - v \rangle \quad \forall \hat{v} \in Z. \end{aligned}$$

Let us point out that our 1-homogeneity assumption on $\Psi(z, \cdot)$ entails that a solution to (1.1) remains so if the time variable is rescaled.

As a matter of fact, (1.1) models *rate-independent* processes, occurring in plasticity, phase transformations in elastic solids, dry friction on surfaces, and several other fields of continuum mechanics (see the survey [4]). In this framework, z is referred to as the *state variable* of the process, while the functionals $\mathcal{E}$ and Ψ are respectively related to the *potential energy* and to the *dissipation*. Rate-independence means that such processes are insensitive to changes in the time-scales, which is also connected to the fact that they may display a hysteresis behavior.

The energetic formulation. A new modelling approach for rate-independent problems has been recently proposed in the seminal papers [5, 8, 7], however dealing with the simpler case of a *state-independent* functional Ψ (i.e., $\Psi(z,v) = \Psi(v)$ for all $(z,v) \in Z \times Z$). Such an approach stems from the fact that, in the applied problems arising in, e.g., plasticity, the energy functional $\mathcal{E}$ cannot be expected to be convex or smooth w.r.t. the state variable. Hence, the subdifferential formulation (1.1) appears to be too restrictive. On the other hand, the functional $t \mapsto \mathcal{E}(t,\cdot)$, which takes into account the external loadings, can be assumed smooth. In this spirit, in the paper [6] the following *energetic formulation* of (1.1) has been considered:

$$\mathcal{E}(t,z(t)) \leq \mathcal{E}(t,\hat{z}) + \Psi(z(t),\hat{z}-z(t)) \quad \forall \hat{z} \in Z, \tag{S}$$

$$\mathcal{E}(t,z(t)) + \int_0^t \Psi(z(\tau),\dot{z}(\tau))\,\mathrm{d}\tau = \mathcal{E}(0,z(0)) + \int_0^t \partial_t\mathcal{E}(\tau,z(\tau))\,\mathrm{d}\tau. \tag{E}$$

Note that (S) is a *stability condition*, which states that switching from the state z to the state $\hat{z}$ enforces a release of potential energy $\mathcal{E}(t,z(t)) - \mathcal{E}(t,\hat{z})$, smaller than the dissipated energy $\Psi(z(t),\hat{z}-z(t))$. On the other hand, (E) is an energy balance. Let us stress that the formulation (S)-(E) does not involve the derivative of $\mathcal{E}$ w.r.t. the variable z, but only the assumedly smooth power of the external forces $\partial_t\mathcal{E}$. Actually, it has been shown in [6] that, under suitable conditions on $\mathcal{E}$ and Ψ (see Sec. 3 later on), the two formulations (1.1) and (S)-(E) are equivalent. Moreover, an existence and approximation result (cf. [6, Thm. 4.7]) has been proved for (S)-(E) by passing to the limit in a suitable time-discretization scheme, which we briefly introduce.

Approximation via time-discretization. To introduce this approximation scheme, let us fix a partition of the interval $(0,T)$

$$\mathcal{P}_\tau := \left\{t_\tau^0 = 0 < t_\tau^1 < \cdots < t_\tau^{N-1} < t_\tau^N = T\right\}, \quad \tau := \max_{j=1,\ldots,N}\{t_\tau^j - t_\tau^{j-1}\}, \tag{1.4}$$

and let us introduce the following *time incremental problem*, associated with the *time-continuous* Problem (1.1): *given* $z_\tau^0 := z_0$, *find* $z_\tau^1,\ldots,z_\tau^N \in Z$ *such that*

$$z_\tau^k \in \operatorname{argmin}\{\, \mathcal{E}(t_\tau^k,z) + \Psi(z_\tau^{k-1}, z - z_\tau^{k-1}) \mid z \in Z \,\} \quad \text{for } k = 1,\ldots,N. \tag{\textbf{IP}}$$

It can be shown (cf. Section 3) that, under suitable assumptions on Ψ and $\mathcal{E}$, for any $k = 1,\ldots,N$ (**IP**) has a *unique* solution $\{z_\tau^k\}$, fulfilling the *subdifferential* *inclusion*

$$\partial\Psi(z_\tau^{k-1}, z_\tau^k - z_\tau^{k-1}) + \partial\mathcal{E}(t_\tau^k, z_\tau^k) \ni 0 \quad \forall k = 1, \dots, N, \tag{1.5}$$

the *stability condition*

$$\mathcal{E}(t_\tau^k, z_\tau^k) \le \mathcal{E}(t_\tau^k, \hat{z}) + \Psi(z_\tau^{k-1}, \hat{z} - z_\tau^k) \quad \forall \hat{z} \in Z, \tag{1.6}$$

as well as the *energy inequality*

$$\mathcal{E}(t_\tau^k, z_\tau^k) + (t_\tau^k - t_\tau^{k-1})\Psi\left(z_\tau^{k-1}, \frac{z_\tau^k - z_\tau^{k-1}}{t_\tau^k - t_\tau^{k-1}}\right) \le \mathcal{E}(t_\tau^{k-1}, z_\tau^{k-1}) + \int_{t_\tau^{k-1}}^{t_\tau^k} \partial_t \mathcal{E}(r, z_\tau^{k-1})\,\mathrm{d}r. \tag{1.7}$$

Indeed, (1.5), (1.6) and (1.7) are nothing but the *discrete* versions of (1.1), (S) and (E). In this setting, the approximate solutions of (1.1) are then constructed as suitable interpolants of the discrete solutions $\{z_\tau^k\}_{k=1}^N$. Yet, in [6] the passage to the limit procedure is performed *only* in the case of *uniform* time-step partitions.

Our main result. Nonetheless, it has been suggested in [6, Rem. 4.10] that it should also be possible to pass to the limit in an approximation scheme constructed with a *variable* time-step partition of $[0, T]$, somehow "adjusted" to the data of the problem, cf. (3.19) later on. In the what follows, we will indeed prove this conjecture, thus obtaining our main existence and approximation result, Theorem 3.8. The proof of this result relies on a (version of) the compactness result for Young measures (which we briefly present in Sec. 2), and on a *chain rule* argument. However, let us point out that the problem of the convergence of a *generic* variable time-step approximation scheme for (S)-(E) remains open.

Error estimates and uniqueness. As for the analysis of the rate of convergence of the approximate solutions, this issue is actually tightly connected to the problem of uniqueness of the solutions to (1.1). In fact, the strong convergence of the approximation scheme in the *state-independent* case has been achieved in [7], by means of the same *energetic estimates* exploited for proving a result of continuous dependence on the initial data. However, the latter type of result is definitely harder to prove in the state-dependent case, due to the quasivariational character of the problem. Uniqueness and continuous dependence theorems for (1.1) have indeed been obtained in [6] under more restrictive assumptions on Ψ and $\mathcal{E}$ than the ones for existence, by carefully combining the *energetic* method of [7] with the *convex analysis* arguments of [2]. Up to now, it has turned out to be arduous to adapt this complex uniqueness proof for proving the strong convergence of the approximate solutions, even in the case of a *uniform* time-step partition.

2. The fundamental theorem for Young measures for the weak topology

Notation. Let X be a separable reflexive Banach space. We denote by $\mathcal{B}(X)$ the Borel σ-algebra of X, while $\mathcal{L}$ is the σ-algebra of the Lebesgue measurable subsets of $(0, T)$ ($|\cdot|$ stands for the Lebesgue measure on $(0, T)$), and $\mathcal{L} \otimes \mathcal{B}(X)$ is the product σ-algebra on $(0,T)\times X$. We say that a $\mathcal{L}\otimes\mathcal{B}(X)$-measurable functional $h:(0,T)\times X\to(-\infty,+\infty]$ is a *weakly normal integrand* if

$$v\mapsto h_t(v)=h(t,v) \text{ is sequentially weakly l.s.c. for a.e. } t\in(0,T). \tag{2.1}$$

We also recall that a sequence $\{u_n\}\subset L^1(0,T;X)$ is *uniformly integrable* if

$$\forall\varepsilon>0\ \exists\delta>0 \qquad \forall J\subset(0,T)\ |J|\le\delta \quad \Rightarrow \sup_{n\in\mathbb{N}}\int_J\|u_n(t)\|\,dt\le\varepsilon. \tag{2.2}$$

Definition 2.1 (Young measures). A *Young measure* (or *parametrized measure*) in X is a family $\boldsymbol{\nu}:=\{\nu_t\}_{t\in(0,T)}$ of Borel probability measures on X such that

$$t\in(0,T)\mapsto\nu_t(B)\quad\text{is}\quad\mathcal{L}\text{-measurable}\quad\forall B\in\mathcal{B}(X). \tag{2.3}$$

We denote by $\mathcal{Y}(0,T;X)$ the set of all Young measures.

The following result, which is a version of the so-called fundamental compactness result for Young measures (see, e.g., [1, Thm.1]) for the *weak topology*, has been proved in [10] (cf. Thm. 3.2 therein).

Theorem 2.2 (The fundamental theorem for weak topologies). *Let $\{v_n\}_{n\in\mathbb{N}}$ be a* bounded *sequence in $L^p(0,T;X)$, for some $p\ge1$; if $p=1$, let $\{v_n\}$ also be uniformly integrable. Then, there exists a subsequence $k\mapsto v_{n_k}$ and a Young measure $\boldsymbol{\nu}=\{\nu_t\}_{t\in(0,T)}\in\mathcal{Y}(0,T;X)$ such that for a.e. $t\in(0,T)$*

$$\nu_t \text{ is concentrated on the set } L(t):=\bigcap\nolimits_{p=1}^{\infty}\overline{\{v_{n_k}(t):k\ge p\}}^{w} \tag{2.4}$$

of the weak limit points *of $\{v_{n_k}(t)\}$, and*

$$\liminf_{k\to\infty}\int_0^T h(t,v_{n_k}(t))\,\mathrm{d}t\ge\int_0^T\left(\int_X h(t,\xi)\,\mathrm{d}\nu_t(\xi)\right)\mathrm{d}t \tag{2.5}$$

for every weakly normal *integrand h such that $\{h^-(\cdot,v_{n_k}(\cdot))\}$ is uniformly integrable (h^- denoting the negative part of h). As a consequence, setting $v(t):=\int_X\xi\,\mathrm{d}\nu_t(\xi)$, we have*

$$v_{n_k}\rightharpoonup v \text{ in } L^p(0,T;X) \text{ if } p<\infty,\quad v_{n_k}\overset{*}{\rightharpoonup}v \text{ in } L^\infty(0,T;X). \tag{2.6}$$

Remark 2.3. Note that our L^p-boundedness assumption (together with the uniform integrability for $p=1$) yields by itself that, up to a subsequence, $\{v_n\}$ weakly converges in $L^p(0,T;X)$ (for $p=1$, this is guaranteed by the *Dunford-Pettis criterion* for L^1 vector-valued functions, see, e.g., [3, Thm. IV.2.1]). Hence, (2.6) (which is a straightforward consequence of the general lower semicontinuity inequality (2.5)) has rather to be interpreted as an identification of the weak-L^p limit in terms of a suitable Young measure $\boldsymbol{\nu}$. Such a measure actually retains some information on the *pointwise* (weak) limiting behavior of the sequence, see (2.4).

3. The main result

First of all, let us enlist all the assumptions on the functionals $\mathcal{E}$ and Ψ which will come into play in the proof of our main theorem (i.e., Thm. 3.8) and of the related intermediate results.

Statement of the assumptions

Besides (1.2), we assume that $\mathcal{E}$ complies with

$$\begin{gathered}\partial_t\mathcal{E}(\cdot,z):[0,T]\to\mathbb{R}\quad\text{is measurable }\forall z\in Z,\text{ and}\\ \exists C_0>0\ \exists\lambda_0\in L^1(0,T;[0,\infty))\ \forall z\in Z:\ |\partial_t\mathcal{E}(t,z)|\le\lambda_0(t)(\mathcal{E}(t,z)+C_0);\end{gathered}\tag{3.1}$$

$$\begin{gathered}\exists\,\lambda_1\in L^1(0,T;[0,\infty))\quad\text{for a.e.}\,t\in(0,T),\quad\forall z,\hat z\in Z:\\ |\partial_t\mathcal{E}(t,z)-\partial_t\mathcal{E}(t,\hat z)|\le\lambda_1(t)\|z-\hat z\|;\end{gathered}\tag{3.2}$$

$$\text{for a.e.}\,t\in(0,T)\text{ the map }z\mapsto\partial_t\mathcal{E}(t,z)\text{ is weakly continuous on }Z;\tag{3.3}$$

$$\partial\mathcal{E}\subset[0,T]\times Z\times Z'\text{ is closed in the strong-weak-weak topology.}\tag{3.4}$$

Finally, we require $z\mapsto\mathcal{E}(t,z)$ to be uniformly convex in the z variable, with a modulus of convexity κ independent of $t\in[0,T]$, i.e., (setting $z_\theta:=(1-\theta)z_0+\theta z_1$):

$$\begin{gathered}\exists\kappa>0\quad\forall z_0,\ z_1\in Z,\quad\forall t\in[0,T],\quad\forall\theta\in[0,1]:\\ \mathcal{E}(t,z_\theta)\le(1-\theta)\mathcal{E}(t,z_0)+\theta\mathcal{E}(t,z_1)-\frac{\kappa}{2}\theta(1-\theta)\|z_0-z_1\|^2.\end{gathered}\tag{3.5}$$

Remark 3.1. Indeed, (3.1) ensures that $\mathcal{E}$ is bounded from below and absolutely continuous in time (see also [4, Sect. 3]): namely, $\forall\,t,s\in[0,T]$ and $\forall\,z\in Z$ we have

$$\mathcal{E}(t,z)\ge-C_0,\quad\text{and}\quad\mathcal{E}(t,z)+C_0\le(\mathcal{E}(s,z)+C_0)\exp\left(\Big|\int_s^t\lambda_0(\tau)\,\mathrm{d}\tau\Big|\right).\tag{3.6}$$

Remark 3.2. Under the assumptions (1.2), (3.1) and (3.2) on $\mathcal{E}$, the following chain rule for the subdifferential $\partial\mathcal{E}$ of $\mathcal{E}$ holds (for the proof, see [6, Prop. 2.6]): *for any curve $z\in W^{1,1}(0,T;Z)$ such that there exists a selection g with*

$$g(t)\in\partial\mathcal{E}(t,z(t))\quad\text{for a.e. }t\in(0,T)\quad\text{and}\quad g\in L^\infty(0,T;Z'),\tag{3.7}$$

then the map $t\mapsto\mathcal{E}(t,z(t))$ is absolutely continuous on $(0,T)$, and for every measurable selection $\zeta(t)\in\partial\mathcal{E}(t,z(t))$ we have the identity

$$\tfrac{\mathrm{d}}{\mathrm{d}t}\mathcal{E}(t,z(t))=\langle\zeta(t),\dot z(t)\rangle+\partial_t\mathcal{E}(t,z(t))\quad\text{for a.e. }t\in(0,T).\tag{3.8}$$

As for the dissipation functional Ψ, in addition to (1.3) we impose:

$$\exists\,C_\Psi>0\ \forall(z,v)\in Z\times Z:\ \Psi(z,v)\le C_\Psi\|v\|;\tag{3.9}$$

$$\begin{gathered}\exists\,\psi^*>0\quad\forall\,z,\hat z,v,\in Z:\quad|\Psi(z,v)-\Psi(\hat z,v)|\le\psi^*\|v\|\|z-\hat z\|,\\ \text{with}\quad\psi^*<\kappa;\end{gathered}\tag{3.10}$$

$$\Psi:Z\times Z\to[0,\infty)\text{ is sequentially weakly lower semicontinuous;}\tag{3.11}$$

$$\forall v\in Z:\ \Psi(\cdot,v):Z\to[0,\infty)\text{ is sequentially weakly continuous.}\tag{3.12}$$

We refer to [6, Sec. 4.2] for a non trivial example of functionals Ψ and $\mathcal{E}$ complying with all the above assumptions.

Remark 3.3. Loosely speaking, the requirement $\psi^* < \kappa$ in (3.10) means that the variations of Ψ with respect to z are weak enough so that the uniform convexity of $\mathcal{E}$ is able to compensate for them. In fact, a simple example (see [6, Sec. 3]) shows that, when $\psi^* \geq \kappa$, the Cauchy problem for (1.1) might not possess absolutely continuous solutions.

Remark 3.4. It is possible to show (cf. [6, Lemma 4.1]) that, under the present conditions on Ψ, (3.12) is equivalent to the fact that the map $z \mapsto \partial\Psi(z,0)$ has a sequentially closed graph in the weak-weak topology of $Z \times Z'$.

Remark 3.5. Assumption (1.3) has an interesting *geometrical* interpretation: i.e., that for every $z \in Z$ there exists a non-empty, closed and convex set $C(z) \subset Z'$, with $\Psi(z,v) := \sup\{ \langle \sigma, v\rangle \mid \sigma \in C(z) \}$ for all $v \in Z$ (i.e., for all $z \in Z$ $\Psi(z,\cdot)$ is the *support function* of the set $C(z)$). Standard convex analysis results (see [9]) ensure that

$$\partial\Psi(z,v) = \operatorname{argmax}\{ \langle \sigma, v\rangle \mid \sigma \in C(z) \} \subset C(z) = \partial\Psi(z,0) \quad \forall\, v, z \in Z.$$

Besides, it is easy to check that, due to (3.9), for all $z \in Z$

$$C(z) \subset B'_{C_\Psi}(0),$$

$B'_{C_\Psi}(0)$ being the ball of Z' centered at 0, with radius C_Ψ.

Remark 3.6. It has been verified (see [6, Prop. 2.7]) that, under the present assumptions, for a.e. $t \in [0,T]$ the stability condition (S) is equivalent to

$$\partial\Psi(z(t),0) + \partial\mathcal{E}(t,z(t)) \ni 0. \tag{3.13}$$

Approximation

Let us a fix a partition $\mathcal{P}_\tau$ (1.4) of $[0,T]$, and consider the discrete time incremental problem (**IP**), with an initial datum $z_0 \in Z$. In [6] (cf. Lemma 4.4 and Cor. 4.5 therein), it has been proved that, under (some of) the assumptions (1.2)–(1.3) and (3.1)–(3.4), problem (**IP**) admits a unique solution, complying with (1.5)–(1.7), as well as with the following *variational inequality*: $\forall \hat{z} \in Z$

$$\frac{\kappa}{2}\|z_\tau^k - \hat{z}\|^2 \leq \mathcal{E}(t_\tau^k, \hat{z}) - \mathcal{E}(t_\tau^k, z_\tau^k) + \Psi(z_\tau^{k-1}, \hat{z} - z_\tau^{k-1}) - \Psi(z_\tau^{k-1}, z_\tau^k - z_\tau^{k-1}). \tag{3.14}$$

Furthermore, note that (1.5) may be rephrased in this way: for all $k = 1, \ldots, N$ there exist $\xi_k \in \partial\mathcal{E}(t_\tau^k, z_\tau^k)$, $\omega_k \in \partial\Psi(z_\tau^{k-1}, z_\tau^k - z_\tau^{k-1})$ fulfilling $\xi_k + \omega_k = 0$.

We can now introduce the piecewise constant interpolants $\overline{Z}_\tau$, $\underline{Z}_\tau : [0,T] \to Z$ and the piecewise linear interpolant $\widehat{Z}_\tau : [0,T] \to Z$ of the discrete solutions $\{z_\tau^k\}_{k=0}^N$ of Problem (**IP**), defined by

$$\overline{Z}_\tau(t) := z_\tau^k \text{ for } t \in (t_\tau^{k-1}, t_\tau^k], \quad \underline{Z}_\tau(t) := z_\tau^{k-1} \text{ for } t \in [t_\tau^{k-1}, t_\tau^k),$$
$$\widehat{Z}_\tau(t) = \frac{t - t_\tau^{k-1}}{t_\tau^k - t_\tau^{k-1}} z_\tau^k + \frac{t_\tau^k - t}{t_\tau^k - t_\tau^{k-1}} z_\tau^{k-1}, \quad \text{for } t \in [t_\tau^{k-1}, t_\tau^k].$$

Analogously, we introduce the (left-continuous piecewise constant) interpolants $\bar{\xi}_\tau$ and $\bar{\omega}_\tau$ of $\{\xi_k\}_{k=1}^N$ and $\{\omega_k\}_{k=1}^N$. Also, let $\bar{\mathsf{t}}_\tau : [0,T] \to [0,T]$ be defined by $\bar{\mathsf{t}}_\tau(0) := 0$ and $\bar{\mathsf{t}}_\tau(t) := t_\tau^k$ for $t \in (t_\tau^{k-1}, t_\tau^k]$. Of course, for every $t \in [0,T]$ we have $\bar{\mathsf{t}}_\tau(t) \downarrow t$ as $\tau \searrow 0$.

Hence, on account of Remark 3.5, (1.5) yields for every $t \in [0,T]$

$$\bar{\xi}_\tau(t) \in \partial\mathcal{E}(\bar{\mathsf{t}}_\tau(t), \overline{Z}_\tau(t)), \quad \overline{\omega}_\tau(t) \in \partial\Psi(\underline{Z}_\tau(t), 0), \quad \overline{\omega}_\tau(t) + \bar{\xi}_\tau(t) = 0. \tag{3.15}$$

On the other hand, summing up (1.7) on each subinterval of the partition $\mathcal{P}_\tau$, we end up with the energy estimate

$$\int_{\mathsf{s}}^{\mathsf{t}} \Psi(\underline{Z}_\tau(r), \widehat{Z}'_\tau(r))\,\mathrm{d}r + \mathcal{E}(\mathsf{t}, \overline{Z}_\tau(\mathsf{t})) \le \mathcal{E}(\mathsf{s}, \overline{Z}_\tau(\mathsf{s})) + \int_{\mathsf{s}}^{\mathsf{t}} \partial_t\mathcal{E}(r, \underline{Z}_\tau(r))\,\mathrm{d}r \tag{3.16}$$

for every pair of nodes $\mathsf{s}, \mathsf{t} \in \mathcal{P}_\tau$, with $\mathsf{s} < \mathsf{t}$.

A priori estimates

We will always denote by the symbol C any positive constant occurring in the following estimates, without specifying the quantities C may depend on.

Claim 1. *There exists a positive constant C such that for all $\tau > 0$*

$$\max\left\{|\mathcal{E}(t, \overline{Z}_\tau(t))|, |\mathcal{E}(t, \underline{Z}_\tau(t))|\right\} \le C \quad \forall\, t \in [0,T]. \tag{3.17}$$

Indeed, using that $\overline{Z}_\tau(\mathsf{t}) = \underline{Z}_\tau(\mathsf{t})$ for every $\mathsf{t} \in \mathcal{P}_\tau$, taking $\mathsf{s} = 0$ and $\mathsf{t} = \bar{\mathsf{t}}_\tau(t)$ in (3.16) we end up with

$$\mathcal{E}(\bar{\mathsf{t}}_\tau(t), \underline{Z}_\tau(t)) \le \mathcal{E}(0, z_0) + \int_0^{\bar{\mathsf{t}}_\tau(t)} \partial_t\mathcal{E}(r, \underline{Z}_\tau(r))\,\mathrm{d}r.$$

Hence, by (3.1) we have

$$\mathcal{E}(\bar{\mathsf{t}}_\tau(t), \underline{Z}_\tau(\mathsf{t})) \le \mathcal{E}(0, z_0) + \textstyle\int_0^{\bar{\mathsf{t}}_\tau(t)} \lambda_0(r)\left(\mathcal{E}(r, \underline{Z}_\tau(r)) + C_0\right)\,\mathrm{d}r.$$

Then, by the Gronwall Lemma and the first of (3.6) we conclude

$$0 \le \mathcal{E}(\bar{\mathsf{t}}_\tau(t), \underline{Z}_\tau(\mathsf{t})) + C_0 \le (\mathcal{E}(0, z_0) + C_0)\exp(\textstyle\int_0^{\bar{\mathsf{t}}_\tau(t)} \lambda_0(r)\,\mathrm{d}r),$$

so that the sequence $\{\mathcal{E}(\bar{\mathsf{t}}_\tau(\cdot), \underline{Z}_\tau(\cdot))\}$ is bounded in $L^\infty(0,T)$. Exploiting the second of (3.6), we conclude a L^∞-bound for the sequence $\{\mathcal{E}(\cdot, \underline{Z}_\tau(\cdot))\}$. We argue likewise for $\{\mathcal{E}(\cdot, \overline{Z}_\tau(\cdot))\}$, and (3.17) is proved.

Claim 2. *There exists a positive constant C such that for all $\tau > 0$*

$$\|\overline{Z}_\tau\|_{L^\infty(0,T;Z)},\ \|\underline{Z}_\tau\|_{L^\infty(0,T;Z)},\ \|\widehat{Z}_\tau\|_{L^\infty(0,T;Z)} \le C. \tag{3.18}$$

This estimate is a straightforward consequence of (3.17) and of the uniform convexity assumption (3.5) on $\mathcal{E}$.

Claim 3. *The sequences $\{\bar{\xi}_\tau\}$ and $\{\bar{\omega}_\tau\}$ are bounded in $L^\infty(0,T;Z')$.*
This is due to (3.15) and to (1.3), (3.9): indeed, in view of Remark 3.5 we have

$$\bar{\omega}_\tau(t) \in \partial\Psi(\underline{Z}_\tau(t), 0) \subset B'_{C_\Psi}(0) \quad \forall t \in [0,T].$$

Claim 4. *Set* $\Lambda_* := \int_0^T \lambda_1(s)\,\mathrm{ds}$, *and, for every* $N \geq 1$, *let* $\mathcal{P}_{\tau N}$ *be the partition:*

$$0 = t_\tau^0 < \cdots \leq t_\tau^k \leq \cdots < t_\tau^N = T, \quad \int_0^{t_\tau^k} \lambda_1(s)\,ds = k\Lambda_*/N, \quad k = 1, \ldots, N. \tag{3.19}$$

Then, the following estimates hold:

$$\|\widehat{Z}_{\tau N} - \overline{Z}_{\tau N}\|_{L^\infty(0,T;Z)} \leq \|\overline{Z}_{\tau N} - \underline{Z}_{\tau N}\|_{L^\infty(0,T;Z)} \leq \frac{\Lambda_*}{(\kappa - \psi^*)N} \quad \forall N \geq 1. \tag{3.20}$$

The first inequality in (3.20) ensues from trivial calculations. As for the second one, we first choose $\hat{z} := z_\tau^{k-1}$ in (3.14), thus obtaining

$$\tfrac{\kappa}{2}\|z_\tau^k - z_\tau^{k-1}\|^2 \leq \mathcal{E}(t_\tau^k, z_\tau^{k-1}) - \mathcal{E}(t_\tau^k, z_\tau^k) - \Psi(z_\tau^{k-1}, z_\tau^k - z_\tau^{k-1}). \tag{3.21}$$

Secondly, we write (3.14) at the $(k-1)$th step and plug in $\hat{z} := z_\tau^k$. Adding the resulting inequality and (3.21), and exploiting (3.1)–(3.2), (3.5), (3.9)–(3.10) (see [6, Prop. 4.7] for details), we get

$$\begin{aligned}
&\kappa\|z_\tau^k - z_\tau^{k-1}\|^2 \\
&\leq \mathcal{E}(t_\tau^k, z_\tau^{k-1}) - \mathcal{E}(t_\tau^k, z_\tau^k) + \mathcal{E}(t_\tau^{k-1}, z_\tau^k) - \mathcal{E}(t_\tau^{k-1}, z_\tau^{k-1}) \\
&\quad + \Psi(z_\tau^{k-2}, z_\tau^k - z_\tau^{k-1}) - \Psi(z_\tau^{k-1}, z_\tau^k - z_\tau^{k-1}) \\
&\leq \int_{t_\tau^{k-1}}^{t_\tau^k} (\partial_t\mathcal{E}(\tau, z_\tau^{k-1}) - \partial_t\mathcal{E}(\tau, z_\tau^k))\,d\tau + \psi^*\|z_\tau^{k-1} - z_\tau^{k-2}\|\|z_\tau^k - z_\tau^{k-1}\| \\
&\leq \|z_\tau^k - z_\tau^{k-1}\| \int_{t_\tau^{k-1}}^{t_\tau^k} \lambda_1(\tau)\,d\tau + \psi^*\|z_\tau^{k-1} - z_\tau^{k-2}\|\|z_\tau^k - z_\tau^{k-1}\|.
\end{aligned}$$

On the other hand, by (3.19) we have $\int_{t_\tau^{k-1}}^{t_\tau^k} \lambda_1(\tau)\,d\tau = \Lambda_*/N$. Thus, dividing both sides of the above inequality by $\kappa\delta_k$, with $\delta_k := \|z_\tau^k - z_\tau^{k-1}\|$ for $k = 1, \ldots, N$ and $\delta_0 := 0$, we end up with the recurrence relation

$$\delta_k \leq \frac{\Lambda_*}{\kappa N} + \frac{\psi^*}{\kappa}\delta_{k-1} \quad \forall k = 1, \ldots, N,$$

and the second inequality in (3.20) easily follows.

Claim 5. *Assume* (3.19). *Then, the sequence* $\{\widehat{Z}'_{\tau N}\}_N \subset L^1(0,T;Z)$ *is bounded and uniformly integrable.*

As a consequence of (3.20), we have for all $k = 1, \ldots, N$

$$\int_{t_\tau^{k-1}}^{t_\tau^k} \|\widehat{Z}'_{\tau N}(s)\|\,\mathrm{ds} \leq \frac{\Lambda_*}{(\kappa - \psi^*)N} = \frac{1}{\kappa - \psi^*}\int_{t_\tau^{k-1}}^{t_\tau^k} \lambda_1(s)\,\mathrm{ds}, \tag{3.22}$$

and the first assertion is proved upon summing on the subintervals of the partition. To check the uniform integrability, we preliminarily note that, for any fixed $N \geq 1$, and for any $s, t \in [0,T]$, with $t_{h-1} < s < t_h \le t_{j-1} \le t < t_j$, there holds

$$\int_s^t \|\widehat{Z}'_{\tau_N}(r)\| \, dr \le \int_{t_{h-1}}^{t_h} \|\widehat{Z}'_{\tau_N}(r)\| \, dr + \int_{t_h}^{t_{j-1}} \|\widehat{Z}'_{\tau_N}(r)\| \, dr + \int_{t_{j-1}}^{t_j} \|\widehat{Z}'_{\tau_N}(r)\| \, dr$$
$$\le \frac{2\Lambda_*}{(\kappa - \psi^*)N} + \int_{t_h}^{t_{j-1}} \lambda_1(r) \, dr \le \frac{2\Lambda_*}{(\kappa - \psi^*)N} + \int_s^t \lambda_1(r) \, dr,$$

in view of (3.22). Then, let us fix $\varepsilon > 0$, and hence $N_* > 1$ such that, for $N > N_*$, $2\Lambda_*/((\kappa - \psi^*)N) < \varepsilon/4$. Let us also pick $\delta_* > 0$ such that

$$s, t \in [0,T], \ |s-t| \le \delta_* \Longrightarrow \begin{cases} \int_s^t \|\widehat{Z}'_{\tau_N}(r)\| \, dr \le \frac{\varepsilon}{4} & \text{for } 1 \le N \le N_*, \\ \int_s^t \lambda_1(r) \, dr \le \frac{\varepsilon}{4}. \end{cases}$$

Combining the two estimates above, we finally end up with

$$|t-s| \le \delta_* \Rightarrow \sup_N \int_s^t \|\widehat{Z}'_{\tau_N}(r)\| \, dr \le \sup_{N \le N_*} \int_s^t \|\widehat{Z}'_{\tau_N}(r)\| \, dr$$
$$+ \sup_{N > N_*} \int_s^t \|\widehat{Z}'_{\tau_N}(r)\| \, dr \le \frac{3\varepsilon}{4}.$$

Remark 3.7. Let us point out that Claims 1–3 hold for *any* variable time-step partition $\mathcal{P}_\tau$, whereas we have been able to prove Claims 4 and 5 just for the partition $\mathcal{P}_{\tau_N}$ defined by (3.19), which is in fact tailored to the function λ_1, cf. (3.2). On the other hand, it can be proved that, if $\lambda_1 \in L^\infty(0,T)$ and if *uniform* time-step partitions are considered, then a stronger version of (3.20) holds, namely

$$\|\widehat{Z}'_\tau\|_{L^\infty(0,T;Z)} \le \frac{\|\lambda_1\|_{L^\infty(0,T)}}{\kappa - \psi^*} \quad \text{for } \tau > 0,$$

see [6, Prop. 4.8]

Our existence and approximation result

Theorem 3.8. *Assume* (1.2)–(1.3), (3.1)–(3.5), (3.9)–(3.12). *Let* $z_0 \in Z$ *fulfil the stability condition*

$$\mathcal{E}(0, z_0) \le \mathcal{E}(0, \hat{z}) + \Psi(z_0, \hat{z} - z_0) \quad \forall \hat{z} \in Z. \tag{3.23}$$

Then, the sequence of partitions $\{\mathcal{P}_{\tau_N}\}_N$ *of* $[0,T]$ *defined by* (3.19) *admits a subsequence (not relabeled), for which there exists a curve* $z \in W^{1,1}(0,T;Z)$ *such that* $z(0) = z_0$, *and the following convergences hold as* $N \nearrow \infty$:

$$\forall t \in [0,T]: \quad \widehat{Z}_{\tau_N}(t), \ \overline{Z}_{\tau_N}(t), \ \underline{Z}_{\tau_N}(t) \rightharpoonup z(t) \quad \text{in } Z, \tag{3.24}$$
$$\widehat{Z}'_{\tau_N} \rightharpoonup \dot{z} \quad \text{in } L^1(0,T;Z). \tag{3.25}$$

Moreover, z *fulfils the energetic formulation* (S)–(E) *of* (1.1) *for all* $t \in [0,T]$.

Proof. It follows from Claim 5 that the sequence $\{\widehat{Z}_{\tau_N}\} \subset \mathrm{C}^0([0,\mathrm{T}];\mathrm{Z})$ is equicontinuous. Hence, the estimate (3.18), combined with the Ascoli-Arzelà compactness theorem in the framework of the weak topology of (the reflexive space) Z, ensures that there exists a limit curve $z \in L^\infty(0,T;Z)$ such that, along a subsequence, the convergence for $\widehat{Z}_{\tau_N}$ in (3.24) holds. We trivially have $z(0) = z_0$. Thanks to (3.20), we also deduce the convergences for $\overline{Z}_{\tau_N}$ and $\underline{Z}_{\tau_N}$. Besides, owing to Claim 5 and the Dunford-Pettis criterion [3, Thm. IV.2.1], we conclude that $z \in W^{1,1}(0,T;Z)$ and that (3.25) holds up to a further extraction.

Proof of (S). In view of Theorem 2.2 and of Claim 3, up to a further extraction the sequences $\{\bar{\xi}_{\tau_N}\}$ and $\{\bar{\omega}_{\tau_N}\}$ have two limiting Young measures $\{\zeta_t\}_{t\in(0,T)} \in \mathcal{Y}(0,T;Z')$ and, respectively, $\{\mu_t\}_{t\in(0,T)} \in \mathcal{Y}(0,T;Z')$ such that for a.e. $t \in (0,T)$ the measure ζ_t (μ_t, respectively) is concentrated on the set $L_\xi(t)$ ($L_\omega(t)$, resp.) of the weak limit points of the sequence $\{\bar{\xi}_{\tau_N}(t)\}$ ($\{\overline{\omega}_{\tau_N}(t)\}$, resp.). Further,

$$\begin{aligned} \bar{\xi}_{\tau_N} &\overset{*}{\rightharpoonup} \xi \quad \text{in } L^\infty(0,T;Z'), \quad \text{with } \xi(t) := \textstyle\int_{Z'} v \, d\zeta_t(v) \quad \text{for a.e. } t \in (0,T), \\ \bar{\omega}_{\tau_N} &\overset{*}{\rightharpoonup} \omega \quad \text{in } L^\infty(0,T;Z'), \quad \text{with } \omega(t) := \textstyle\int_{Z'} v \, d\mu_t(v) \quad \text{for a.e. } t \in (0,T). \end{aligned}$$

Passing to the limit in (3.15), we have

$$\xi(t) + \omega(t) = 0 \quad \text{for a.e.}\, t \in (0,T). \tag{3.26}$$

On the other hand, thanks to (3.4), (3.11)–(3.12), Remark 3.4, (3.15) and (3.24), we readily conclude that

$$L_\xi(t) \subset \partial\mathcal{E}(t, z(t)), \quad L_{\omega(t)} \subset \partial\Psi(z(t), 0) \quad \text{for a.e. } t \in (0,T).$$

Thus, a convexity argument yields that $\xi(t) \in \partial\mathcal{E}(t,z(t))$ and $\omega(t) \in \partial\Psi(z(t),0)$ for a.e. $t \in (0,T)$. Combining this with (3.26), we end up with (3.13). Therefore, by Remark 3.6, (S) holds for a.e. $t \in (0,T)$, hence for all $t \in [0,T]$ due to (3.23) and by a continuity argument. By the way, let us stress that the above computations yield that there exists $\xi \in L^\infty(0,T;Z')$ fulfilling $\xi(t) \in \partial\mathcal{E}(t,z(t)) \cap (-\partial\Psi(z(t),0))$ for a.e. $t \in (0,T)$. Recalling that $\Psi(z(t),0) = 0$, we conclude that

$$\Psi(z(r), \dot{z}(r)) + \langle \xi(r), \dot{z}(r) \rangle \geq 0 \quad \text{for a.e. } r \in (0,t), \tag{3.27}$$

Proof of (E). First, we establish the one-sided estimate

$$\textstyle\int_0^t \Psi(z(r), \dot{z}(r))\,\mathrm{d}r + \mathcal{E}(t, z(t)) \leq \mathcal{E}(0, z_0) + \int_0^t \partial_t\mathcal{E}(r, z(r))\,\mathrm{d}r \tag{3.28}$$

by passing to the limit as $N \nearrow \infty$ in the discrete energy inequality

$$\begin{aligned} \int_0^{\bar{\mathfrak{t}}_{\tau_N}(t)} \Psi(\underline{Z}_{\tau_N}(r), \widehat{Z}'_{\tau_N}(r))\,\mathrm{d}r + \mathcal{E}(\bar{\mathfrak{t}}_{\tau_N}(t), \overline{Z}_{\tau_N}(\bar{\mathfrak{t}}_{\tau_N}(t))) \\ \leq \mathcal{E}(0, z_0) + \int_0^{\bar{\mathfrak{t}}_{\tau_N}(t)} \partial_t\mathcal{E}(r, \underline{Z}_{\tau_N}(r))\,\mathrm{d}r \end{aligned} \tag{3.29}$$

for all $t \in [0,T]$. Owing to the assumptions (3.1)–(3.3), it is easy to pass to the limit in the terms involving $\mathcal{E}$ and $\partial_t\mathcal{E}$, see the proof of [6, Thm. 4.6] for further details. In fact, we end up with

$$\partial_t\mathcal{E}(\cdot, \underline{Z}_{\tau_N}(\cdot)) \to \partial_t\mathcal{E}(\cdot, z(\cdot)) \quad \text{in } L^1(0,T), \tag{3.30}$$

$$\liminf_{N\to\infty}(\mathcal{E}(\bar{\mathfrak{t}}_{\tau_N}(t), \overline{Z}_\tau(\bar{\mathfrak{t}}_{\tau_N}(t)) \geq \mathcal{E}(t, z(t))), \tag{3.31}$$

In order to pass to the limit in the first integral term on left-hand side of (3.29), we remark that, by Claims 2 and 5, the sequence $\{(\underline{Z}_{\tau_N}, \widehat{Z}'_{\tau_N})\}$ is bounded in

$L^1(0,T;Z\times Z)$. Thus, applying Theorem 2.2 in the space $X := Z \times Z$, we conclude that, up to a subsequence, $(\underline{Z}_{\tau_N}, \widehat{Z}'_{\tau_N})$ generates a limiting Young measure $\{\nu_t\}_{t\in(0,T)} \in \mathcal{Y}(0,T;Z\times Z)$. Recalling that Ψ is a weakly normal integrand (cf. Section 2) on the space $(0,T)\times Z\times Z$, we thus obtain

$$\liminf_{N\to\infty} \int_0^{\bar{\mathrm{t}}_{\tau_N}(t)} \Psi(\underline{Z}_\tau(r), \widehat{Z}'_{\tau_N}(r))\,\mathrm{d}r \geq \int_0^t \left(\int_{Z\times Z} \Psi(z,v)\,\mathrm{d}\nu_r(z,v)\right)\mathrm{d}r. \tag{3.32}$$

On the other hand, in view of (3.24), (3.25), (2.4) and (2.6), for a.e. $t\in(0,T)$ we have $\nu_t = \delta_{z(t)}\otimes\sigma_t$, with $\{\sigma_t\}_{t\in(0,T)} \in \mathcal{Y}(0,T;Z)$ and $\dot z(t) = \int_Z v\,\mathrm{d}\sigma_t(v)$ for a.e. $t\in(0,T)$. Therefore, also by the Jensen inequality we conclude

$$\begin{aligned}\int_0^t \left(\int_{Z\times Z} \Psi(z,v)\,\mathrm{d}\nu_r(z,v)\right)\mathrm{d}r &= \int_0^t \left(\int_{Z} \Psi(z(r),v)\,\mathrm{d}\sigma_r(v)\right)\mathrm{d}r \\ &\geq \int_0^t \Psi(z(r),\dot z(r))\,\mathrm{d}r. \end{aligned} \tag{3.33}$$

Collecting (3.30), (3.31), (3.32) and (3.33), we conclude (3.28). To obtain the opposite inequality, we combine (3.27) with the chain rule formula (3.8) (see Remark 3.2), applied to the L^∞-selection ξ of $\partial\mathcal{E}(\cdot, z(\cdot))$ previously retrieved. Therefore, we find

$$\tfrac{\mathrm{d}}{\mathrm{d}t}\mathcal{E}(t,z(t)) + \Psi(z(t),\dot z(t)) \geq \partial_t\mathcal{E}(t,z(t)) \text{ for a.e. } t\in(0,T).$$

Integration of this inequality leads to the converse inequality of (3.28), and (E) ensues. □

Acknowledgment

I would like to thank Professor Alexander Mielke for all he has taught to me about rate-independent problems.

References

[1] E.J. Balder, *A general approach to lower semicontinuity and lower closure in optimal control theory.* SIAM J. Control Optim. **22** (1984), 570–598.

[2] M. Brokate, P. Krejčí, and H. Schnabel, *On uniqueness in evolution quasivariational inequalities.* J. Convex Analysis **11** (2004), 111–130.

[3] J. Diestel and J.J. Uhl, Jr., *Vector Measures.* American Mathematical Society, Providence, R.I., 1977. Mathematical Surveys, No. 15.

[4] A. Mielke, *Evolution in rate-independent systems.* In In C. Dafermos and E. Feireisl, editors, *Handbook of Differential Equations. Evolutionary Equations*, volume 2, pages 461–559, Elsevier B.V., 2005.

[5] A. Mielke and F. Theil, *A mathematical model for rate–independent phase transformations with hysteresis.* In H.-D. Alber, R. Balean, and R. Farwig, editors, *Proceedings of the Workshop on "Models of Continuum Mechanics in Analysis and Engineering"*, pages 117–129. Shaker–Verlag, 1999.

[6] A. Mielke and R. Rossi, *Existence and uniqueness results for general rate-independent problems.* Quaderno del Seminario Matematico di Brescia n. 21/2005, 2005. Math. Models Methods Appl. Sci. (to appear).
[7] A. Mielke and F. Theil, *On rate–independent hysteresis models.* NoDEA Nonl. Diff. Eqns. Appl. **11** (2004), 151–189.
[8] A. Mielke, F. Theil, and V. Levitas, *A variational formulation of rate-independent phase transformations using an extremum principle.* Arch. Rational Mech. Anal. **162** (2002), 137–177.
[9] R. Rockafellar, *Convex Analysis.* Princeton University Press, 1970.
[10] R. Rossi and G. Savaré, *Gradient flows of non convex functionals in Hilbert spaces and applications.* ESAIM Control Optim. Calc. Var. **12** (2006), 564–614.

Riccarda Rossi
Dipartimento di Matematica dell'Università di Brescia
Via Valotti 9
I-25133 Brescia, Italy
e-mail: riccarda.rossi@ing.unibs.it

International Series of Numerical Mathematics, Vol. 154, 381–390

Global Attractors for the Quasistationary Phase Field Model: a Gradient Flow Approach

Antonio Segatti

Abstract. In this note we summarize some results of a forthcoming paper (see [15]), where we examine, in particular, the long time behavior of the so-called quasistationary phase field model by using a gradient flow approach. Our strategy in fact, is inspired by recent existence results which show that gradient flows of suitable non-convex functionals yield solutions to the related quasistationary phase field systems. Thus, we firstly present the long-time behavior of solutions to an abstract non-convex gradient flow equation, by carefully exploiting the notion of *generalized semiflows* by J.M. Ball and we provide some sufficient conditions for the existence of the global attractor for the solution semiflow. Then, the existence of the global attractor for a proper subset of all the solutions to the quasistationary phase field model is obtained as a byproduct of our abstract results.

Mathematics Subject Classification (2000). 80A22, 37L30, 35K55.

Keywords. Quasistationary phase field models, gradient flow equation, generalized semiflows, global attractor.

1. Introduction

In this paper we are interested in the study of the asymptotic stability from the point of view of the global attractor of the so-called quasistationary phase field system

$$\partial_t(\vartheta + \chi) - \Delta\vartheta = g, \tag{1.1}$$

$$-\Delta\chi + W'(\chi) = \vartheta, \tag{1.2}$$

in $\Omega \times (0,T)$, where Ω is a bounded domain, occupied by a medium liable to phase transition in the time interval $(0,T)$, for $T > 0$. Here, ϑ is the *relative* temperature of the system, and χ is the phase variable. Further, the function W' is the derivative of the double well potential (e.g., $W(\chi) := (\chi^2 - 1)^2/4$, but in our analysis we can cover also the case in which W presents some singular parts) and g

is a source term that will be taken independent of time. The system (1.1)–(1.2) can be formally obtained from the standard parabolic phase field system (firstly discussed by CAGINALP in [6]) by suppressing the time derivative of χ in the equation for the order parameter. However, proving the convergence of solutions of the phase field system to the solutions of its quasistationary version is still an open and apparently difficult problem. Indeed, the problem of existence of solutions to (1.1)–(1.2) is intrinsically difficult, because of its mixed elliptic-parabolic nature, i.e., the lack of the term $\partial_t \chi$ in (1.2), which prevents from directly controlling the variation in time of the order parameter. Thus, standard approximation arguments are not straightforwardly available. However, the existence of a suitable solution to (1.1)–(1.2) has been proved by PLOTNIKOV & STAROVOITOV in [11] and by SCHÄTZLE [16] in the technically different cases of Dirichlet and of Neumann boundary condition for ϑ. In their approach, the proof of the convergence of the discrete approximation to the solution to (1.1)–(1.2) relies on Holmgren uniqueness continuation theorem or on refined spectral analysis tools and in both cases essentially depends on the particular shape and regularity of the double well potential $W = (\chi^2 - 1)^2/4$. More recently, ROSSI AND SAVARÉ in [13, 14] obtained an existence results by a procedure which somehow exploits the underlying physics of the system. In particular, their analysis relies on the crucial observation that (1.1)–(1.2) stems as a gradient flow for a non convex function strictly related to the entropy of the system. Thus the existence of solutions to (1.1)–(1.2) is obtained as a by product of the general existence theory for gradient flows for non convex functional developed in [13, 14]. We recall that also the quasistationary version of the Penrose-Fife model for phase transitions has recently received a good deal of interest, as the paper [8] shows.

Our approach to the analysis of the long-time behavior of (1.1)–(1.2) actually follows from the existence analysis of [13, 14], which we briefly recall. For later convenience, we recast (1.1)–(1.2) by introducing the *internal energy* (*or enthalpy*) variable $u := \vartheta + \chi$, thus obtaining

$$\partial_t u - \Delta(u - \chi) = g, \text{ in } \Omega \times (0, T) \tag{1.3}$$

$$-\Delta\chi + W'(\chi) = u - \chi, \text{ in } \Omega \times (0, T) \tag{1.4}$$

$$u - \chi = 0, \quad \partial_n \chi = 0 \text{ on } \partial\Omega \times (0, T). \tag{1.5}$$

In [13, 14] relation (1.4) is interpreted as the Euler-Lagrange equation for the minimization, with respect to χ and for fixed u, of the functional

$$\mathscr{F}(u, \chi) := \int_\Omega \left(\frac{1}{2}|u - \chi|^2 + \frac{1}{2}|\nabla\chi|^2 + W(\chi) \right) dx,$$

whose gradient flow with respect to the variable u also yields (1.3). Namely, we turn to the system ($f := (-\Delta^{-1})g$)

$$\begin{cases} (-\Delta^{-1})\partial_t u - \frac{\delta \mathscr{F}}{\delta u} = f & \text{in } \Omega \times (0, T), \\ \frac{\delta \mathscr{F}}{\delta \chi} = 0 & \text{in } \Omega \times (0, T), \end{cases} \tag{1.6}$$

which has a clear variational structure. In fact, in [14] it has been rigorously proved that (1.6) may be interpreted as the *gradient flow* equation (for a suitable notion of subdifferential which we introduce below), in the Hilbert space $H^{-1}(\Omega)$ (recall the conditions (1.5) on $u - \chi$), of the functional defined by $\phi : H^{-1}(\Omega) \to (-\infty, +\infty]$

$$\phi(u) := \inf_{\chi \in H^1(\Omega)} \int_\Omega \left(\frac{1}{2}|u-\chi|^2 + \frac{1}{2}|\nabla\chi|^2 + W(\chi) \right) dx, \quad \text{with } D(\phi) = L^2(\Omega). \tag{1.7}$$

Let us point out that ϕ is a concave perturbation of a quadratic functional, hence it is *non-convex*. In [14] (see also Theorem 3.1 in this paper) existence and approximation results have been obtained for the abstract Cauchy problem

$$u'(t) + \partial_s\phi(u(t)) \ni f \quad \text{a.e. in } (0,T), \quad u(0) = u_0, \tag{GF}$$

for a given initial datum $u_0 \in D(\phi)$ and source term f. The term with $\partial_s\phi$ is a suitable limiting version (cf. the forthcoming Section 3 for the rigorous definition) of the Fréchet subdifferential of the (general) proper and lower semicontinuous functional $\phi : \mathscr{H} \to (-\infty, +\infty]$, not necessarily convex, defined in a (separable) Hilbert space $\mathscr{H}$ with scalar product $\langle\cdot,\cdot\rangle$ and norm $|\cdot|_{\mathscr{H}}$ (which we will often simply denote by $|\cdot|$). Such techniques have been then applied in order to deduce existence and approximation results for the quasistationary phase field evolution problem (1.3)–(1.4), supplemented with Dirichlet boundary conditions. Thus, in order to study the long time behavior of (1.3)–(1.4), we are naturally led to study the long time behavior of (GF) with the potential ϕ given by (1.7) in the Hilbert space $H^{-1}(\Omega)$. For the investigation of the long time behavior of a gradient flow equation for a fairly general non-convex function, the reader is referred to the forthcoming paper [15].

The long time dynamics of gradient flow equations of the type (GF) when the potential ϕ is a convex and lower semicontinuous function (thus $\partial_s\phi$ reduces to the subdifferential of the convex analysis) is rather well known. In particular, the existence of the global attractor has been proved and the long-time convergence to single stationary states investigated even in the non-autonomous situation (see, among the others, [4] and [19]). When ϕ is non convex nor a smooth perturbation of a convex function (as (1.7)), things are remarkably more difficult. In fact, due to the non convexity of the potential ϕ, the uniqueness of the solutions is no longer to be expected (as in the concrete case of the quasistationary phase field system (1.3)–(1.4)). Hence, (GF) does not generate a semigroup, and we cannot rely on the theory of [20] for the study of the long-term dynamics of the solutions.

In recent years, several approaches have been developed in order to address the asymptotic behavior of solutions of differential problems without uniqueness. Without any claim of completeness, we may refer the reader to, e.g., the results by Sell [17] (but see also Chepyzhov & Vishik [7]), Melnik & Valero [10], and, especially, to the work of J.M. Ball, [1, 2]. In particular, we will especially focus here on the theory of *generalized semiflows* proposed in [1]. By definition, a generalized semiflow is a family of functions on $[0, +\infty)$ taking values in a given phase space (we have to think for instance to the solutions to a given differential problem), and complying with suitable existence, stability for time translation, concatenation, and upper semicontinuity axioms. Within this setting, it is possible to introduce a suitable notion of *global attractor*, and to characterize the existence of such an attractor in terms of suitable boundedness and compactness properties on the generalized semiflow. We refer the reader to Section 2 for an overview of these general notions and results, which we have exploited in connection with (GF) in the framework of the metric space (see (3.4))

$$\mathcal{X} = D(\phi), \quad d_{\mathcal{X}} = |u - v| + |\phi(u) - \phi(v)| \ \ \forall u, v \in \mathcal{X}.$$

Indeed, we define the phase space in terms of the energy functional ϕ (see [12, 18] for some analogous choices), which turns out to be a *Lyapunov function* for the system.

We have shown that the set of the solutions to (GF) on the half-line $[0, +\infty)$ is a generalized semiflow, cf. Theorem 3.3 later on, and that it possesses a global attractor, cf. Theorem 3.4. Referring to the forthcoming paper [15] for all the details, we just stress that the energy identity and a proper chain rule for the potential ϕ combined with the compactness of the sublevels of ϕ (see Theorem 3.1) will play a key role both in the proof of the upper-semicontinuity axiom, and of the boundedness and compactness properties of the trajectories. Eventually, we apply our abstract results Theorems 3.3 and 3.4 to the investigation of the long-time behavior of the *energy solutions* to (1.3)–(1.4), i.e., the solutions deriving from the related gradient flow equation (GF), for the functional ϕ in (1.7). Thus, we show that, under suitable conditions, the energy solutions to (1.3)–(1.4) form a generalized semiflow, which admits the global attractor. Let us stress that this gradient flow approach does not provide the description of the long-term behavior of the *whole* set of solutions to (1.3)–(1.4), but it is rather concerned with a proper subclass of trajectories (i.e., the solutions to the gradient flow).

2. Generalized semiflows

Suppose we are given a metric space (not necessarily complete) $\mathcal{X}$ with metric $d_{\mathcal{X}}$. If C is a subset of $\mathcal{X}$ and b is a point in $\mathcal{X}$, we set $\rho(b, C) := \inf_{c \in C} d_{\mathcal{X}}(b, c)$; consequently, if $C \subset \mathcal{X}$ and $B \subset \mathcal{X}$, we set $\operatorname{dist}(B, C) := \sup_{b \in B} \rho(b, C)$.

Definition 2.1. A *generalized semiflow* $\mathfrak{F}$ on $\mathcal{X}$ is a family of maps $u : [0, +\infty) \to \mathcal{X}$, called solutions, satisfying the following hypotheses:

(H1) (*Existence*) For each $v \in \mathcal{X}$ there exists at least one $u \in \mathfrak{F}$ with $u(0) = v$.

(H2) (*Translates of solutions are still solutions*) If $u \in \mathfrak{F}$ and $\tau \geq 0$, then $u^{\tau} \in \mathfrak{F}$ where $u^{\tau}(t) := u(t + \tau)$, $t \in (0, +\infty)$.

(H3) (*Concatenation*) If $u, v \in \mathfrak{F}$, and $t \geq 0$ with $u(t) = v(0)$ then the function w defined by

$$w(\tau) := \begin{cases} u(\tau) & \text{for } 0 \leq \tau \leq t, \\ v(\tau - t) & \text{for } t < \tau, \end{cases}$$

belongs to $\mathfrak{F}$.

(H4) (*Upper semi-continuity with respect to initial data*) If $u_n \in \mathfrak{F}$ with $u_n(0) \to v$, then there exist a subsequence u_{n_k} of u_n and $u \in \mathfrak{F}$ with $u(0) = v$ such that $u_{n_k}(t) \to u(t)$ for each $t \geq 0$.

It is possible to extend to *generalized semiflows* the standard concepts concerning absorbing sets, ω-limit sets and attractors given for semiflows and semigroups (see [1]). In particular, for a given *generalized semiflows* $\mathfrak{F}$ any $t \geq 0$, we define

$$T(t)E = \{u(t) : u \in \mathfrak{F} \text{ with } u(0) \in E\}, \tag{2.1}$$

where $E \subset \mathcal{X}$. It is clear that $T(t) : 2^{\mathcal{X}} \to 2^{\mathcal{X}}$, denoting by $2^{\mathcal{X}}$ the space of all subsets of $\mathcal{X}$. Moreover, thanks to $(H2)$ and $(H3)$, $\{T(t)\}_{t\geq 0}$ defines a semigroup on $2^{\mathcal{X}}$. On the other hand, $(H4)$ implies that $T(t)z$ is compact for any $z \in \mathcal{X}$. We say that the subset $\mathcal{U} \subset \mathcal{X}$ *attracts* a set E if $\operatorname{dist}(T(t)E, \mathcal{U}) \to 0$ as $t \to +\infty$.

We say that $\mathcal{U}$ is *invariant* if $T(t)\mathcal{U} = \mathcal{U}$ for all $t \geq 0$.

The subset $\mathcal{U} \subset \mathcal{X}$ is a global attractor if $\mathcal{U}$ is compact, invariant, and attracts all bounded sets.

$\mathfrak{F}$ is *eventually bounded* if, given any bounded $B \subset \mathcal{X}$, there exists $\tau \geq 0$ with $\gamma^\tau(B)$ bounded.

$\mathfrak{F}$ is *point dissipative* if there exists a bounded set B_0 such that, for any $u \in \mathfrak{F}$, $u(t) \in B_0$ for all sufficiently large $t \geq 0$.

$\mathfrak{F}$ is *asymptotically compact* if for any sequence $u_n \in \mathfrak{F}$ with $u_n(0)$ bounded, and for any sequence $t_n \nearrow +\infty$, the sequence $u_n(t_n)$ has a convergent subsequence. We will also make use of the notion of *Lyapunov function*, which can be introduced starting from the following definitions: we say that a complete orbit $g \in \mathfrak{F}$ is *stationary* if there exists $x \in \mathfrak{F}$ such that $g(t) = x$ for all $t \in \mathbb{R}$ – such x is then called a *rest point*. We denote the set of rest points of $\mathfrak{F}$ by $Z(\mathcal{X})$. A function $V : \mathcal{X} \to \mathbb{R}$ is said to be a *Lyapunov function* for $\mathfrak{F}$ if: V is continuous, $V(g(t)) \leq V(g(s))$ for all $g \in \mathfrak{F}$ and $0 \leq s \leq t$ (i.e., V decreases along solutions), and, whenever the map $t \mapsto V(g(t))$ is constant for some complete orbit g, then g is a stationary orbit. Finally, we say that a global attractor A for $\mathfrak{F}$ is *Lyapunov stable* if for any $\varepsilon > 0$ there exists $\delta > 0$ such that for any $E \subset \mathfrak{F}$ with $\operatorname{dist}(E, A) \leq \delta$, then $\operatorname{dist}(T(t)E, A) \leq \varepsilon$ for all $t \geq 0$. The following Theorem (see [1, Theorem 5.1]) gives sufficient conditions for a generalized semiflow to have a global attractor.

Theorem 2.2 (Ball 1997). *Assume that each element $u \in \mathfrak{F}$ is continuous from $(0, +\infty) \to \mathcal{X}$ and that $\mathfrak{F}$ is asymptotically compact. Suppose further that there exists a Lyapunov function V for $\mathfrak{F}$ and that the sets of its rest points $Z(\mathcal{X})$ is bounded. Then, $\mathfrak{F}$ is also point dissipative, and thus admits a global attractor A.*

Moreover for all trajectories $u \in \mathfrak{F}$, the limit sets $\alpha(u), \omega(u)$ are connected subsets of $Z(\mathcal{X})$ on which V is constant. When $Z(\mathcal{X})$ is totally disconnected the limits

$$z_- = \lim_{t \searrow -\infty} u(t), \qquad z_+ = \lim_{t \nearrow +\infty} u(t) \tag{2.2}$$

exist and z_-, z_+ are rest points; moreover, $v(t)$ tends to a rest point as $t \nearrow +\infty$ for every $v \in \mathfrak{F}$.

3. Abstract gradient flows in Hilbert spaces and their long time behavior

In this section we briefly recall the existence theorem of Rossi and Savaré in [14] and we state, without proofs, our main results. The interested reader is referred to the forthcoming paper [15] for the proofs and for some related remarks.

First of all, we have to introduce the notion of subdifferential we aim to use in our analysis. Since the function ϕ is *non convex*, a preliminary choice could be the so called Fréchet subdifferential defined by

$$\xi \in \partial\phi(v) \quad \Leftrightarrow \quad v \in D(\phi), \ \liminf_{w \to v} \frac{\phi(w) - \phi(v) - \langle \xi, w - v \rangle}{|w - v|} \geq 0. \tag{3.1}$$

It is easy to see that the Fréchet subdifferential reduces to the usual one as soon as ϕ is convex. Unfortunately, (3.1) has some drawbacks. In particular, easy finite dimensional examples show that the graph of the Fréchet subdifferential may not be strongly-weakly closed, which is one of the major features of the convex case. We thus define the *strong limiting subdifferential* $\partial_s\phi$ at a point $v \in D(\phi)$ as the set of the vectors ξ such that there exists sequences

$$v_n, \xi_n \in \mathscr{H} \quad \text{with} \quad \xi_n \in \partial\phi(v_n), \ v_n \to v, \ \xi_n \to \xi, \ \phi(v_n) \to \phi(v), \tag{3.2}$$

as $n \uparrow +\infty$. Of course, $\partial_s\phi$ reduces to the usual subdifferential $\partial\phi$ whenever ϕ is convex. Let us now recall one of the existence results proved in [14] for the Cauchy problem (GF).

Theorem 3.1 (Rossi-Savaré 04). *Let $\phi : \mathscr{H} \to (-\infty, +\infty]$ be a proper and lower semicontinuous function which complies with the* coercivity *assumption*

$$\exists\, \kappa \geq 0 : \quad v \mapsto \phi(v) + \kappa|v|^2 \ \text{ has compact sublevels,} \tag{COMP}$$

and with the Chain Rule *condition for any bounded interval (a, b)*

$$\begin{gathered} \text{if } v \in H^1(a,b;\mathscr{H}),\ \xi \in L^2(a,b;\mathscr{H}),\ \xi \in \partial_s\phi(v) \ \textit{a.e. in } (a,b), \\ \textit{and } \phi \circ v \textit{ is bounded, } \text{then } \phi \circ v \in AC(a,b) \textit{ and} \\ \frac{d}{dt}\phi(v(t)) = \langle \xi(t), v'(t) \rangle \quad \textit{for a.e. } t \in (a,b). \end{gathered} \tag{CHAIN}$$

Then, for any $u_0 \in D(\phi)$, $T > 0$ and $f \in \mathscr{H}$, the Cauchy problem

$$u'(t) + \partial_s\phi(u(t)) \ni f \quad \textit{a.e. in } (0,T), \quad u(0) = u_0,$$

admits a solution $u \in H^1(0,T;\mathscr{H})$. *Moreover, there holds the* energy identity

$$\int_s^t |u'(\sigma)|^2\, d\sigma + \phi(u(t)) = \phi(u(s)) + \int_s^t \langle f, u'(\sigma)\rangle\, d\sigma, \quad \forall s,t \in [0,T]. \tag{3.3}$$

3.1. Long time behavior of (GF)

The assumption $u_0 \in D(\phi)$ in the existence Theorem 3.1, suggests to investigate the long time behavior of (GF) in the phase space (not complete in general)

$$\mathcal{X} := D(\phi), \quad \text{with} \quad d_{\mathcal{X}}(u,v) := |u-v| + |\phi(u) - \phi(v)| \quad \forall u,v \in \mathcal{X}. \tag{3.4}$$

Definition 3.2. *We denote by* $\mathcal{S}$ *the set of all solutions* $u \in H^1_{loc}(0,+\infty;\mathscr{H})$ *to the gradient flow equation*

$$u'(t) + \partial_s\phi(u(t)) \ni f \quad \text{for a.e. } t \in (0,+\infty). \tag{3.5}$$

We can now state our main results (see [15] for the proofs).

Theorem 3.3 (The generalized semiflow). *Let* ϕ *comply with the assumptions of Theorem* 3.1. *In addition, assume that*

$$\exists K_1, K_2 \geq 0: \quad \phi(u) \geq -K_1|u| - K_2 \quad \forall u \in \mathscr{H}. \tag{3.6}$$

Then, $\mathcal{S}$ *is a generalized semiflow on* $(\mathcal{X}, d_{\mathcal{X}})$.

In order to study the long time behavior for our gradient flow equation, we assume some additional continuity property for the potential ϕ, that is

$$v_n \to v, \quad \sup_n \left(|(\partial_s\phi(v_n))^\circ|, \phi(v_n) \right) < +\infty \ \Rightarrow \phi(v_n) \to \phi(v), \tag{CONT}$$

where $|(\partial_s\phi(v))^\circ| := \inf_{\xi \in \partial_s\phi(v)} |\xi|$. The latter is indeed a natural request. In fact, (CONT) is readily fulfilled by lower semicontinuous convex functionals. We thus have

Theorem 3.4 (The global attractor). *Let* ϕ *fulfill the above assumptions of Theorem* 3.1, (CONT) *and*

$$\exists J_1, J_2 > 0: \quad \phi(u) \geq J_1|u| - J_2 \quad \forall u \in \mathscr{H}. \tag{3.7}$$

Further, let $\mathcal{D}$ *be a non-empty subset of* $\mathcal{X}$ *satisfying*

$$\begin{gathered} \mathcal{T}(t)\mathcal{D} \subset \mathcal{D} \quad \forall t \geq 0, \\ Z(\mathcal{S}) \cap \mathcal{D} := \{u \in D(\partial_s\phi) \,:\, 0 \in \partial_s\phi(u) - f\} \cap \mathcal{D} \text{ is bounded in } (\mathcal{X}, d_{\mathcal{X}}). \end{gathered} \tag{3.8}$$

Then, there exists a unique, Lyapunov stable attractor A *for* $\mathcal{S}$ in $\mathcal{D}$, *given by* $A := \cup\{\omega(D) \,:\, D \subset \mathcal{D} \text{ bounded}\}$.

4. Long time behavior of the quasistationary phase field system

First of all, we have to specify the class of solutions of the quasistationary phase field model (1.3)–(1.4) for which we construct the global attractor. We introduce the following

Definition 4.1 (Energy solutions). We say that a function

$$u \in H^1_{loc}(0,+\infty; H^{-1}(\Omega)) \cap L^\infty_{\mathrm{loc}}(0,+\infty; L^2(\Omega))$$

is an *energy solution* to Problem (1.3)–(1.4) with the boundary conditions (1.5) if u solves the gradient flow equation

$$u'(t) + \partial_s \phi(u(t)) \ni f \quad \text{for a.e. } t \in (0,+\infty),$$

in the Hilbert space $\mathscr{H} := H^{-1}(\Omega)$, for the functional

$$\phi(u) := \inf_{\chi \in H^1(\Omega)} \int_\Omega \left(\frac{1}{2}|u-\chi|^2 + \frac{1}{2}|\nabla \chi|^2 + W(\chi) \right) dx, \quad u \in L^2(\Omega). \tag{4.1}$$

We denote by $\mathcal{E}$ the set of all energy solutions.

The set $\mathcal{E}$ is not empty thanks to Theorem 3.1. In fact, in [14] it has been proved that the potential ϕ in (1.7) is proper and lower semicontinuous and satisfies the chain rule (CHAIN) and the coercivity condition (COMP) in the Hilbert space $\mathscr{H} = H^{-1}(\Omega)$. As a by product of our main results Theorem 3.3 and Theorem 3.4 we thus have the following result ($D(W)$ is the realization in $L^2(\Omega)$ of the domain of W) in the framework of the phase space (see (4.1))

$$\mathcal{X} = L^2(\Omega) \quad d_{\mathcal{X}}(u,v) = \|u-v\|_{H^{-1}(\Omega)} + |\phi(u) - \phi(v)| \ \ \forall u,v \in L^2(\Omega) \tag{4.2}$$

Theorem 4.2. *Let the double well potential W in (1.4) be such that: there exist constants $\kappa_1,\ \kappa_2 > 0$ such that for all $v \in H^1(\Omega) \cap D(W)$*

$$\int_\Omega W(v)dx \geq \kappa_1 \|v\|^2_{L^2(\Omega)} - \kappa_2, \tag{4.3}$$

and either one of the following

1. *the set $H^1(\Omega) \cap D(W')$ is bounded in $(L^2(\Omega), d_{\mathcal{X}})$,* (4.4)
2. *there exist two positive constants κ_3, κ_4 such that for all $v \in H^1(\Omega) \cap D(W')$*

$$\int_\Omega W'(v)v \geq \kappa_3 \|v\|_{L^2(\Omega)} - \kappa_4. \tag{4.5}$$

Then, the set $\mathcal{E}$ of all the energy solutions to Problem (1.3)–(1.4) is a generalized semiflow in the phase space $(L^2(\Omega), d_{\mathcal{X}})$ (see (4.2)). Moreover, $\mathcal{E}$ possesses a unique global attractor $A_{\mathcal{E}}$, which is Lyapunov stable. Finally, for any trajectory $u \in \mathcal{E}$ and for any $u_\infty \in \omega(u)$, we have

$$\begin{aligned} -\Delta u_\infty + W'(u_\infty) &= f, \\ \partial_n u_\infty &= 0 \end{aligned} \tag{4.6}$$

Meaningful examples of potentials W satisfying the coercivity assumption (4.3) and (4.2) (or(4.5)) are the standard double-well potential

$$W(\chi) := \frac{(\chi^2 - 1)^2}{4}, \tag{4.7}$$

but also

$$W(\chi) := I_{[-1,1]}(\chi) + (1 - \chi)^2; \tag{4.8}$$

$$W(\chi) := c_1\left((1+\chi)\ln(1+\chi) + (1-\chi)\ln(1-\chi)\right) - c_2\chi^2 + c_3\chi + c_4, \tag{4.9}$$

with $c_1, c_2 > 0$ and $c_3, c_4 \in \mathbb{R}$ (see, e.g., [5, 4.4, p. 170] for (4.9), [3, 21] for (4.8)). In particular, the term with $I_{[-1,1]}$ is the indicator function of $[-1, 1]$, thus forcing χ to lie between -1 and 1.

Remark 4.3. We stress that the question of the convergence of *all* the trajectory $u(t)$ to a single solution of equation (4.6) is a nontrivial one and is not answered by the preceding Theorem. This problem would have an affirmative answer if the set of all the solution would be totally disconnected (see Theorem 2.2). Unfortunately, it is well known (see [9]) that problem (4.6) may well admit a continuum of solutions.

Remark 4.4 (The Neumann-Neumann boundary condition case). If one replace the first in (1.5) with $\partial_n(u-\chi) = 0$ (i.e., homogeneous Neumann boundary conditions for the temperature ϑ), we get the so-called quasistationary phase field model with Neumann-Neumann boundary condition. This situation is very delicate since with this type of (non coercive) boundary conditions problem (1.3)–(1.4) does not have a gradient flow structure (see [14]). In [14] however, the existence of solutions has been deduced by means of a suitable approximation with more regular problems of gradient flow type. This kind of approximation has been reconsidered in [15] from the point of view of the long time dynamics. More precisely, in [15] we show that the set of all the solutions to (1.3)–(1.4) obtained with the above mentioned approximation still retain a (kind of) generalized semiflow structure. In particular this set, named $\mathcal{E}_N$, does not satisfy the concatenation property, but complies with some substantial properties, which allow us to prove the existence of a suitable weak notion of global attractor $A_{\mathcal{E}_N}$. Here weak means that this subset of the phase space is no longer *invariant* but only *quasi invariant* in the sense that for any $v \in A_{\mathcal{E}_N}$ there exists a complete orbit w with $w(0) = v$ and $w(t) \in A_{\mathcal{E}_N}$ for all $t \in \mathbb{R}$.

References

[1] J.M. Ball, *Continuity properties and global attractors of generalized semiflows and the Navier-Stokes equations*, J. Nonlinear Sci. **7** (1997), 475–502.

[2] ———, *Global attractors for damped semilinear wave equations*, Discrete Continuous. Dynam. Systems, **10**(1) (2004), 31–52.

[3] J.F. Blowey and C.M. Elliott, *The Cahn-Hilliard gradient theory for the phase separations with nonsmooth free energy. I*, European J. Appl. Math, **2**(3) (1991), 233–280.

[4] H. Brézis, *Opérateurs maximaux monotones et semi-groupes de contractions dans les espaces de Hilbert*, North-Holland Publishing Co., Amsterdam, 1973, North-Holland Mathematics Studies, No. 5. Notas de Matemática (50). MR 50 #1060

[5] M. Brokate and J. Sprekels, *Hysteresis and phase transitions*, Appl. Math. Sci., 121, Springer, New York, 1996.

[6] G. Caginalp, *An analysis of a phase field model of a free boundary*, Arch. Rational Mech. Anal., **92**, (1986), 205–245.

[7] V.V. Chepyshoz and M.I. Vishik, *Attractors for equations of mathematical physics*, American Mathematical Society Colloquium Publications, 49, American Mathematical Society, Providence, RI, 2002.

[8] P. Colli and P.I. Plotnikov, *Global solution to a quasistationary Penrose-Fife model*, Indiana Univ. Math. J., **54** (2) (2005), 349–382.

[9] A. Haraux, *Systèmes dynamiques dissipatifs et applications*, Recherches en Mathématiques Appliquées, Masson, Paris 1991.

[10] V.S. Melnik and J. Valero, *On attractors of multivalued semi-flows and differential inclusions*, Set-Valued Anal. **6** (4) (1998), 83–111.

[11] P.I. Plotnikov and V.N. Starovoitov, *The Stefan problem with surface tension as the limit of a phase field model*, Differential Equations **29** (1993), 395–404.

[12] E. Rocca and G. Schimperna, *Universal attractor for some singular phase transition systems*, Phys. D **192** (3-4) (2004), 279–307.

[13] R. Rossi and G. Savaré, *Existence and approximation results for gradient flows*, "Rend. Mat. Acc. Lincei", **15** (2004), 183–196.

[14] R. Rossi and G. Savaré, *Gradient flows of non convex functionals in Hilbert spaces and applications*, preprint IMATI-CNR n. 7-PV (2004) 1-45. To appear on *ESAIM Control Optim. Calc. Var.*

[15] R. Rossi, A. Segatti and U. Stefanelli, *Attractors for gradient flows of non convex functionals and applications*, (2005). Submitted.

[16] R. Schätzle, *The quasistationary phase field equations with Neumann boundary conditions*, J. Differential Equations, **162** (2) (2000), 473–503. MR **2001b**:35143

[17] G.R.Sell, *Differential equations without uniqueness and classical topological dynamics*, J. Differential Equations **14** (1973), 42–56.

[18] A. Segatti, *Global attractor for a class of doubly nonlinear abstract evolution equations*, Discrete Contin. Dyn. Syst. **14** (4) (2006), 801–820.

[19] K. Shirakawa, A. Ito, N. Yamazaki, and N. Kenmochi, *Asymptotic stability for evolution equations governed by subdifferentials*. Recent developments in domain decomposition methods and flow problems (Kyoto, 1996; Anacapri, 1996), GAKUTO Internat. Ser. Math. Sci. Appl., **11**, 287–310, Gakkōtosho, Tokyo, 1998.

[20] R. Temam, *Infinite dimensional mechanical systems in mechanics and physics*. Applied Mathematical Sciences 68, Springer-Verlag, New York, 1988. 1988.

[21] A. Visintin, *Models of phase transitions*, Progress in Nonlinear Differential Equations and Their Applications, vol. 28, Birkhäuser, Boston, 1996.

Antonio Segatti
Department of Mathematics "F. Casorati", Università di Pavia
Via Ferrata 1, I-27100 Pavia, Italy
e-mail: `antonio.segatti@unipv.it`

International Series of Numerical Mathematics, Vol. 154, 391–401

Aleksandrov and Kelvin Reflection and the Regularity of Free Boundaries

Henrik Shahgholian and Georg S. Weiss

Abstract. The first part of this paper is an announcement of a result to appear. We apply the Aleksandrov reflection to obtain regularity and stability of the free boundaries in the *two-dimensional* problem

$$\Delta u = \frac{\lambda_+}{2}\chi_{\{u>0\}} - \frac{\lambda_-}{2}\chi_{\{u<0\}} ,$$

where $\lambda_+ > 0$ and $\lambda_- > 0$.

In the second part we show that the Kelvin reflection can be used in a similar way to obtain regularity of the classical obstacle problem

$$\Delta u = \chi_{\{u>0\}}$$

in higher dimensions.

Mathematics Subject Classification (2000). Primary 35R35, Secondary 35J60.

Keywords. Free boundary, singular point, branch point, obstacle problem, regularity, Kelvin transform, global solution, blow-up, monotonicity formula.

1. Part I

From the physical point of view, the problem

$$\Delta u = \frac{\lambda_+}{2}\chi_{\{u>0\}} - \frac{\lambda_-}{2}\chi_{\{u<0\}} \qquad \text{in } \Omega, \tag{1.1}$$

where $\lambda_+ > 0, \lambda_- > 0$ and $\Omega \subset \mathbf{R}^n$ (cf. Fig. 1) arises for example as the "two-phase membrane problem": consider an elastic membrane touching the planar phase boundary between two liquid/gaseous phases with constant densities $\rho_1 > \rho_2$ in a gravity field, for example water and air. If the constant density ρ_m of the membrane

H. Shahgholian has been partially supported by the Swedish Research Council.

G.S. Weiss has been partially supported by a Grant-in-Aid for Scientific Research, Ministry of Education, Japan. He wishes to thank the Max Planck Institute for Mathematics in the Sciences for the hospitality during his stay in Leipzig and the Göran Gustafsson Foundation for visiting appointments to the Royal Inst. of Technology in Stockholm.

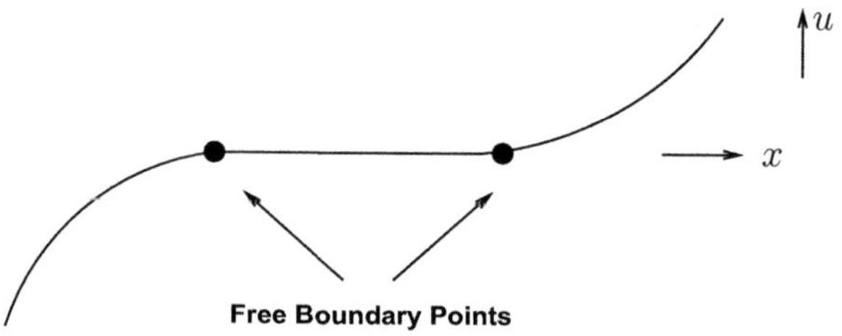

FIGURE 1. Example of a one-dimensional membrane

satisfies $\rho_1 > \rho_m > \rho_2$, then the membrane is being buoyed up in the phase with higher density and pulled down in the phase with lesser density, so the equilibrium state can be described by equation (1.1). Notice that (1.1) arises also as limiting case in the model of temperature control through the interior described in [7, 2.3.2] as $h_1, h_2 \to 0$.

Properties of the solution etc. have been derived by the G.S. Weiss in [19] and by N. Uraltseva in [16]. Moreover, in [15], H. Shahgholian-N. Uraltseva-G.S. Weiss gave a complete characterization of global two-phase solutions satisfying a quadratic growth condition at the two-phase free boundary point 0 and at infinity. It turned out that each such solution coincides after rotation with the one-dimensional solution $u(x) = \frac{\lambda_+}{4}\max(x_n, 0)^2 - \frac{\lambda_-}{4}\min(x_n, 0)^2$. In particular this implies that each blow-up limit u_0 at so-called "branch points" (see Fig. 2), $\Omega \cap \partial\{u > 0\} \cap \partial\{u < 0\} \cap \{\nabla u = 0\}$, is after rotation of the form $u_0(x) = \frac{\lambda_+}{4}\max(x_n, 0)^2 - \frac{\lambda_-}{4}\min(x_n, 0)^2$.

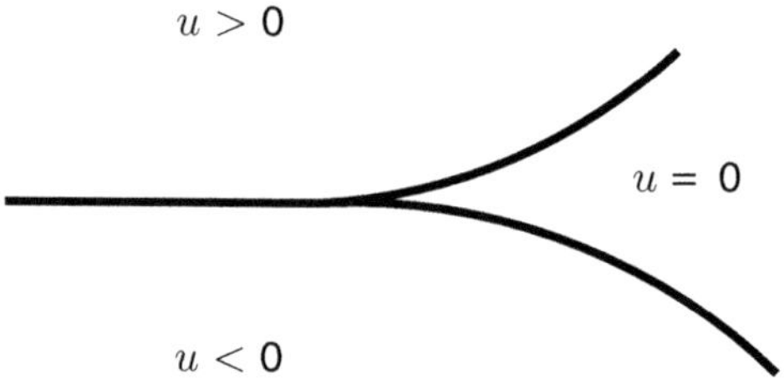

FIGURE 2. Example of a branch point

In this paper we prove (cf. Theorem 1.5) that *in two dimensions* the free boundary is in a neighborhood of each branch point the union of (at most) two C^1-graphs. As application we obtain the following stability result: If the free boundary contains no singular one-phase point for certain boundary data (B_0), then for boundary data (B) close to (B_0) the free boundary consists of C^1-arcs converging to those of (B) (cf. Theorem 1.7).

Let $\lambda_+ > 0$ and $\lambda_- > 0$, $n \geq 2$, let Ω be a bounded open subset of $\mathbf{R}^n$ with Lipschitz boundary and assume that $u_D \in W^{1,2}(\Omega)$. From [19] we know then that there exists a "solution", i.e., a function $u \in W^{2,2}(\Omega)$ solving the strong equation $\Delta u = \frac{\lambda_+}{2}\,\chi_{\{u>0\}} - \frac{\lambda_-}{2}\,\chi_{\{u<0\}}$ a.e. in Ω, and attaining the boundary data u_D in L^2. The boundary condition may be replaced by other, more general boundary conditions.

The tools at our disposition include two powerful monotonicity formulae. One is the monotonicity formula introduced in [18] by G.S. Weiss for a class of semilinear free boundary problems (see also [17]). The second monotonicity formula has been introduced by H.W. Alt-L.A. Caffarelli-A. Friedman in [1]. What we are actually going to apply in Theorem 1.4 is a stronger statement than the one in [1].

For the sake of completeness let us state both monotonicity formulae here.

Theorem 1.1 (Weiss's Monotonicity Formula). *Suppose that $B_\delta(x_0) \subset \Omega$. Then for all $0 < \rho < \sigma < \delta$ the function*

$$\Phi_{x_0}(r) := r^{-n-2} \int_{B_r(x_0)} \Big(|\nabla u|^2 + \lambda_+ \max(u,0) + \lambda_- \max(-u,0) \Big)$$

$$- 2\, r^{-n-3} \int_{\partial B_r(x_0)} u^2 \, d\mathcal{H}^{n-1} ,$$

defined in $(0,\delta)$, satisfies the monotonicity formula

$$\Phi_{x_0}(\sigma) - \Phi_{x_0}(\rho) = \int_\rho^\sigma r^{-n-2} \int_{\partial B_r(x_0)} 2 \left(\nabla u \cdot \nu - 2\,\frac{u}{r} \right)^2 d\mathcal{H}^{n-1}\, dr \geq 0 .$$

For a proof see [18].

In Theorem 1.4 we are going to need the following stronger version of the Alt-Caffarelli-Friedman monotonicity formula.

Theorem 1.2 (Alt-Caffarelli-Friedman Monotonicity Formula). *Let h_1 and h_2 be continuous non-negative subharmonic $W^{1,2}$-functions in $B_R(z)$ satisfying $h_1 h_2 = 0$ in $B_R(z)$ as well as $h_1(z) = h_2(z) = 0$.*

Then for

$$\Psi_z(r, h_1, h_2) := r^{-4} \int_{B_r(z)} \frac{|\nabla h_1(x)|^2}{|x-z|^{n-2}}\, dx \int_{B_r(z)} \frac{|\nabla h_2(x)|^2}{|x-z|^{n-2}}\, dx ,$$

and for $0 < \rho < r < \sigma < R$, we have $\Psi_z(\rho) \leq \Psi_z(\sigma)$. Moreover, if equality holds for some $0 < \rho < r < \sigma < R$ then one of the following is true:

(A) *$h_1 = 0$ in $B_\sigma(z)$ or $h_2 = 0$ in $B_\sigma(z)$,*

(B) *for $i = 1, 2$, and $\rho < r < \sigma$, supp $(h_i) \cap \partial B_r(z)$ is a half-sphere and $h_i \Delta h_i = 0$ in $B_\sigma(z) \setminus B_\rho(z)$ in the sense of measures.*

For a proof of this version of monotonicity see [15]. We also refer to [1], for the original proof.

It is noteworthy that

$$\Psi_z(r, (\partial_e u)^+, (\partial_e u)^-) = \Psi_0(1, (\partial_e u_r)^+, (\partial_e u_r)^-) \text{ and } \Phi_z(r, u) = \Phi_0(1, u_r),$$

where

$$u_r(x) = \frac{u(rx+z)}{r^2}.$$

It is in fact possible to apply Theorem 1.2 to the positive and negative part of directional derivatives of u: due to N. Uraltseva, the functions $\max(\partial_e u, 0)$ and $-\min(\partial_e u, 0)$ are subharmonic in Ω (see Lemma 2 in [16]).

A quadratic growth estimate near the set $\Omega \cap \{u=0\} \cap \{\nabla u = 0\}$ had already been proved in [19] for more general coefficients λ_+ and λ_-, but local $W^{2,\infty}$- or $C^{1,1}$-regularity of the solution has been shown for the first time in [16]. See also [14]. So we know that

$$u \in W^{2,\infty}_{\text{loc}}(\Omega). \tag{1.2}$$

The following lemma relates the value of the density of the ACF-monotonicity formula to the structure of the singularity:

Lemma 1.3. *Let u be a solution of (1.1) in B_1 and suppose that the origin is a free boundary point. Then the following statements are equivalent:*

1) *Either $\nabla u(0) \neq 0$, or $\lim_{r\to 0} \Psi_0(r, (\partial_e u)^+, (\partial_e u)^-) = 0$ for each direction e.*
2) *Either $\nabla u(0) \neq 0$, or each blow-up limit*

$$u_0(x) = \lim_{m\to\infty} \frac{u(r_m x)}{r_m^2}$$

is after rotation of the form

$$u_0(x) = a_1 \frac{\lambda_+}{4} \max(x_1, 0)^2 - a_2 \frac{\lambda_-}{4} \min(x_1, 0)^2$$

where $a_1, a_2 \in \{0,1\}$ and $a_1 + a_2 \neq 0$.

3) *Either $\nabla u(0) \neq 0$, or at least one blow-up limit*

$$u_0(x) = \lim_{m\to\infty} \frac{u(r_m x)}{r_m^2}$$

is after rotation of the form

$$u_0(x) = a_1 \frac{\lambda_+}{4} \max(x_1, 0)^2 - a_2 \frac{\lambda_-}{4} \min(x_1, 0)^2$$

where $a_1, a_2 \in \{0,1\}$.

4) *The origin is not a one-phase singular free boundary point, i.e., no blow-up limit*
$$u_0(x) = \lim_{m\to\infty} \frac{u(r_m x)}{r_m^2}$$
is allowed to be a non-negative/non-positive homogeneous polynomial of degree 2.

Let us now define the class

$$\begin{aligned} M^* := \{u : B_1(0) \to \mathbf{R} \ : \\ u(x_1,\dots,x_n) = \beta_1\left(\frac{\lambda_+}{4}\max(x_1,0)^2 - \frac{\lambda_-}{4}\min(x_1-\tau,0)^2\right) + \beta_2 x_1, \quad (1.3) \\ \text{where } \tau\in[-1,0], 0\le\beta_1\le a, 0\le\beta_2\le b, 0<c\le\beta_1+\beta_2, \\ \text{and } \beta_2\neq 0 \text{ implies } \tau=0\}. \end{aligned}$$

The class M is then defined as all rotated elements of M^*, i.e.,

$$M := \{u : B_1(0) \to \mathbf{R} \ : u = v\circ U \text{ where } U \text{ is a rotation}, v\in M^*\}. \quad (1.4)$$

Observe that singular one-phase solutions are excluded from M.

Theorem 1.4. *Let $(u_\alpha)_{\alpha\in I}$ be a family of solutions of* (1.1) *in B_1 that is bounded in $W^{2,\infty}(B_1)$, and suppose that $0\in\Omega\cap(\partial\{u_{\alpha_0}>0\}\cup\partial\{u_{\alpha_0}<0\})$ for some $\alpha_0\in I$, and either $\nabla u_{\alpha_0}(0)\neq 0$ or $\lim_{r\to 0}\Psi_0(r,(\partial_e u_{\alpha_0})^+,(\partial_e u_{\alpha_0})^-) = 0$ for each direction e; this means by Lemma* 1.3 *that* 0 *is not a singular one-phase free boundary point. Define further S_r by*
$$r^{n-1}S_r^2(y,u_\alpha) = \int_{\partial B_r(y)} u_\alpha^2,$$
Then, if $u_\alpha \to u_{\alpha_0}$ in $L^1(B_1)$ as $\alpha\to\alpha_0, \partial\{u_\alpha>0\}\ni\ y\to 0$ and $r\to 0$, all possible limit functions of the family
$$\frac{u_\alpha(y+r\cdot)}{S_r(y,u_\alpha)},$$
belong to M for some a, b, c as above.

The following theorem contains our main result, i.e., regularity at branch points. Unfortunately the known techniques seem to be insufficient to do a conclusive analysis at branch points. One reason is that the density of the monotonicity formula by H.W. Alt-L.A. Caffarelli-A. Friedman takes the value 0 at branch points.

The situation is complicated by the fact that the limit manifold of all possible blow-ups at branch points (including the case of varying centers) is not a one-dimensional or even smooth manifold, but has a more involved structure. Also the convergence to blow-up limits is close to the branch-point *not uniform!* Here we use an intersection-comparison approach based on the Aleksandrov reflection to show that – although the flow with respect to the limit manifold may not slow down when blowing up – the free boundaries are still *uniformly graphs* (see Proposition

1.6). The approach in Proposition 1.6 uses – apart from the reflection invariance – very little information about the underlying PDE and so yields a general approach to the regularity of free boundaries in two space dimensions provided that there is some information on the blow-up limits.

The Aleksandrov reflection has been recently used to prove regularity in geometric parabolic PDE ([10], [11], [12]). In contrast to those results, where structural conditions for the initial data are preserved under the flow, our results are completely local.

Theorem 1.5. *Let $n = 2$, let $(u_\alpha)_{\alpha\in I}$ be a family of solutions of* (1.1) *in B_1 that is bounded in $W^{2,\infty}(B_1)$, and suppose that for some $\alpha_0 \in I$, a blow-up limit*

$$\lim_{m\to\infty} \frac{u_{\alpha_0}(r_m\cdot)}{r_m^2}$$

is contained in M^*.

Then, if $u_\alpha \to u_{\alpha_0}$ in $L^1(B_1)$ as $\alpha \to \alpha_0$, $B_{r_0}\cap\partial\{u_\alpha > 0\}$ and $B_{r_0}\cap\partial\{u_\alpha < 0\}$ are C^1-graphs uniformly in $\alpha\in N_\kappa(\alpha_0)$ for some $r_0>0$ and $\kappa>0$; here the direction of every graph is the same, and $N_\kappa(\alpha_0)$ is a given open neighborhood of α_0.

The crucial tool in the proof of the theorem is the following proposition which uses an Aleksandrov reflection approach.

Proposition 1.6. *Let $n = 2$, let $(u_\alpha)_{\alpha\in I}$ be a family of solutions of* (1.1) *in B_1 that is bounded in $W^{2,\infty}(B_1)$, and suppose that for some $\alpha_0 \in I$, a blow-up limit*

$$\lim_{m\to\infty} \frac{u_{\alpha_0}(r_m\cdot)}{r_m^2}$$

is contained in M^*.

Then, if $u_\alpha \to u_{\alpha_0}$ in $L^1(B_1)$ as $\alpha \to \alpha_0$, there exist for given $\epsilon \in (0, 1/8)$ positive κ, δ and ρ such that for $\alpha \in N_\kappa(\alpha_0), y \in B_\delta \cap \partial\{u_\alpha > 0\}$ and $r \in (0, \rho)$, the scaled function

$$u_r(x) = \frac{u_\alpha(rx + y)}{S_r(y, u_\alpha)} \tag{1.5}$$

satisfies

$$\operatorname{dist}(u_r, M^*) = \inf_{v\in M^*} \sup_{B_1(0)} |v(x) - u_r(x)| < \epsilon.$$

The idea of the proof is to reflect the solution at a plane as in Fig. 3 and to compare the reflected solution to the original solution. As a consequence we obtain the following stability result:

Theorem 1.7. *Let $\Omega \subset \mathbf{R}^2$ be a bounded Lipschitz domain and assume that for given Dirichlet data $u_D \in W^{1,2}(\Omega)$ the free boundary does not contain any one-phase singular free boundary point (cf. Lemma* 1.3*).*

Then for $K \subset\subset \Omega$ and $\tilde u_D \in W^{1,2}(\Omega)$ satisfying $\sup_{\partial\Omega} |u_D - \tilde u_D| < \delta_K$, there is $\omega > 0$ such that the free boundary is for every $y \in K$ in $B_\omega(y)$ the union of (at most) two C^1-graphs which approach those of the solution with respect to boundary data u_D as $\sup_{\partial\Omega} |u_D - \tilde u_D| \to 0$.

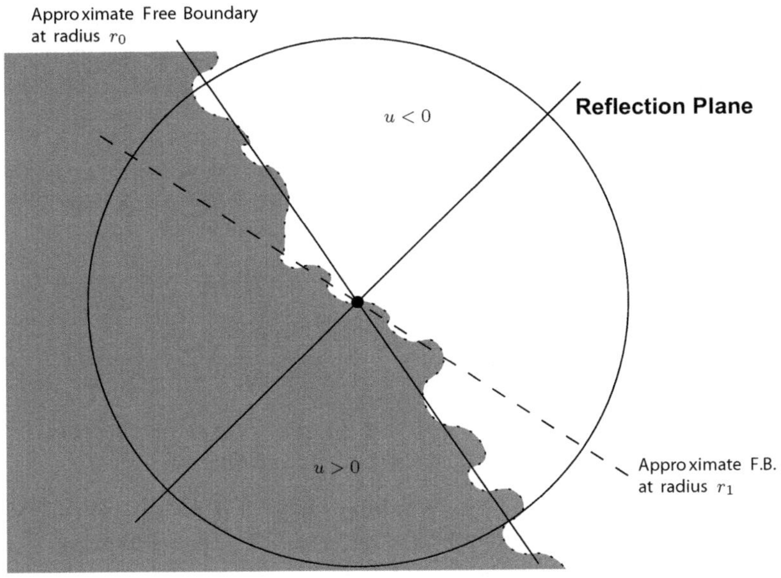

FIGURE 3. Turning free boundary

2. Part II

We are going to give a sketch of how a similar approach can be applied to the classical obstacle problem in higher dimensions.

The solution of the classical obstacle problem u is non-negative and satisfies

$$\Delta u = \frac{1}{2}\chi_{\{u>0\}} \, . \tag{2.1}$$

Classical results include local $C^{1,1}$-regularity (see [8]) and non-degeneracy of the solution. Regularity of the free boundary in higher dimensions has first been proved by L.A. Caffarelli in [4]. Here we give an alternative proof of the fact that the free boundary is close to regular points the graph of a differentiable function. To this end we define the class

$$M^* := \{u : B_1(0) \to \mathbf{R} \, : \, u(x_1, \ldots, x_n) = \frac{1}{4}\max(x_1, 0)^2 \,\}. \tag{2.2}$$

The class M is then defined as all rotated elements of M^*, i.e.,

$$M := \{u : B_1(0) \to \mathbf{R} \, : u = v \circ U \text{ where } U \text{ is a rotation, } v \in M^*\}. \tag{2.3}$$

Moreover for any $\gamma \in (0, \pi)$ the class M^γ is defined as

$$\begin{aligned} M^\gamma := \{u : B_1(0) \to \mathbf{R} \, : u = v \circ U \text{ where } U \text{ is a rotation,} \\ v \in M^*, \text{ and } \sup_{e \in \partial B_1} |\arccos((Ue) \cdot e)| \le \gamma\}. \end{aligned} \tag{2.4}$$

Lemma 2.1. *Let u be a $W^{2,\infty}(B_1)$-solution of (2.1). If one blow-up limit u_0 of the blow-up sequence $u(x_0 + r_j\cdot)/r_j^2$ as $r_j \to 0$ is contained in M then all blow-up limits of $u(x_k + r_k\cdot)/r_k^2$ as $x_k \to x_0$ and $r_k \to 0$ are contained in M.*

Proof. This follows from the upper semicontinuity of the density $x \mapsto \Phi_x(0+)$, from the consequent homogeneity of blow-up limits of $u(x_k + r_k\cdot)/r_k^2$ and from the known fact that each non-trivial homogeneous solution of degree 2 is either contained in M or a quadratic homogeneous polynomial (cf. [4]). □

Theorem 2.2. *Let u be a $W^{2,\infty}(B_1)$-solution and suppose that a blow-up limit*

$$\lim_{m\to\infty} \frac{u(r_m\cdot)}{r_m^2}$$

is contained in M^*.

Then the free boundary $\partial\{u > 0\}$ is in some open neighborhood of x_0 the graph of a differentiable, Lipschitz continuous function.

Theorem 2.2 follows from the combination of Lipschitz continuity (see the following Proposition) and flatness (see Lemma 2.1). The following Proposition based on the Kelvin transform is crucial.

Proposition 2.3. *Let u be a $W^{2,\infty}(B_1)$-solution and suppose that a blow-up limit*

$$\lim_{m\to\infty} \frac{u(r_m\cdot)}{r_m^2}$$

is contained in M^*.
Then there exist for $\epsilon \in (0,1)$ positive δ and ρ such that for $y \in B_\delta \cap \partial\{u > 0\}$ and $r \in (0,\rho)$, the scaled function

$$u_r(x) = \frac{u(y + rx)}{r^2} \tag{2.5}$$

satisfies

$$\mathrm{dist}(u_r, M^{\gamma_0}) = \inf_{v\in M^{\gamma_0}} \sup_{B_1(0)} |v(x) - u_r(x)| < \epsilon,$$

where $\gamma_0 = \pi/2 - 1/10$.

Proof. First, by continuity and by Lemma 2.1, for any $\tilde\epsilon > 0$ there are positive $\tilde\kappa, \tilde\delta$ and $\tilde\rho$ such that

$$\mathrm{dist}(u_{\tilde\rho}, M^*) < \tilde\epsilon \text{ for } \alpha \in N_{\tilde\kappa}(\alpha_0) \text{ and } y \in \partial\{u_\alpha > 0\} \cap B_{\tilde\delta}$$
$$\text{and } \mathrm{dist}(u_r, M) < \tilde\epsilon \text{ for } \alpha \in N_{\tilde\kappa}(\alpha_0), y \in \partial\{u_\alpha > 0\} \cap B_{\tilde\delta} \text{ and } r \in (0,\tilde\rho).$$

Now if the statement of the theorem does not hold, then there are positive r_0 and a rotation U_{θ_0} satisfying $\arccos((U_{\theta_0}e)\cdot e) \geq \pi/2 - \gamma_0 - c_1\epsilon > 0$ as well as

$$\mathrm{dist}(u_{r_0} \circ U_{\theta_0}, M^*) \leq \tilde\epsilon;$$

here c_1 is a constant depending on $(a, b, \lambda_+, \lambda_-)$. It is important for what follows that α and y are the same for u_{r_0} and $u_{\tilde\rho}$. In the remainder of the proof α and y are

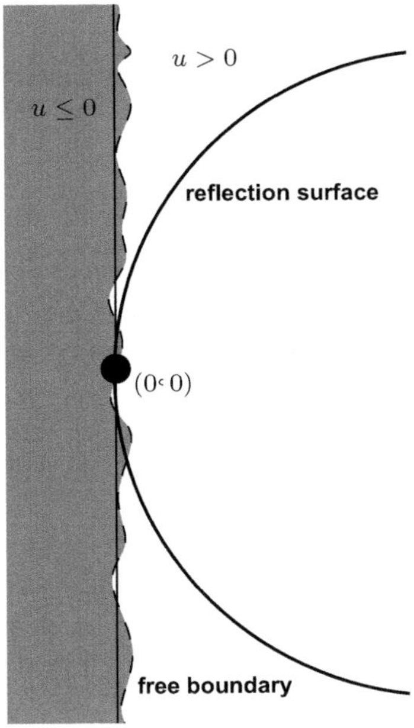

FIGURE 4. Reflection at a sphere cap

fixed. Let us now take an arbitrary rotation V such that $\sup_{e\in\partial B_1} |\arccos((Ve)\cdot e)| \le 2(\pi/2-\gamma_0)$, let $w := u_{\tilde\rho}\circ V$ and define v define by the Kelvin transform (cf. [13, Theorem 4.13], i.e.,

$$v(x) := |x-2e_1|^{2-n} w(\frac{x-2e_1}{|x-2e_1|^2})$$

(see Fig. 4). The function v satisfies in B_1 the equation

$$\Delta v(x) = |x-2e_1|^{-n-2}\chi_{\{v>0\}} \,.$$

By the C^1-closeness of $u_{\tilde\rho}$ to M^* we know that $w\ge v$ on $\partial(B_1(0)\cap B_1(2e_1))$. Thus

$$\int_{B_1(0)\cap B_1(2e_1)} |\nabla \max(v-w,0)|^2 = \int_{B_1(0)\cap B_1(2e_1)} \Delta(w-v)\max(v-w,0)$$

$$= \int_{B_1(0)\cap B_1(2e_1)} \max(v-w,0)\left(\chi_{\{w>0\}} - |x-2e_1|^{-n-2}\chi_{\{v>0\}}\right) \le 0,$$

implying that $w\ge v$ in $B_1(0)\cap B_1(2e_1)$. Consequently

$$u_{r_0} \ge |\frac{r_0}{\tilde\rho}x-2e_1|^{2-n} u_{r_0}(V(\frac{x-(\tilde\rho/r_0)2e_1}{|(r_0/\tilde\rho)x-2e_1|^2})) \text{ in } B_1(0)\cap B_1(2e_1)\,,$$

a contradiction to

$$\operatorname{dist}(u_{r_0}\circ U_{\theta_0}, M^*) \le \tilde\epsilon$$

in view of $\arccos((U_{\theta_0}e)\cdot e) \ge \pi/2-\gamma_0-c_1\epsilon > 0$ and the arbitrary choice of V. □

References

[1] H.W. Alt, L.A. Caffarelli, A. Friedman, *Variational problems with two phases and their free boundaries,* Trans. Amer. Math. Soc. 282 (1984), 431–461.

[2] M.-F. Bidaut-Véron, V. Galaktionov, P. Grillot, L. Véron, Singularities for a 2-dimensional semilinear elliptic equation with a non-Lipschitz nonlinearity. J. Differential Equations 154 (1999), 318–338.

[3] H. Brézis, D. Kinderlehrer, The smoothness of solutions to nonlinear variational inequalities. Indiana Univ. Math. J. 23 (1973/74), 831–844.

[4] L.A. Caffarelli, *The regularity of free boundaries in higher dimensions.* Acta Math. 139 (1977), 155–184.

[5] L.A. Caffarelli, L. Karp, H. Shahgholian, *Regularity of a free boundary in potential theory with application to the Pompeiu problem,* Ann. of Math. (2) 151 (2000), 269–292.

[6] X.-Y. Chen, Uniqueness of the ω-limit point of solutions of a semilinear heat equation on the circle. Proc. Japan Acad. Ser. A Math. Sci. 62 (1986), 335–337.

[7] G. Duvaut, J.-L. Lions, Inequalities in mechanics and physics. Translated from the French by C. W. John. Grundlehren der Mathematischen Wissenschaften, 219. Springer-Verlag, Berlin-New York, 1976.

[8] J. Frehse, *On the regularity of the solution of a second order variational inequality,* Boll. Un. Mat. Ital. (4) 6 (1972), 312–315.

[9] X.-Y. Chen, H. Matano, L. Véron, Anisotropic singularities of solutions of nonlinear elliptic equations in $\mathbf{R}^2$. J. Funct. Anal. 83 (1989), 50–97.

[10] B. Chow, R. Gulliver, Aleksandrov reflection and geometric evolution of hypersurfaces. Comm. Anal. Geom. 9 (2001), 261–280.

[11] B. Chow, Geometric aspects of Aleksandrov reflection and gradient estimates for parabolic equations. Comm. Anal. Geom. 5 (1997), 389–409.

[12] B. Chow, R. Gulliver, Aleksandrov reflection and nonlinear evolution equations. I. The n-sphere and n-ball. Calc. Var. Partial Differential Equations 4 (1996), 249–264.

[13] D. Gilbarg, N.S. Trudinger, Elliptic partial differential equations of second order. Second edition. Springer-Verlag, Berlin, 1983

[14] H. Shahgholian, *$C^{1,1}$-regularity in semilinear elliptic problems,* Comm. Pure Appl. Math. 56 (2003), 278–281.

[15] H. Shahgholian, N.N. Uraltseva, G.S. Weiss, *Global solutions of an obstacle-problem-like equation with two phases.* Monatsh. Math. 142 (2004), 27–34.

[16] N.N. Uraltseva, *Two-phase obstacle problem.* Problems in Math.Analysis, v. 22, 2001, 240–245 (in Russian. English translation: Journal of Math. Sciences, v.106, N 3, 2001, pp. 3073–3078).

[17] G.S. Weiss, *A Homogeneity Improvement Approach to the Obstacle Problem,* Inv. Math. 138 (1999), 23–50.

[18] G.S. Weiss, *Partial Regularity for Weak Solutions of an Elliptic Free Boundary Problem,* Commun. Partial Differ. Equations 23 (1998), 439–455.

[19] G.S. WEISS, *An Obstacle-Problem-Like Equation with Two Phases: Pointwise Regularity of the Solution and an Estimate of the Hausdorff Dimension of the Free Boundary,* Interfaces and Free Boundaries 3 (2001), 121–128.

Henrik Shahgholian
Department of Mathematics
Royal Institute of Technology
S-100 44 Stockholm, Sweden
e-mail: `henriksh@math.kth.se`
URL: `http://www.math.kth.se/~henriksh/`

Georg S. Weiss
Graduate School of Mathematical Sciences
University of Tokyo
3-8-1 Komaba, Meguro-ku
Tokyo-to, 153-8914 Japan
and
Guest of the Max Planck Institute
for Mathematics in the Sciences
Inselstr. 22
D-04103 Leipzig, Germany
e-mail: `gw@ms.u-tokyo.ac.jp`
URL: `http://www.ms.u-tokyo.ac.jp/~gw/`

International Series of Numerical Mathematics, Vol. 154, 403–412

Solvability for a PDE Model of Regional Economic Trend

Ken Shirakawa, Akio Ito and Atsushi Kadoya

Abstract. The aim of this work is to develop a simulation method focused on regional economic trend. In this light, an original model, formulated by partial differential equations, will be proposed. Consequently, the existence of time-local solutions of our mathematical model will be concluded, as a transitional report in the research.

Mathematics Subject Classification (2000). Primary 91B62; Secondary 35K45.

Keywords. Regional economic trends, Mathematical model by PDEs, Solvability.

1. Introduction

This work is motivated to represent the dynamics of economic trends in local communities, such as cities, towns and villages. Generally, economic conditions have been estimated through the observation of an economic parameter, called "production", and it has been said that two kinds of parameters, respectively called "labor force" and "capital", are mainly involved in the movement of production. In the field of macro economics, these parameters are classified as "flow variable", and each of them is supposed to be some representative value, which indicates a subaccumulation of economical quantity circulating in a certain period (usually one year). So in the actual research activities, flow variables are often observed as discrete data.

However, since economic data actually vary the whole times, it is conceivable that discrete data are not enough to foresee some accidental movements, such as price escalation and heavy falls in price. Also, at the level of local communities, the regional disparities of economic conditions, coming from urban congestion and under population etc., will not be negligible.

The research for this paper is partially supported by a grant from Institute for Advanced Research Hiroshima Shudo University, 2004.

In view of this, we set the main purpose of this paper to propose and to study a mathematical model described in a continuous framework. To this end, let us denote the space-time coordinate system by a product space $Q_T := \Omega \times (0,T)$ of a two or three dimensional domain Ω with a smooth boundary $\Gamma := \partial\Omega$, and an open interval $(0,T)$ with a constant $T > 0$. As a natural consequence, the mathematical model should somehow represent the movements of flow variables, such as production, labor force and capitals. But, these flow variables would be unsuitable to adopt as unknown parameters of our model, because flow variables are mathematically supposed to map a subset $A \subset \Omega$ and an interval $I \subset [0,T]$, into an economical quantity at the area A, accumulated during the period I. Hence, we first introduce the following virtual parameter, named as "density", for each flow variable.

Definition 1.1. Let F be any flow variable. Let us denote by $\mathbb{R}_+$ the half interval $[0,+\infty)$ of nonnegative values, and for any Borel set $A \subset \Omega$ and any open interval $I \subset [0,T]$, let us denote by $F(A,I) \in \mathbb{R}_+$ the value of flow F at the subarea A and the period I. Then, a nonnegative function $f : \Omega \times [0,T] \longrightarrow \mathbb{R}_+$ is called the density of flow F, if and only if:

$$F(A,I) = \int_I \int_A f(x,t)\,dx\,dt$$

for any Borel set $A \subset \Omega$ and any open interval $I \subset [0,T]$.

Throughout this paper, we denote by p, u, $\boldsymbol{w} = {}^{\mathrm{t}}(w_1,\ldots,w_N)$ with $N \in \mathbb{N}$, the density of production, the density of labor force, and the vectorial parameter of densities of N-kinds of capitals respectively, and assume the following equality for these densities:

$$\begin{aligned} p = c\,u^{\alpha_0} w_1^{\alpha_1} \cdots w_N^{\alpha_N} \text{ for some constants prescribed to satisfy} \\ c > 0,\ 0 < \alpha_i < 1\ (i = 0,1,\ldots,N) \text{ and } \sum_{i=0}^{N} \alpha_i = 1. \end{aligned} \tag{1.1}$$

Equality (1.1) is known as the relational expression of Cobb-Douglas type, and in the context, the constant c is a scaling coefficient, and the exponent α_0 and N-exponents α_i $(i = 1,\ldots,N)$ are respectively the ratios that labor force u and N-capitals w_i $(i = 1,\ldots,N)$ contribute to economic growth. Thus, the conservation law $\sum_{i=0}^{N} \alpha_i = 1$ is usually assumed for these exponents. Incidentally, it is empirically said that $\alpha_0 \approx 0.75$ and $\sum_{i=1}^{N} \alpha_i \approx 0.25$.

Now, our mathematical model is derived in the track of the modelling method as in [4, section 3]. In [4], the authors have tentatively ruled out spatial variation of densities, and have proposed a mathematical model described by the following ordinary differential equations:

$$u'(t) = n_0(\log p)'(t)u(t),\ t \in (0,T], \tag{1.2}$$

$$\boldsymbol{w}'(t) = p(t)\boldsymbol{s} - \boldsymbol{D}\,\boldsymbol{w}(t),\ t \in (0,T]; \tag{1.3}$$

subject to the relational expression (1.1) and appropriate initial conditions, where $0 < n_0 < 1$ is a constant, $\boldsymbol{s} := {}^{\mathrm{t}}(s_1,\ldots,s_N)$ is a vector consisted of positive constants s_i $(i = 1, \ldots, N)$, and $\boldsymbol{D} := \begin{pmatrix} \delta_1 & & 0 \\ & \ddots & \\ 0 & & \delta_N \end{pmatrix}$ is a $N \times N$-diagonal matrix with positive constants δ_i $(i = 1, \ldots, N)$.

The first equation (1.2) is a kinetic equation of labor force, and it is derived on the basis of Malthus's law with a time-dependent growth rate $n_0(\log p)'(t)$. Here, since the term $(\log p)'(t)$ is nothing but the so-called "economic growth rate", this equation also implies that the growth rate of labor force directly depends on the economic vibrancy at each time (see [4, Remark 3.1], for details). Furthermore, under the relational expression (1.1), it is easily checked that equation (1.2) is equivalent to:

$$(\log u)'(t) = \sum_{i=1}^{N} \nu_i (\log w_i)'(t), \ t \in (0, T]; \tag{1.4}$$

where $\nu_i := n_0 \alpha_i / (1 - n_0 \alpha_0)$ $(i = 1, \ldots, N)$. Notice that N-constants ν_i $(i = 1, \ldots, N)$ satisfy $0 < \nu_i < 1$ $(i = 1, \ldots, N)$ and $\sum_{i=1}^{N} \nu_i < 1$.

On the other hand, the second equation (1.3) is a balance equation among the variation $\boldsymbol{w}'(t)$ of capitals in time, the positive contribution $p(t)\boldsymbol{s}$ from the production and the negative contribution $-\boldsymbol{D}\,\boldsymbol{w}(t)$ such as depreciations. The coupled system $\{(1.1),(1.3)\}$ is known as "Solow model", in honor of the name of proposer (cf. [10]), and the original model is formulated by ordinary differential equations in scalar-valued framework of $\boldsymbol{w}$. Nevertheless, since the category of capitals is not singleton in general, the study of Solow model has enhanced into vectorial framework, as in (1.3), by some follow-on researchers (e.g. [4, 7]).

In this paper, we first modify the system $\{(1.1)–(1.3)\}$ into the framework to enable the representation of spatial (regional) disparities of parameters. Hence, our mathematical model will be described by a system of appropriate partial differential equations that will be based on the system $\{(1.1)–(1.3)\}$. But, because of a technical reason in mathematical treatment, the parameter $\boldsymbol{w}$ (capitals) will be considered under scalar-valued setting. Consequently, the existence of time-local solutions of the original system will be reported in the main theorem of this paper.

Notation. For any Banach space X, we denote by $|\cdot|_X$ the norm in X, throughout this paper. Also, for any Hilbert space H, we denote by $(\cdot,\cdot)_H$ the inner product in H.

2. Statement of the main result

First of all, let us start with proposing the following coupled system of two partial differential equations:

$$\begin{cases} (\log u)_t - \kappa_0 \Delta(\log u) = \nu_1 (\log w)_t \text{ in } Q_T, \\ \dfrac{\partial}{\partial n}(\log u) = 0 \text{ on } \Sigma_T := \Omega \times (0,T), \ \ u(\cdot, 0) = u_0 \text{ in } \Omega; \end{cases} \tag{2.1}$$

$$\begin{cases} w_t - \kappa_1 \Delta w = s_1 c\, u^{\alpha_0} w^{1-\alpha_0} - \delta_1 w \text{ in } Q_T, \\ \dfrac{\partial w}{\partial n} = 0 \text{ on } \Sigma_T, \ \ w(\cdot, 0) = w_0 \text{ in } \Omega; \end{cases} \tag{2.2}$$

as a possible mathematical model of regional economic trends.

This system consists of two Cauchy problems (2.1) and (2.2) of parabolic equations subject to homogeneous Neumann boundary conditions, which are respectively derived from the equivalent formula (1.4) of the kinetic equation (1.2) of labor force, and Solow model {(1.1), (1.3)} under scalar-valued setting of $\boldsymbol{w}$. Therefore, unknown parameters of this model are the density u of labor force and the (scalar-valued) density w of a capital, and in the context, κ_i $(i = 0, 1)$ are diffusion coefficients of (small) positive constants.

Now, let us denote by σ_1 the constant $s_1 c$, as in (2.2), for simplicity. Also, let us set $D_N := \{ \ z \in H^2(\Omega) \mid \frac{\partial z}{\partial n} = 0 \text{ a.e. on } \Gamma \ \}$, and let us denote by Δ_N the operator $\Delta_N : D_N \longrightarrow L^2(\Omega)$, defined as $\Delta_N z = \Delta z$ for any $z \in D_N$. As is well known, operator $-\Delta_N$ forms a maximal monotone graph in the product space $L^2(\Omega) \times L^2(\Omega)$.

In this paper, we mainly focus on the existence of solutions of the system {(2.1),(2.2)}, prescribed as follows.

Definition 2.1. A pair $\{u, w\}$ of functions $u, w : Q_T \longrightarrow \mathbb{R}_+$ is called a solution of system {(2.1),(2.2)}, if and only if:

(s1) $u, w \in W^{1,2}(0, T; L^2(\Omega)) \cap L^\infty(0, T; H^1(\Omega))$, $v := \log u \in L^\infty(Q_T)$, and $\log w \in L^\infty(Q_T)$;

(s2) $u(0) = u_0 \in H^1(\Omega)$, $v_0 := v(0) = \log u_0 \in L^\infty(\Omega)$,

$$v_t(t) - \kappa_0 \Delta_N v(t) = \nu_1 (\log w)_t(t) \text{ in } L^2(\Omega), \text{ a.e. } t \in (0, T];$$

(s3) $w(0) = w_0 \in D_N$, $\log w_0 \in L^\infty(\Omega)$,

$$w_t(t) - \kappa_1 \Delta_N w(t) = \sigma_1 e^{\alpha_0 v(t)} w(t)^{1-\alpha_0} - \delta_1 w(t) \text{ in } L^2(\Omega), \text{ a.e. } t \in [0, T].$$

On the basis of the above definition, the following theorem will be concluded as the main result of this paper.

Theorem 2.2. System {(2.1), (2.2)} admits at least one time-local solution.

Remark 2.3. As it is mentioned in introduction, the main objective of the research is to represent the dynamics of regional economic trends. So the study on the dynamical system generated by system {(2.1), (2.2)} (including the solvability) will be a basilicus assessment criterion for the adequacy of mathematical models and simulation data. From this viewpoint, Theorem 2.2 can be said a report on interim step yet, while it gives some positive answer for the adequacy of our model.

However, since the original Solow model generates a dynamical system having quite strong stability, the major focus in next challenges will be to know whether the dynamical systems generated by PDE models, such as system {(2.1), (2.2)}, also show similar stability as in the case of ODE models.

3. Auxiliary problems

The main theorem will be proved by means of Schauder's fixed point argument, seen in [5, section 3], or others (e.g., [2, 8]). To this end, let us put:

$$M_0 := |\log u_0|_{L^\infty(\Omega)} + 1;$$

and let us fix two functions:

$$\tilde{v}, \tilde{w} \in W^{1,2}(0,T; L^2(\Omega)) \text{ with } |\tilde{v}|_{L^\infty(Q_T)} \leq 2M_0 \text{ and } \log \tilde{w} \in L^\infty(Q_T); \tag{3.1}$$

to consider the following Cauchy problems:

$$\begin{cases} w_t(t) - \kappa_1 \Delta_N w(t) + \delta_1 w(t) = \sigma_1 e^{\alpha_0 \tilde{v}(t)} |w(t)|^{1-\alpha_0} \\ \qquad\qquad\qquad\qquad \text{in } L^2(\Omega), \text{ a.e. } t \in (0,T], \\ w(0) = w_0 \in H^1(\Omega), \text{ a.e. in } \Omega; \end{cases} \tag{3.2}$$

$$\begin{cases} v_t(t) - \kappa_0 \Delta_N v(t) = \nu_1 (\log \tilde{w})_t(t) \text{ in } L^2(\Omega), \text{ a.e. } t \in (0,T], \\ v(0) = v_0 \in H^1(\Omega) \cap L^\infty(\Omega), \text{ a.e. in } \Omega. \end{cases} \tag{3.3}$$

This section is devoted to show some lemmas concerned with key properties of the above auxiliary problems.

Let φ be a proper l.s.c. and convex functional on $L^2(\Omega)$, defined as:

$$\varphi(z) := \begin{cases} \dfrac{\kappa_1}{2} \displaystyle\int_\Omega |\nabla z|^2 \, dx + \dfrac{\delta_1}{2} \displaystyle\int_\Omega |z|^2 \, dx, \text{ if } z \in H^1(\Omega), \\ +\infty, \text{ otherwise;} \end{cases}$$

and for any $t \in [0,T]$, let $g(t,\cdot)$ be a continuous operator from $L^2(\Omega)$ into itself, defined as:

$$g(t,z) := -\sigma_1 e^{\alpha_0 \tilde{v}(t,\cdot)} |z|^{1-\alpha_0} \text{ for any } z \in L^2(\Omega).$$

Then, Cauchy problem (3.2) is equivalently reformulated to that for the following evolution equation:

$$w_t(t) + \partial\varphi(w(t)) + g(t, w(t)) \ni 0 \text{ in } L^2(\Omega), \text{ a.e. } t \in [0,T]; \tag{3.4}$$

subject to the subdifferential $\partial\varphi$ of φ in $L^2(\Omega)$, and a non-Lipschitz perturbation $g(t,\cdot)$ $(t \in [0,T])$.

In recent years, this type of evolution equation has been studied by a lot of mathematicians, such as Brézis [1], from various viewpoint. So applying some of those theories (e.g., [9]), we immediately see the existence of solutions of Cauchy problem (3.2). But the uniqueness is hard to say at this moment, because the property of the perturbation $g(t,\cdot)$ $(t \in [0,T])$ is too weak to satisfy conditions for uniqueness, in any general theory.

In this light, the following lemma will be keypoint to solve the barriers for the fixed point argument, including the uniqueness.

Lemma 3.1. (L^∞-estimates for solutions of (3.2)) Let us fix positive constants r and R to satisfy:

$$\delta_1 r < \sigma_1 e^{-2\alpha_0 M_0} r^{1-\alpha_0} \text{ and } \delta_1 R > \sigma_1 e^{2\alpha_0 M_0} R^{1-\alpha_0}; \tag{3.5}$$

respectively, and let us assume that:

$$0 < r \leq w_0(x) \leq R < +\infty \text{ a.e. } x \in \Omega. \tag{3.6}$$

Then, for any solution w of Cauchy problem (3.2), we see that

$$r \leq w(x,t) \leq R \text{ a.e. } (x,t) \in Q_T.$$

Proof. First, let us prove the nonnegativeness of w in Q_T. Multiplying the both sides of (3.4) by $-w^-(t)$, we have

$$\frac{1}{2}\frac{d}{dt}|w^-(t)|^2_{L^2(\Omega)} \leq -\sigma_1 \left(e^{\alpha_0 \tilde{v}(t)}|w(t)|^{1-\alpha_0}, w^-(t)\right)_{L^2(\Omega)} \leq 0, \text{ a.e. } t \in [0,T]. \tag{3.7}$$

So, integrating the both sides of (3.7) over any interval $[0,t] \subset [0,T]$,

$$|w^-(t)|^2_{L^2(\Omega)} \leq |w_0^-|^2_{L^2(\Omega)} = 0 \text{ for any } t \in [0,T],$$

which implies $w \geq 0$ a.e. in Q_T.

Next, we show that $r \leq w \leq R$ a.e. in Q_T. Let us put

$$\underline{w}_r(t) := r - w(t), \ \overline{w}_R(t) := w(t) - R \text{ in } L^2(\Omega), \text{ for any } t \in [0,T].$$

Then, it is easily seen from (3.5) that

$$(\underline{w}_r)_t - \kappa_1 \Delta \underline{w}_r + \delta_1 \underline{w}_r < \sigma_1(e^{-2\alpha_0 M_0} r^{1-\alpha_0} - e^{\alpha_0 \tilde{v}}|w|^{1-\alpha_0}) \text{ a.e. in } Q_T \tag{3.8}$$

$$(\overline{w}_R)_t - \kappa_1 \Delta \overline{w}_R + \delta_1 \overline{w}_R < \sigma_1(e^{\alpha_0 \tilde{v}}|w|^{1-\alpha_0} - e^{2\alpha_0 M_0} R^{1-\alpha_0}) \text{ a.e. in } Q_T \tag{3.9}$$

$$\frac{\partial \underline{w}_r}{\partial n} = \frac{\partial \overline{w}_R}{\partial n} = 0 \text{ a.e. on } \Sigma_T. \tag{3.10}$$

Let us multiply the both sides of (3.8) and (3.9) by $\underline{w}_r^+$ and $\overline{w}_R^+$, respectively. Then, on account of (3.10), the nonnegativeness of w and the monotonicity of the function $\exp(\,\cdot\,)$ in $\mathbb{R}$, integrating the both sides of the results over Ω yields that

$$\frac{1}{2}\frac{d}{dt}|\underline{w}_r^+(t)|^2_{L^2(\Omega)} \leq \sigma_1 e^{-2\alpha_0 M_0} \left((r^{1-\alpha_0} - w(t)^{1-\alpha_0})^+, \underline{w}_r^+(t)\right)_{L^2(\Omega)}, \tag{3.11}$$

$$\frac{1}{2}\frac{d}{dt}|\overline{w}_R^+(t)|^2_{L^2(\Omega)} \leq \sigma_1 e^{2\alpha_0 M_0} \left((w(t)^{1-\alpha_0} - R^{1-\alpha_0})^+, \overline{w}_R^+(t)\right)_{L^2(\Omega)}, \tag{3.12}$$

a.e. $t \in [0,T]$.

Here, since:

$$\frac{r^{1-\alpha_0} - \omega^{1-\alpha_0}}{r - \omega} \le \frac{1}{r^{\alpha_0}}$$
for any $\omega \in (0, r)$ (see Fig.1),

$$\frac{\omega^{1-\alpha_0} - R^{1-\alpha_0}}{\omega - R} \le \frac{1-\alpha_0}{R^{\alpha_0}}$$
for any $\omega \in (R, +\infty)$ (see Fig.2);

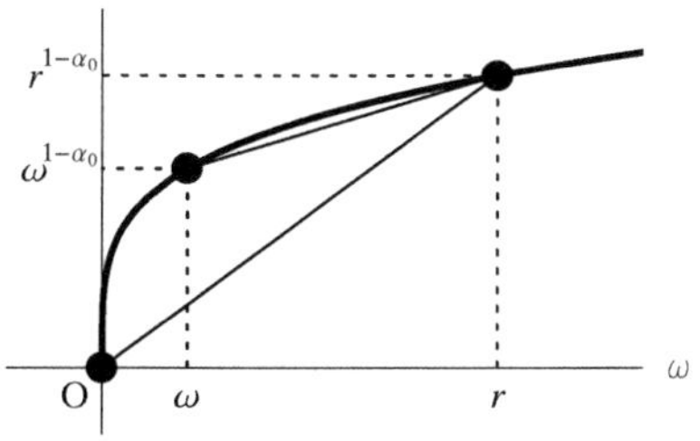

Fig.1

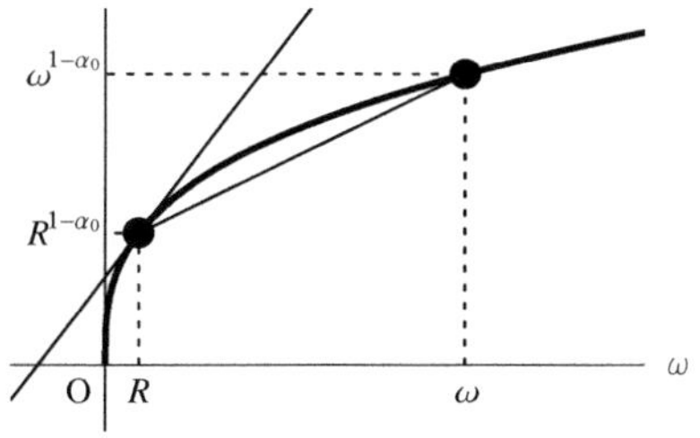

Fig.2

we see from (3.11) and (3.12) that

$$\frac{d}{dt}|\underline{w}_r^+(t)|^2_{L^2(\Omega)} \le \frac{2\sigma_1 e^{-2\alpha_0 M_0}}{r^{\alpha_0}}|\underline{w}_r^+(t)|^2_{L^2(\Omega)}, \text{ and}$$
$$\frac{d}{dt}|\overline{w}_R^+(t)|^2_{L^2(\Omega)} \le \frac{2\sigma_1(1-\alpha_0)e^{2\alpha_0 M_0}}{R^{\alpha_0}}|\overline{w}_R^+(t)|^2_{L^2(\Omega)};$$

a.e. $t \in [0, T]$, respectively. Thus, applying Gronwall's lemma for the above inequalities, we conclude that $r \le w \le R$ a.e. in Q_T. □

Corollary 3.2. (Uniqueness of solutions of (3.2)) For any initial value $w_0 \in H^1(\Omega)$ satisfying (3.6), solutions of Cauchy problem (3.2) have the uniqueness.

Proof. By Lemma 3.1, the perturbation $g(t, \cdot)$ ($t \in [0, T]$) can be regarded as Lipschitz perturbation with a time-independent Lipschitz constant. Thus, the uniqueness immediately follows from the general theory as in [1]. □

Next, let us look towards another Cauchy problem (3.3). On account of (3.1), this Cauchy problem just turns out that for a heat equation, subject to a forcing term $(\log \tilde{w})_t = \tilde{w}_t/\tilde{w} \in L^2(0, T; L^2(\Omega))$ and homogeneous Neumann boundary condition. Therefore, by usual energy estimates, solution v has the boundedness in topologies of $W^{1,2}(0, T; L^2(\Omega))$ and $L^\infty(0, T; H^1(\Omega))$. However, referring to the technique as in Ladyženskaja-Solonnikov-Ural'ceva [6], we further obtain the following L^∞-estimate of solutions.

Lemma 3.3. (L^∞-estimates for solutions of (3.3)) In addition of (3.1), let us assume that:

$$(\log \tilde{w})_t = \frac{\tilde{w}_t}{\tilde{w}} \in L^\infty(0, T; L^2(\Omega)). \tag{3.13}$$

Then, there exists a time $T_0 = T_0(|v_0|_{L^\infty(\Omega)}, |(\log \tilde{w})_t|_{L^\infty(0,T;L^2(\Omega))}) \in (0, T]$, depending on $|v_0|_{L^\infty(\Omega)}$ and $|(\log \tilde{w})_t|_{L^\infty(0,T;L^2(\Omega))}$, such that

$$|v(t)|_{L^\infty(\Omega)} \le 2M_0 = 2(|v_0|_{L^\infty(\Omega)} + 1) \text{ for any } t \in [0, T_0].$$

Proof. Since this lemma is obtained just as in [6, Theorem 7.1 in Chapter III], we omit to show the detailed proof. Additionally, the thread of the proof can be found in [3, Lemma 2.1], too. □

In the proof of the main theorem, the above lemma will be required to complete the iteration, associated with solution operators for problems (3.2) and (3.3).

4. Proof of the main theorem

Now we are on the stage to prove the main theorem. Throughout this section, we use the same notations as in the previous sections.

For any $T \in (0,1]$, let us set a compact and convex subset $X_0(T)$ in $C([0,T];L^2(\Omega))$, by putting:

$$X_0(T) := \left\{ \begin{array}{l} \xi \in W^{1,2}(0,T;L^2(\Omega)) \\ \cap L^\infty(0,T;H^1(\Omega)) \end{array} \middle| \begin{array}{l} \int_0^t |\xi_t(\tau)|^2_{L^2(\Omega)}\,d\tau + \kappa_0 |\nabla \xi(t)|^2_{L^2(\Omega)^n} \\ \leq \kappa_0 |\nabla v_0|^2_{L^2(\Omega)^n} + 1, \text{ and} \\ |\xi(t)|_{L^\infty(\Omega)} \leq 2M_0, \text{ for any } t \in [0,T] \end{array} \right\};$$

and for any $\tilde{v} \in X_0(T)$, let us consider Cauchy problem (3.2) with the initial value $w_0 \in D_N$ satisfying (3.6). Then, as is seen in the last section, problem (3.2) admits a unique solution $w \in W^{1,2}(0,T;L^2(\Omega)) \cap L^\infty(0,T;H^1(\Omega))$, and we find a positive constant $\rho_0 = \rho_0(|v_0|_{L^\infty(\Omega)}, |w_0|_{H^1(\Omega)}, R)$, independent of $T \in (0,1]$, such that

$$|w|_{W^{1,2}(0,T;L^2(\Omega))} + |w|_{L^\infty(0,T;H^1(\Omega))} \leq \rho_0.$$

Also, taking the time-derivative of the both sides of (3.4) and multiplying the both sides of the result by $w_t(t)$, we further find a positive constant $\rho_1 = \rho_1(|v_0|_{L^\infty(\Omega)}, |v_0|_{H^1(\Omega)}, |w_0|_{H^2(\Omega)}, r, R)$, independent of $T \in (0,1]$, such that

$$\rho_1 \geq \rho_0 \text{ and } |w|_{C^1([0,T];L^2(\Omega))} \leq \rho_1.$$

Here, for any $T \in (0,1]$, let us introduce a compact and convex subset $Y_0(T)$, by putting:

$$Y_0(T) := \left\{ \eta \in C^1([0,T];L^2(\Omega)) \middle| \begin{array}{l} r \leq \eta \leq R \text{ a.e. in } Q_T, \text{ and} \\ |\eta|_{C^1([0,T];L^2(\Omega))} + |\eta|_{L^\infty(0,T;H^1(\Omega))} \leq \rho_1 \end{array} \right\};$$

to define an operator $\mathcal{P}_T : X_0(T) \longrightarrow Y_0(T)$, which maps any function $\tilde{v} \in X_0(T)$ into the solution $w \in Y_0(T)$ of Cauchy problem (3.2). Then, for each $T \in (0,1]$, the operator $\mathcal{P}_T$ is continuous in the sense that:

$$\begin{array}{l} w_i := \mathcal{P}_T \tilde{v}_i \to w := \mathcal{P}_T \tilde{v} \text{ in } C([0,T];L^2(\Omega)) \text{ as } i \to +\infty, \\ \text{if } \{\tilde{v}_i\} \subset X_0(T),\ \tilde{v} \in X_0(T) \text{ and } \tilde{v}_i \to \tilde{v} \text{ in } C([0,T];L^2(\Omega)) \text{ as } i \to +\infty. \end{array} \tag{4.1}$$

The continuous dependence (4.1) is checked by applying Gronwall's lemma to the following differential inequality:

$$\begin{aligned}\frac{1}{2}\frac{d}{dt}|(w_i-w)(t)|^2_{L^2(\Omega)} &\le \sigma_1\Big(e^{\alpha_0\tilde{v}_i(t)}w_i(t)^{1-\alpha_0}-e^{\alpha_0\tilde{v}(t)}w(t)^{1-\alpha_0},(w_i-w)(t)\Big)_{L^2(\Omega)}\\ &\le \sigma_1 e^{2\alpha_0 M_0}\left(R^{1-\alpha_0}+\frac{1}{r^{\alpha_0}}\right)\left(|(w_i-w)(t)|^2_{L^2(\Omega)}+|(\tilde{v}_i-\tilde{v})(t)|^2_{L^2(\Omega)}\right);\end{aligned}$$

for a.e. $t\in[0,T]$, that is obtained in a standard way with use of a similar technique as in the proof of Lemma 3.1.

Next, for any $T\in(0,1]$ and any $\tilde{w}\in Y_0(T)$, let us consider Cauchy problem (3.3). Then, since:

$$|(\log\tilde{w})_t(t)|^2_{L^2(\Omega)}=\left|\frac{\tilde{w}_t(t)}{\tilde{w}(t)}\right|^2_{L^2(\Omega)}\le\frac{\rho_1^2}{r^2}\ \text{ for any } t\in[0,T];\tag{4.2}$$

the following energy estimate:

$$\begin{aligned}&\int_0^t|v_t(\tau)|^2_{L^2(\Omega)}\,d\tau+\kappa_0|\nabla v(t)|^2_{L^2(\Omega)^n}\\ &\le\kappa_0|\nabla v_0|^2_{L^2(\Omega)^n}+\frac{\nu_1^2\rho_1^2}{r^2}\cdot t,\ \text{ for any } t\in[0,T];\end{aligned}\tag{4.3}$$

is easily obtained by multiplying the both sides of the (heat) equation by $v_t(t)$.

Now, for each $T\in(0,1]$, let us define a solution operator $\mathcal{Q}_T : Y_0(T)\longrightarrow C([0,T];L^2(\Omega))$, which maps any $\tilde{w}\in Y_0(T)$ into the solution $v\in W^{1,2}(0,T;L^2(\Omega))\cap L^\infty(0,T;H^1(\Omega))$. Then, on account of the uniqueness of solutions and the demi-closedness of the maximal monotone operator, the operator $\mathcal{Q}_T$ shows the following continuity:

$$\begin{aligned}&v_i:=\mathcal{Q}_T\,\tilde{w}_i\to v:=\mathcal{Q}_T\,\tilde{w}\text{ in } C([0,T];L^2(\Omega))\text{ as } i\to+\infty,\\ &\text{if }\{\tilde{w}_i\}\subset Y_0(T),\ \tilde{w}\in Y_0(T)\text{ and }\tilde{w}_i\to\tilde{w}\text{ in }C([0,T];L^2(\Omega))\text{ as } i\to+\infty;\end{aligned}\tag{4.4}$$

for any $T\in(0,1]$.

Furthermore, on account of (4.2), (4.3) and Lemma 3.3, we find a sufficiently small constant $T_*\in(0,1]$ such that

$$(\mathcal{Q}_T\tilde{w})|_{[0,T_*]}\in X_0(T_*)\text{ for any } T\in[T_*,1]\text{ and any }\tilde{w}\in Y_0(T),$$

where the notation "$|_{[0,T_*]}$" denotes the restriction of functions in $C([0,T];L^2(\Omega))$, with $T\in[T_*,1]$, onto the compact interval $[0,T_*]$. Therefore, by the compactness of $X_0(T_*)$ in $C([0,T_*];L^2(\Omega))$, the operator $\mathcal{Q}_{T_*}:Y_0(T_*)\longrightarrow X_0(T_*)$ is well defined and continuous in the sense of (4.4) under $T=T_*$.

Consequently, we figure out that the composition $\mathcal{S}_{T_*}:=\mathcal{P}_{T_*}\circ\mathcal{Q}_{T_*}:Y_0(T_*)\longrightarrow Y_0(T_*)$ forms a well-defined and continuous operator in the topology of $C([0,T_*];L^2(\Omega))$. Since $Y_0(T_*)$ is a compact and convex subset in $C([0,T_*];L^2(\Omega))$, Schauder's fixed point theorem enables us to conclude the existence of a fixed point $w_*\in Y(T_*)$ for the operator (iteration) $\mathcal{S}_{T_*}$. Here, putting $v_*:=\mathcal{Q}_{T_*}w_*$ and $u_*:=e^{v_*}$, the pair $\{u_*,w_*\}$ fulfills all conditions (s1)–(s3) under $T=T_*$. This entails the existence of time-local solutions of our system $\{(2.1),(2.2)\}$. □

Acknowledgment. The research for this paper is partially supported by a grant from Institute for Advanced Research Hiroshima Shudo University, 2004.

References

[1] H. Brézis, *Opérateurs maximaux monotones et semigroupes de contractions dans les espace de Hilbert,* North-Holland, Amsterdam, 1973.
[2] P. Colli and K. Shirakawa, *Attractors for the one-dimensional Frémond model of shape memory alloys,* Asymptot. Anal., **40** (2004), no. 2, 109–135.
[3] T. Fukao, *Weak solutions for Stefan problems with convection in non-cylindrical domains,* Nonlinear Partial Differential Equations and Their Applications, 28–47, GAKUTO Internat. Ser. Math. Sci. Appl., **20**, Gakkōtosho, 2004.
[4] A. Kadoya, K. Shirakawa and A. Ito, *Mathematical models for describing the regional economic trend,* submitted to Mathematical Approach to Nonlinear Phenomena; Modelling, Analysis and Simulations, GAKUTO Internat. Ser. Math. Sci. Appl., Gakkōtosho.
[5] N. Kenmochi and M. Niezgódka, *Evolution systems of nonlinear variational inequalities arising from phase change problems,* Nonlinear Anal., **22** (1994), no. 9, 1163–1180.
[6] O.A. Ladyženskaja, V.A. Solonnikov and N.N. Ural'ceva, *Linear and Quasi-linear Equations of Parabolic Type,* Translations of Mathematical Monographs, **23**, Amer. Math. Soc., 1968.
[7] N.G. Mankiw, D. Romer and D.N. Weil, *A contribution to the empirics of economic growth,* The Quarterly Journal of Economics, **107** (1992), no. 2, 407–437.
[8] K. Shirakawa, *Large-time behavior for a phase field system associated with total variation energy,* Adv. Math. Sci. Appl. **15** (2005), no. 1, 1–27.
[9] K. Shirakawa, A. Ito, N. Yamazaki and N. Kenmochi, *Asymptotic stability for evolution equations governed by subdifferentials,* Recent Developments in Domain Decomposition Methods and Flow Problems, 287–310, GAKUTO Internat. Ser. Math. Sci. Appl., **11**, Gakkōtosho, 1998.
[10] R.M. Solow, *A contribution to the theory of economic growth,* The Quarterly Journal of Economics, **70** (1956), no. 1, 65–94.

Ken Shirakawa
Department of Applied Mathematics, Faculty of Engineering, Kobe University
1-1 Rokkodai, Nada, Kobe, 657-8501, Japan
e-mail: `sirakawa@cs.kobe-u.ac.jp`

Akio Ito
Department of Electronic Engineering and Computer Sciences, School of Engineering
Kinki University, 1 Takayaumenobe, Higashihiroshimashi, Hiroshima, 739-2116, Japan
e-mail: `aito@hiro.kindai.ac.jp`

Atsushi Kadoya
Department of Economic Informatics, Faculty of Economic Sciences,
Hiroshima Shudo University, 1-1-1 Otsuka, Asaminami-ku
Hiroshimashi, Hiroshima, 731-3195, Japan
e-mail: `kadoya@shudo-u.ac.jp`

International Series of Numerical Mathematics, Vol. 154, 413–423

Surface Energies in Multi-phase Systems with Diffuse Phase Boundaries

Björn Stinner

Abstract. A Ginzburg–Landau type functional for a multi-phase system involving a diffuse interface description of the phase boundaries is presented with the following calibration property: Prescribed surface energies (possibly anisotropic) of the phase transitions are correctly recovered in the sense of a Γ-limit as the thickness of the diffuse interfaces converges to zero. Possible applications are grain boundary motion and solidification of alloys on which numerical simulations are presented.

Mathematics Subject Classification (2000). 82B26, 74N20, 74E10.

Keywords. Phase transitions, phase field model, sharp interface model, surface energy, anisotropy.

1. Introduction

Consider a domain $\Omega \subset \mathbb{R}^d$, $d \in \{1,2,3\}$, which is subdivided into several (not necessarily connected) regions Ω_α, $\alpha \in \{1,\dots,\mathcal{M}\}$, $\mathcal{M} \in \mathbb{N}$, separated by hypersurfaces $\Gamma_{\alpha\beta}$, $1 \leq \alpha < \beta \leq \mathcal{M}$. The interfaces are supposed to carry energy obtained by integrating surface densities which may depend on the orientation of the hypersurface. The total energy of the system has the form

$$\mathcal{F}_{SI} = \sum_{\alpha<\beta} \int_{\Gamma_{\alpha\beta}} \sigma_{\alpha\beta}(\nu_{\alpha\beta}) d\mathcal{H}^{d-1}. \tag{1.1}$$

To avoid wetting effects is is assumed that $\sigma_{\alpha\beta} + \sigma_{\beta\delta} > \sigma_{\alpha\delta} > 0$ for each triple of phases α, β, δ.

Energies as in (1.1) can be approximated by Ginzburg–Landau energies of the form

$$\mathcal{F}_{PF} = \int_\Omega \Big(\varepsilon a(\phi, \nabla\phi) + \frac{1}{\varepsilon} w(\phi)\Big) d\mathcal{L}^d. \tag{1.2}$$

Here, $\phi = (\phi_1, \ldots, \phi_{\mathcal{M}}) : \Omega \to \Sigma^{\mathcal{M}}$ with

$$\Sigma^{\mathcal{M}} := \Big\{ v \in \mathbb{R}^{\mathcal{M}} : \sum_{\alpha=1}^{\mathcal{M}} v_\alpha = 1 \text{ and } 0 \le v_\alpha \le 1 \text{ for all } \alpha = 1, \ldots, \mathcal{M} \Big\}$$

is a set of phase field variables. For each α, ϕ_α is assigned to one of the phases (labelled α and corresponding to Ω_α) and describes its presence in a point of Ω. The function $a(\phi, \nabla\phi)$ is a non-negative gradient term, and $w(\phi)$ is a multi-well potential with $\mathcal{M}$ minima corresponding to the phases. It is well known that such Ginzburg–Landau energies cause transition regions where the phase fields vary from one of the minima of w, i.e., from one of the phases, to another one. The thickness of the interfacial layers is of the order ε, a small length scale.

Bellettini et al. [2] showed that the Γ-limit of (1.2) as $\varepsilon \to 0$ has the form (1.1), and they proved a relation between the $\sigma_{\alpha\beta}$ and the functions a and w. Using matched asymptotic expansions, Sternberg [10] for the isotropic case and Garcke et al. [5] for the general case found the slightly simpler relation

$$\sigma_{\alpha\beta}(\nu) = \inf_p \Big\{ \int_{-1}^{1} \sqrt{w(p) a(p, p' \otimes \nu)} dy, \\ p \in C^{0,1}([-1,1]; \Sigma^{\mathcal{M}}),\ p(-1) = e_\alpha,\ p(1) = e_\beta \Big\} \quad (1.3)$$

where e_α and e_β are the corners of the simplex $\Sigma^{\mathcal{M}}$ corresponding to the phases α and β, i.e., $e_\eta = (\delta_{\eta\theta})_{\theta=1}^{\mathcal{M}}$ with the Kronecker symbols $\delta_{\eta\theta}$. Using numerical simulations they showed that this formula holds true for a large class of anisotropies. Vice versa, it is a non-trivial task to construct functions a and w such that the surface energies obtained via (1.3) coincide with given surface energies (which, in applications, may be known from experiments).

A possible solution has been found by the author in collaboration with H. Garcke and R. Haas (see [7]) and will be presented in the following sections. First some facts on the minimization problem (1.3), afterwards the precise aims are stated. Some additional conditions on a and w are imposed:

Definition 1.1. Let

$$T\Sigma^{\mathcal{M}} := \Big\{ v = (v_1, \ldots, v_{\mathcal{M}}) \in \mathbb{R}^{\mathcal{M}} : \sum_{\alpha=1}^{\mathcal{M}} d_\alpha = 0 \Big\}.$$

The function $a : \Sigma^{\mathcal{M}} \times (T\Sigma^{\mathcal{M}})^d \to \mathbb{R}$ is an admissible gradient term if

$$a(\phi, X) \ge 0, \quad a(\phi, \eta X) = \eta^2 a(\phi, X) \quad \forall \phi \in \Sigma^{\mathcal{M}}, X \in (T\Sigma^{\mathcal{M}})^d, \eta \in \mathbb{R}, \quad (1.4)$$

$$a_{\alpha\beta}(\nu) := a(s e_\beta + (1-s) e_\alpha, (e_\beta - e_\alpha) \otimes \nu) \quad (1.5)$$

does not depend on $s \in [0,1]$ $\forall \alpha, \beta \in \{1, \ldots, \mathcal{M}\}$.

The function $w : \Sigma^{\mathcal{M}} \to \mathbb{R}$ is an admissible multi-well potential if

$$w(\phi) \geq 0 \ \forall \phi \in \Sigma^{\mathcal{M}}, \quad w(\phi) = 0 \Leftrightarrow \phi \in \{e_1, \dots, e_N\}, \tag{1.6}$$
$$w(se_\beta + (1-s)e_\alpha) = w_{\alpha\beta} s^2 (1-s)^2, \quad \forall s \in [0,1],\ \alpha, \beta \in \{1, \dots, \mathcal{M}\}. \tag{1.7}$$

A minimizer of the problem in (1.3) fulfills after appropriate rescaling to some function $\phi : \mathbb{R} \to \Sigma^{\mathcal{M}}$ (cf. [5] for this procedure)

$$\begin{aligned} &w_{,\phi}(\phi) + a_{,\phi}(\phi, \partial_z \phi \otimes \nu) - \frac{d}{dz}\Big(a_{,X}(\phi, \partial_z \phi \otimes \nu)\nu\Big) = \boldsymbol{\lambda}, \\ &\lim_{z \to \infty} \phi(z) = e_\beta, \qquad \lim_{z \to -\infty} \phi(z) = e_\alpha, \end{aligned} \tag{1.8}$$

where $\lambda = \frac{1}{\mathcal{M}} \sum_{i=1}^{\mathcal{M}} w_{,\phi_i}(\phi) + a_{,\phi_i}(\phi, \partial_z \phi \otimes \nu) - \frac{d}{dz}\big(a_{,X_i}(\phi, \partial_z \phi \otimes \nu)\nu\big)$ is a Lagrange multiplier and $\boldsymbol{\lambda} = \lambda(1, \dots, 1) \in \mathbb{R}^{\mathcal{M}}$. In particular, the minimization problem in (1.3) reads now

$$\begin{aligned} \sigma_{\alpha\beta}(\nu) = \inf_{\phi} \Big\{ &\int_{\mathbb{R}} \Big(a(\phi, \partial_z \phi \otimes \nu) + w(\phi)\Big) dz, \\ &\phi \in C^{0,1}(\mathbb{R}; \Sigma^{\mathcal{M}}),\ \lim_{z \to -\infty} \phi(z) = e_\alpha,\ \lim_{z \to \infty} \phi(z) = e_\beta \Big\}. \end{aligned} \tag{1.9}$$

Lemma 1.2. Consider a function of the form

$$\phi(z) = \chi(z) e_\beta + (1 - \chi(z)) e_\alpha. \tag{1.10}$$

Then ϕ solves (1.8) for admissible functions a and w if

$$\chi(z) = \frac{1}{2}\Big(1 + \tanh\big(\sqrt{\tfrac{w_{\alpha\beta}}{a_{\alpha\beta}(\nu)}}\, \frac{z}{2}\big)\Big) \quad \text{and if} \tag{1.11}$$

$$\lambda = w_{,\phi_i} + |\chi'|^2 a_{,\phi_i} - \frac{d}{dz}\Big(\chi' a_{,X_i} \nu\Big) \tag{1.12}$$

for some λ independent of $i \in \{1, \dots, \mathcal{M}\}$ where $w_{,\phi_i}$ is evaluated at $\phi(z)$ and the derivatives of a at $(\phi(z), (e_\beta - e_\alpha) \otimes \nu)$.

Moreover, if ϕ is a solution to (1.9) then the surface energy is

$$\sigma_{\alpha\beta}(\nu) = \frac{1}{3}\sqrt{a_{\alpha\beta}(\nu) w_{\alpha\beta}} \tag{1.13}$$

Proof. Given ϕ as in (1.10) obviously $\phi(z) \to e_\beta \Leftrightarrow \chi(z) \to 1 \Leftrightarrow z \to \infty$ and $\phi(z) \to e_\alpha \Leftrightarrow \chi(z) \to 0 \Leftrightarrow z \to -\infty$ whence the second line of (1.8) follows. By assumption (1.4) $a_{,\phi}$ is two-homogeneous and $a_{,X}$ is one-homogeneous with respect to the second variable. Since $\partial_z \phi(z) \otimes \nu = \chi'(z)(e_\beta - e_\alpha)$

$$\begin{aligned} a_{,\phi}(\phi(z), \partial_z \phi(z) \otimes \nu) &= |\chi'(z)|^2 a_{,\phi}(\phi(z), (e_\beta - e_\alpha) \otimes \nu), \\ a_{,X}(\phi(z), \partial_z \phi(z) \otimes \nu) &= \chi'(z) a_{,X}(\phi(z), (e_\beta - e_\alpha) \otimes \nu). \end{aligned}$$

The first line of (1.8) then follows directly from (1.12).

A straightforward calculation shows the identities

$$|\chi'|^2 = \frac{w_{\alpha\beta}}{a_{\alpha\beta}} \chi^2 (1-\chi)^2, \quad \chi'' = \frac{w_{\alpha\beta}}{a_{\alpha\beta}} \chi(1-\chi)(1-2\chi). \tag{1.14}$$

By assumption (1.5) $a(\phi(z),(e_\beta - e_\alpha)\otimes\nu) = a_{\alpha\beta}(\nu)$. Therefore, if ϕ solves (1.9) then the surface energy is (substituting $\frac{z}{2}\sqrt{\frac{a_{\alpha\beta}(\nu)}{w_{\alpha\beta}}} = t$)

$$\begin{aligned}\sigma_{\alpha\beta}(\nu) &= \int_{\mathbb{R}} \Big(a(\phi,\partial_z\phi\otimes\nu) + w(\phi)\Big)dz \\ &= \int_{\mathbb{R}} \Big(|\chi'(z)|^2 a_{\alpha\beta}(\nu) + w_{\alpha\beta}\chi(z)^2(1-\chi(z))^2\Big)dz \\ &= \int_{\mathbb{R}} 2w_{\alpha\beta}(1+\tanh(t))^2(1-\tanh(t))^2\sqrt{\frac{a_{\alpha\beta}(\nu)}{w_{\alpha\beta}}}dt \\ &= \frac{1}{3}\sqrt{a_{\alpha\beta}(\nu)w_{\alpha\beta}}.\end{aligned}$$

□

Now it is possible to state the aim precisely:

Task. *Construct admissible functions a and w such that*

1. *the values $\frac{1}{3}\sqrt{a_{\alpha\beta}(\nu)w_{\alpha\beta}}$ coincide with given surface energies $\sigma_{\alpha\beta}(\nu)$,*
2. *the function $\phi(z) = \chi(z)e_\beta + (1-\chi(z))e_\alpha$ with χ given by* (1.11) *fulfills condition* (1.12), *and*
3. *this function $\phi(z)$ solves* (1.9).

2. On the solvability of the task

In this section, an admissible gradient potential a will be presented which is based on the irreducible representations $\phi_\alpha\nabla\phi_\beta - \phi_\beta\nabla\phi_\alpha$ first used in [8] and later in [11] in the context of the phase field approach. Polynomial multi-well potentials w will also be presented such that the second point of the task set in the previous section is fulfilled and a critical point of the minimization problem (1.9) has the desired structure (1.10). In view of the first point, the values on the right-hand side of (1.13) can be determined in terms of coefficients of a and w. In particular, the presented potentials allow for a large class of anisotropic surface energies $\sigma_{\alpha\beta}(\nu)$. It remains to examine whether the third point of the given task is satisfied. For this purpose, numerical simulations of test problems have been performed indicating when a critical point of the form (1.10) indeed solves the minimization problem (1.9). More results and other examples for gradient terms a and multi-well potentials w with the demanded properties including results of further numerical simulations will be published in a forthcoming paper [7].

Definition 2.1. For each $\alpha \neq \beta$, $\alpha,\beta \in \{1,\dots,\mathcal{M}\}$, let $s_{\alpha\beta} : \mathbb{R}^d \to \mathbb{R}$ be one-homogeneous functions which are positive on the unit sphere $S^{d-1} = \{\nu \in \mathbb{R}^d : \|\nu\|_2 = 1\}$, even and smooth (except in zero), and let $g_{\alpha\beta}$ positive coefficients. In addition the symmetries $g_{\alpha\beta} = g_{\beta\alpha}$ and $s_{\alpha\beta}(\nu) = s_{\beta\alpha}(\nu)$ for all $\nu \in S^{d-1}$, $1 \le \alpha,\beta \le N$, are assumed to hold. The gradient potential in (1.2) is defined by

$$a(\phi,\nabla\phi) = \sum_{\alpha<\beta} g_{\alpha\beta}\big(s_{\alpha\beta}(\phi_\alpha\nabla\phi_\beta - \phi_\beta\nabla\phi_\alpha)\big)^2. \tag{2.1}$$

The homogeneity and the symmetry of the $s_{\alpha\beta}(\nu)$ yield $\nabla s_{\alpha\beta}(\nu)\cdot\nu = s_{\alpha\beta}(\nu)$ and $\nabla s_{\alpha\beta}(-\nu) = -\nabla s_{\alpha\beta}(\nu)$. Since

$$a(se_\beta + (1-s)e_\alpha), (e_\beta - e_\alpha)\otimes\nu) = g_{\alpha\beta}\big(s_{\alpha\beta}(\nu)\big)^2 = a_{\alpha\beta}(\nu) \tag{2.2}$$

is independent of $s \in [0,1]$ a fulfills the assumptions of Definition 1.1 and is admissible.

Proposition 2.2. *The multi-well potential*

$$w(\phi) = 9\sum_{\alpha<\beta} g_{\alpha\beta}\phi_\alpha^2\phi_\beta^2\Big(1 + 8\sum_{\delta\neq\alpha,\beta}\phi_\delta\Big) \tag{2.3}$$

satisfies the assumptions of Definition 1.1, *and, given* $a(\phi,\nabla\phi)$ *by* (2.1), *the condition* (1.12) *of Lemma* 1.2 *is satisfied.*

Proof. It is easy to derive that (1.6) is fulfilled. Since $w(se_\beta + (1-s)e_\alpha) = 9g_{\alpha\beta}s^2(1-s)^2$ condition (1.7) is fulfilled, too, with $w_{\alpha\beta} = 9g_{\alpha\beta}$.

When evaluating the derivatives of a at $(\chi e_\beta + (1-\chi)e_\alpha, (e_\beta - e_\alpha)\otimes\nu)$ short calculations show that

$$\begin{aligned}
a_{,\phi_i} &= 2g_{\alpha\beta}(s_{\alpha\beta}(\nu))^2, \quad i = \alpha,\beta,\\
a_{,X_i} &= \begin{cases} 2g_{\alpha\beta}s_{\alpha\beta}(\nu)\nabla s_{\alpha\beta}(\nu)(-\chi), & i=\alpha,\\ 2g_{\alpha\beta}s_{\alpha\beta}(\nu)\nabla s_{\alpha\beta}(\nu)(1-\chi), & i=\beta,\end{cases}\\
\frac{d}{dz}\Big(a_{,X_i}\nu\chi'\Big) &= \begin{cases} 2g_{\alpha\beta}(s_{\alpha\beta}(\nu))^2(-\chi\chi'' - |\chi'|^2), & i=\alpha,\\ 2g_{\alpha\beta}(s_{\alpha\beta}(\nu))^2((1-\chi)\chi'' - |\chi'|^2), & i=\beta,\end{cases}
\end{aligned}$$

and if $i \neq \alpha,\beta$ it holds that $a_{,\phi_i} = 0$ and $a_{,X_i} = 0$, whence $\frac{d}{dz}(a_{,X_i}\nu\chi') = 0$. For the derivatives of w evaluated at $\phi = \chi e_\beta + (1-\chi)e_\alpha$ on can derive

$$w_{,\phi_i}(\chi e_\beta + (1-\chi)e_\alpha) = \begin{cases} 18g_{\alpha\beta}\chi^2(1-\chi), & i=\alpha,\\ 18g_{\alpha\beta}\chi(1-\chi)^2, & i=\beta,\\ 72g_{\alpha\beta}\chi^2(1-\chi)^2, & i\neq\alpha,\beta.\end{cases} \tag{2.4}$$

It follows then from (1.12) for $i \neq \alpha,\beta$ necessarily

$$\lambda = 72g_{\alpha\beta}\chi^2(1-\chi)^2 = 8w_{\alpha\beta}\chi^2(1-\chi)^2. \tag{2.5}$$

For $i = \alpha$ the right-hand side of (1.12) reads using the identities (1.14) and (2.2)

$$\begin{aligned}
&w_{,\phi_\alpha} + |\chi'|^2 a_{,\phi_\alpha} - \frac{d}{dz}\Big(\chi' a_{,X_\alpha}\nu\Big)\\
&= 2w_{\alpha\beta}\chi^2(1-\chi) + \frac{w_{\alpha\beta}}{a_{\alpha\beta}(\nu)}\chi^2(1-\chi)^2\,2a_{\alpha\beta}(\nu)\\
&\quad - 2a_{\alpha\beta}(\nu)\Big(-\chi\frac{w_{\alpha\beta}}{a_{\alpha\beta}(\nu)}\chi(1-\chi)(1-2\chi) - \frac{w_{\alpha\beta}}{a_{\alpha\beta}(\nu)}\chi^2(1-\chi)^2\Big)
\end{aligned}$$

$$= 2w_{\alpha\beta}\chi^2(1-\chi) + w_{\alpha\beta}\chi^2(1-\chi)^2$$
$$+ w_{\alpha\beta}(\nu)\Big(4\chi^2(1-\chi)^2 - 2\chi^2(1-\chi) + 2\chi^2(1-\chi)^2\Big)$$
$$= 8w_{\alpha\beta}\chi^2(1-\chi)^2 = \lambda,$$

hence condition (1.12) holds for $i = \alpha$ with λ given by (2.5). Similarly this can be shown for $i = \beta$. □

Theorem 2.3. *The multi-well potential*

$$w(\phi) = 9\sum_{\alpha<\beta} g_{\alpha\beta}\phi_\alpha^2\phi_\beta^2\Big(1 + 8\sum_{\delta\neq\alpha,\beta}\phi_\delta\Big) + \sum_{\alpha<\beta<\delta} g_{\alpha\beta\delta}\phi_\alpha^2\phi_\beta^2\phi_\delta^2 \tag{2.6}$$

satisfies the assumptions of Definition 1.1. *Given* $a(\phi,\nabla\phi)$ *by* (2.1) *the second point of the set task is satisfied.*

Proof. For the additional term $w_{add}(\phi) := \sum_{\alpha<\beta<\delta} g_{\alpha\beta\delta}\phi_\alpha^2\phi_\beta^2\phi_\delta^2$ it holds that $(w_{add})_{,\phi}(\chi e_\beta + (1-\chi)e_\alpha) = 0$. Therefore, (2.4) is not changed. Hence, the same arguments as in the proof of the preceding proposition followed by Lemma 1.2 can be applied to show the theorem. □

Observe that $a_{\alpha\beta}(\nu) = g_{\alpha\beta}(s_{\alpha\beta}(\nu))^2$ and $w_{\alpha\beta} = 9g_{\alpha\beta}$ yield

$$\sigma_{\alpha\beta}(\nu) = g_{\alpha\beta}s_{\alpha\beta}(\nu)$$

from Lemma 1.2 if, in addition, also the third point of the given task is satisfied. In order to recover a prescribed surface energy the idea is that $g_{\alpha\beta}$ determines its typical size involving the physical units (one may choose the mean value of $\sigma_{\alpha\beta}(\nu)$ on the unit sphere) and that $s_{\alpha\beta}(\nu)$, a dimensionless multiplier, models the deviations from this typical value depending on the direction ν, i.e., the anisotropy. In spite of the constraints imposed on the $g_{\alpha\beta}$ and the $s_{\alpha\beta}(\nu)$ in Definition 2.1 a wide class of anisotropic surface tensions can be recovered by that procedure. Hence, there is a good chance that the first point of the set task can be fulfilled provided the third point is satisfied for the presented potentials. Numerical experiments for test problems have been performed and are presented in the following section indicating that this is in fact the case.

3. Numerical tests

In the following let $\alpha = 1$ and $\beta = 2$. In one space dimension a sharp transition relaxed under a gradient flow of the energy (1.2) (cf. (4.1) in the following section), i.e., a jump of ϕ from e_1 to e_2, expecting a stable form which approximates a solution to (1.9). Let $M = 3$, $\varepsilon = 0.1$ and consider the domain $D = [0,1]$ discretized with a uniform grid of mesh size $\Delta x = 0.01$. Homogeneous boundary conditions were imposed which were checked to have no influence on the results. For the numerical method based on finite differences [6] is an appropriate reference.

Choosing $g_{\alpha\beta} = 1$ and $s_{\alpha\beta}(\nu) = 1$ for all pairs (α,β), very small contributions of the order 10^{-4} of phase 3 in the transition region were found after relaxation.

Refining the grid, the contributions became even smaller indicating that discretization errors had come in. Also the case of different surface energies was examined defining g_{12}, g_{13} and g_{23} by a permutation of the values $\frac{2}{\sqrt{3}}$, 1.0 and $\frac{1}{\sqrt{3}}$. If $g_{12} = \frac{2}{\sqrt{3}}$ the largest contributions from phase 3 of order 10^{-3} were observed. As in the case of equal surface energies they became smaller when refining the grid.

For $M = 4$ the multi-well potential (2.3) converges to $-\infty$ if $\xi \to \infty$ and $\phi = \xi(1, 1, -1, -1)$. During the relaxation of the sharp phase transition it was indeed observed that ϕ_3 and ϕ_4 were equal and grew such that no equilibrium was reached. Adding a coercive term as in (2.6) the blow up could be avoided. Analogously, if $M > 4$ the additional term in w turned out to be necessary.

Let $g_{\alpha\beta} = 1$ and $s_{\alpha\beta}(\nu) = 1$ as before and use the potential (2.6) where $g_{\alpha\beta\delta} = C$ for all triples α, β, δ with a positive constant C. In the case $M = 4$, for $C \leq 120.0$ we found large contributions of the phases 3 and 4 of order 10^{-1} in the transition region. The energy (1.2) can be approximated by the sum

$$\mathcal{F}_{\text{Approx}} := \Delta x \sum_{i=0}^{N} \varepsilon\, a(\phi(x_i), \partial_x^h \phi(x_i)) + \frac{1}{\varepsilon} w(\phi(x_i)),$$

where $\partial_x^h \phi(x_i) = \frac{1}{\Delta x}(\phi(x_{i+1}) - \phi(x_i))$ and the $\{x_i\}_{i=0}^N$ are the grid points. Computing this sum gives energies of $\mathcal{F}_{\text{Approx}} = 0.970590$ if $C = 5.0$, i.e., in the presence of third phase contributions, and $\mathcal{F}_{\text{Approx}} = 0.999242$ if $C = 200.0$ without. This indicates that if C is to small then the solution to (1.9) has not the form (1.10). Moreover, the approximation of the surface energy, here $\sigma_{\alpha\beta}(\nu) = 1$, is worse in that case.

To examine whether a given surface energy is correctly recovered by the phase field model the contraction of a sphere by a curvature flow in two space dimensions is a good test problem since an analytical solution is known. Using matched asymptotic expansions, it is shown in [5] that the phase field model yields a model with free boundaries moving according to a curvature flow in the sharp interface limit as $\varepsilon \to 0$. For a sphere in 2D, this curvature flow can be expressed in terms of the radius $r(t)$ being a function of the time t and reads

$$\dot{r}(t) = -\frac{\sigma}{r(t)}, \quad r(0) = r_0. \tag{3.1}$$

For $\sigma = 1$ the exact solution is $r(t) = \sqrt{r_0^2 - 2t}$.

On the domain $D = [0, 2]^2$ phase 1 initially occupied a quarter of a ball with radius $r_0 = 1.4$ centered in $(0, 0)$, phase 2 occupied the remaining part of D, and another phase 3 is considered but initially nowhere present. The initial data were chosen as follows:

$$\phi_{1,ic} = \tfrac{1}{2}\big(1 - \tanh(\tfrac{3(r-r_0)}{2\varepsilon})\big), \quad \phi_{2,ic} = 1 - \phi_{1,ic}, \quad \phi_{3,ic} = 0.$$

The other parameters were $\varepsilon = 0.2$, $\Delta x = 0.01$, $g_{\alpha\beta} = 1$ and $s_{\alpha\beta}(\nu) = 1$ for all pairs $\alpha < \beta$. Homogeneous Neumann boundary conditions were imposed. Two numerical simulations were performed, a first one with the straightforward generalization

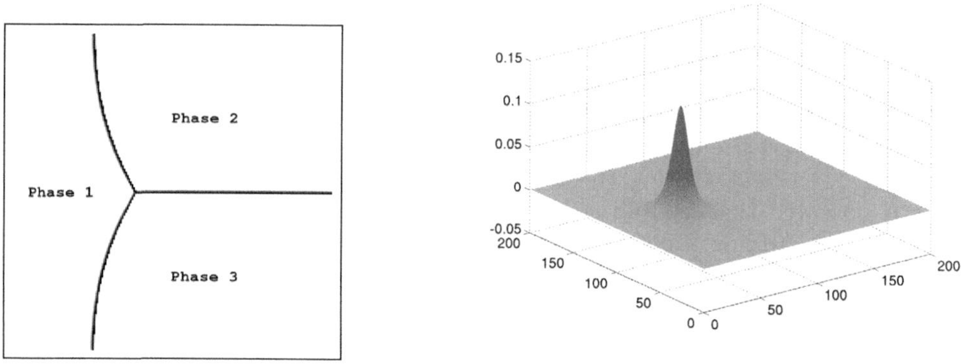

FIGURE 1. On the left: Simulated triple junction. On the right: Contribution of phase 4 in the triple junction.

$w(\phi) = 9\sum_{\alpha<\beta} g_{\alpha\beta}\phi_\alpha^2\phi_\beta^2$ of the standard multi-well potential and a second one with the new potential (2.3).

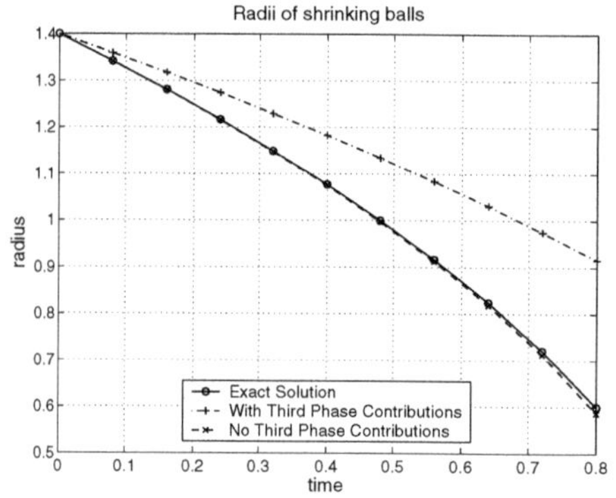

During the first simulation a strong contribution of ϕ_3 in the interfacial layer was observed in contrast to the second simulation. The figure shows the radii of the sets where $\phi_1 = \phi_2$ in comparison with the exact solution over the time. If no third phase contributions are present the simulation remarkably good approximates the exact solution while in the other case the velocity of the shrinking circle is too small. This is consistent with the results in [6]. There for the above choice of w a surface energy $\sigma < 1$ was measured which, by (3.1), leads to a slower motion.

When simulating with $\mathcal{M} = 4$ or more phases, phase field variables are present in the region of a triple junction which do not correspond to the adjacent phases. Fig. 1 shows a triple junction on the domain $D = [0,1]^2$. Thanks to an appropriate choice of the $g_{\alpha\beta\delta}$, on the phase transitions no contributions of other phase field variables than corresponding to the adjacent phases are observed. But in the triple junction a fourth phase has developed which was not present initially ($\phi_4(t=0) = 0$) and whose height turned out to be independent of ε and Δx. Nevertheless, the angles in the triple point still agree remarkably well with the theoretically predicted values of 120° (see [3]) in spite of the presence of the artificial fourth order contributions.

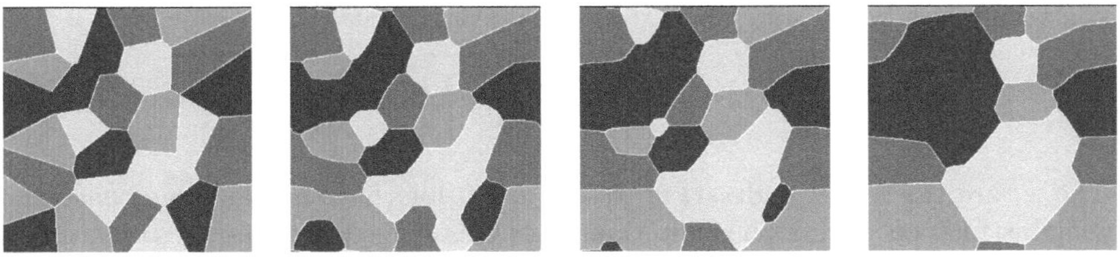

FIGURE 2. Growth of grains. From left to right and from top to bottom the situation at the times t =0.0, 0.0036, 0.0108, and 0.9.

4. Applications

As a first application example grain growth involving four grains of different structure was simulated, modeled by an L^2-gradient flow of the energy (1.2). It reads

$$\omega\partial_t\phi_\alpha = \varepsilon\big(\nabla\cdot a_{,X_\alpha}(\phi,\nabla\phi) - a_{,\phi_\alpha}(\phi,\nabla\phi)\big) - w_{,\phi_\alpha} - \lambda \tag{4.1}$$

with an appropriate Lagrange multiplier λ ensuring $\partial_t\phi \in T\Sigma^{\mathcal{M}}$ and a calibration constant ω. For the surface energy anisotropies the crystalline functions

$$s_{\alpha\beta}(\nu) = \max\{\eta_{\alpha\beta}^{(k)}\cdot\nu,\, k=1,\dots,6\},$$

were chosen where the $\{\eta_{\alpha\beta}^{(k)}\}_k$ are the edges of a hexagonal crystal. For every phase transition the same surface energy was chosen, namely $g_{\alpha\beta}=1$ and

$$\left\{\eta_{\alpha\beta}^{(k)}\right\}_{k=1}^{6} = \left\{\begin{pmatrix}1\\0\end{pmatrix}, \frac{1}{2}\begin{pmatrix}1\\\sqrt{3}\end{pmatrix}, \frac{1}{2}\begin{pmatrix}-1\\\sqrt{3}\end{pmatrix}, \begin{pmatrix}-1\\0\end{pmatrix}, \frac{1}{2}\begin{pmatrix}-1\\-\sqrt{3}\end{pmatrix}, \frac{1}{2}\begin{pmatrix}1\\-\sqrt{3}\end{pmatrix}\right\}$$

for all $\alpha<\beta$. On the domain $D=[0,1]^2$ the system of parabolic partial differential equations (4.1) was discretized explicitly in time using a finite difference method (gradients were replaced by forward differences, the divergence by a backward difference, cf. [6]). The values for the phase field variables were initialized using a Voronoj diagram. The grid size of the uniform grid was $\Delta x = 0.0033$, the time step $\Delta t = 1.5\cdot 10^{-6}$. Setting $g_{\alpha\beta\delta} = 250.0$ for all triples $\alpha<\beta<\delta$ third phase contributions on the phase boundaries could be avoided. Periodic boundary conditions were imposed. Fig. 2 shows the evolution of the grains. During the relaxation, the angles remarkably good approximate the predicted values of 120° and the phase boundaries are oriented according to the preferred directions (cf. [1] for the crystalline curvature flow resulting from the gradient flow of (1.1) with such crystalline surface energies).

As a second example the gradient flow (4.1) for three phases (two solid phases and liquid one) was coupled to parabolic evolution equations balancing the mass respectively the concentrations of two components of a eutectic alloy (cf. [4] for a description of a general model for solidification of alloys on which the present one is based). The concentrations are denoted by c_1 and c_2 and satisfy the algebraic constraint $c_1 + c_2 = 1$ and the partial differential equations

$$\partial_t c_i = -\nabla \cdot J_i = -\nabla \cdot \sum_{j=1}^{2} L_{ij} \nabla \frac{-\mu_j}{T}, \quad 1 \leq i \leq 2.$$

The J_i are the fluxes and satisfy $J_1 + J_2 = 0$ for that the algebraic constraint remains fulfilled during the evolution. The Onsager coefficients L_{ij} constitute a symmetric matrix and were chosen to be

$$L_{11} = L_{22} = -L_{12} = -L_{21} = D^{(\alpha)} c_1 c_2$$

with diffusion coefficients D^α depending on the phase but independent of the components. In the liquid phase D^α was set to 1.5 while in the two solid phases to 0.015. The chemical potentials μ_j are the derivatives of the free energies of the possible phases with respect to the concentrations, $\mu_j = \partial_{c_j} f$. The free energies were chosen of the form

$$f^\alpha(c, \phi) := \sum_{\alpha=1}^{3} \sum_{i=1}^{2} \left(c_i L_i^\alpha \frac{T - T_i^\alpha}{T_i^\alpha} h(\phi_\alpha) \right) + \sum_{i=1}^{2} T c_i \ln(c_i).$$

The function h satisfying $h(0) = 0$ and $h(1) = 1$ is a monotone function. The quantity L_i^α is the latent heats per unit volume of the phase transition from phase α to the liquid phase and of the pure component i, and T_i^α is the melting temperature of the ith component in phase α. Indexing the solid phases with α and β and the liquid one with l the following values were chosen:

$$L_i^\alpha = L_i^\beta = 10.5467, \; L_i^l = 0, \; \forall i, \quad T_1^\alpha = T_2^\beta = 2.4, \; T_2^\alpha = T_1^\beta = 1.6.$$

These choices yield a symmetric eutectic phase diagram with a eutectic concentration of $c_E = 0.5$, a eutectic temperature of about $T_E \approx 2.11686$, and equilibrium concentrations $c^\alpha = 0.1$ and $c^\beta = 0.9$ of the solid phases at eutectic temperature. The above choice of the Onsager coefficients together with this free energy density yields linear diffusion equations in the pure phases, cf. [4].

The system temperature T was kept constant during the simulation and set to 2.01686 which means that the liquid is slightly undercooled. The evolution of the phase field variables is coupled to the thermodynamic quantities be adding a term $-\frac{1}{T} f_{,\phi_\alpha}(c, \phi)$ to the right-hand side of (4.1). The remaining values were set to $\varepsilon = 0.4$ and $\omega = 0.1$. Initially, a situation as in Fig. 3 on the left was considered on the domain $[4.8, 9.6]$. The concentrations in the phases were chosen according to the equilibrium values at eutectic temperature. To the left and to the right periodic boundary conditions were imposed while to the top and to the bottom homogeneous Neumann boundary conditions. The initial situation relaxed on a uniform grid with spacing $\Delta x = 0.04$ and time stepping $\Delta t = 0.0002$. After a while, a self-similar lamellar growth of the solid phases into the liquid one was observed (see Fig. 3). For more numerical simulations and results on eutectic alloys [9] is an appropriate reference.

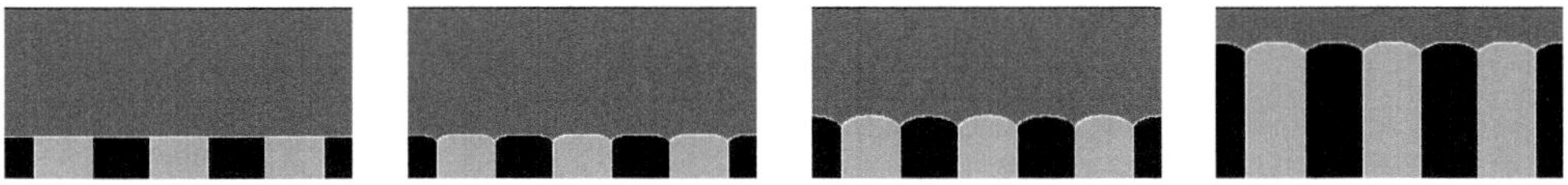

FIGURE 3. Eutectic solidification, lamellar solid structure growing self-similarly (after a while) into an undercooled melt. From the left to the right the situation at $t = 0.0, 0.06, 1.2, 6.0$.

Acknowledgement. The author gratefully acknowledges the financial support by the German Research Foundation (DFG) within the priority research program (SPP) "Analysis, Modeling and Simulation of Multiscale Problems" and the scholarship granted by the University of Coimbra for visiting the Mathematics Department on the occasion of the FBP2005 conference.

References

[1] F. Almgren and J. Taylor, *Flat flow is motion by crystalline curvature for curves with crystalline energies.* J. Differential Geom. **42** (1995), 1–22.

[2] G. Bellettini, A. Braides, and G. Riey, *Variational Approximation of Anisotropic Functionals on Partitions.* Ann. Mat. **184** (2002), 75–93.

[3] L. Bronsard and F. Reitich, *On three-phase boundary motion and the singular limit of a vector-valued Ginzburg-Landau equation.* Arch. Rat. Mech. Anal. **124** (1993), 355–379.

[4] H. Garcke, B. Nestler, and B. Stinner, *A Diffuse Interface Model for Alloys with Multiple Components and Phases.* SIAM J. Appl. Math. **64** (2004), 775–799.

[5] H. Garcke, B. Nestler, and B. Stoth, *On anisotropic Order Parameter Models for Multi Phase Systems and their Sharp Interface Limits.* Physica D **115** (1998), 87–108.

[6] H. Garcke, B. Nestler, and B. Stoth, *A Multi Phase Field Concept: Numerical Simulations of Moving Phase Boundaries and Multiple Junctions.* SIAM J. Appl. Math. **60** (1999), 295–315.

[7] H. Garcke and R. Haas and B. Stinner, *On Ginzburg-Landau Type Free Energies for Multi-phase Systems.* In preparation.

[8] L.D. Landau and E.M. Lifschitz, *Lehrbuch der theoretischen Physik, Band 5, Statistische Physik Teil I.* Akademie Verlag Berlin (1979).

[9] B. Nestler and A.A. Wheeler, *A Multi-phase-field Model of Eutectic and Peritectic Alloys: Numerical Simulation of Growth Structures.* Phys. D **138** (2000), 114–133.

[10] P. Sternberg, *Vector-valued local minimizers of nonconvex variational problems.* Rocky Mountain J. Math. **21** (1991), 799–807.

[11] I. Steinbach et al., *A Phase Field Concept for Multiphase Systems.* Physica D **94** (1996), 135–147.

Björn Stinner
NWF I - Mathematik, Universität Regensburg, D-93040 Regensburg, Germany
e-mail: `bjoern.stinner@mathematik.uni-regensburg.de`

International Series of Numerical Mathematics, Vol. 154, 425–434

High-order Techniques for Calculating Surface Tension Forces

M. Sussman and M. Ohta

Abstract. In this paper we develop further the "height fraction" technique for computing curvature directly from volume fractions. In particular we, (1) develop a systematic approach for calculating curvature from volume fractions which is accurate to any order, and (2) we test the second-order "height fraction" technique on the following two-phase problems: (1) the break-up of a cylindrical column of liquid due to Rayleigh-capillary instability, (2) surface tension induced droplet oscillations and (3) the steady motion of gas bubbles rising in liquid.

Keywords. Curvature, volume-of-fluid method, level set, multiphase flow.

1. Introduction

Numerical algorithms for calculating surface tension have been presented from the perspective of front tracking algorithms [14, 26, 25, 28, 16], Volume-of-Fluid algorithms [3, 7, 1, 18, 8], level set methods [24, 12, 13], and various hybrid methods [23, 20, 6]. The ability to accurately calculate surface tension can be important for modeling the impact of drops on surfaces, contact line dynamics, bubble motion, and the break-up of liquid jets. In our previous work [20], a second-order coupled level set and volume of fluid method was presented for calculating bubble growth and collapse. In that work, the "height fraction" technique[9] was employed to accurately calculate curvature directly from volume fractions. In this paper, we present additional calculations further validating the second-order method originally proposed in [20]. Furthermore, we demonstrate that the "height fraction" technique can be extended to calculate curvature to *any* order of accuracy.

Conventional wisdom would have it that only a level set representation of an interface is capable of having a very high-order accurate method for extracting

Work supported in part by NSF grant number 0242524 (U.S. Japan Cooperative Science).
Work supported in part by JSPS.

the interface curvature. For example, a "spectral" level set approach was presented by[21]. Previous methods for extracting curvature from volume fractions have been proposed by Chorin [4] (osculating circle technique), Poo and Ashgriz [15], Aleinov and Puckett [1] (convolution technique), Williams et al. [27], Renardy et al. [17] ("PROST"), and the "height fraction" approach [9, 20, 7, 5]. The challenge in accurately calculating curvature from the volume-of-fluid function F is the fact that F transitions sharply from 0 (computational cell containing only gas) to 1 (computational cell containing only liquid). Of all the approaches for calculating curvature from volume fractions, the "height fraction" approach is the most direct, and, unlike convolution techniques, it is "localized."

In this paper we present a systematic approach to extending the "height fraction" approach to any order of accuracy (i.e., higher than second order).

2. Curvature discretization using "height fractions"

The curvature of an interface is computed to second- or fourth-order accuracy directly from the volume fractions. The method is based on reconstructing the "height" function directly from the volume fractions [9]. Without loss of generality, we assume that the free surface is oriented more horizontal than vertical. For a second-order curvature algorithm, a $3 \times 3 \times 7$ stencil of volume fractions is constructed about cell (i, j, k). For the fourth-order algorithm, a $5 \times 5 \times 13$ stencil of volume fractions is constructed about cell (i, j, k). The 3×3 (5×5 for fourth-order) vertical sums, $F_{i',j'}$, $i' = -1 \ldots 1$, $j' = -1 \ldots 1$, are exact integrals of the height function $h(x, y)$ (up to a constant); i.e.,

$$F_{i',j'} \Delta x \Delta y \Delta z = \int_{x_{i+i'-1/2}}^{x_{i+i'+1/2}} \int_{y_{j+j'-1/2}}^{y_{j+j'+1/2}} h(x, y) dx dy + C. \tag{2.1}$$

It can be shown that $\Delta z(F_{1,0} - F_{-1,0})/(2\Delta x)$ is a second-order approximation to $h_x(x_i, y_j)$ and that $\Delta z(F_{1,0} - 2F_{0,0} + F_{-1,0})/\Delta x^2$ is a second-order approximation to $h_{xx}(x_i, y_j)$. In general, one expands $h(x, y)$ in a Taylor series,

$$\begin{aligned} h(x, y) &= h(x_i, y_j) + (x - x_i)h_x(x_i, y_j) + \frac{1}{2}(x - x_i)^2 h_{xx}(x_i, y_j) \\ &+ \frac{1}{6}(x - x_i)^3 h_{xxx}(x_i, y_j) + \frac{1}{24}(x - x_i)^4 h_{xxxx}(x_i, y_j) + (y - y_j)h_y(x_i, y_j) \\ &+ \frac{1}{2}(y - y_j)^2 h_{yy}(x_i, y_j) + \frac{1}{6}(y - y_j)^3 h_{yyy}(x_i, y_j) + \frac{1}{24}(y - y_j)^4 h_{yyyy}(x_i, y_j) \\ &+ (x - x_i)(y - y_j)h_{xy}(x_i, y_j) + \frac{1}{2}(x - x_i)^2(y - y_j)h_{xxy}(x_i, y_j) \\ &+ \frac{1}{2}(x - x_i)(y - y_j)^2 h_{xyy}(x_i, y_j) + \frac{1}{4}(x - x_i)^2(y - y_j)^2 h_{xxyy}(x_i, y_j) \\ &+ \text{higher-order terms.} \end{aligned}$$

After integrating the Taylor series expansion of $h(x, y)$ and using (2.1), one has

$$F_{i',j'}\Delta z = h(x_{i+i'}, y_{j+j'}) + \frac{1}{24}\Delta x^2 h_{xx}(x_{i+i'}, y_{j+j'}) + \frac{1}{(16)(120)}\Delta x^4 h_{xxxx}(x_{i+i'}, y_{j+j'}) + \frac{1}{24}\Delta y^2 h_{yy}(x_{i+i'}, y_{j+j'}) + \frac{1}{(16)(120)}\Delta y^4 h_{yyyy}(x_{i+i'}, y_{j+j'}) + \frac{1}{(24)(24)}\Delta x^2 \Delta y^2 h_{xxyy}(x_{i+i'}, y_{j+j'}) + \text{higher-order terms.} \quad (2.2)$$

For a horizontally orientated surface, the curvature is written as,

$$\kappa = \nabla \cdot \boldsymbol{n}$$

where,

$$\boldsymbol{n} = \left(-\frac{h_x}{\sqrt{1+h_x^2+h_y^2}}, -\frac{h_y}{\sqrt{1+h_x^2+h_y^2}}, \frac{1}{\sqrt{1+h_x^2+h_y^2}} \right).$$

For a fourth-order approximation to the curvature, we must approximate h_x, h_y, h_{xx}, h_{yy} and h_{xy} with fourth-order accuracy. We assume the discretization to each of these terms has the form,

$$\frac{\partial^{l+m} h(x_i, y_j)}{\partial x^l \partial y^m} = \Delta z \sum_{i'=-2}^{2} \sum_{j'=-2}^{2} A_{i',j'}^{l,m} F_{i',j'}.$$

The coefficients, $A^{l,m}$ ($l = 0, 1, 2$ and $m = 0, 1, 2$), are determined by the "method of undetermined coefficients" in which one uses the relation (2.2) to relate $F_{i',j'}$ to h, and the fact that our discretization for the derivatives of h should be exact for the polynomials $h(x, y) = (x - x_i)^l (y - y_j)^m$ where $l = 0 \dots 4$ and $m = 0 \dots 4$. As a result, one constructs a matrix system of equations with 25 equations and 25 unknowns. For example, in 2d, one has the following fourth-order approximations,

$$\frac{\partial h(x_i)}{\partial x} \approx \frac{\Delta y}{\Delta x} \left(\frac{5}{48}(F_{-2} - F_2) + \frac{17}{24}(F_1 - F_{-1}) \right)$$

$$\frac{\partial^2 h(x_i)}{\partial x^2} \approx \frac{\Delta y}{\Delta x^2} \left(\frac{-1}{8}(F_{-2} + F_2) + \frac{3}{2}(F_1 + F_{-1}) - \frac{11}{4}F_0 \right).$$

Remark: For the fourth-order algorithm, a $5 \times 5 \times 13$ stencil is used. A possible concern here is that in underresolved regions, the interface might pass through the stencil more than one time resulting in an erroneous approximation to the curvature. A simple patch (not implemented for any of the results presented in this paper) for this problem would be to locate the interface crossing, in each 13 cell column of data, closest to z_k and then delete other interface crossings by looking at where the divided difference,

$$Df_{i',j',k'} = \frac{f_{i',j',k'} - f_{i',j',k'-1}}{\Delta z}$$

TABLE 1. Convergence study for computing curvatures from volume fractions of a unit circle in 2*d*. Results for the second-order and fourth-order discretizations are reported.

Δx	max error (2nd)	max error (4th)	avg. error (2nd)	avg. error (4th)
1/16	0.0031	0.00104	0.0019	0.00016
1/32	0.0007	4.21E-5	0.0005	7.9E-6

TABLE 2. Convergence study for computing curvatures from volume fractions of a unit sphere in 3*d*. Results for the second-order and fourth-order discretizations are reported.

Δx	max error (2nd)	max error (4th)	avg. error (2nd)	avg. error (4th)
1/16	0.050	0.03431	0.0035	0.00081
1/32	0.010	0.00060	0.0009	2.78E-5

changes sign. All volume fractions in the stencil, $f_{i',j',k''}$, in which $k'' \geq k' \geq k$ are set to $f_{i',j',k'-1}$.

3. Numerical validation of curvature discretization for a circle

We check our curvature discretization algorithm for a circle in 2d or a sphere in 3d. In 2d, we have a unit circle located at the point (2,2) in a 4×4 domain. In 3d, we have a unit sphere located at the origin in a 2×2 domain. Symmetric boundary conditions are used at the borders of the domain. As demonstrated by Tables 1 and 2, we get the appropriate order of accuracy for our high-order height fraction curvature discretization schemes.

4. Parasitic currents

In this section we test our implementation of surface tension for the problem of a static 2d drop with surface tension. The exact solution for such a problem is that the velocity $\boldsymbol{u}$ is identically zero. In terms of the Ohnesorge number,

$$Oh = \frac{\mu}{\sqrt{\sigma \rho D}},$$

and assuming constant density and constant viscosity in the drop, the Navier Stokes equations are,

$$\frac{D\boldsymbol{u}}{Dt} = -\nabla p + \frac{1}{Oh}\Delta \boldsymbol{u} - \frac{1}{Oh}\kappa \nabla H.$$

We assume the drop is surrounded by a constant pressure void. The numerical simulation uses the second-order coupled levelset and volume-of-fluid (CLSVOF) algorithm described in [20]. We investigate the maximum velocity of our numerical

TABLE 3. Convergence study for static droplet with surface tension (parasitic currents test). Maximum velocity at $t = 250$ is shown. $Oh = 12000$. $\rho_L = 1$ and $\rho_G = 0$. Results for the second-order discretization of curvature and the fourth-order discretization are reported.

Δx	max. velocity (2nd)	max. velocity (4th)
1.0/32	1.4E-6	5.5E-7
1.0/64	1.5E-7	2.9E-8

method for varying grid resolutions at the dimensionless time $t = 250$. The dimensions of our computational grid are 1×1 with symmetric boundary conditions at all boundaries. A one diameter drop is placed at the origin of our domain. Our tolerance for the pressure solver and viscous solver is $1.0E-12$. In Table 3 we display results of our grid refinement study for $Oh = 12000$. We used both the second-order height fraction algorithm and the fourth-order height fraction algorithm for calculating curvature. Our results indicate at least second-order convergence using both approaches. We remark that the overall results are not expected to be 4th-order accurate when using the 4th-order height fraction curvature discretization since the underlying Navier-Stokes solver is second order. For a reference of previous results for parasitic currents, we refer the reader to work by [16, 7]. Our results using the fourth-order accurate curvature discretization algorithm crush any doubt about the ability to calculate surface tension using the volume-of-fluid method.

5. 2d axisymmetric test problems

In this section we validate the second-order height fraction curvature discretization scheme for the problem of (1) surface tension driven drop oscillations, (2) Rayleigh-capillary instability, and (3) steady bubble motion. The governing equations are the Navier-Stokes equations for two phase flows,

$$\rho \frac{D\boldsymbol{U}}{Dt} = \nabla \cdot (-pI + 2\mu D) + \rho g \hat{\boldsymbol{z}} - \sigma \kappa \nabla H \tag{5.1}$$

$$\nabla \cdot \boldsymbol{U} = 0$$

$$\frac{D\phi}{Dt} = 0$$

$$\rho = \rho_L H(\phi) + \rho_G(1 - H(\phi))$$

$$\mu = \mu_L H(\phi) + \mu_G(1 - H(\phi))$$

$$H(\phi) = \begin{cases} 1 & \phi \geq 0 \\ 0 & \phi < 0 \end{cases}$$

TABLE 4. Convergence study for zero gravity drop oscillations $\gamma = 1/2$.

Δr	$E^{avg}_{Amplitude}$	$E^{max}_{amplitude}$
3/64	N/A	N/A
3/128	0.00073	0.00174
3/256	0.00021	0.00054

We either use the second-order "single-phase" method described in [20] ($\rho_G = 0$) or the sharp interface "two-phase" method described in [11, 22] ($\rho_G = 0.001$).

For the first problem (1), surface tension driven drop oscillations, we compute the evolution of a drop in a void with a surface tension coefficient $\sigma = 1/2$ and initial perturbation of $\epsilon = 0.05$. Table 4 gives the successive errors in amplitude as one refines the computational grid. In Figure 1, we plot the minor amplitude versus time for the three different grid resolutions.

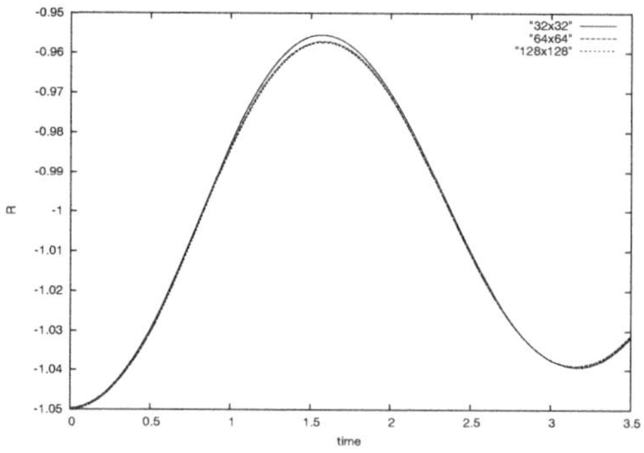

FIGURE 1. Perturbation in minor amplitude for zero gravity drop oscillations. $\mu_L = 1/50$, $\gamma = 1/2$.

For the second problem (2), we calculate the break-up of a liquid jet (in a void) due to capillary instability. The surface tension coefficient is $\sigma = 1$ and the viscosity coefficient is $\mu = 1/200$. In Figure 2, we display the results of our computations for the capillary jet as it breaks up. In Table 5, we measure the relative errors for the interface and velocity field for grid resolutions ranging from 16x32 to 64x128.

For our third test problem (3), we compute the steady state shapes of a gas bubble rising in a viscous Newtonian liquid. For comparison, we use the experimental results found in [2] and [10] and computational results in [19].

As in [2] and [10], we present our computational results in terms of the following dimensionless groups. The Reynolds number R, the Eötvös number Eo, and the Morton number Mo are defined as follows

$$R = \frac{\rho L U}{\eta_L} \qquad Eo = \frac{g L^2 U}{\sigma} \qquad Mo = \frac{g \eta_L^4}{\rho \sigma^3}. \tag{5.2}$$

TABLE 5. Convergence study for the Rayleigh capillary instability problem. Two-phase sharp interface method. $t = 80$.

grid	$E_{\text{interface}}$	$E^{\text{avg}}_{\text{Liquid}}$	$E^{\text{max}}_{\text{Liquid}}$
16x32	N/A	N/A	N/A
32x64	3.87	2.77	0.014
64x128	0.62	0.84	0.003

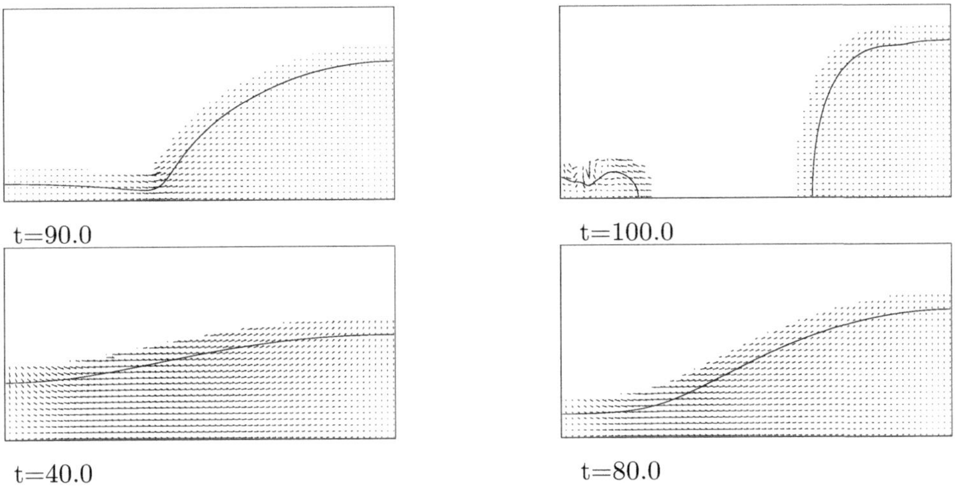

FIGURE 2. Capillary Instability. Grid resolution is 64×128.

ρ is the liquid density, L is the bubble diameter, U is a characteristic velocity, η_L is the liquid viscosity, σ is the surface tension, and g is the acceleration of gravity. A comparison of computed terminal bubble shapes versus previous computational and experimental results are reported in Figure 3. Our comparisons include oblate ellipsoidal cap bubbles studied by [2] ($Eo = 243$, $Mo = 266$, and $R = 7.77$ for bubble figure 2(d) and $Eo = 116$, $Mo = 5.51$, and $R = 13.3$ for bubble figure 3(d)), spherical cap bubbles studied by Hnat & Buckmaster [10] ($R = 9.8$, $Mo = 0.065$, and $C = 4.95$, where $C = \frac{r}{(\nu^2/g)^{1/3}}$), and a disk-bubble studied by Ryskin & Leal [19] ($R = 100$ and $We = 10$).

6. Conclusions

The "height fraction" approach for deriving curvature from volume fractions was extended from second-order accurate to fourth-order accurate. The improved accuracy was verified both analytically and through numerical tests. When applied to the "parasitic currents" test, there was a factor of 5 improvement of the fourth-order method over the second-order method. Besides developing a fourth-order

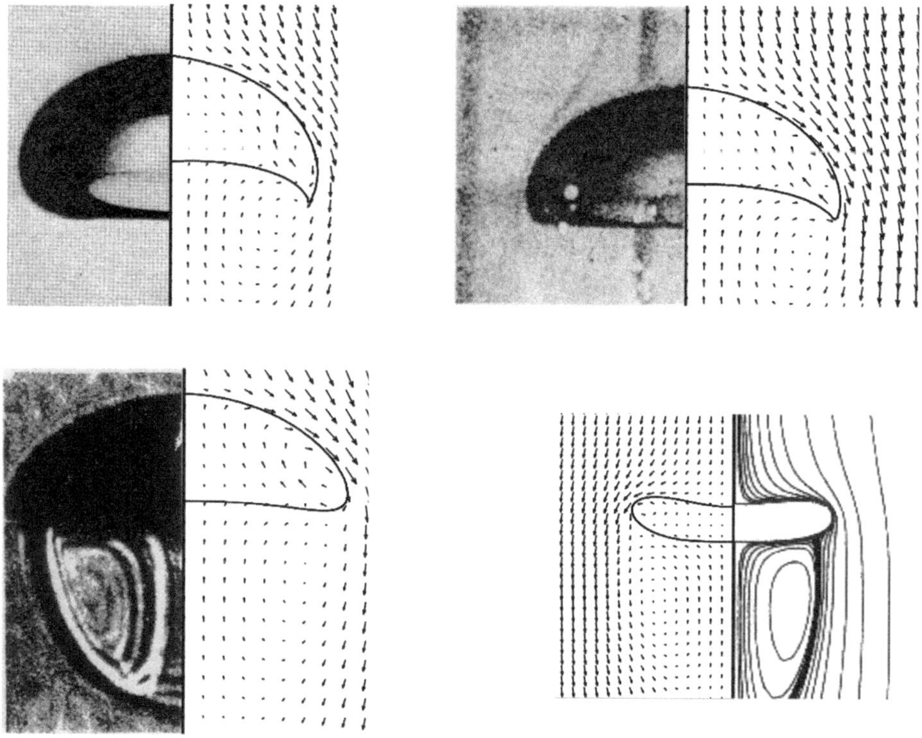

FIGURE 3. Comparison of numerical results with experimental results and previous computational results. Upper left: Bhaga & Weber (figure 2, bubble (d)). Upper right: Bhaga & Weber (figure 3, bubble (d)). Lower left: Hnat & Buckmaster. Lower right: Ryskin & Leal.

height fraction technique for finding curvature, we also verified further the robustness of using the second-order height fraction technique together with a 2nd-order Navier-Stokes solver for problems such as the pinch-off of a liquid jet and the steady rise of bubbles.

Acknowledgments

We thank Mr. D. Kikuchi and Mr. S. Yamaguchi for their help in preparing this manuscript.

References

[1] I. Aleinov and E.G. Puckett. Computing surface tension with high-order kernels. In *Proceedings of the 6th International Symposium on Computational Fluid Dynamics*, Lake Tahoe, CA, 1995.

[2] D. Bhaga and M.E. Weber. Bubbles in viscous liquids: Shapes, wakes and velocities. *J. Fluid Mech.*, 105:61–85, 1981.

[3] J.U. Brackbill, D.B. Kothe, and C. Zemach. A continuum method for modeling surface tension. *J. Comput. Phys.*, 100:335–353, 1992.

[4] A.J. Chorin. Curvature and solidification. *J. Comput. Phys.*, 57:472–490, 1985.

[5] S. Cummins, M. Francois, and D. Kothe. Estimating curvature from volume fractions. *Computers and Structures*, 83:425–434, 2005.

[6] D. Enright, S. Marschner, and R. Fedkiw. Animation and rendering of complex water surfaces. In *SIGGRAPH 2002*, volume ACM TOG 21, pages 736–744, 2002.

[7] M. Francois, S. Cummins, E. Dendy, D. Kothe, J. Sicilian, and M. Williams. A balanced-force algorithm for continuous and sharp interfacial surface tension models within a volume tracking framework. *J. Comput. Phys.*, 2005. in press.

[8] D. Gueyffier, J. Li, A. Nadim, R. Scardovelli, and S. Zaleski. Volume of fluid interface tracking with smoothed surface stress methods for three-dimensional flows. *J. Comput. Phys.*, 152:423–456, 1999.

[9] J. Helmsen, P. Colella, and E.G. Puckett. Non-convex profile evolution in two dimensions using volume of fluids. LBNL technical report LBNL-40693, Lawrence Berkeley National Laboratory, 1997.

[10] J.G. Hnat and J.D. Buckmaster. Spherical cap bubbles and skirt formation. *Physics of Fluids*, 19 (2):182–194, 1976.

[11] E. Jimenez, M. Sussman, and M. Ohta. A computational study of bubble motion in newtonian and viscoelastic fluids. *Fluid Dynamics and Materials Processing*, 2005. (to appear).

[12] M. Kang, R. Fedkiw, and X.-D. Liu. A boundary condition capturing method for multiphase incompressible flow. *J. Sci. Comput.*, 15:323–360, 2000.

[13] H. Liu, S. Krishnan, S. Marella, and H.S. Udaykumar. Sharp interface cartesian grid method ii: A technique for simulating droplet interactions with surfaces of arbitrary shape. *J. Computational Physics*, 210(1):32–54, 2005.

[14] M.R. Nobari, Y.J. Jan, and G. Tryggvason. Head on collision of drops; a numerical investigation. Technical Report ICOMP-93-45, NASA ICOMP Lewis Research Center, November 1993.

[15] J.Y. Poo and N. Ashgriz. A computational method for determining curvatures. *J. Comput. Phys.*, 84:483–491, 1989.

[16] S. Popinet and S. Zaleski. A front-tracking algorithm for accurate representation of surface tension. *International Journal for Numerical Methods in Fluids*, 30(6):775–793, 1999.

[17] Y. Renardy and M. Renardy. Prost: A parabolic reconstruction of surface tension for the volume-of-fluid method. *J. Comput. Phys.*, 183(2):400–421, 2002.

[18] M. Rudman. A volume tracking method for interfacial flows with large density variations. *Int. J. Numer. Methods Fluids*, 28:357–378, 1998.

[19] G. Ryskin and L.G. Leal. Numerical solution of free boundary problems in fluid mechanics. part 2 buoyancy-driven motion of a gas bubble through a quiescent liquid. *J. Fluid Mech.*, 148:19–35, 1984.

[20] M. Sussman. A second-order coupled levelset and volume of fluid method for computing growth and collapse of vapor bubbles. *Journal of Computational Physics*, 187:110–136, 2003.

[21] M. Sussman and M.Y. Hussaini. A discontinuous spectral element method for the level set equation. *J. Scientific Computing*, 19:479–500, 2003.

[22] M. Sussman, M.Y. Hussaini, K. Smith, R. Zhi-Wei, and V. Mihalef. A second-order adaptive sharp interface method for incompressible multiphase flow. In *Proceedings of the 3rd international conference on Computational Fluid Dynamics*, Toronto, Canada, 2004. to appear.

[23] M. Sussman and E.G. Puckett. A coupled level set and volume of fluid method for computing 3d and axisymmetric incompressible two-phase flows. *J. Comp. Phys.*, 162:301–337, 2000.

[24] M. Sussman, P. Smereka, and S.J. Osher. A level set approach for computing solutions to incompressible two-phase flow. *J. Comput. Phys.*, 114:146–159, 1994.

[25] H.S. Udaykumar, M.M. Rao, and W. Shyy. Elafint – a mixed Eulerian-Lagrangian method for fluid flows with complex and moving boundaries. *Int. J. Numer. Meths. Fluids.*, 22(8):691–704, 1996.

[26] S.O. Unverdi and G. Tryggvason. A front-tracking method for viscous, incompressible, multi-fluid flows. *J. Comput. Phys.*, 100:25–37, 1992.

[27] M. Williams, D. Kothe, and E.G. Puckett. Convergence and accuracy of kernel-based continuum surface tension models. In *Proceedings of the Thirteenth U.S. National Congress of Applied Mechanics*, Gainesville, FL, June 16-21 1998.

[28] T. Ye, W. Shyy, and J.N. Chung. A fixed-grid, sharp interface method for bubble dynamics and phase change. *J. Comp. Phys.*, 174:781–815, 2001.

M. Sussman
Department of Mathematics
Florida State University
Tallahassee, FL 32306, USA
e-mail: `sussman@math.fsu.edu`

M. Ohta
Department of Applied Chemistry
Muroran Institute of Technology
27-1, Mizumotocho, Muroran-shi
Hokkaido 050-8585, Japan
e-mail: `mohta@mmm.muroran-it.ac.jp`

International Series of Numerical Mathematics, Vol. 154, 435–444

Simulation of a Model of Tumors with Virus-therapy

Youshan Tao and Qian Guo

Abstract. We consider a procedure for cancer therapy which consists of injecting replication-competent viruses into the tumor. The viruses infect tumor cells, replicate inside them, and eventually cause their death. As infected cells die, the viruses inside them are released and then proceed to infect adjacent tumor cells. However, a major factor influencing the efficacy of virus agents is the immune response that may limit the replication and spread of the replication-competent virus. The immune response is cytokine-mediated. The expression of viruses in tumor cells sensitize cells to lysis by the TNF (tumor necrosis factor) cytokine. The competition between tumor cells, a replication-competent virus and an immune response is modelled as a free boundary problem for a nonlinear system of partial differential equations, where the free boundary is the surface of the tumor. In this model, the immune response equation is a non-standard parabolic equation due to the *chemotaxis* (spatial gradients of diffusible chemicals) of the immune response. The purpose of this paper is to give the numerical methods for solving this kind of free boundary problems. Several simulation results are also given.

Mathematics Subject Classification (2000). 35R35; 92A15.

Keywords. Tumors, replication-competent virus, immune response, chemotaxis, free boundary problems.

1. The model

In the past three decades, the gene therapy approaches for cancer treatment have been studied and the corresponding mathematical models of tumor growth have been developed (for example, see [1–6, 10, 12–14]). These models have generated valuable insights into cancer treatment. However, one of the obstacles in developing efficient gene therapy to cancer is in the delivery process. The macromolecules used as gene delivery carriers are too large to be transported into, and diffuse within, the tumor (see Jain [7]). Recently, replication-competent viruses have been proposed as an approach to bypass the delivery problem. The virus is engineered to selectively bind to receptors on the tumor cell surface (but not to the surface of normal healthy cells). The virus particles then gain entry by endocytosis and proceed to proliferate exponentially within the tumor cell, eventually causing death (lysis). Thereupon the newly reproduced virus particles are released and then proceed to infect adjacent cancer cells.

A mathematical model which describes the evolution of tumors under viral injection was recently developed by Wu, Byrne, Kirn and Wein [16]. They computed and compared the evolution of the tumor under different initial conditions using the *simplified* version of their model. Friedman and Tao [5] presented a somewhat different model and made a rigorous mathematical analysis on their model. The PDE for the virus is the main difference between the FT model and the WBKW model. The WBKW model does not include diffusion term and their mathematical system is not a well posed problem.

However, a major factor influencing the efficacy of virus agents is the immune response. New clinical data ([8], [9]) revealed an innate immune response to virus that may mitigate the effects of treatment. The immune response is cytokine-mediated. The expression of virus gene in tumor cells sensitize cells to lysis by the TNF (tumor necrosis factor) cytokine (see [11] and references therein). The binding of TNF to death receptors on the tumor cell surface induces apoptosis. TNF preferentially induces apoptosis of viral-infected tumor cells, whereas the uninfected tumor cells are generally resistant to TNF-induced killing.

In order to improve the efficacy of oncolytic viruses, we therefore need to better understand the dynamics between tumor growth, a replication-competent virus and an immune response. Recently, Wein, Wu and Kirn [15] incorporated an immune response into their earlier model [16]. In [17] the authors made some analysis on the WWK model [15] using ODEs which are *simplified* approximation to their PDE model, whereas Tao and Guo [11] made a rigorous mathematical analysis on the PDE model. Furthermore, TG model [11] includes a diffusion term in the immune response equation for the consideration of mathematical well-posedness.

Clinical results suggest that the immune response is cytokine-mediated (see Section 1 in [11]). The immune response migrates towards cytokines. The movement of the immune response is sometimes an active process induced by chemoattractants. Therefore, the immune response equation should include not only the *random diffusion* but also the *chemotaxis* as done in [10].

The physical variables are assumed as follows:

$\hat{x}$ = density of uninfected tumor cells,
$\hat{y}$ = density of infected tumor cells,
$\hat{n}$ = density of necrotic cells,
$\hat{v}$ = density of free virus, i.e., virus in the extracellular tissue,
$\hat{z}$ = density of the immune response, and
u = the velocity field within the tumor.

The tumor volume is modelled as an incompressible fluid, through which the cells travel via a convective field whose velocity is u. The velocity field is a result of the spatio-temporal variation due to the proliferation of uninfected cells and the removal of necrotic cells. We assume that the problem is radially symmetric, so that all the unknown functions depend only on (r, t) where r is the distance from the center of the tumor. The model were derived by applying the principle of mass conservation to each of the variables. The model consists of the following system of equations:

$$\begin{aligned}\frac{D\hat{x}}{Dt} &\equiv \frac{\partial \hat{x}(r,t)}{\partial t} + \frac{1}{r^2}\frac{\partial}{\partial r}\Big(r^2 u(r,t)\hat{x}(r,t)\Big) \\ &= \lambda \hat{x}(r,t) - \beta \hat{x}(r,t)\hat{v}(r,t),\end{aligned} \tag{1.1}$$

$$\begin{aligned}\frac{D\hat{y}}{Dt} &\equiv \frac{\partial \hat{y}(r,t)}{\partial t} + \frac{1}{r^2}\frac{\partial}{\partial r}\Big(r^2 u(r,t)\hat{y}(r,t)\Big) \\ &= \beta \hat{x}(r,t)\hat{v}(r,t) - \delta \hat{y}(r,t) - k\hat{y}(r,t)\hat{z}(r,t),\end{aligned} \tag{1.2}$$

$$\begin{aligned}\frac{D\hat{n}}{Dt} &\equiv \frac{\partial \hat{n}(r,t)}{\partial t} + \frac{1}{r^2}\frac{\partial}{\partial r}\Big(r^2 u(r,t)\hat{n}(r,t)\Big) \\ &= \delta \hat{y}(r,t) + k\hat{y}(r,t)\hat{z}(r,t) - \mu \hat{n}(r,t),\end{aligned} \tag{1.3}$$

$$\frac{\partial \hat{v}(r,t)}{\partial t} = N\delta \hat{y}(r,t) - \gamma \hat{v}(r,t) + D_1 \frac{1}{r^2}\frac{\partial}{\partial r}\Big(r^2 \frac{\partial \hat{v}(r,t)}{\partial r}\Big), \quad \frac{\partial \hat{v}(0,t)}{\partial r} = 0, \tag{1.4}$$

$$\begin{aligned}\frac{\partial \hat{z}(r,t)}{\partial t} &= l\hat{z}(r,t)\hat{y}(r,t) - \omega[\hat{z}(r,t)]^2 + D_2 \frac{1}{r^2}\frac{\partial}{\partial r}\Big(r^2 \frac{\partial \hat{z}(r,t)}{\partial r}\Big) \\ &\quad - \frac{\chi}{r^2}\frac{\partial}{\partial r}\Big(r^2 \hat{z}(r,t)\frac{\partial \hat{y}(r,t)}{\partial r}\Big), \quad D_2 \frac{\partial \hat{z}(0,t)}{\partial r} - \chi \frac{\partial \hat{y}(0,t)}{\partial r} = 0.\end{aligned} \tag{1.5}$$

In (1.1), λ is the proliferation rate of the uninfected cancer cells and β is the infection rate of the uninfected cells; in (1.2), δ is the death rate of the infected cells by lysis and k is the killing rate of the infected cells killed by the immune response; in (1.3), μ is the removal rate of the necrotic cells; in (1.4), γ is the removal (or clearance) rate of virus ($1/\gamma$ is the mean lifetime of free virus), $N\delta$ is the virus release rate (N is the burst size of virus at the death of a cell) and D_1 is the diffusion coefficient of virus; in (1.5), l is the productive rate of the immune response, ω is the second-order clearance rate of the immune response, D_2 is the diffusion coefficient of the immune response and χ is the chemotactic coefficient; the last equation in (1.4) or (1.5) is a consequence of the radial symmetry. The readers may refer to [5, 11, 15–17] for more detailed biological explanations of the equations (1.1)–(1.5) (except the chemotaxis term in (1.5)).

We finally assume that all cells have the same size and density, and that they are uniformly distributed in the tumor (see [13]), so that

$$\hat{x} + \hat{y} + \hat{n} \equiv \text{const.} \equiv \theta. \tag{1.6}$$

Summing equations (1.1)–(1.3) and invoking the assumption (1.6) we get

$$\frac{\theta}{r^2}\frac{\partial}{\partial r}\Big(r^2 u(r,t)\Big) = \lambda \hat{x}(r,t) - \mu \hat{n}(r,t). \tag{1.7}$$

The boundary conditions, at the moving boundary, are

$$\frac{\partial}{\partial r}\hat{v}(R(t),t) = D_2\frac{\partial}{\partial r}\hat{z}(R(t),t) - \chi\hat{z}(R(t),t)\frac{\partial}{\partial r}\hat{y}(R(t),t) = 0, \tag{1.8}$$

$$\frac{dR(t)}{dt} = u(R(t),t). \tag{1.9}$$

(1.8) means that virus particles and the immune response do not cross the boundary, for $t > 0$, and (1.9) is the equation of continuity: the velocity of the free surface is the same as the velocity u at the surface.

We note that Equation (1.3) is a consequence of Equations (1.1), (1.2), (1.6) and (1.7), so that in the sequel we may drop this equation and replace $\hat{n}$ by $\theta - \hat{x} - \hat{y}$ in (1.7).

We also note that since the velocity field is radially symmetric,

$$u(0,t) = 0. \tag{1.10}$$

To complete the model, we impose the following initial conditions

$$\begin{array}{l} R(0) \quad \text{is prescribed,} \\ \hat{x}(r,0) = \hat{x}_0(r), \hat{y}(r,0) = \hat{y}_0(r), \hat{v}(r,0) = \hat{v}_0(r), \hat{z}(r,0) = \hat{z}_0(r) \\ \text{where } \hat{z}_0(r) > 0 \text{ and } \hat{x}_0(r), \hat{y}_0(r), \hat{v}_0(r) \text{ are nonnegative} \\ \text{functions with } \hat{x}_0(r) + \hat{y}_0(r) \le \theta, \text{ for } 0 \le r \le R(0). \end{array} \tag{1.11}$$

This paper studies numerically the model (1.1)–(1.11). We shall study analytically in a subsequent paper the global existence of the solutions of the model (1.1)–(1.11).

Remark 1.1. The model (1.1)–(1.11) was originally developed by Wein, Wu and Kirn (see [15], [17]). However, in the WWK model there does not appear a diffusion term in the virus equation (1.4) and there do not appear a diffusion term and a chemotaxis term for the immune response equation (1.5). Without a diffusion term for virus density or for immune response density, the WWK model is, mathematically, not a well posed problem, as explained in Section 3 of [5]. As said before, the immune response migrates towards cytokines. The movement of the immune response is sometimes an active process induced by chemoattractants. So we incorporate a chemotaxis term into the immune response equation (1.5). Mathematically, with a chemotaxis term, the equation (1.5) is *non-standard* parabolic equation. Biologically, the chemotaxis may have effects on the tumor growth and on the spatio-temporal distributions of tumor cells, viruses and the immune response.

The structure of this paper is as follows. First, in Section 2 we transform the problem (1.1)–(1.11) into a problem in a fixed region and we introduce a variable change to simplify the difficulty which is due to the chemotaxis term. In Section 3 and 4, we study the model using numerical methods. The numerical methods are given in Section 3, whereas the simulation results are presented in Section 4.

2. Transformation

We introduce the variables

$$\tilde{x} = \frac{\hat{x}}{\theta}, \quad \tilde{y} = \frac{\hat{y}}{\theta}, \quad \tilde{v} = \frac{\hat{v}}{\theta N}, \quad \tilde{z} = \hat{z}, \quad \tilde{u} = u$$

and the quantity

$$p_0 = \frac{\beta N \theta}{\gamma}.$$

The parameter p_0 is called the *basic reproductive ratio* in the epidemic modeling. It represents the mean number of virus particles released by one virus.

It will be convenient to transform the region $\{0 \le r \le R(t)\}$ into the fixed region $\{0 \le \rho \le 1\}$ by

$$\rho = \rho(r,t) = \frac{r}{R(t)}.$$

We further introduce the variable change:

$$\begin{aligned} x(\rho,t) &= \tilde{x}(r,t), \quad y(\rho,t) = \tilde{y}(r,t), \quad z(\rho,t) = \tilde{z}(r,t) e^{-\frac{\chi}{D_2}\tilde{y}(r,t)}, \\ v(\rho,t) &= \tilde{v}(r,t), \quad u(\rho,t) = \tilde{u}(r,t)/R(t). \end{aligned}$$

In terms of the new variable, the system (1.1)–(1.11) takes the following form:

$$\begin{aligned} &\frac{\partial x}{\partial t} + \big[u(\rho,t) - \rho u(1,t)\big]\frac{\partial x}{\partial \rho} \\ &= \lambda x - p_0 \gamma x v - \big[-\mu + (\lambda+\mu)x + \mu y\big]x, \end{aligned} \tag{2.1}$$

$$\begin{aligned} &\frac{\partial y}{\partial t} + \big[u(\rho,t) - \rho u(1,t)\big]\frac{\partial y}{\partial \rho} \\ &= p_0 \gamma x v - \delta y - k y z e^{\frac{\chi}{D_2} y} - \big[-\mu + (\lambda+\mu)x + \mu y\big]y, \end{aligned} \tag{2.2}$$

$$\frac{\partial v}{\partial t} - \frac{D_1}{R^2(t)}\frac{1}{\rho^2}\frac{\partial}{\partial \rho}\Big(\rho^2 \frac{\partial v}{\partial \rho}\Big) - \rho u(1,t)\frac{\partial v}{\partial \rho} = \delta y - \gamma v, \quad v_\rho(0,t) = 0, \tag{2.3}$$

$$\begin{aligned} &\frac{\partial z}{\partial t} - \frac{D_2}{R^2(t)}\frac{1}{\rho^2}\frac{\partial}{\partial \rho}\Big(\rho^2 \frac{\partial z}{\partial \rho}\Big) - \Big(\frac{\chi}{R^2(t)}\frac{\partial y}{\partial \rho} + \rho u(1,t)\Big)\frac{\partial z}{\partial \rho} \\ &= \Big[-\frac{\chi}{D_2}\Big(\frac{\partial y}{\partial t} - \rho u(1,t)\frac{\partial y}{\partial \rho}\Big) + l\theta y\Big]z - \omega e^{\frac{\chi}{D_2} y} z^2, \quad z_\rho(0,t) = 0, \end{aligned} \tag{2.4}$$

$$u(\rho,t) = \frac{1}{\rho^2}\int_0^\rho s^2\big[-\mu + (\lambda+\mu)x(s,t) + \mu y(s,t)\big]ds \tag{2.5}$$

in $\{0 < \rho < 1, t > 0\}$,

$$\dot{R}(t) = R(t)u(1,t), \quad R(0) \text{ is given}, \tag{2.6}$$

$$v_\rho(1,t) = z_\rho(1,t) = 0, \tag{2.7}$$

$$x(\rho,0) = x_0(\rho), y(\rho,0) = y_0(\rho), v(\rho,0) = v_0(\rho), z(\rho,0) = z_0(\rho), \tag{2.8}$$

$$x_0(\rho) \ge 0, y_0(\rho) \ge 0, \quad x_0(\rho) + y_0(\rho) \le 1, \quad z_0(\rho) > 0. \tag{2.9}$$

We shall also assume that

$$x_0(\rho), y_0(\rho), v_0(\rho) \text{ and } z_0(\rho) \text{ belong to } C^1[0,1], \text{ and } \frac{\partial v_0}{\partial \rho}(1) = \frac{\partial z_0}{\partial \rho}(1) = 0. \tag{2.10}$$

In this paper we focus on numerically solving the system (2.1)–(2.10). The numerical methods are given in Section 3 and the simulation results are presented in Section 4. The global existence of the solutions to the system (2.1)–(2.10) will be studied in a forthcoming paper.

3. Numerical methods

We solve the system (2.1)–(2.10) by finite difference methods (for simplicity, we take $\delta y = \gamma v$). We give a spatial mesh, with mesh points $s_i = ih, h = \frac{1}{N_0}$ for $0 \le i \le N_0$, which partition the interval $[0,1]$. And we set time step $\tau, t_m = m\tau, m = 0, 1, 2, \dots$.

We denote

$$R^{(m)} \approx R(t_m), \qquad u^{(i,m)} \approx u(s_i, t_m), \qquad v^{(i,m)} \approx v(s_i, t_m),$$
$$x^{(i,m)} \approx x(s_i, t_m), \qquad y^{(i,m)} \approx y(s_i, t_m), \qquad z^{(i,m)} \approx z(s_i, t_m).$$

By the composite trapezoid rule, we discretize (2.5)

$$\begin{aligned} u^{(i,m)} &= -\frac{\mu s_i}{3} + \frac{(\lambda+\mu)}{s_i^2} h \sum_{j=0}^{i-1} \frac{1}{2}\left(s_j^2 x^{(j,\, m-1)} + s_{j+1}^2 x^{(j+1,\, m-1)}\right) \\ &\quad + \frac{\mu}{s_i^2} h \sum_{j=0}^{i-1} \frac{1}{2}\left(s_j^2 y^{(j,\, m-1)} + s_{j+1}^2 y^{(j+1,\, m-1)}\right). \end{aligned} \tag{3.1}$$

and, $u^{(0,m)} = 0$.

By the backward Euler method, we discretize (2.6)

$$R^{(m)} = \frac{R^{(m-1)}}{1 - \tau * u^{(N_0,m)}}. \tag{3.2}$$

Using the implicit centered difference formula, we discretize equations (2.1), (2.2) and (2.4) with initial value conditions (2.17)

$$\begin{aligned} &x^{(i,m)} - x^{(i,m-1)} + \frac{\tau}{h}\left[u^{(i,m)} - s_i \times u^{(N_0,m)}\right]\left(x^{(i,m)} - x^{(i-1,m)}\right) \\ &-\tau\left\{\lambda - p_0\gamma v^{(i,m-1)} - \left[-\mu + (\lambda+\mu)x^{(i,m)} + \mu y^{(i,m)}\right]\right\}x^{(i,m)} = 0 \end{aligned} \tag{3.3}$$

where $i = 1, \dots, N_0 - 1$;

$$x^{(i,m)} - x^{(i,m-1)} - \tau\left\{\lambda - p_0\gamma v^{(i,m-1)} - \left[-\mu + (\lambda+\mu)x^{(i,m)} + \mu y^{(i,m)}\right]\right\}x^{(i,m)} = 0$$

where $i = 0, N_0$ and we have used the fact that $[u(\rho,t) - \rho u(1,t)]|_{\rho=0,1} = 0$;

$$y^{(i,m)} - y^{(i,m-1)} + \frac{\tau}{h}\big[u^{(i,m)} - s_i \times u^{(N_0,m)}\big]\big(y^{(i,m)} - y^{(i-1,m)}\big) \qquad (3.4)$$
$$-\tau\{p_0\gamma v^{(i,m-1)}x^{(i,m)} - \delta y^{(i,m)} - k y^{(i,m)} z^{(i,m-1)} e^{\frac{\chi}{D_2} y^{(i,m)}}$$
$$-[-\mu + (\lambda+\mu)x^{(i,m)} + \mu y^{(i,m)}]y^{(i,m)}\} = 0$$

where $i = 1, \ldots, N_0 - 1$;

$$y^{(i,m)} - y^{(i,m-1)} - \tau\{p_0\gamma v^{(i,m-1)}x^{(i,m)} - \delta y^{(i,m)} - k y^{(i,m)} z^{(i,m-1)} e^{\frac{\chi}{D_2} y^{(i,m)}}$$
$$-[-\mu + (\lambda+\mu)x^{(i,m)} + \mu y^{(i,m)}]y^{(i,m)}\} = 0$$

where $i = 0, N_0$ and we have used the fact that $[u(\rho,t) - \rho u(1,t)]|_{\rho=0,1} = 0$;

$$z^{(i,m)} - z^{(i,m-1)} + \tau\Big\{ - \frac{D_2}{R^{(m)2}}\Big[\frac{2}{s_i}\frac{z^{(i,m)} - z^{(i-1,m)}}{h} \qquad (3.5)$$
$$+ \frac{z^{(i+1,m)} - 2\, z^{(i,m)} + z^{(i-1,m)}}{h^2}\Big]$$
$$-\Big[\frac{\chi}{R^{(m)2}}\frac{y^{(i,m)} - y^{(i-1,m)}}{h} + s_i\, u^{(N_0,m)}\Big]\frac{z^{(i,m)} - z^{(i-1,m)}}{h}$$
$$-\Big[l\theta\, y^{(i,m)} - \frac{\chi}{D_2}\Big(\frac{y^{(i,m)} - y^{(i,m-1)}}{\tau} - s_i u^{(N_0,m)}\frac{y^{(i,m)} - y^{(i-1,m)}}{h}\Big)\Big]z^{(i,m)}$$
$$+\omega e^{\frac{\chi}{D_2} y^{(i,m)}} z^{(i,m)^2}\Big\} = 0$$

where $i = 1, \ldots, N_0 - 1$;

$$z^{(i,m)} - z^{(i,m-1)} + \tau\Big\{ - \frac{D_2}{R^{(m)2}}\Big[\frac{z^{(i+1,m)} - 2\, z^{(i,m)} + z^{(i+1,m)}}{h^2}\Big]$$
$$-\Big[l\theta\, y^{(i,m)} - \frac{\chi}{D_2}\frac{y^{(i,m)} - y^{(i,m-1)}}{\tau}\Big]z^{(i,m)} + \omega e^{\frac{\chi}{D_2} y^{(i,m)}} z^{(i,m)^2}\Big\} = 0$$

where $i = 0$,

$$z^{(i,m)} - z^{(i,m-1)} + \tau\Big\{ - \frac{D_2}{R^{(m)2}}\Big[\frac{z^{(i-1,m)} - 2\, z^{(i,m)} + z^{(i-1,m)}}{h^2}\Big]$$
$$-\Big[l\theta\, y^{(i,m)} - \frac{\chi}{D_2}\frac{y^{(i,m)} - y^{(i,m-1)}}{\tau}\Big]z^{(i,m)} + \omega e^{\frac{\chi}{D_2} y^{(i,m)}} z^{(i,m)^2}\Big\} = 0$$

where $i = N_0$.

Here, we have used the boundary conditions $\frac{\partial y}{\partial \rho}(1,t) = \frac{\partial z}{\partial \rho}(0,t) = \frac{\partial z}{\partial \rho}(1,t) = 0$.

We propose the following iteration algorithm:

(1) Initialization: Given corresponding parameters and $m_{\max}$, set $m = 1$, $x^{(i,0)} = x_0(i)$, $y^{(i,0)} = y_0(i)$, $i = 0, 1, \ldots, N$, $R^{(0)} = 1$;
(2) Solve (3.1) for $u^{(i,m)}$, $i = 0, 1, \ldots, N$;
(3) Solve (3.2) for $R^{(m)}$;

(4) Solve (3.3) and (3.4) for $x^{(i,m)}$, $y^{(i,m)}$, $i = 0, 1, \ldots, N$;
(5) Evaluate (3.5) at $\rho = s_i$, $t = t_m$ for $z^{(i,m)}$, $i = 0, 1, \ldots, N$;
(6) $v^{(i,m)} = \frac{\delta\ y^{(i,m)}}{\gamma}$, $i = 0, 1, \ldots, N$;
(7) STOP if $m > m_{\max}$; otherwise set $m := m + 1$, GOTO (2).

4. Simulation results

The typical parameter values for our numerical work are $\lambda = 3.2 \times 10^{-4}$; $\mu = 1/72$; $\omega = 1600$; $\delta = 1/48$; $p_0 = 3.73$; $\kappa = 15.3$; $\gamma = 1$; $\theta = 1e6$; $l = 0.0016$; $D_2 = 0.0005$; $xs = 9.7531 \times 10^{-1}$; $ys = 3.8298 \times 10^{-3}$; $zs = 3.0 \times 10^{-1}$; $x_0 = xs + xs * (\rho^3/3 - \rho^2/2)/10$; $y_0 = ys + ys * (\rho^3/3 - \rho^2/2)/50$; $z_0 = zs + zs * (\rho^3/3 - \rho^2/2) * 2$, (see [13]).

Figure 1 shows that the tumor growth is monotonous increasing in chemotaxis coefficient χ if $t \geq 20$ (hours) and the diffusivity D_2 is small. However, the tumor growth is insensitive to the chemotaxis coefficient χ if the diffusivity D_2 is large.

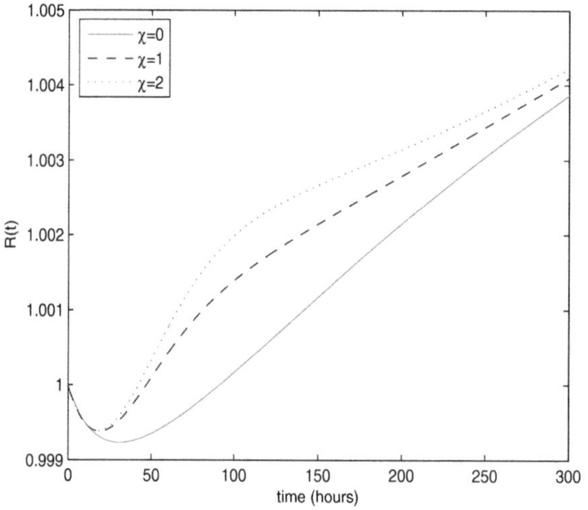

FIGURE 1. The effect of the chemotaxis coefficient χ on tumor growth.

Figure 2 shows that the effects of the chemotaxis coefficient χ on the spatial distribution of the infected cell density $y(\rho, t)$, which varies with time.

The stronger the chemotaxis is, the more the immune response migrates towards the infected cells, and therefore more virus may be killed. Overall, the strong chemotaxis may mitigate the effects of treatment as shown in Figure 1.

At the early stage of tumor growth, the density of the infected cells is increasing in the spatial direction (see the spatial distribution of the density $y(\rho, t)$ of the

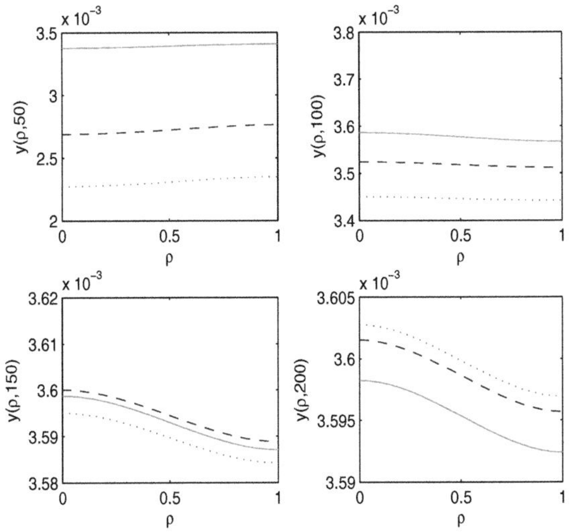

FIGURE 2. The spatial distributions of the density $y(\rho, t)$ of the infected cells at times $t = 50, 100, 150, 200$.

infected cells at time $t = 50$ in Figure 2). However, the infected cells can secrete chemoattractants, which can induce the movement of the immune response (which in turn kills the infected cells). The spatial gradient of the infected cells directs this movement. Hence, as time t increasing, the above monotony of $y(\rho, t)$ in the spatial direction changes. The density of the infected cells may become decreasing in the spatial direction (see the spatial distributions of the density $y(\rho, t)$ of the infected cells at time $t = 100, 150, 200$ in Figure 2). Our simulation also shows that the density of the infected cells depends on the chemotactic coefficient χ. However, this dependence is *not* monotonous (see Figure 2), which may indicate the complexity of the competition between tumor cells, virus and an immune response.

Acknowledgement

The authors are supported by NSFC 10571023. The first author is also supported by Shanghai Educational Development Foundation.

References

[1] H.M. Byrne and M.A.J. Chaplain, *Growth of nonnecrotic tumors in the presence and absence of inhibitors*, Mathematical Biosciences, 181 (1995), 130–151.

[2] H. Byrne and M. Chaplain, *Growth of necrotic tumors in the presence and absence of inhibitors*, Math. Biosci. 135 (1996), 187–216.

[3] M.A.J. Chaplain, V.A. Kuznetsov, Z.H. James and L.A. Stepanova, *Spatio-temporal dynamics of the immune system response to cancer.* Proceedings of the mathematical

Models in Medical and Health Sciences Conference (eds. M.A. Horn, G. Simonett, G. Webb), Vanderbilt University Press, 1998, ISBN 0-8265-1310-7.

[4] A. Friedman and F. Reitich, *Analysis of a mathematical model for the growth of tumors*, J. Math. Biol. 38 (1999), 262–284.

[5] A. Friedman and Y. Tao, *Analysis of a model of a virus that selectively in tumor cells*, J. Math. Biol. 47 (2003), 391–423.

[6] T.L. Jackson and H.M. Byrne, *A mathematical model to study the effects of drug resistance and vasculature on the response of solid tumors to chemotherapy.* Math. Biosci. 164 (2000), 17–38.

[7] R. Jain, *Barriers to drug delivery in solid tumors,* Sci. Am. 271 (1994), 58–65.

[8] J. Nemunaitis, I. Ganly, F. Khuri, J. Arseneau, J. Kuhn, T. McCarty, S. Landers, P. Maples, L. Romel, B. Randley, T. Reid, S. Kaye and D. Kirn, *Selective replication and oncolysis in p53 mutant tumors with ONYX-015, an E1B-55kD gene-deleted adenovirus, in patients with advanced head and neck cancer: a phase II trial*, Cancer Res. 60 (2000), 6359–6366.

[9] J. Nemunaitis, C. Cunningham, A. Buchanan, A. Blackburn, G. Edelman, P. Maples, G. Netto, A. Tong, B. Randley, S. Olson and D. Kirn, *Intravenous infusion of a replication-selective adenovirus (ONYX-015) in cancer patients: safety, feasibility and biological activity*, Gene Therapy 8 (2001), 746–759.

[10] M. Owen and J.A. Sherratt, *Pattern formation and spatio-temporal irregularity in a model for macrophage-tumor interactions*, J. Theor. Biol. 189 (1997), 63–80.

[11] Y. Tao and Q. Guo, *The competitive dynamics between tumor cells, a replication-competent virus and an immune response*, J. Math. Biol. 51 (2005), 37–74.

[12] Y. Tao, N. Yoshida and Q. Guo, *Nonlinear analysis of a model of vascular tumor growth and treatment*, Nonlinearity 17 (2004), 867–895.

[13] J. P. Ward and J. R. King, *Mathematical modelling of avascular-tumor growth*, IMA J. Math. Appl. Med. Biol. 14 (1997), 39–69.

[14] J.P. Ward and J.R. King, *Mathematical modelling of drug transport in tumor multi-cell spheroids and monolayer cultures*, Math. Biosci. 181 (2003), 177–207.

[15] L.M. Wein, J.T. Wu and D.H. Kirn, *Validation and analysis of a mathematical model of a replication-competent oncolytic virus for cancer treatment: implications for virus design and delivery*, Cancer Res. 63 (2003), 1317–1324.

[16] J.T. Wu, H.M. Byrne, D.H. Kirn and L.M. Wein, *Modeling and analysis of a virus that replicates selectively in tumor cells,* Bull. Math. Biol., 63 (2001), 731–768.

[17] J.T. Wu, D.H. Kirn and L.M. Wein, *Analysis of a three-way race between tumor growth, a replication-competent virus and an immune response,* Bull. Math. Biol. 66 (2004), 605–625.

Youshan Tao and Qian Guo
Department of Applied Mathematics
Dong Hua University,
Shanghai 200051, P.R. China
e-mail: `taoys@dhu.edu.cn`
e-mail: `guoqian@dhu.edu.cn`

International Series of Numerical Mathematics, Vol. 154, 445–455

Asymptotic Behavior of a Hyperbolic-parabolic Coupled System Arising in Fluid-structure Interaction

Xu Zhang and Enrique Zuazua

Abstract. In this paper we summarize some recent results on the asymptotic behavior of a linearized model arising in fluid-structure interaction, where a wave and a heat equation evolve in two bounded domains, with natural transmission conditions at the interface. These conditions couple, in particular, the heat unknown with the velocity of the wave solution. First, we show the strong asymptotic stability of solutions. Next, based on the construction of ray-like solutions by means of Geometric Optics expansions and a careful analysis of the transfer of the energy at the interface, we show the lack of uniform decay of solutions in general domains. Finally, we obtain a polynomial decay result for smooth solutions under a suitable geometric assumption guaranteeing that the heat domain envelopes the wave one. The system under consideration may be viewed as an approximate model for the motion of an elastic body immersed in a fluid, which, in its most rigorous modeling should be a nonlinear free boundary problem, with the free boundary being the moving interface between the fluid and the elastic body.

Mathematics Subject Classification (2000). Primary: 37L15; Secondary: 35B37, 74H40, 93B07.

Keywords. Fluid-structure interaction, Strong asymptotic stability, Non-uniform decay, Gaussian Beams, Polynomial decay, Weakened observability inequality.

1. Introduction

Let $\Omega \subset \mathbb{R}^n$ ($n \in \mathbb{N}$) be a bounded domain with C^2 boundary $\Gamma = \partial\Omega$. Let Ω_1 be a sub-domain of Ω and set $\Omega_2 = \Omega \setminus \overline{\Omega}_1$. We denote by γ the interface, $\Gamma_j = \partial\Omega_j \setminus \overline{\gamma}$

The work is supported by the Grant MTM2005-00714 of the Spanish MEC, the DOMINO Project CIT-370200-2005-10 in the PROFIT program of the MEC (Spain), the SIMUMAT projet of the CAM (Spain), the EU TMR Project "Smart Systems", and the NSF of China under grants 10371084 and 10525105.

$(j = 1, 2)$, and ν_j the unit outward normal vector of Ω_j $(j = 1, 2)$. We assume $\gamma \neq \emptyset$ and γ is of class C^1 (unless otherwise stated). Denote by $\Box$ the d'Alembert operator $\partial_{tt} - \Delta$. Consider the following hyperbolic-parabolic coupled system:

$$\begin{cases} y_t - \Delta y = 0 & \text{in } (0, \infty) \times \Omega_1, \\ \Box z = 0 & \text{in } (0, \infty) \times \Omega_2, \\ y = 0 & \text{on } (0, \infty) \times \Gamma_1, \\ z = 0 & \text{on } (0, \infty) \times \Gamma_2, \\ y = z_t, \quad \frac{\partial y}{\partial \nu_1} = -\frac{\partial z}{\partial \nu_2} & \text{on } (0, \infty) \times \gamma, \\ y(0) = y_0 & \text{in } \Omega_1, \\ z(0) = z_0, \quad z_t(0) = z_1 & \text{in } \Omega_2. \end{cases} \tag{1.1}$$

This is a simplified and linearized model for fluid-structure interaction. In system (1.1), y may be viewed as the velocity of the fluid; while z and z_t represent respectively the displacement and velocity of the structure. This system consists of a wave and a heat equation coupled through an interface with transmission conditions. More realistic models should involve the Stokes (*resp.* the elasticity) equations instead of the heat (*resp.* the wave) ones. In [7] and [11], the same system was considered but for the transmission condition $y = z$ on the interface instead of $y = z_t$. From the point of view of fluid-structure interaction, the transmission condition $y = z_t$ in (1.1) is more natural since y and z_t represent velocities of the fluid and the elastic body, respectively. On the other hand, in the most rigorous formulation the model should consist on a free boundary problem, with the free boundary being the moving interface between the fluid and the elastic body. After linearization around the trivial solution the interface is kept fixed in time. Our analysis concerns this later linearized formulation.

Put $H^1_{\Gamma_1}(\Omega_1) \triangleq \left\{ h|_{\Omega_1} \,\middle|\, h \in H^1_0(\Omega) \right\}$ and $H^1_{\Gamma_2}(\Omega_2) \triangleq \left\{ h|_{\Omega_2} \,\middle|\, h \in H^1_0(\Omega) \right\}$. As we shall see, system (1.1) is well posed in the Hilbert space

$$H \triangleq L^2(\Omega_1) \times H^1_{\Gamma_2}(\Omega_2) \times L^2(\Omega_2).$$

The space H is asymmetric with respect to the wave and heat components since the regularity differs in one derivative from one side to the other.

When Γ_2 is a non-empty open subset of the boundary, the following is an equivalent norm on H:

$$|f|_H = \sqrt{|f_1|^2_{L^2(\Omega_1)} + |\nabla f_2|^2_{(L^2(\Omega_2))^n} + |f_3|^2_{L^2(\Omega_2)}}, \qquad \forall\, f = (f_1, f_2, f_3) \in H.$$

This simplifies the dynamical properties of the system in the sense that the only stationary solution is the trivial one. The analysis is simpler as well. The same can be said when Γ_2 has positive capacity since, then, the Poincaré inequality holds. Note that when $\Gamma_2 = \emptyset$ or, more generally, when Cap Γ_2, the capacity of Γ_2, vanishes, $|\cdot|_H$ is no longer a norm on H. In this case, there are non-trivial stationary solutions of the system. Thus, the asymptotic behavior is more complex and one should rather expect the convergence of each individual trajectory to a specific stationary solution. Therefore, to simplify the presentation of this paper, we shall assume Cap $\Gamma_2 \neq 0$ in what follows.

Define the energy of system (1.1) by

$$E(t) \triangleq E(y, z, z_t)(t) = \frac{1}{2}|(y(t), z(t), z_t(t))|_H^2.$$

By means of the classical energy method, it is easy to check that

$$\frac{d}{dt}E(t) = -\int_{\Omega_1} |\nabla y|^2 dx. \tag{1.2}$$

Therefore, the energy of (1.1) is decreasing as $t \to \infty$. First of all, we show that $E(t) \to 0$ as $t \to \infty$, without any geometric conditions on the domains Ω_1 and Ω_2. Note however that, due to the lack of compactness of the resolvent of the generator of the underlying semigroup of system (1.1) for $n \geq 2$, one can not use directly the LaSalle's invariance principle to prove this result. Instead, using the "relaxed invariance principle" ([9]), we conclude first that the first and third components of every solution (y, z, z_t) of (1.1), y and z_t, tend to zero strongly in $L^2(\Omega_1)$ and $L^2(\Omega_2)$, respectively; while its second component z tends to zero weakly in $H^1_{\Gamma_2}(\Omega_2)$ as $t \to \infty$. Then, we use the special structure of (1.1) and the key energy dissipation law (1.2) to "recover" the desired strong convergence of z in $H^1_{\Gamma_2}(\Omega_2)$.

The main goal of this paper is to summarize the results we have obtained in the analysis of the longtime behavior of $E(t)$. Especially, we study whether or not the energy of solutions of system (1.1) tends to zero uniformly as $t \to \infty$, i.e., whether there exist two positive constants C and α such that

$$E(t) \leq CE(0)e^{-\alpha t}, \qquad \forall\, t \geq 0 \tag{1.3}$$

for every solution of (1.1).

According to the energy dissipation law (1.2), the uniform decay problem (1.3) is equivalent to showing that: there exists $T > 0$ and $C > 0$ such that every solution of (1.1) satisfies

$$|(y_0, z_0, z_1)|_H^2 \leq C\int_0^T \int_{\Omega_1} |\nabla y|^2 dxdt, \qquad \forall\, (y_0, z_0, z_1) \in H. \tag{1.4}$$

Inequality (1.4) can be viewed as an observability estimate for equation (1.1) with observation on the heat subdomain.

Note however that, as indicated in [10], there is no uniform decay for solutions of (1.1) even in one space dimension. The analysis in [10] exhibits the existence of a hyperbolic-like spectral branch such that the energy of the eigenvectors is concentrated in the wave domain and the eigenvalues have an asymptotically vanishing real part. This is obviously incompatible with the exponential decay rate. The approach in [10], based on spectral analysis, does not apply to multidimensional situations. But the $1-d$ result in [10] is a warning in the sense that one may not expect (1.4) to hold.

Exponential decay property also fails in several space dimensions, as the $1-d$ spectral analysis suggests. For this purpose, following [7], we analyze carefully the interaction of the wave and heat-like solutions on the interface for general geometries. The main idea is to use Gaussian Beams ([6] and [5]) to construct approximate solutions of (1.1) which are highly concentrated along the generalized rays of the d'Alembert operator $\Box$ in the wave domain Ω_2 and are almost completely reflected on the interface γ. Due to the asymmetry of the energy space H, the same construction in [7] does not give the desired estimate. One has to compute higher order corrector terms on the phases and amplitudes of the wave-like solutions to recover an accurate description.

In view of the above analysis, it is easy to see that, one can only expect a polynomial stability property of smooth solutions of (1.1) even under the Geometric Control Condition (GCC for short, *see* [1]), i.e., when the heat domain where the damping of the system is active is such that all rays of Geometric Optics propagating in the wave domain touch the interface in an uniform time. To verify this, we need to derive a weakened observability inequality by viewing the whole system as a perturbation of the wave equation in the whole domain Ω. This technique was applied in the simpler model analyzed in [7]. However, as before, some efforts are necessary to treat the asymmetric structure of the energy space H.

We refer to [12] for the details of the proofs of the results in this paper and other results in this context (especially for the analysis without the technical assumption Cap $\Gamma_2 \neq 0$).

2. Some preliminary results

In this section, we shall present some preliminary results.

Define an unbounded operator $\mathcal{A} : D(\mathcal{A}) \subset H \to H$ by $\mathcal{A}Y = (\Delta Y_1, Y_3, \Delta Y_2)$, where $Y = (Y_1, Y_2, Y_3) \in D(\mathcal{A})$, and

$$D(\mathcal{A}) = \Big\{(Y_1, Y_2, Y_3) \in H \ \Big|\ \Delta Y_1 \in L^2(\Omega_1),\ \Delta Y_2 \in L^2(\Omega_2),\ Y_3 \in H^1(\Omega_2),$$
$$Y_1|_{\Gamma_1} = Y_3|_{\Gamma_2} = 0,\ Y_1|_\gamma = Y_3|_\gamma,\ \frac{\partial Y_1}{\partial \nu_1}\Big|_\gamma = -\frac{\partial Y_2}{\partial \nu_2}\Big|_\gamma\Big\}.$$

Remark 1. Obviously, in one space dimension, i.e., $n = 1$, we have $D(\mathcal{A}) = \Big\{(Y_1, Y_2, Y_3) \in H \ \Big|\ Y_1 \in H^2(\Omega_1),\ Y_2 \in H^2(\Omega_2),\ Y_3 \in H^1(\Omega_2),\ Y_1|_{\Gamma_1} = Y_3|_{\Gamma_2} = 0,\ Y_1|_\gamma = Y_3|_\gamma,\ \frac{\partial Y_1}{\partial \nu_1}\Big|_\gamma = -\frac{\partial Y_2}{\partial \nu_2}\Big|_\gamma\Big\} \subset H^2(\Omega_1) \times H^2(\Omega_2) \times H^1(\Omega_2)$. But this is not longer true in several space dimensions.

It is easy to see that system (1.1) can be re-written as an abstract Cauchy problem in H: $X_t = \mathcal{A}X$ for $t > 0$ with $X(0) = X_0$, where $X = (y, z, z_t)$ and $X_0 = (y_0, z_0, z_1)$. We have the following result.

Theorem 1. *The operator $\mathcal{A}$ is the generator of a contractive C_0-semigroup in H, and $0 \in \rho(\mathcal{A})$, the resolvent of $\mathcal{A}$.*

Remark 2. When $n = 1$, in view of the embedding in Remark 1, it is easy to check that $\mathcal{A}^{-1}$ is compact. However, $\mathcal{A}^{-1}$ is not guaranteed to be compact in several space dimensions, i.e., $n \geq 2$. Indeed, for any $F = (F_1, F_2, F_3) \in H$, the second component Y_2 of $\mathcal{A}^{-1}F$ belongs to $H^1_{\Gamma_2}(\Omega_2)$, which has the same regularity as the second component F_2 of F. (According to the regularity theory of elliptic equations, this regularity property for Y_2 is sharp as we shall see.)

The following result shows that $\mathcal{A}^{-1}$ is not compact.

Proposition 1. *In dimensions $n \geq 2$, the domain $D(\mathcal{A})$ is noncompact in H.*

The proof of Proposition 1 is due to Thomas Duyckaerts. The main idea is as follows: It suffices to show that there exists a sequence of $\{(Y_1^k, Y_2^k, Y_3^k)\}_{k=1}^{\infty} \subset D(\mathcal{A})$ such that $(Y_1^k, Y_2^k, Y_3^k) \rightharpoonup 0$ in $D(\mathcal{A})$ as $k \to \infty$ and $\inf_{k\in\mathbb{N}} |(Y_1^k, Y_2^k, Y_3^k)|_H \geq c$ for some constant $c > 0$. For this purpose, for any nonempty open subset Γ_0 of Γ, we denote by $H_{\Gamma_0}^{-1/2}(\Gamma)$ the completion of $C(\overline{\Gamma_0})$ with respect to the norm:

$$|u|_{H_{\Gamma_0}^{-1/2}(\Gamma)} = \sup\left\{ \frac{|\int_\Gamma uf d\Gamma|}{|f|_{H^{1/2}(\Gamma)}} \;\middle|\; f \in H^{1/2}(\Gamma) \setminus \{0\} \text{ and } f = 0 \text{ on } \Gamma \setminus \overline{\Gamma_0} \right\}.$$

Since $H_\gamma^{-1/2}(\partial\Omega_1)$ can be identified with $H_\gamma^{-1/2}(\partial\Omega_2)$ (algebraically and topologically), we denote them simply by $H_\gamma^{-1/2}$. It is easy to see that $H_\gamma^{-1/2}$ is an infinite-dimensional separable Hilbert space whenever $n \geq 2$. Hence there is a sequence $\{\beta^k\}_{k=1}^{\infty} \subset H_\gamma^{-1/2}$ such that $|\beta^k|_{H_\gamma^{-1/2}} = 1$ for each k and $\beta^k \rightharpoonup 0$ in $H_\gamma^{-1/2}$ as $k \to \infty$.

We solve the following two systems

$$\begin{cases} \Delta Y_1^k = 0 & \text{in } \Omega_1, \\ Y_1^k = 0 & \text{on } \Gamma_1, \\ \frac{\partial Y_1^k}{\partial \nu_1} = -\beta^k & \text{on } \gamma, \end{cases} \qquad \begin{cases} \Delta Y_2^k = 0 & \text{in } \Omega_2, \\ Y_2^k = 0 & \text{on } \Gamma_2, \\ \frac{\partial Y_2^k}{\partial \nu_2} = \beta^k & \text{on } \gamma \end{cases}$$

to get $Y_i^k \in H^1_{\Gamma_i}(\Omega_i)$ $(i = 1, 2)$, and then solve

$$\begin{cases} \Delta Y_3^k = 0 & \text{in } \Omega_2, \\ Y_3^k = 0 & \text{on } \Gamma_2, \\ Y_3^k = Y_1^k & \text{on } \gamma \end{cases}$$

to get $Y_3^k \in H^1_{\Gamma_2}(\Omega_2)$. This produces the desired $\{(Y_1^k, Y_2^k, Y_3^k)\}_{k=1}^{\infty}$.

Remark 3. Noting the structure of $D(\mathcal{A})$, it is easy to see that

$$D(\mathcal{A}) \subset H^1_{\Gamma_1}(\Omega_1) \times H^1_{\Gamma_2}(\Omega_2) \times H^1_{\Gamma_2}(\Omega_2). \tag{2.1}$$

This, at least, produces H^1-regularity for the heat and wave components of system (1.1) whenever its initial datum belongs to $D(\mathcal{A})$. One may need the H^2-regularity for the heat and wave components of system (1.1) when the initial data are smooth. For this to be true it is not sufficient to take the initial data in $D(\mathcal{A})$ since generally $D(\mathcal{A}) \not\subset (H^2(\Omega_1) \cap H^1_{\Gamma_1}(\Omega_1)) \times (H^2(\Omega_2) \cap H^1_{\Gamma_2}(\Omega_2)) \times H^1_{\Gamma_2}(\Omega_2)$ unless $n = 1$.

In order to prove the existence of smooth solutions of (1.1), we introduce the following Hilbert space:

$$V = \left\{ (y_0, z_0, z_1) \in D(\mathcal{A}) \;\middle|\; y_0 \in H^2(\Omega_1), z_0 \in H^2(\Omega_2) \right\} \subset D(\mathcal{A}),$$

with the canonical norm. Note however that, according to Proposition 1, $D(\mathcal{A}^k)$ is not necessarily a subspace of V even if $k \in \mathbb{N}$ is sufficiently large.

We have the following regularity result:

Theorem 2. *Let $\Gamma \cap \gamma = \emptyset$ and $\gamma \in C^2$. Then for any $(y_0, z_0, z_1) \in V$, the solution of (1.1) satisfies $(y, z, z_t) \in C([0, \infty); V)$, and for any $T \in (0, \infty)$, there is a constant $C_T > 0$ such that*

$$|(y, z, z_t)|_{C([0,T];V)} \leq C_T |(y_0, z_0, z_1)|_V.$$

The main idea to show Theorem 2 is as follows: We first take the tangential derivative of the system and show that the tangential derivative of the solution is of finite energy and then by using the original equation, one obtains the regularity of the other derivatives.

3. Asymptotic behavior

First of all, we show the strong asymptotic stability of (1.1) without the GCC.

Theorem 3. *For any given $(y_0, z_0, z_1) \in H$, the solution (y, z, z_t) of (1.1) tends to 0 strongly in H as $t \to \infty$.*

To prove Theorem 3, by density, it suffices to assume $(y_0, z_0, z_1) \in D(\mathcal{A})$. As we said above, we apply the relaxed invariance principle, using the energy as Lyapunov function. This yields the strong convergence to zero of the components y and z_t of the solution in the corresponding spaces. But this argument fails to give strong convergence to zero of z in $H^1_{\Gamma_2}(\Omega_2)$, because of the lack of compactness of the embedding from $D(A)$ into H. This argument, in principle, only yields the weak convergence of z. We need a further argument to show that the convergence of z holds in the strong topology of $H^1_{\Gamma_2}(\Omega_2)$. The key point is that, in view of the energy dissipation law (1.2), one has

$$\nabla y \in L^2(0, \infty; (L^2(\Omega_1))^n). \tag{3.1}$$

Also, by the standard semigroup theory and (2.1) in Remark 3, we see that $\nabla y \in C([0, \infty); (L^2(\Omega_1))^n)$. Therefore, (3.1) implies that there is a sequence $\{s_n\}_{n=1}^{\infty}$ which tends to ∞ such that

$$\nabla y(s_n) \to 0 \;\text{ strongly in }\; (L^2(\Omega_1))^n \;\text{ as }\; n \to \infty.$$

With this, we can deduce that

$$\nabla z(s_n) \to 0 \;\text{ strongly in }\; (H^1(\Omega_2))^n \;\text{ as }\; n \to \infty,$$

and, using the decreasing character of the energy of the system, we may conclude that the convergence holds along all the continuous one parameter family $z(s)$ as s tends to infinity.

Next, we analyze the non-uniform decay of solutions to (1.1). For this purpose, we recall that a *null bicharacteristic* for $\Box$ in $\mathbb{R}^n$ is defined to be a solution of the ODE:

$$\begin{cases} \dot{x}(t) = 2\xi(t), & \dot{\xi}(t) = 0, \\ x(0) = x^0, & \xi(0) = \xi^0, \end{cases}$$

where the initial data ξ^0 are chosen such that $|\xi^0| = 1/2$. Clearly, $(t, x(t))$, the projection of the null bicharacteristic to the physical time-space, traces a line in $\mathbb{R}^{1+n}$ (starting from $(0, x^0)$), which is called a *ray* for $\Box$ in the sequel. Sometimes, we also refer to $(t, x(t), \xi(t))$ as the ray. Obviously, rays for $\Box$ in $\mathbb{R}^n$ are simply straight lines.

In the presence of boundaries, rays, when reaching the boundary, are reflected following the usual rules of Geometric Optics. More precisely, for a $T > 0$ and a bounded domain $M \subset \mathbb{R}^n$ with piecewise C^1 boundary ∂M, the singular set being localized on a closed (topological) sub-manifold S with $\dim S \leq n-2$, we introduce the following definition of multiply reflected rays.

Definition 1. A continuous parametric curve: $[0, T] \ni t \mapsto (t, x(t), \xi(t)) \in C([0, T] \times \overline{M} \times \mathbb{R}^n)$, with $x(0) \in M$ and $x(T) \in M$, is called a multiply reflected ray for the operator $\Box$ in $[0, T] \times \overline{M}$ if there exist $m \in \mathbb{N}$, $0 < t_0 < t_1 < \cdots < t_m = T$ such that each $(t, x(t), \xi(t))|_{t_i < t < t_{i+1}}$ is a ray for $\Box$ $(i = 0, 1, 2, \ldots, m-1)$, which arrives at $\partial M \setminus S$ at time $t = t_{i+1}$, and is reflected by $(t, x(t), \xi(t))|_{t_{i+1} < t < t_{i+2}}$ by the usual geometric optics law whenever $i < m - 1$.

In view of [7, Lemma 2.2 and Remark 2.4], we have the following geometric lemma.

Lemma 1. *For each $T > 0$, there is a multiply reflected ray for the operator $\Box$ in M which meets $\partial M \setminus S$ transversally and non-normally.*

We have the following key result.

Theorem 4. *Let the boundary $\partial\Omega_2$ of the wave domain Ω_2 be of class C^4. For any $T > 0$, let $[0, T] \ni t \mapsto (t, x(t), \xi(t)) \in C([0, T] \times \overline{\Omega_2} \times \mathbb{R}^n)$ be a multiply reflected ray for the operator $\Box$ in Ω_2, which meets the boundary $\partial\Omega_2$ transversally and non-normally. Then there is a family of solutions $\{(y_\varepsilon, z_\varepsilon)\}_{\varepsilon>0}$ of system (1.1) in $(0, T)$ (the initial conditions being excepted), such that*

$$|\nabla y_\varepsilon|^2_{(L^2((0,T)\times\Omega_1))^n} = O(\varepsilon), \qquad E_\varepsilon(0) = E(y_\varepsilon, z_\varepsilon, \partial_t z_\varepsilon)(0) \geq c_0,$$

where $c_0 > 0$ is a constant, independent of ε.

Now, combining Lemma 1 and Theorem 4, one obtains the following non-uniform decay result:

Theorem 5. *Let the boundary $\partial\Omega_2$ of the wave domain Ω_2 be of class C^4. Then*

i) *For any given $T > 0$, there is no constant $C > 0$ such that (1.4) holds for all solutions of (1.1);*

ii) *The energy $E(t)$ of solutions of system (1.1) does not decay exponentially as $t \to \infty$.*

Finally, we analyze the long time behavior of solutions of system (1.1) in several space dimensions under suitable geometric assumptions.

We introduce the following internal observability assumption for the wave equation in Ω:

(H) There exist $T_0 > 0$ such that for some constant $C > 0$, all solutions of the following system

$$\begin{cases} \Box\zeta = 0 & \text{in} \quad (0, T_0) \times \Omega, \\ \zeta = 0 & \text{on} \quad (0, T_0) \times \Gamma, \\ \zeta(0) = \zeta_0, \quad \zeta_t(0) = \zeta_1 & \text{in} \quad \Omega \end{cases}$$

satisfy

$$|\zeta_0|^2_{H^1_0(\Omega)} + |\zeta_1|^2_{L^2(\Omega)} \leq C \int_0^{T_0} \int_{\Omega_1} |\zeta_t|^2 dxdt, \quad \forall\, (\zeta_0, \zeta_1) \in H^1_0(\Omega) \times L^2(\Omega).$$

It is well known that assumption (H) holds when T_0 and Ω_1 satisfy the *Geometric Optics Condition* (GCC) introduced in [1]. This condition asserts that all rays of Geometric Optics propagating in Ω and bouncing on the boundary enter the control domain Ω_1 in a uniform time $T_0 > 0$. A relevant particular case in which the GCC is satisfied is when the heat domain Ω_1 envelopes the wave domain Ω_2. This simple case can be handled by the multiplier method ([4]).

Now, we may state our polynomial decay result for system (1.1) as follows.

Theorem 6. *Let T_0 and Ω_1 satisfy (H). Then there is a constant $C > 0$ such that for any $(y_0, z_0, z_1) \in D(\mathcal{A})$, the solution of* (1.1) *satisfies*

$$|(y(t), z(t), z_t(t))|_H \leq \frac{C}{t^{1/6}} |(y_0, z_0, z_1)|_{D(\mathcal{A})}, \quad \forall\, t > 0.$$

Remark 4. Theorem 6 is not sharp for $n = 1$ since in [10] we have proved that the decay rate is $1/t^2$. However, similar to [7], the WKB asymptotic expansion for the flat interface allows to show that it is impossible to expect the same decay rate for several space dimensions. This suggests that the rate of decay in the multidimensional case is slower than in the one dimensional one. According to Remark 5 and the possible sharp weakened observability inequality (3.4) below, it seems reasonable to expect $1/t$ to be the sharp polynomial decay rate for smooth solutions of (1.1) with initial data in $D(\mathcal{A})$. But this is an open problem. We refer to [3] for an interesting partial solution to this problem with a decay rate of the order of $1/t^{1-\delta}$ for all $\delta > 0$ but under stronger assumptions on the geometry that Ω is of C^∞ and $\overline{\Gamma}_1 \cap \overline{\Gamma}_2 = \emptyset$.

The proof of Theorem 6 is based on the following key weakened observability inequality for equation (1.1):

Theorem 7. *Let T_0 and Ω_1 satisfy (H). Then there exist two constants T_0 and $C > 0$ such that for any $(y_0, z_0, z_1) \in D(\mathcal{A}^3)$, and any $T \geq T_0$, the solution of* (1.1) *satisfies*

$$|(y_0, z_0, z_1)|_H \leq C |\nabla y|_{H^3(0,T;(L^2(\Omega_1))^n)}. \tag{3.2}$$

The main idea to prove Theorem 7 is as follows: Setting $w = y\chi_{\Omega_1} + z_t\chi_{\Omega_2}$, noting (1.1) and recalling that $\partial z_t/\partial\nu_2 = -\partial y_t/\partial\nu_1$ on $(0,T)\times\gamma$, and by $(y_0, z_0, z_1) \in D(\mathcal{A}^2)$, one sees that $w \in C([0,T]; H_0^1(\Omega)) \cap C^1([0,T]; L^2(\Omega))$ satisfies

$$\begin{cases} \Box w = (y_{tt} - y_t)\chi_{\Omega_1} + \left(\frac{\partial y}{\partial\nu_1} - \frac{\partial y_t}{\partial\nu_1}\right)\delta_\gamma & \text{in } (0,T)\times\Omega, \\ w = 0 & \text{on } (0,T)\times\Gamma, \\ w(0) = y_0\chi_{\Omega_1} + z_1\chi_{\Omega_2},\ w_t(0) = (\Delta y_0)\chi_{\Omega_1} + (\Delta z_0)\chi_{\Omega_2} & \text{in } \Omega. \end{cases} \tag{3.3}$$

Then, by means of the energy method and assumption (H), one concludes Theorem 7.

Remark 5. Note that (3.2) is, indeed, a weakened version of (1.4), in which we do not only use the norm of ∇y on $(L^2((0,T)\times\Omega_1))^n$ to bound the total energy of solutions but the stronger one on $H^3(0,T;(L^2(\Omega_1))^n)$. Nevertheless, inequality (3.2) is very likely not sharp. One can expect, under assumption (H), the following stronger inequality to hold:

$$|(y_0, z_0, z_1)|_H \le C|\nabla y|_{H^{1/2}(0,T;(L^2(\Omega_1))^n)}. \tag{3.4}$$

This is also an open problem.

4. Open problems

This subject is full of open problems. Some of them seem to be particularly relevant and could need important new ideas and further developments:

• *Logarithmic decay without the GCC.* Inspired on [8], it seems natural to expect a logarithmic decay result for system (1.1) without the GCC. However, there is a difficulty to do this. In [7] we show this decay property for system (1.1) but with the interface condition $y = z_t$ replaced by $y = z$. The key point is to apply the known very weak observability inequalities for the wave equations without the GCC ([8]) to a perturbed wave equation similar to (3.3), and use the crucial fact that the generator of the underlying semigroup has compact resolvent. It is precisely the lack of compactness for (1.1) in multi-dimensions that prevents us from showing the logarithmic decay result in the present case.

• *More complex and realistic models.* In the context of fluid-structure interaction, it is more physical to replace the wave equation in system (1.1) by the system of elasticity and the heat equation by the Stokes system, and the fluid-solid interface γ by a free boundary. It would be interesting to extend the present analysis to these situations. But this remains to be done.

• *Nonlinear models.* A more realistic model for fluid-structure interaction would be to replace the heat and wave equations in system (1.1) by the Navier-Stokes and elasticity systems coupled through a moving boundary. To the best of our knowledge, very little is known about the well-posedness and the long time behavior for the solutions to the corresponding equations (We refer to [2] for some existence results of weak solutions in two space dimensions).

• *Control problems.* In [10], we analyze the null controllability problem for system (1.1) in one space dimension by means of spectral methods. It is found that the controllability results depend strongly on whether the control enters the system through the wave component or the heat one. This problem is completely open in several space dimensions.

Acknowledgment

The authors thank Dr. Thomas Duyckaerts for fruitful discussions.

References

[1] C. Bardos, G. Lebeau and J. Rauch, Sharp sufficient conditions for the observation, control and stabilization of waves from the boundary, *SIAM J. Control Optim.*, **30** (1992), 1024–1065.

[2] M. Boulakia, Existence of weak solutions for the motion of an elastic structure in an incompressible viscous fluid, *C. R. Math. Acad. Sci. Paris,* **336** (2003), 985–990.

[3] T. Duyckaerts, *Optimal decay rates of the solutions of hyperbolic-parabolic systems coupled by an interface*, Preprint, 2005.

[4] J.L. Lions, *Contrôlabilité Exacte, Stabilisation et Perturbations de Systèmes Distribués. Tome 1: Contrôlabilité Exacte*, RMA 8, Masson, Paris, 1988.

[5] F. Macià and E. Zuazua, On the lack of observability for wave equations: a Gaussian beam approach, *Asymptot. Anal.,* **32** (2002), 1–26.

[6] J. Ralston, Gaussian beams and the propagation of singularitie, in *Studies in Partial Differential Equations,* Edited by W. Littman, MAA Studies in Mathematics **23**, Washington, DC, 1982, 206–248.

[7] J. Rauch, X. Zhang and E. Zuazua, Polynomial decay of a hyperbolic-parabolic coupled system, *J. Math. Pures Appl.,* **84** (2005), 407–470.

[8] L. Robbiano, Fonction de coût et contrôle des solutions des équations hyperboliques, *Asymptotic Anal.,* **10** (1995), 95–115.

[9] M. Slemrod, Weak asymptotic decay via a "relaxed invariance principle" for a wave equation with nonlinear, nonmonotone damping, *Proc. Roy. Soc. Edinburgh Sect. A,* **113** (1989), 87–97.

[10] X. Zhang and E. Zuazua, Control, observation and polynomial decay for a coupled heat-wave system, *C.R. Math. Acad. Sci. Paris,* **336** (2003), 823–828.

[11] X. Zhang and E. Zuazua, Polynomial decay and control of a $1-d$ hyperbolic-parabolic coupled system, *J. Differential Equations,* **204** (2004), 380–438.

[12] X. Zhang and E. Zuazua, Long time behavior of a coupled heat-wave system arising in fluid-structure interaction, in submission.

Xu Zhang
School of Mathematics
Sichuan University
Chengdu 610064, China
and
Departamento de Matemáticas
Facultad de Ciencias
Universidad Autónoma de Madrid
E-28049 Madrid, Spain
e-mail: xu.zhang@uam.es

Enrique Zuazua
Departamento de Matemáticas
Facultad de Ciencias
Universidad Autónoma de Madrid
E-28049 Madrid, Spain
e-mail: enrique.zuazua@uam.es

International Series of Numerical Mathematics, Vol. 154, 457–460

List of Participants

Laouar ABDELHAMID – University of Annaba, Algery
Toyohiko AIKI – Gifu University, Japan
Fuensanta ANDREU – University of Valencia, Spain
Stanislav ANTONTSEV – University of Beira Interior, Portugal
Ioannis ATHANASOPOULOS – University of Crete, Greece
Assis AZEVEDO – University of Minho, Portugal
Caterine BANDLE – University of Basel, Switzerland
Sílvia BARBEIRO – University of Coimbra, Portugal
Andres BARREA – National University of Cordoba, Argentina
Raquel BARREIRA – Portugal
John BARRETT – Imperial College, UK
Zakaria BELHACHMI – University of Metz, France
Marino BELLONI – University of Parma, Italy
Henri BERESTYCKI – EHESS Paris, France
Piotr BILER – University of Wroclaw, Poland
Adrien BLANCHET – CERMICS - ENPC, France
Ivan BLANK – Worcester Polytechnical Institute, USA
Giovanni BONFANTI – University of Brescia, Italy
Mahdi BOUKROUCHE – University of Saint-Etienne, France
Xavier BRESSON – Federal Polytechnical School of Lausanne, Switzerland
Martin BROKATE – Technical University of Munich, Germany
Antonio BRU – University Complutense, Spain
Dorin BUCUR – University of Metz, France
Vincent CASELLES – University Pompeu Fabra, Spain
Hector CENICEROS – University of California Santa Barbara, USA
John CHADAM – University of Pittsburgh, USA
Fabio CHALUB – CMAF/University of Lisbon, Portugal
Sagun CHANILLO – Rutgers University, USA
Etienne CHEVALIER – University of Marne-la-Vallée, France
Pierluigi COLLI – University of Pavia, Italy
Luisa CONSIGLIERI – University of Lisbon, Portugal
Panagiota DASKALOPOULOS – University of Columbia, USA
Klaus DECKELNICK – University of Magdeburg, Germany

Irina DENISOVA – Institute of Problems of Mechanical Engineering, Russia
Wolfgang DREYER – Weierstrass Institute, Germany
Marc DROSKE – University of Duisburg, Germany
Michela ELEUTERI – University of Trento, Italy
Charles ELLIOTT – University of Sussex, UK
Antonio FASANO – University of Florence, Italy
Isabel FIGUEIREDO – University of Coimbra, Portugal
Alessandro FONDA – University of Trieste, Italy
Michel FRÉMOND – Lagrange Laboratory, France
Avner FRIEDMAN – Ohio State University, USA
Takesi FUKAO – Gifu National College of Technology, Japan
Gonzalo GALIANO – University of Oviedo, Spain
Harald GARCKE – University of Regensburg, Germany
Maria Luisa GARZON – University of Oviedo, Spain
Yves van GENNIPE – Netherlands
Frederic GIBOU – University of California, USA
Martin GLICKSMAN – Rensselaer Polytechnic Institute, USA
César GONÇALVES – Polytechnical Institute of Guarda, Portugal
Frank HAUSSER – Crystal Growth Group Research Center CAESER, Germany
Eurica HENRIQUES – University of Trás-os-Montes-e-Alto-Douro, Portugal
Danielle HILHORST – University of Paris-Sud, France
Matthieu HILLAIRET – ENS of Lyon, France
Dietmar HÖMBERG – Technical University of Berlin and WIAS, Germany
Noureddine IGBIDA – University of Picardie Jules Verne, France
Akio ITO – Kinki University 1, Japan
Karim IVAZ – Shabestar Islamic Azad University, Iran
Ashley JAMES – University of Minnesota, USA
Jörg KAMPEN – University of Heidelberg, Germany
Bernd KAWOHL – University of Cologne, Germany
Nobuyuki KENMOCHI – Chiba University, Japan
John KING – University of Nottingham, UK
Tero KILPELÄINEN – University of Jyväskylä, Finland
Theodore KOLOKOLNIKOV – Free University of Brussels, Belgium
Chekhonin KONSTANTIN – Far Eastern State Transport University Russia, Russia
Jana KOPFOVA – Silesian University in Opava, Czech Republic
Petra KORDULOVA – Silesian University in Opava, Czech Republic
Marianne KORTEN – Kansas State University, USA
Pavel KREJČI – Academy of Sciences of the Czech Republic, Czech Republic
Thomas LACHAND–ROBERT – University of Savoie, France
Omar LAKKIS – University of Sussex, UK
Carlos LEAL – University of Coimbra, Portugal
Ki–Ahm LEE – Seoul National University, Korea

Ana Cristina LEMOS – School of Technology and Management of Leiria, Portugal
Chiara LEONE – University “Federico II” of Naples, Italy
John LOWENGRUB – University of California at Irvine, USA
Riccardo MARCH – Institute for Aplication of Calculus C.N.R., Italy
Peter MARKOWITHC – University of Vienna, Austria
Sébastien MARTIN – INSA of Lyon, France
Ramon Escobedo MARTINEZ – University “Carlos III” of Madrid, Spain
Simon MASNOU – University of Paris VI, France
Norayr MATEVOSYAN – University of Vienna, Austria
José MAZÓN – University of Valencia, Spain
Wilhelm MERZ– University Erlangen-Nürnberg, Germany
Emmanouil MILAKIS – University of Crete, Greece
Masayasu MIMURA – Meiji University, Japan
Fernando MIRANDA– University of Minho, Portugal
Régis MONNEAU – Cermics–ENPC, France
Andreas MÜNCH – Humboldt University Berlin, Germany
Adrian MUNTEAN – University of Bremen, Germany
Carlo NITSCH – University “Federico II” of Naples, Italy
Robert NÜRNBERG – Imperial College, UK
Hermenegildo OLIVEIRA – University of Algarve, Portugal
Arshak PETROSYAN – Purdue University, USA
Marco PICASSO – Federal Polytechnique School of Lausanne, Switzerland
Konstantin PILECKAS – Shiauliai University, Lithuania
Cecília PINTO – Polytechnical Institute of Viseu, Portugal
Angela PISTOIA – University of Rome “La Sapienza”, Italy
Sorin POP – Technical University of Eindhoven, Netherlands
Mario PRIMICERIO – University of Florence, Italy
Gregory RAPUCH – EHESS Paris, France
Vincenzo RECUPERO – University of Trento, Italy
Wolfgang REICHEL – University of Zürich, Switzerland
Elisabetta ROCCA – University of Milan, Italy
José-Francisco RODRIGUES – Univ. of Lisbon and CMUC/Univ. of Coimbra, Portugal
Matthias ROEGER –Technical University of Eindhoven, Netherlands
Riccarda ROSSI – University of Brescia, Italy
Martin RUMPF – University of Bonn, Germany
Piotr RYBKA – Warsaw University, Poland
Lisa SANTOS – University of Minho and CMAF/University of Lisbon, Portugal
Giuseppe SAVARÉ – University of Pavia, Italy
Mohamed SEBIH – University of Tlemcen, Algerie
Antonio SEGATTI – University of Pavia, Italy
Henrik SHAHGHOLIAN – Royal Institute of Technology, Sweden
Farshbaf SHAKER – Weierstrass Institute, Germany

Ken SHIRAKAWA – Kobe University, Japan
Sergey SHMAREV – University of Oviedo, Spain
Luis SILVESTRE – University of Texas, USA
Vsevolod SOLONNIKOV – Steklov Institute of Mathematics, Russia
Jürgen SPREKELS – Weierstrass Institute, Germany
Bjorn STINNER – University of Regensburg, Germany
Yoshie SUGIYAMA – Max Planck Institute, Germany
Mark SUSSMAN – Florida State University, USA
Xue-Cheng TAI – University of Bergen, Norway
Youshan TAO – Dong Hua University, China
Domingo TARZIA – University Austral, Argentina
Rodica TOADER – University of Udine, Italy
Julián TOLEDO – University of Valencia Burjassot, Spain
German TORRES – National University of Cordoba, Argentina
Marius TUSCNAK – University of Nancy 1, France
Nina URALTSEVA – St. Petersburg State University, Russia
José Miguel URBANO – University of Coimbra, Portugal
Carlos VÁZQUEZ – University of Coruña, Spain
Juan Luis VAZQUEZ – University Autonoma of Madrid, Spain
Stella Piro VERNIER – University of Cagliari, Italy
Luis Nunes VICENTE – University of Coimbra, Portugal
Walter VILLANUEVA – KTH Mechanics, Sweden
Augusto VISINTIN – University of Trento, Italy
Axel VOIGT – Crystal Growth Group Research Center CAESER, Germany
Alfred WAGNER – RWTH Aachen Institute of Mathematics, Germany
Barbara WAGNER – Weierstrass Institute, Germany
Michael WARD – University of British Columbia, Canada
Stephen WATSON – Nortwestern University, USA
Juncheng WEI – Chinese University of Hong Kong, China
Goerg WEISS – University of Tokyo, Japan
Richard WELFORD – University of Sussex Falmer, UK
Fahuai YI – South China Normal University, China
Noriaki YAMAZAKI – Muroran Institute of Technology, Japan
Boris ZALTZMAN – Ben-Gurion University of the Negev, Israel
Xu ZHANG – University Autonoma of Madrid, Spain
Yong-Wei ZHANG – National University of Singapore, Singapore

GPSR Compliance

The European Union's (EU) General Product Safety Regulation (GPSR) is a set of rules that requires consumer products to be safe and our obligations to ensure this.

If you have any concerns about our products, you can contact us on ProductSafety@springernature.com

In case Publisher is established outside the EU, the EU authorized representative is:

Springer Nature Customer Service Center GmbH
Europaplatz 3
69115 Heidelberg, Germany

Batch number: 10732470

Printed by Printforce, the Netherlands